PACIFIC SYMPOSIUM ON
BIOCOMPUTING 2012

PACIFIC SYMPOSIUM ON BIOCOMPUTING 2012

Kohala Coast, Hawaii, USA
3–7 January 2012

Edited by

Russ B. Altman
Stanford University, USA

A. Keith Dunker
Indiana University, USA

Lawrence Hunter
University of Colorado Health Sciences Center, USA

Tiffany Murray
Stanford University, USA

Teri E. Klein
Stanford University, USA

World Scientific

NEW JERSEY • LONDON • SINGAPORE • BEIJING • SHANGHAI • HONG KONG • TAIPEI • CHENNAI

Published by

World Scientific Publishing Co Pte Ltd
5 Toh Tuck Link, Singapore 596224
USA office: 27 Warren Street, Suite 401-402, Hackensack, NJ 07601
UK office: 57 Shelton Street, Covent Garden, London WC2H 9HE

British Library Cataloguing-in-Publication Data
A catalogue record for this book is available from the British Library.

BIOCOMPUTING 2012
Proceedings of the Pacific Symposium

ISBN-13 978-981-4596-37-4
ISBN-10 981-4596-37-X

PACIFIC SYMPOSIUM ON BIOCOMPUTING 2012

2012 marks the 17th Pacific Symposium on Biocomputing. The proceedings and sessions over the last 17 years are available online at http://psb.stanford.edu/prevmeet.html. It is remarkable to watch how the field and the meeting topics evolve. The PSB proceedings have published more than 800 articles, all of which are listed in PubMed (the PSB meeting is abbreviated **Pac Symp Biocomput)**.

We would like to thank our keynote speakers. Dr. Elaine Mardis, Professor of Genetics, Washington University St. Louis will talk about recent progress in genome sequencing for clinical applications. Our keynote in the area of Ethical, Legal and Social Implications of Technology will be James Heywood, Co-Founder and Chairman of Patientslikeme.com.

PSB provides sessions focusing on emerging areas in biomedical computation. These sessions are often conceived at the previous PSB meeting as people discuss the opportunities for new and exciting sessions. The efforts of a dedicated group of leaders has produced an outstanding set of sessions, with associated introductory tutorials. These organizers provide the scientific core of PSB and their sessions are as follows:

Identification of Aberrant Pathway and Network Activity from High-Throughput Data
Rachel Karchin, Michael Ochs, Josh Stuart, Joel Bader

Intrinsically Disordered Proteins: Analysis, Prediction, Simulation, and Biology
Jianhan Chen, Jianlin Cheng, Keith Dunker

Microbiome studies: Analytical Tools and Techniques
James A. Foster, Jason Moore, Jack Gilbert, John Bunge

Modeling host-pathogen interactions: Computational biology and bioinformatics for infectious disease research
Jason McDermott, Pascal Braun, Daniel Hyduke, Richard Bonneau

Personalized Medicine: from genotypes and molecular phenotypes towards computed therapy
Jennifer Listgarten, Oliver Stegle, Fritz Roth, Quaid Morris

Text and knowledge mining for pharmacogenomics: Genotype-phenotype-drug relationships
Kevin Bretonnel Cohen, Yael Garten, Nigham Shah, Udo Hahn

We are also pleased to present three workshops in which investigators with a common interest come together to exchange results and new ideas in a format that is more informal than the peer-reviewed sessions. For this year, the workshops and their organizers are:

The structure and function of chromatin and chromosomes
William Stafford Noble, C. Anthony Blau, Job Dekker, Zhi-jun Duan, Eric Garcia

Law, Bioethics and the Current Status of Ownership, Privacy, Informed Consent in the Genomic Age
Greg Hampikian, Eric M. Meslin

Systems Pharmacogenomics -- Bridging the Gap
Marylyn Ritchie, Nancy Cox, Cheng Cheng, Scott Weiss, Teri Klein, Russ Altman

The Systems Pharmacogenomics workshop is jointly sponsored by the NIH Pharmacogenomics Research Network (PGRN) Statistical Analysis Resource (P-STAR) and Systems Biology Group, as well as the PharmGKB.

Tiffany Murray manages the peer review process and assembly of the proceedings, and also plays a key role in many other aspects of the meeting. We also want to give a special thanks to the assistance of Blanca Pineda this year in helping with the manuscript review and notification process. We also thank the National Institutes of Health and the International Society for Computational Biology (ISCB) for travel grant support. We are particularly grateful to BJ McKay-Morrison at ISCB for her assistance. We also acknowledge the many busy researchers who reviewed the submitted manuscripts on a very tight schedule. The partial list following this preface does not include many who wished to remain anonymous, and of course we apologize to any who may have been left out by mistake.

We look forward to a great meeting once again.

Aloha!

Pacific Symposium on Biocomputing Co-Chairs,
October 14, 2011

Russ B. Altman
Departments of Bioengineering, Genetics & Medicine, Stanford University

A. Keith Dunker
Department of Biochemistry and Molecular Biology, Indiana University School of Medicine

Lawrence Hunter
Department of Pharmacology, University of Colorado Health Sciences Center

Teri E. Klein
Department of Genetics, Stanford University

Thanks to the reviewers...

Finally, we wish to thank the scores of reviewers. PSB aims for every paper in this volume be reviewed by three independent referees. Since there is a large volume of submitted papers, paper reviews require a great deal of work from many people. We are grateful to all of you listed below and to anyone whose name we may have accidentally omitted or who wished to remain anonymous.

Alexander Alekseyenko
Gokmen Altay
Michael Bada
Chris Baker
Elisar Barbar
Olivier Bodenreider
Sarah Bondos
Richard Bonneau
Karsten Borgwardt
Pascal Braun
Celeste Brown
Patrick Cai
Mauro Castro
Jianhan Chen
Jianlin Cheng
Kenny Chua
Melissa Clilne
Courtney Corley
Richard Cowper-Sallari
Trevor Creamer
Rahul Das
Nanjie Deng
Xin Deng
Rahul Deo
Tom Doak
Zsuzsanna Dosztanyi
Michael Ferris
Russ Finley
Satoski Fukuchi
Monika Fuxreiter
Suzanne Gallagher
Don Geman
Todd Gibson
Anna Goldenberg
Graciela Gonzalez
Sara Gosline
Alexander Hartemink
Sampsa Hauteniemi
Jianfeng He
Douglas Hill
Ryan Hoffman
Jeremy Horst
Wolfgang Huber
Daniel Hyduke
Takashi Ishida
Junji Iwahara
Nebojsa Jojic
David Jones
Lars Kangas
Carl Kingsford
Suzanne Komili
Christina Leslie
Jilong Li
Simon Lin
Christopher Lippert
Haibin Liu
Song Liu
Zhirong Liu
Kevin Livingston
Ray Luo
Juliette Madan
Florian Markowetz
Steve Matallo
Jason McDermott
Joe Mellor
Tanja Mittag
Sara Mostafavi
Gabe Musso
Tyrell Nelson
Aurelie Neveol
Micheal Ng
Bill Noble
Motonori Ota
Pier Palamara
Rohit Pappu
Pankaj Pasricha
Frances Pearl
Sampo Pyysalo
Gerald Quon
Predreg Radivojac
Habtom Ressom
Fabio Rinaldi
Thomas Rindflesch
Andrey Rzhetsky
Uwe Sauer
Neethu Shah
Kana Shimizu
Larry Smith
Tobin Sosnic
Alfred Spormann
Adam Szalkowski
Chuan Yi Tang
Haixu Tang
Murat Tasan
Christopher Taylor
Alden Timme
Peter Tompa
Silvio Tosatto
Yoshimasa Tsuruoka
Svitlana Tyekucheva
Karin Verspoor
Xinglong Wang
Zheng Wang
Kejun Wang,
Paul Whitford
Chris Wiggins
John Wilbur
Anthony Wright
LC Xue
Yinyin Yuan
Jinfeng Zhang
Yaoqi Zhou
Blaz Zupan

MICROBIOME STUDIES: ANALYTICAL TOOLS AND TECHNIQUES

IDENTIFICATION OF ABERRANT PATHWAY AND NETWORK ACTIVITY FROM HIGH-THROUGHPUT DATA

RACHEL KARCHIN
Department of Biomedical Engineering and Institute for Computational Medicine
Johns Hopkins University
Baltimore, MD 21218, USA
Email: karchin{at}jhu.edu

MICHAEL F. OCHS
Department of Oncology and Division of Oncology, Biostatistics and Bioinformatics
Sidney Kimmel Comprehensive Cancer Center
Johns Hopkins University
Baltimore, MD 21205, USA
Email: mfo{at}jhu.edu

JOSHUA M. STUART
Biomolecular Engineering
University of California Santa Cruz
Santa Cruz, CA 95064, USA
Email: jstuart{at}soe.ucsc.edu

JOEL S. BADER
Department of Biomedical Engineering and High-Throughput Biology Center
Johns Hopkins University
Baltimore, MD 21218, USA
Email: joel.bader{at}jhu.edu

Introduction

Biological functions are often explained in terms of networks and pathways. Innumerable college students have memorized canonical metabolic pathways, and signal transduction pathways such as MAPK, JAK/STAT and WNT have entered the biological lexicon. Inasmuch as these pathways provide a useful abstraction of biology, they can be used as a framework for understanding how mutations affect life processes and potentially cause disease.

Much work has been invested in mapping these and other pathways in human and simpler model organisms. Data sets are not pathways *per se*, but rather measurements of the individual interactions between proteins, genes, metabolites, drugs, and other species that define a network. In model organisms these networks can be perturbed by directed experiments. The first series of papers in this session explores how the experimental data sets can be analyzed and explained.

Directed perturbations of human networks *in vivo* are not possible, but genotype and phenotype data sets from individuals present a rich spectrum of information. The session concludes with analyses of the effects of primarily somatic mutations in cancer, focusing on the identification of subnetworks that may be responsible for disease phenotypes.

Genetic interaction networks in model organisms

Genetic epistatic interactions occur when mutations in two genes simultaneously create a phenotype different from the expectation from the two individual mutations. Genetic interactions are distinct from gene regulatory interactions, which refer to physical interactions between transcription factors and promoters. Large data sets of genetic interactions have been generated for yeast using a variety of experimental methods, which has motivated several contributions investigating the properties of these networks.

Bandyopadhyay *et al.* use gene expression data to assist in the prediction of genetic epistatic interactions in "SSLPred : Predicting Synthetic Sickness Lethality." The concept explored in this work is the feasibility of using network information to guide the generation of features to use in a predictive regression model. This work illustrates a recent theme in regression in a large feature space, the use of L1 regularization known as the lasso [1]. Lasso regression adds a penalty term that is related to the absolute value of parameter estimates. Quadratic programming and other methods permit fast identification of the global optimum, and, unlike ridge regression using L2 regularization, features can have regression coefficients that are set strongly to zero.

While Bandyopadhyay *et al.* provide global predictions for a genome-scale network, Carter *et al.* focus on quantitative understanding of a single network in "Predicting The Effects Of Copy-Number Variation In Double And Triple Mutant Combinations," investigate the contribution of genetic interactions to quantitative traits. The problem is important because of current interest in what has been called "missing heritability" in the context of human genome-wide association studies, where the contributions of known genetic factors remain below the total genetic contributions, which can be estimated accurately from epidemiology [2]. One possibility is that most analyses consider only additive models, in part because populations are not yet sufficiently large to have power to detect interaction terms. This work by Carter *et al.* is valuable in investigating non-additive contributions to a quantitative trait, defined by gene expression levels within the context of a detailed model of filamentous growth in yeast. The authors build on their earlier model of the gene regulatory network for filamentous growth [3] and use singular-value decomposition to investigate the dominant modes of transcriptional output [4]. The computational models are able to predict transcriptional phenotypes for perturbations including deletions, hypomorph alleles, and copy number variations, all important for human studies. Predictions for novel mutant combinations are in general concordance with experimental results. This manuscript points to the complexity of predicting phenotypes for multi-mutant combinations in yeast and illuminates the challenges ahead for predicting phenotypes from personal human genome data.

The subject of "Role of Synthetic Genetic Interactions in Understanding Functional Interactions Among Pathways" by Mohammadi *et al.* is the interpretation of genetic interactions in the context of other network data. Previous efforts have described genetic interactions in terms of within-pathway models (genetic interaction partners reside within the same pathway) and between-pathway models (genetic interactions occur between pathways with compensating or overlapping function). Comparisons with physical interactions [5, 6] and metabolic pathways [7] have typically favored the between-pathway model, particularly for deletion mutations that destroy the functionality of a linear pathway. The current contribution reexamines this problem using more recent data sets and pathways from KEGG to suggest functional connections between biological pathways in yeast.

Human data and local subnetworks

Data sets revealing network activity in human cancer are becoming available through The Cancer Genome Atlas (TCGA). These data sets provide an important test of the ability of methods developed for model organisms, primarily yeast, to scale to much more complex human systems involving more genes, larger networks, and distinct cell types.

In "Integrative Network Analysis to Identify Aberrant Pathway Networks in Ovarian Cancer," Chen *et al.* investigate the challenge of predicting cancer survival from genome-scale TCGA data involving copy number alterations, gene expression changes, and protein interactions. The rationale of TCGA is the hypothesis that genome-scale data will indeed result in improved methods for detecting, grading, and treating cancers. Demonstrations of effectiveness, including this work, are crucial for making progress and justifying future efforts related to TCGA. The authors approach this problem by identifying subnetworks of genes linked by protein-protein interactions that are effective at classification, then using support vector machines (SVMs) to perform the classification. The subnetwork learning illustrates an important theme in biological network analysis known as the "active subnetwork problem," in which a subset of genes or proteins strongly implicated in a process are linked together parsimoniously [8]. This problem can be expressed as a version of a classic problem in computer science known as the prize-collecting Steiner tree, known to be NP-hard, but for which efficient solvers have been recently invented [9, 10, 11].

Vandin *et al.* also analyze TCGA data in "Discovery of Mutated Subnetworks Associated with Clinical Data in Cancer," but with the goal of identifying subnetworks of genes with high mutation rates whose proteins are known to interact. This method adds to exploratory tools for investigating data from next-generation sequencing of tumors. The methods make use of graph diffusion kernels, which provide robust measures of association or similarity for graphs [12] and are the basis of the Google PageRank algorithm [13]. Graph kernels are a method of choice for extracting features for machine learning from network data, with applications ranging from predictions of protein similarity [14] to genetic epistasis [15].

The local subnetwork problem attacked by Chen *et al.* and Vandin *et al.* in these contributions has an analogous global subnetwork problem, carving a large network into subnetworks that reflect discrete biological processes or functions. Algorithms developed for clustering in other contexts, including powerful spectral clustering techniques [16, 17], have not performed well for biological networks or social networks. Modularity scores that maximize the enrichment of edges within groups are closely related to spectral clustering [18] and suffer from resolution limit problems in which small groups are merged inappropriately into larger groups [19]. Important recent progress has come from probabilistic models for hierarchical networks, essentially hierarchical stochastic block models. These methods can identify groups through Monte Carlo searches for small networks [20] and by approximate greedy algorithms for genome-size networks [21]. These methods have the advantage of extending readily to joint analysis of physical, genetic, gene regulatory, metabolic, and other types of interactions.

Converging problems and challenges

These examples highlight the use of networks as a framework for interpreting genome-scale data and predicting the response to new mutations. Important new challenges are motivated by the growing ability to obtain personal genome sequences, identifying variants that are unique to individuals. The future may involve a synthesis of methods designed for common variants, such as the copy number variants seen across TCGA samples, with methods designed for rare variants, leading to predictions of personal pathway activities and therapeutic options.

References

[1] Robert Tibshirani. Regression Shrinkage and Selection via the Lasso. *Journal of the Royal Statistical Society. Series B (Methodological)*, 58(1):267–288, 1996.

[2] Teri A Manolio, Francis S Collins, Nancy J Cox, David B Goldstein, Lucia A Hindorff, David J Hunter, Mark I Mccarthy, Erin M Ramos, Lon R Cardon, Aravinda Chakravarti, Judy H Cho, Alan E Guttmacher, Augustine Kong, Leonid Kruglyak, Elaine Mardis, Charles N Rotimi, Montgomery Slatkin, David Valle, Alice S Whittemore, Michael Boehnke, Andrew G Clark, Evan E Eichler, Greg Gibson, Jonathan L Haines, Trudy F C Mackay, Steven A Mccarroll, and Peter M Visscher. Finding the missing heritability of complex diseases. *Nature*, 461(7265):747, October 2009.

[3] Gregory W Carter, Susanne Prinz, Christine Neou, J Patrick Shelby, Bruz Marzolf, Vesteinn Thorsson, and Timothy Galitski. Prediction of phenotype and gene expression for combinations of mutations. *Molecular systems biology*, 3:96, 2007.

[4] O Alter, P O Brown, and D Botstein. Singular value decomposition for genome-wide expression data processing and modeling. *Proceedings of the National Academy of Sciences of the United States of America*, 97(18):10101–10106, August 2000.

[5] Ryan Kelley and Trey Ideker. Systematic interpretation of genetic interactions using protein networks. *Nature Biotechnology*, 23(5):561–566, May 2005.

[6] Ping Ye, Brian D Peyser, Xuewen Pan, Jef D Boeke, Forrest A Spencer, and Joel S Bader. Gene function prediction from congruent synthetic lethal interactions in yeast. *Molecular systems biology*, 1:2005.0026, 2005.

[7] Daniel Segrè, Alexander Deluna, George M Church, and Roy Kishony. Modular epistasis in yeast metabolism. *Nature genetics*, 37(1):77–83, January 2005.

[8] Trey Ideker, Owen Ozier, Benno Schwikowski, and Andrew F Siegel. Discovering regulatory and signalling circuits in molecular interaction networks. *Bioinformatics*, 18 Suppl 1:S233–40, 2002.

[9] I Ljubić, R Weiskircher, U Pferschy, and G Klau. Solving the prize-collecting Steiner tree problem to optimality. *Proceedings of ALENEX*, 2005.

[10] Marcus T Dittrich, Gunnar W Klau, Andreas Rosenwald, Thomas Dandekar, and Tobias Müller. Identifying functional modules in protein-protein interaction networks: an integrated exact approach. *Bioinformatics*, 24(13):i223–31, July 2008.

[11] M Bailly-Bechet, C Borgs, A Braunstein, J Chayes, A Dagkessamanskaia, J-M François, and R Zecchina. Finding undetected protein associations in cell signaling

by belief propagation. *Proceedings of the National Academy of Sciences of the United States of America*, 108(2):882–887, January 2011.

[12] Risi Imre Kondor and John Lafferty. Diffusion kernels on graphs and other discrete input spaces. *The Nineteenth International Conference on Machine Learning (ICML-2002)*, 2002.

[13] S Brin and L Page. The anatomy of a large-scale hypertextual Web search engine. *Computer Networks And Isdn Systems*, 30(1-7):107–117, 1998.

[14] Jason Weston, Andre Elisseeff, Dengyong Zhou, Christina S Leslie, and William Stafford Noble. Protein ranking: from local to global structure in the protein similarity network. *Proceedings of the National Academy of Sciences of the United States of America*, 101(17):6559–6563, April 2004.

[15] Yan Qi, Yasir Suhail, Yu-yi Lin, Jef D Boeke, and Joel S Bader. Finding friends and enemies in an enemies-only network: a graph diffusion kernel for predicting novel genetic interactions and co-complex membership from yeast genetic interactions. *Genome research*, 18(12):1991–2004, December 2008.

[16] Alex Pothen, Horst D. Simon, and Kang-Pu Liou. Partitioning sparse matrices with eigenvectors of graphs. *SIAM Journal on Matrix Analysis and Applications*, 11(3):430–452, July 1990.

[17] U Von Luxburg. A tutorial on spectral clustering. *Statistics and Computing*, 17(4):395–416, 2007.

[18] M E J Newman. Modularity and community structure in networks. *Proceedings of the National Academy of Sciences of the United States of America*, 103(23):8577–8582, June 2006.

[19] Santo Fortunato and Marc Barthélemy. Resolution limit in community detection. *Proceedings of the National Academy of Sciences of the United States of America*, 104(1):36–41, 2007.

[20] Aaron Clauset, Cristopher Moore, and M E J Newman. Hierarchical structure and the prediction of missing links in networks. *Nature*, 453(7191):98–101, May 2008.

[21] Yongjin Park and Joel S Bader. Resolving the structure of interactomes with hierarchical agglomerative clustering. *BMC Bioinformatics*, 12 Suppl 1:S44, 2011.

SSLPRED : PREDICTING SYNTHETIC SICKNESS LETHALITY*

NIRMALYA BANDYOPADHYAY†, SANJAY RANKA, AND TAMER KAHVECI

CISE Department, University of Florida,
Gainesville, FL 32611, USA
E-mail: {nirmalya†, ranka, tamer}@cise.ufl.edu
www.cise.ufl.edu

Two genes in an organism have a *Synthetic Sickness Lethality* (SSL) interaction, if their joint deletion leads to a lower than expected fitness. Synthetic Gene Array (SGA) is a technique that helps in identifying SSL values for pairs of genes in a given set of genes. SSL interactions are useful to discover the co-expressed gene groups in the regulatory and signaling networks. Also, they are used to unravel the pair of pathways (subset of physically interacting genes) that substitute the functions of each other. Generating an SGA entry is costly as it requires producing and monitoring a *double mutant* (a progeny with two mutated genes). Generating a comprehensive SGA can be very expensive as the number of gene pairs is quadratic in the number of genes of the corresponding organism.

In this paper, we develop a new method *SSLPred* to predict the SSL interactions in an organism. Our method is built on the concept of *Between Pathway Models (BPM)*, where majority of the SSL pairs span across the two functionally complementing pathways. We develop a regression based approach that learns the mapping between the gene expressions of single deletion mutant to the corresponding SGA entries. We compare our method to the one by Hescott et al. for predicting the GI (Genetic Interaction) score of *Saccharomyces cerevisiae* (*S. cerevisiae*) on four benchmark datasets. On different experimental setups, on average SSLPred performs significantly better compared to the other method.

Keywords: Synthetic Sickness Lethality, Regression, Essential Genes

1. Introduction

Analysis of gene essentiality is a crucial problem to understand the roles of different genes at the molecular and genetic levels. A gene is defined *essential* if it is required for proper growth and sustenance of that organism. Essential genes have been thoroughly investigated using techniques such as single gene deletion screening for some low level organisms such as *Escherichia coli (E. coli)*.[1] Though identification of essential genes enlightens us about the functions of individual genes in an organism, it provides little conclusive information about the nature of their genetic relationships in gene regulatory and signaling networks. Recently, studies on *Synthetic Sickness Lethality (SSL)* opened up new directions in the areas of functional genomics. Two *non-essential* genes follow an SSL interaction if their joint deletion leads to a less than expected *fitness* for the organism. Here *fitness* denotes the growth and sustenance rate of an organism. An expected fitness corresponds to that of a double mutant when the two knocked out genes are not in an SSL interaction. Note that the fitness of an organism due to an SSL interaction can be less than (aggravating) or more than (alleviating) the expected fitness.[2] A genome wise catalog of SSL interactions enables in-depth molecular analysis, by creating a functional map of the cell, predicting functions and relations of the genes and deciphering complex regulatory relations from the global genetic network.[3]

The Synthetic Genetic Array (SGA)[4] and diploid-based synthetic lethality analysis on microarray (dSLAM)[5] are two pioneering approaches that enable systematic identification of SSL

*This work was supported partially by NSF under grants CCF-0829867 and IIS-0845439.

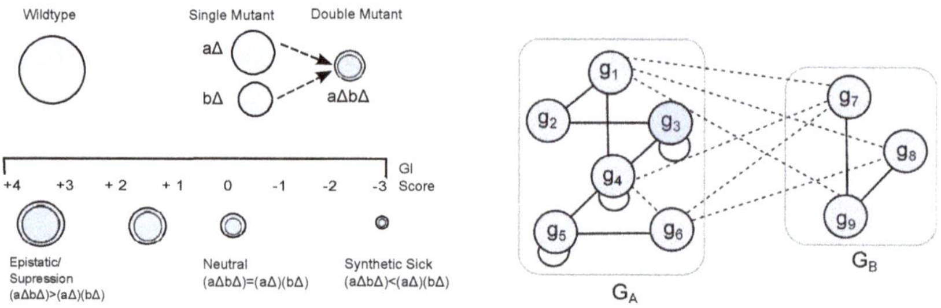

(a) Synthetic Sickness Lethality

(b) Between Pathway Models

Fig. 1. Figure 1(a) illustrates the concepts of synthetic sickness and lethality. A double mutant produced from the cross of two single mutants can have a specific fitness in a range of GI scores based of the relationship between the two genes. The single and double circles represent single and double mutants respectively. Here, the size of a circle corresponds to the observed fitness of the corresponding mutant. Based on whether the two genes have a epistatic, neutral of SSL interaction, the observed fitness of the double mutant can have more than, equal to or less than the expected fitness. In those cases, the GI score can be a significantly large positive, close to zero or significantly large negative number, respectively. Figure 1(b) depicts the concept of Between Pathway Models (BPM). The hypothetical BPM consists of two sub-networks (also called pathways) G_A and G_B who are functionally independent and complementing. The solid lines denote physical interactions, while the dashed directed lines stand for the SSL interactions. It is evident that the number of SSL edges between G_A and G_B is higher compared to the ones within the two groups.

interactions. Both methods require generation of double mutant strains and monitoring their growth. Each entry of SGA is a triple that consists of two genes and a GI (genetic interaction) score for those two genes. A score close to zero indicates that there is no SSL interaction between the two genes. For a gene-pair, a negative GI with a large magnitude indicates an aggravating SSL interaction. A significant large positive number denotes a higher chance of being alleviating relationship.[2] Figure 1(a) illustrates the concepts of synthetic sickness and lethality relationship. The EMAP strategy exploits the SGA technique by enabling colony sizes to be measured in an array format, thus quantifying genetic interactions in a high-throughput fashion.

Both SGA and dSLAM are costly techniques as for a pair of genes they require creation of two single mutant strains and crossing between them to produce a double mutant strain. For an organism with N genes, we need to generate and monitor the growth of $\frac{N(N-1)}{2}$ different double mutants. As a result, millions of double mutants need to be produced to tabulate all the genetic interaction scores for an organism that consists of thousands of genes. Creating such double mutants in wet-lab is an expensive and time consuming process. Therefore, we need an efficient method to predict whether there exists a synthetic lethality relation between two genes. Briefly, we can describe the problem considered in this paper as follows: *given two genes g_A and g_B what is the GI score between them?*

In order to predict the GI score between two genes, we incorporate the *genetic profile* of single mutant strains. This is a promising strategy for the number of single mutants can not be more than the number of genes. First, we elaborate on the term genetic profile. Consider single gene knockout dataset (also termed as single gene mutant data). Here, in each experiment a non-essential gene is knocked out from an organism. For each gene, expressions are obtained before and after the knock-out and ratio of this after to before expression is calculated. Finally, the logarithm of that ratio is computed and tabulated. If the magnitude of a logarithm is large, it indicates that the expression of the corresponding gene changed significantly after the knockout of the non-essential gene under consideration. The genetic profile for a single mutant or a single gene knockout experiment consists of entries for all genes computed in the way described above.

In this paper our objective is to learn the GI scores of gene pairs with the help of *genetic profile* of single mutants. Formally we solve the following problem.

Problem *Let* $\mathcal{V} = \{g_1, g_2, \cdots, g_M\}$ *denote the set of genes in an organism. Assume that we are given the genetic profiles of* K *single mutant genes.* $\mathcal{X}$ *is a* $K \times M$ *matrix, where each row corresponds to the genetic profile of a single mutant. Let us represent the GI score of gene pairs* g_a *and* g_b *with* $t_{a,b}$*. Let* $\mathcal{T}$ *denote the set of all the available GI scores for that organism. For any gene pair* (g_i, g_j) *such that* $g_i \in \mathcal{V}, g_j \in \mathcal{V}$*, we would like to predict the GI score.* □

Before discussing our contribution in this paper, we summarize the Between Pathway Models (BPM), which is a building block of our model.[6] A BPM consists of two gene subnetworks (also called pathways) G_A and G_B, such that there are few SSL interactions within G_A and within G_B, but many of those between G_A and G_B. The opposite holds for the physical interaction edges. That is, many physical interactions exist within G_A and G_B, but few of them exist between G_A and G_B. Figure 1(b) depicts a hypothetical BPM. According to Kelley and Ideker, the two pathways in a BPM are functionally compensating due to the orientation of genetic and physical edges.[6] Now that we have introduced all the relevant building blocks, we discuss our contribution in this paper.

Contribution In this paper, we develop a new method *SSLPred* to predict the GI scores. To our knowledge, our method is the first one to predict GI scores using a mathematical machine learning based technique.

In accordance with the concept of BPM, we propose the following conjecture. If there is an SSL interaction between two genes and if these two genes belong to two pathways of a BPM, then knocking out one of them will change the expressions of most of the genes in both of the pathways in that BPM. The pathway containing the mutated gene is directly affected and dysfunctional as most of the consisting genes have a direct connection with the mutated gene through physical edges. The other pathway compensates for its affected pair, and due to the additional activities the genes in it change their expressions noticeably.

In our regression based method SSLPred, we develop a mapping between the genetic profiles of single mutants and the corresponding GI score. For every genetic interaction entry $(g_a, g_b, t_{a,b})$, such that either of g_a and g_b has been mutated in a single mutant gene experiment and $t_{a,b}$ is the GI score for g_a and g_b, we create a training sample. As we have already conjectured in the previous paragraph, if this genetic interaction entry represents an SSL, the mutated gene affects the expressions of all the genes in the corresponding BPM. Thus, we use the gene expression changes only from the pathways of that BPM to extract the features of the training point and correlate it with the corresponding GI score $t_{a,b}$ using a regression model. After we estimate the parameters of SSLPred, we are able to predict the GI score for a new pair of genes.

We compare our method to the one by Hescott et al.[7] in their ability to identify BPMs in the gene networks of *S. cerevisiae* on four benchmark datasets. On average SSLPred performs significantly better compared to the other method. We summarize our contribution as follows:

(1) According to our knowledge, SSLPred is the first *predictive method* to predict the GI score for a pair of genes. All other relevant computational methods are *descriptive*.
(2) The GI scores predicted by SSLPred assume a real value. This is more useful than a binary prediction, since it enables to conduct statistical analysis such as permutation tests and p-Value generation associated with the validation of benchmark BPMs.

The rest of the paper is organized as follows. Section 2 discusses the related work. Section 3 describes our method SSLPred. Section 4 presents the experimental results. Finally, Section 5 concludes the paper.

2. Background

Recent studies on synthetic sickness and lethality analysis opened up new directions in the areas of functional genomics. These works can be classified into two categories, namely experimental and computational. Experimental methods include Synthetic Gene Array (SGA),[4] dSLAM (diploid-based synthetic lethal analysis with microarrays)[5] and EMAP (epistatic mini-array profile).[8] We describe them in detail in Section 1. We summarize the computation methods next.

Kelly and Ideker introduced Between Pathway Models (BPM) by combining synthetic sickness and lethality informations from EMAP data with information on protein-protein, protein-DNA or metabolic networks.[6] Hescott et al.[7] proposed a new method to validate BPMs using single gene deletion microarray data. They evaluated the quality of the BPMs from four different studies and described how their methods might be extended to refine BPM pathways. Kelley and Kingsford developed a new method called Expected Graph Compression to identify compensatory pathways (BPMs) by clustering genes into modules and establishing relationships between those modules.[9]

Though researches in these two avenues enriched our understanding of gene interactions and gene networks, we did not find any predictive model to predict the GI scores. In this paper, we proposed a new regression based method SSLPred to predict GI scores.

3. Methods

In this section, we discuss our method in detail. Section 3.1 describes the notation and formulates the problem. Section 3.2 explains our conjectures which guide our model and the rationale behind it. Section 3.3 discusses the feature extraction and regression model.

3.1. *Problem Formulation and Notation*

In this section, we mathematically formulate the problem, and for that purpose, we describe the relevant notation. We group our notation in three classes based on three related entities. These are gene network, single mutant data and SGA. Here, gene network stands for gene interaction network, specifically the union of gene regulatory and signaling network.

(1) Gene network. The gene network is a union of gene regulatory and signaling networks that can be modeled as a set of genes and the directed edges (i.e., interactions) connecting these genes. Here, an edge between two genes denotes different kinds of genetic interactions such as activation, inhibition and phosphorylation. Let us denote the set of all M genes by $\mathcal{V} = \{g_1, g_2, \cdots, g_M\}$. We denote the set of all edges in the gene network by $\mathcal{W} = \{(g_i, g_j) | g_i \in \mathcal{V}, g_j \in \mathcal{V}\}$, where (g_i, g_j) implies a directed interaction from g_i to g_j. thus, $G = (\mathcal{V}, \mathcal{W})$ defines

the gene network.

(2) SINGLE MUTANT DATASET. In a single mutant, one gene is mutated in an organism and gene expression is obtained before and after the mutation. *Single deletion mutant* (also known as *single gene knockout*) is an important kind of gene mutant, where one gene is knocked out from an organism. In a single mutant dataset, each entry contains the logarithm of the ratio of the expressions of a gene after the gene knockout to that of the same gene before the gene knockout.[10]

Let $e'_{h,j}$ and $e_{h,j}$ denote the expressions of the gene g_j after and before the mutation of g_h respectively. We define the genetic profile of the organism when gene g_h is mutated by $X_h = \{x_{h,j} | x_{h,j} = ln(e'_{h,j}/e_{h,j}), j \in \{1, 2, \cdots, M\}\}$. Let $\mathcal{H} \subseteq \mathcal{G}$ be the set of genes that have been mutated in total. We define the single mutant genetic profile of N genes as $\mathcal{X} = \{X_h | g_h \in \mathcal{H}\}$.

(3) SYNTHETIC GENE ARRAY. An SGA is a set of triples, $\mathcal{T} = \{(g_i, g_j, t_{i,j}) | i, j \in \{1, 2, \cdots, M\}, i \leq j\}$, where $t_{i,j}$ is a real number that corresponds to the ratio of the observed fitness to the expected fitness when the organism has both gene g_i and g_j knocked out. A value with a large magnitude implies a potential SSL edge. A positive and a negative value stand for alleviating and aggravating relations, respectively.

Problem Formulation *Given a gene network $\mathcal{G}$, the single mutant dataset $\mathcal{X}$ and the SGA dataset $\mathcal{T}$, find the mapping $\Upsilon : \mathcal{X}, \mathcal{G} \longrightarrow \mathcal{T}$ which minimizes a predetermined risk function.*

Risk function is a measure of expected miss prediction rate. In this paper, while estimating the mapping function Υ, we minimize least square error in order to minimize expected miss prediction rate. Based on the mapping learned, we would predict the GI score $t_{i,j}$ for a new double mutant whose two genes g_i and g_j haven been mutated.

3.2. *Between Pathway Conjectures*

In this section, we describe our two conjectures that are central to SSLPred and the rationale for them. These two conjectures are built on the concepts of BPMs. Incorporating the structure and properties of BPMs into our model to improve its prediction accuracy was the motivation behind these conjecture.

CONJECTURE 1. Let $\mathcal{B}$ denote a BPM, consisting of two pathways G_A and G_B. Also, consider an SSL edge $\mathcal{S} = \{g_a, g_b\}$ such that $g_a \in G_A$ and $g_b \in G_B$. Then, mutating g_a will significantly alter the expressions of many genes in G_A and G_B. □

Since, g_a is connected to most other genes in G_A through physical interactions, altering the expression level of g_a will affect the expression of all the genes connected to g_a as they regulate each other. This effect will propagate through the gene network and eventually may change the expressions of many genes in G_A. Eventually g_a will severely affect G_A and prohibit it from working properly. Since G_A and G_B constitute a BPM, G_B will compensate this loss by changing the expression of the genes in G_B. Thus, mutating g_a eventually changes the expressions of the genes in both G_A and G_B.

From this conjecture, we conclude that there is a mapping between an SGA entry $(g_a, g_b, t_{a,b})$ and the corresponding single mutant dataset X_h, $g_h \in \{g_a, g_b\}$. This implies a non-trivial mapping, if the SGA entry corresponds to an SSL or epistatic relationship and we have a higher chance to find both g_a and g_b embedded in two pathways of a BPM. In that case, most of the genes in that BPM are supposed to have their expression changed in the single mutant dataset and an appropriate regression method can correlate the changes in the single mutant gene expressions and the corresponding GI score.

Before stating the second conjecture, we define a relevant term, *neighbor*. We say that, gene g_b is an rth layer incoming neighbor of gene g_a in the directed gene network, if the shortest path from g_b to g_a consists of r directed edges. In that case, g_a is a rth layer outgoing neighbor of g_b. Figure 2 depicts the incoming and outgoing neighbors for the two genes in a genetic interaction.

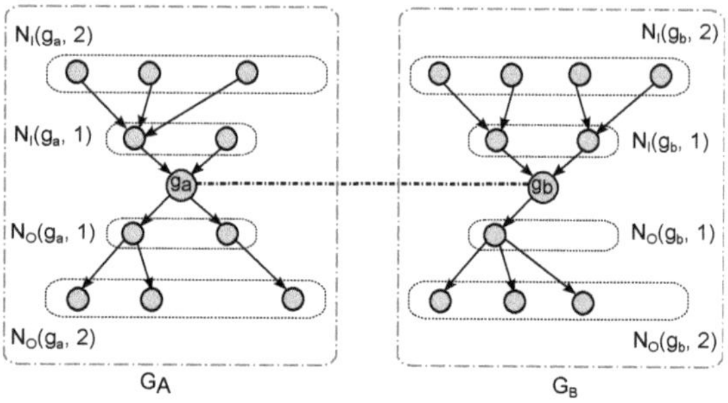

Fig. 2. This figure depicts the layered neighbor structure around a gene interaction edge (g_a, g_b). $N_{IN}(g_h, r)$ denotes the set of incoming neighbors of gene g_h at layer r. The set outgoing neighbor $N_{OUT}(g_h, r)$ is defined similarly. The example contains only up to 2 layers for each direction and gene. The dotted rectangles denote the putative BPM (G_A, G_B) around the gene interaction edge.

CONJECTURE 2. Let $\mathcal{B}$ denote a BPM, consisting of two pathways G_A and G_B. Consider an SSL edge $\mathcal{S} = \{g_a, g_b\}$ such that $g_a \in G_A$ and $g_b \in G_B$. If the expression of g_a changes significantly, then in G_A expression change is most prominent in the first layer of neighbors of g_a and gradually decreases with increasing layers. Similarly, in G_B the effect is most prominent for g_b and gradually decreases with increasing layers of neighbors. □

In brief, our conjecture is that the effect of a gene knockout eventually wanes away through the gene network. The rationale behind this is that the neighbors that are close to g_a and g_b have a higher chance of being connected only to the nodes of $\mathcal{B}$. The genes in the distant neighborhood have a greater possibility to take part in other pathways. Hence, the closer nodes are more susceptible to undergo a major effect, while the distant neighbors are supposed to be partially screened from that affect due to their activity in the other pathways. Based on these two conjectures, we build a regression based model that we describe in the next section.

3.3. *Regression based solution*

This section describes the customized regression based approach that we developed to build the mapping $\Upsilon : \mathcal{X}, \mathcal{G} \longrightarrow \mathcal{T}$, where $\mathcal{X}$, $\mathcal{G}$ and $\mathcal{T}$ denote the single mutant gene expression, gene network and GI score, respectively. Based on the two conjectures in Section 3.2, we extract a set of features for training and testing samples.

We start from the SGA and for each entry $(g_a, g_b, t_{a,b})$, we create a sample point provided either g_a or g_b has been mutated in the single mutant dataset available to us, otherwise we discard that SGA entry. Without losing the generality, assume that g_a has been mutated in this case. Thus, we extract the feature functions from the single mutant data $X_a = \{x_{a,1}, x_{a,2}, \cdots, x_{a,M}\}$. In designing the set of features, we leverage the information from gene networks by incorporating the two conjectures in our solution. According to the first conjecture, the mutated gene is suppose to perturb only the genes in the host BPM. Thus, while processing the SGA entry $(g_a, g_b, t_{a,b})$,

Table 1. The table summarizes the feature functions of the regression model and the corresponding parameters. Feature function represents the set of different features for the regression. A parameter quantifies the strength of the corresponding feature function.

Feature Function	Parameter	Description
$\Psi_{IN}(N_{IN}(g,r))$	$w_{IN}(g,r)$	For incoming neighbors of rth layer for gene g.
$\Psi_{OUT}(N_{OUT}(g,r))$	$w_{OUT}(g,r)$	For outgoing neighbors of rth layer for gene g.
$\Psi(g_b)$	w	For gene g_b when considering (g_a, g_b), g_a is knocked out.
	w_0	A constant representing the bias of the regression.

we consider only the genes from G_A and G_B and discard the ones from $\mathcal{G} - (G_A \cup G_B)$. We use the GI score $t_{a,b}$ as the label of the training sample.

Note that while we create the features for a training point, all the data we have is the single mutant data, GI scores and the gene networks. However, for a specific pair of genes (g_a, g_b) we do not know the set of genes that consists of the putative BPM $\mathcal{B} = (G_A, G_B)$ around the gene pair. Rather, we are suppose to validate that information using our model. In fact, if the SGA entry does not correspond to an SSL, there may not be a *real* BPM for the pair (g_a, g_b). To circumvent this problem, we assume the BPM as part of our model rather an input to the model.

Specifically, we use the concept of *rth layer neighbors*, introduced in Section 3.2. Let R represent the maximum number of layers to construct G_A and G_B. (Usually, R will be set by the user.) Let us denote the rth layer incoming and outgoing neighbors of gene g_h by $N_{IN}(g_h, r)$ and $N_{OUT}(g_h, r)$ respectively. Figure 2 demonstrates the layered structure of incoming and outgoing neighbors for a pair of genes. We define the putative BPM pathway for g_h as the union of the sets of incoming and outgoing neighbors of g_h up to the layer Rth given by,

$$G_H = \bigcup_{r=1}^{R} (N_{IN}(g_h, r) \cup N_{OUT}(g_h, r)) \tag{1}$$

In a comprehensive SGA data each GI score is a real valued number that varies in the range of two small numbers such as -5 to +5. However, if the score $t_{a,b}$ has a small magnitude (close to zero), the gene pair (g_a, g_b) may not have an SSL/epistatic interaction and may not be part of a BPM. Since the GI score $t_{a,b}$, which is the label of the regression model is real valued, we still shall use this sample point to train our model. However, the regression model is expected *not* to discover any interesting pattern of a BPM in the gene expression, and will adjust its parameters accordingly.

To incorporate the second conjecture, we design the features of the regression in a layered approach that directly depends on the concept of layered neighbors introduced in Section 3.2. We denote the feature function associated with the incoming neighbors of layer r of gene g by $\Psi_{IN}(N_{IN}(g,r))$ and the corresponding regression parameter by $w_{IN}(g,r)$. Similarly, the feature function and parameters for the rth layer outgoing neighbor are given by $\Psi_{OUT}(N_{OUT}(g,r))$ and $w_{OUT}(g,r)$, respectively. Thus, for $J \in \{IN, OUT\}$ we state that the feature function $\Psi_J(N_J(g_c, r))$ corresponds to neighbors of gene g_c in direction J at layer r. Given that g_h has been knocked out and we are considering the neighborhood of g_c, $\Psi_J(N_J(g_c, r))$ can be defined as follows,

$$\Psi_J(N_J(g_c, r)) = \frac{\sum_{g_i \in N_J(g_c, r)} |x_{h,i}|}{|N_J(g_c, r)|} \tag{2}$$

We define another feature function for g_b by $\Psi(g_b)$ and the corresponding parameter by w.

However, we do not create any feature function to capture the expression of g_a, since g_a is mutated and its expression may not be available for inspection. Finally, we create the last parameter w_0 that acts as a bias constant in the model. Table 1 summarizes the feature functions and the corresponding parameters. By aggregating all these feature functions, we can fit the SGA entry $(g_a, g_b, t_{a,b})$, where g_a is knocked out in the single mutant data as,

$$y_{a,b} = w_0 + w\Psi(g_b) + \sum_{\substack{r \in \{1,2,\cdots,R\},\\ J \in \{IN,OUT\},\ c \in \{a,b\}}} w_J(g_c, r)\Psi_J(N_J(g_c, r)) \quad (3)$$

Parameter Estimation. When the ratio of number of samples to that of the parameters is small (typically less than 20), the estimated value of the parameters experience high variance due to overfitting of data.[11] To alleviate this problem, we augment a regularization term on top of the regression model. Specifically, we aim to minimize the difference between the parameter values at neighbor levels r and r+1, to smoothen the decaying of gene expression change. Formally, the regularization term can be written as,

$$Q = \sum_{\substack{r \in \{1,2,\cdots,R-1\},\\ J \in \{IN,OUT\},\ c \in \{a,b\}}} |w_J(g_c, r+1) - w_J(g_c, r)| \quad (4)$$

We augment this regularization term with the objective function when estimating the parameters. Using least square error approach, we estimate the parameters of the regression by minimizing the following,

$$E = \sum_{a,b \in \{1,2,\cdots,M\},\ a<b} (t_{a,b} - y_{a,b})^2 + \lambda Q \quad (5)$$

A detailed discussion on the simplification of the regularization term can be found in Section 3 of Tibshirani et al.[12] We use the interior-point method to solve this parameter estimation problem.[13] The value of λ is estimated using five fold cross validation.

After the parameter estimation step of the regression method is complete, SSLPred is prepared to predict the GI score for a test sample. For a pair of test genes, we extract the set of features in the same way as that of a training point. Plugging those extracted features and the estimated parameters in Equation 3, we obtain the predicted GI score.

4. Experiments

In this section, we describe the experiments and discusses the results. Section 4.1 describes the datasets we used for the experiments. Section 4.2 demonstrates SSLPred with another relevant method recently published by Hescott et al.[7,14]

4.1. *Datasets*

As implied in Section 3.1, we classify the datasets into three different categories, namely, single gene mutant data, SGA data and gene networks. We decided on collecting datasets for *S. cerevisiae*, since this is a well researched organism of yeast with extensive datasets available for all these three categories.[3,10,15] We extracted BPMs from four studies, which were also used by

Hescott et al. to validate their method.[14] We employed these BPMs as gold standards in this paper. Next, we describe these datasets in detail.

(1) **Gene network BioGRID.** The dataset maintains one of the most comprehensive gene networks for S. cerevisiae.[15] We collected 155,287 the genetic interactions in total from this database.
(2) **Single mutant data.** In this paper, we collected 287 single gene knockout experiments from the compendium of expression profile of *S. cerevisiae* developed by Hughes et al.[10] Each experiment contains 6,316 entries. Every entry contains the the logarithm of after to before ratio of expressions of a gene as described in Section 3.1.
(3) **SGA data.** Costanzo et al. generated a genome scale SGA profile for *S. cerevisiae* with neatly 5.4 millions of genetic interactions out of nearly 75% genes.[3] Out of this comprehensive profile, we selected GI scores for 370,913 interactions such that for every edge, at least one of the two consisting genes was knocked out in the gene knockout experiments.
(4) **BPMs.** We obtained four sets of BPMs, all are of *S. cerevisiae*, itemized in the following – *Kelley-Ideker*,[6] *Ulitsky-Shamir*,[16] *Brady et al.*,[17] and *Ma et al.*[18] We denote a dataset using the authors' names of the corresponding paper. The numbers of BPMs that contain three or more genes in each pathway in these datasets are 160, 36, 959 and 54 respectively.

4.2. *Comparison with Hescott's Method*

This section describes the comparison between SSLPred and the method proposed by Hescott et al.[7,14] Hescott et al. employs microarray expression data of single gene knockout experiments to identify BPMs. Though their method does not predict GI score, according to our knowledge, this is the only published method that integrates the concept of single gene mutants and between pathway motifs.

Before coming to the main discussion, we describe how we create a matrix of predicted GI scores using five-fold cross validation. We divide the 287 knockout experiments into nearly equal five groups, each of them being a 57×6316 matrix. For each fold of cross validation, we use four out of five groups to create sample training points along with the corresponding GI scores as described in Section 3.3. If for a gene pair the corresponding GI score is not available, we discard the that sample point. After training, we create test points from the left-out one group and predict the test scores for them. Repeating this process in a five fold cross validation fashion, we predict GI scores for all possible pairs of genes from the 287×6316 matrix. Now that we have the predicted GI matrix which we denote by $\mathcal{T}_P$, we discuss how we employ it for comparison between SSLPred and the one proposed by Hescott et al.

Consider a BPM $\mathcal{B} = (G_A, G_B)$ obtained from a known sets of BPMs. Now, consider a gene $g_x \in G_A$. Hescott et al. ranks all the genes $\mathcal{G}$ of the organism with respect to g_x. Let us denote that rank by $\mathcal{G}_\Phi(g_x)$. Then, from that rank, it retrieves G_B and calculate the quality of retrieved G_B by a scoring method called *ClusterRankScore*. We now describe ClusterRankScore which is adapted from Gene Set Enrichment Analysis.[19]

ClusterRankScore accepts an ordered list of genes $\mathcal{L}$ and another set $\mathcal{C}$ as input. Then, it explore the distribution of $\mathcal{C}$ along $\mathcal{L}$. Intuitively, if $\mathcal{C}$ appears at the head or tail of $\mathcal{L}$, it is enriched with the specific properties represented by the ordered list $\mathcal{L}$. In the current context, consider a BPM $\mathcal{B} = \{G_A, G_B\}$. Let us knock out a gene g_a from G_A and measure the change

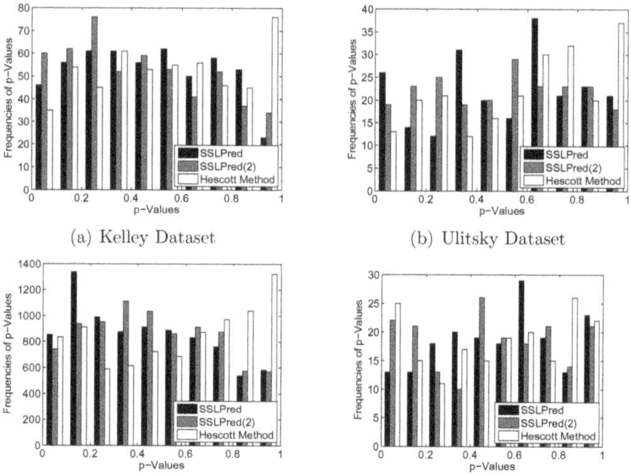

(a) Kelley Dataset (b) Ulitsky Dataset

(c) Brady Dataset (d) Ma Dataset

Fig. 3. Comparison of SSLPred with the method from Hescott et al. SSLPred and SSLPred(2) denote variants of SSLPred with at most one or two layers of neighbors, respectively. The X axis that ranges between zero and one represents the p-Values of the permutation tests. The Y axis represents the frequencies of the BPMs at a particular p-Value of the permutation test. The four sub-figures demonstrate that apart from on Ma dataset, SSLPred outperforms Hescott's method, as it maintains a higher frequency at the p-Value ranges between zero and 0.1.

Table 2. BPMs with p-Values less than 0.1 [%]

Method	BPMs			
	Kelley-Ideker	Ulitsky-Shamir	Brady et al.	Ma et al.
SSLPred	8.74	11.71	9.95	7.02
SSLPred (2)	11.4	8.55	8.64	11.89
Hescott et al.	6.65	5.85	9.75	13.51

of expressions for all the other genes. Hescott et al. now arrays the genes according to its own criteria. Here, this ordered list is $\mathcal{L}$ and the pathway G_B is $\mathcal{C}$. Thus, a correlation between the ordered gene list and the pathway G_B implies that the BPM $\mathcal{B}$ is validated by Hescott et al.

Using SSLPred we create a similar rank as follows. We obtain the predicted GI score of all the gene pairs (g_x, g_y), $g_y \in \mathcal{G}$ from the predicted GI matrix $\mathcal{T}_P$. Then we sort $\mathcal{G}$ in increasing value of the retrieved GI scores of (g_x, g_y). Let us denote the sorted list of genes by $\mathcal{G}_\Psi$. After this we calculate the ClusterRankScore of G_B based on $\mathcal{G}_\Psi$. Let us denote the ClusterRankScore of G_B with respect to Hescott et al. and SSLPred by $CRS_\Phi(g_x, G_B)$ and $CRS_\Psi(g_x, G_B)$, respectively.

To calculate the statistical significance of the two ClusterRankScore, we design separate permutation tests for each of them and calculate p-Values with respect to those permutation tests. Here the null hypothesis can be stated as $\mathcal{B}$ *is not a BPM*. A detailed account of *ClusterRankScore* and the permutation test can be found at Hescott et al.[7,14]

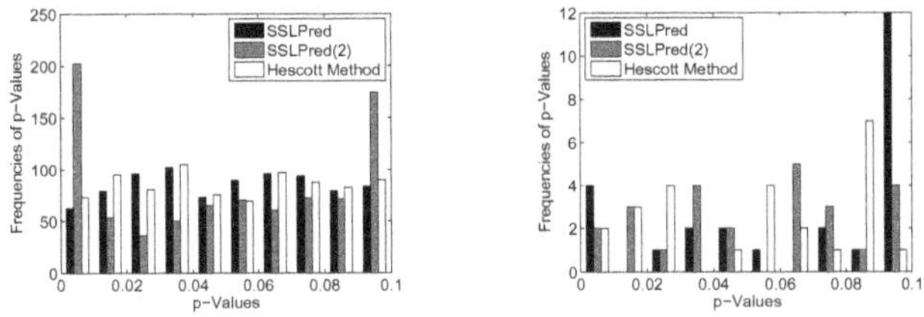

(a) Brady Dataset　　(b) Ma Dataset

Fig. 4. Comparison of SSLPred with the method from Hescott et al. SSLPred and SSLPred(2) denote variants of SSLPred with at most one or two layers of neighbors, respectively. The X axis, that ranges between zero and 0.1, represents the p-Values of the permutation tests. The Y axis represents the frequencies of the pathways at a particular p-Value of the permutation test. We display histograms for the very two datasets for which our method performs similar or worse in Figure 3. The two sub-figures demonstrate that on these two specific datasets, SSLPred maintains a higher frequency at the p-Value ranges between zero and 0.01.

Consider a BPM $\mathcal{B} = (G_A, G_B)$. For all the combinations, $(g_x, G_B), g_x \in G_A$ and $(g_y, G_A), g_y \in G_B$ we calculate the p-Values using the procedure described above. For every dataset, we plot the histograms of those p-Values. Since all the BPMs have been obtained from published literature, we assume them to be equivalent to a gold standard. Hence, in the histogram, an increased frequency of the BPMs with lower p-Values corresponds to a better quality of the BPM retrieval method.

Figure 3 compares SSLPred with the other method. SSLPred and SSLPred (2) denote variants of SSLPred with at most one or two layers of neighbors, respectively. The X axis, that ranges between zero and one, represents the p-Values of the permutation tests. The Y axis represents the frequencies of the BPMs at a particular p-Value of the permutation test. SSLPred outperforms Hescott et al. by 100%, 31%, 2% for Ulitsky, Kelley and Bradly dataset when the p-Value is equal to or smaller than 0.1. For SSLPred (2) the corresponding numbers are 46%, 71% and -12% respectively. For Dataset Ma, Hescott et al. is better by 48% and 12% than SSLPred and SSLPred (2) respectively. If we relax the p-Value to 0.3 we observe that SSLPred (2) performs better than Hescott et al. by 24%, 12%, 10% and 48% for Ulitsky, Brady, Ma and Kelley respectively. It can be concluded that that apart from on Ma dataset, SSLPred outperforms Hescott's method, since it maintains a higher frequency at the p-Value ranges between zero and 0.1. For Kelley and Ulitsky dataset SSLPred outperforms with a high margin between 0 to 0.1 p-Value range. For Ma and Brady datasets, the two methods perform very competitively on an average. Also, its is difficult to compare between two variants of SSLPred. Though these two variants are in close competition, SSLPred (2) has a slightly better advantage over SSLPred. Table 2 summarizes the results in Figure 3 by tabulating the percentage of BPMs with p-Values less than or equal to 0.1.

Figure 4 highlights a special case of the frequency distribution when the p-Value is restricted to be less than or equal to 0.01. We are specifically interested in the very two datasets Brady and Ma for which our method performs similar or worse when we restrict the p-Values to 0.1. Since a lower p-value implies a lower chance of false positive detection, these results are important in determining the superiority of the competing methods. Here also we observe that SSLPred (2) has a better accuracy compared to Hescott et al. in identifying larger number of small p-Value BPMs. For Brady dataset SSLPred (2) outperforms the other method by 177%. For Ma dataset both of them detect two BPMs. This concludes that, our method demonstrates superior accuracy in validating more BPMs with very low p-Values (≤ 0.01). We also conducted a third set of experiments SSLPred (3) with highest layer of neighbors $R = 3$. However, SSLPred (3) performed poorly compared to Hescott et al. We believe that most BPMs are of small sizes with a diameter of less that equal to four edges as indicated in Kelley et al.[9] Hence, assuming a bigger size BPMs with $R = 3$ compromises the accuracy of our method.

Code. All the code developed in this paper is available from `http://bioinformatics.cise.ufl.edu/projects/SSLPred.html`.

5. Conclusion

In this paper, we developed a new method SSLPred to predict SSL interactions in an organism. Our method is built on the concept of Between Pathway Models, where majority of the SSL pairs span across the two functionally complementing pathways. We developed a regression based approach that learns the mapping between the gene expressions of single deletion mutant to the corresponding synthetic gene array.

We compared our method to the one by Hescott et al. for predicting the GI scores of *S. cerevisiae* on four benchmark datasets. On different experimental setups, on average SSLPred performed significantly better compared to the other method.

References

1. T. Baba, T. Ara and M. H. et al., *Mol Syst Biol* **2**, p. 2006.0008 (2006).
2. P. Beltrao, G. Cagney and N. Krogan, *Cell* **141**, 739 (2010).
3. M. Costanzo, A. Baryshnikova and J. B. et al., *Science* **327**, 425 (2010).
4. A. Tong, M. Evangelista and A. P. et al., *Science* **294**, 2364 (2001).
5. X. Pan, D. Yuan and D. X. et al, *Mol Cell* **16**, 487 (2004).
6. R. Kelley and T. Ideker, *Nat Biotechnol* **23**, 561 (2005).
7. B. Hescott, M. Leiserson and L. C. et al., *J Comput Biol* **17**, 477 (2010).
8. S. Collins, K. Miller and N. M. et al., *Nature* **446**, 806 (2007).
9. D. Kelley and C. Kingsford, *J Comput Biol* **18**, 379 (2011).
10. T. R. Hughes, M. J. Marton and A. R. J. et al, *Cell* **102**, 109 (2000).
11. V. S. Cherkassky and F. Mulier, *Learning from Data: Concepts, Theory, and Methods*, 1st edn. (John Wiley & Sons, Inc., New York, NY, USA, 1998).
12. R. Tibshirani, M. Saunders and S. R. et al., *Journal of the Royal Statistical Society Series B* , 91 (2005).
13. M. H. Wright, *Bull. Amer. Math. Soc. (N.S)* **42**, 39 (2005).
14. B. J. Hescott, M. D. M. Leiserson and L. C. et al., Evaluating between-pathway models with expression data, in *RECOMB*, 2009.
15. B. Breitkreutz, C. Stark and T. R. et al., *Nucleic Acids Res* **36**, D637 (2008).
16. I. Ulitsky and R. Shamir, *BMC Syst Biol* **1**, p. 8 (2007).
17. A. Brady, K. Maxwell and N. D. et al., *PLoS One* **4**, p. e5364 (2009).
18. X. Ma, A. Tarone and W. Li, *PLoS One* **3**, p. e1922 (2008).
19. A. Subramanian, P. Tamayo and V. M. et al., *Proc Natl Acad Sci U S A* **102**, 15545 (2005).

PREDICTING THE EFFECTS OF COPY-NUMBER VARIATION IN DOUBLE AND TRIPLE MUTANT COMBINATIONS*

GREGORY W. CARTER†
The Jackson Laboratory, 600 Main Street, Bar Harbor, ME 04609, USA
Email: greg.carter@jax.org

MICHELLE HAYS
Institute for Systems Biology, 401 Terry Avenue North, Seattle, WA 98109, USA
Email: mhays@systemsbiology.org

SONG LI
Institute for Systems Biology, 401 Terry Avenue North, Seattle, WA 98109, USA
Email: soli@systemsbiology.org

TIMOTHY GALITSKI
Millipore Corporation, 290 Concord Road, Billerica, MA 01821, USA, and
Institute for Systems Biology, 401 Terry Avenue North, Seattle, WA 98109, USA
Email: timothy.galitski@merckgroup.com

The study of genetic interactions is a powerful tool in inferring structure and function of biological networks. To date, genetic interaction studies have been dominated by pair-wise gene deletion screens. However, classical genetic analysis and natural genetic variation involve diverse gene forms ranging from null alleles to copy number variants. Moreover, genetic variation is typically multifactorial. Addressing multiple combinatorial genetic variations ranging in gene activity is therefore of critical value. We approach this problem using genetic network modeling that quantitatively encodes how genes influence the activity of one another and phenotype outcomes. A network model was initially inferred from linear decomposition of gene expression data. We used this network to predict the effects of combining multi-copy and deletion mutations of specific gene pairs and a gene triplet. Predicted expression patterns across hundreds of genes were experimentally validated. Prediction success was critically dependent on how a multi-copy gene interacted with other genes in the network model. This strategy provides a template for the inference, prediction, and testing of genetically complex hypotheses involving diverse genetic variation.

* This work was supported by grants P50 GM076547 from NIGMS and by the ISB-University of Luxembourg Program. The content is solely the responsibility of the authors and does not necessarily represent the official views of the NIMGS or the NIH.

† Work partially supported by grant K25 GM079404 from NIGMS.

1. Introduction

Understanding how genetic variation affects phenotype is a central challenge of modern genetics. The hundreds of disease-related genetic variants identified in genome-wide association studies[1,2] typically account for a fraction of the heritable phenotype variation.[3] One likely contributor to this "missing heritability" is that undetected genetic interactions amplify the effects of genetic variations. In natural populations these interacting variants arise from mutations that affect genetic activity in a wide range of hypomorphic and hypermorphic allele forms. Advances in high-throughput sequencing technologies now allow greater experimental detection of structural variation and other types of genetic diversity beyond single-nucleotide polymorphisms.[4] These genetic variants appear in arbitrary combinations within a population. Therefore analytical methods designed to analyze and predict how diverse allele forms combine to influence phenotypes will be a valuable tool in the study of systems genetics.

Studies in model organisms provide evidence that genetic interactions are both prevalent and diverse in their effects. Systems-genetic studies allow the inference of multiple genetic factors[5,6] and their interactions.[7-10] To date, large-scale genetic interaction studies have been dominated by pair-wise gene deletion screens in model organisms.[11-15] However, genetic variation typically results in partial losses or gains of function, and population mixing combines these alleles in combinations with more than pair-wise complexity.[16] Interactions between these variations can result in a rich spectrum of phenotypes[7] including health outcomes.[17] For example, multi-copy suppression (or rescue) is the reversal of the effect of one mutant gene by the overexpression of a second gene. Predicting such effects requires techniques that encompass continuous variation in gene activity rather than simplified binary (on/off) gene states and multi-gene models that allow assessment of arbitrary combinations of two or more perturbations. These polygenic effects are often best accounted for by network models of interacting system elements rather than models of genes that affect phenotypes independently.[18]

To address this problem we extend a previously-developed method of analyzing genetic interactions to construct quantitative network models.[18] We approach genetic interactions as quantitative influences, defined as positive or negative numbers of varying magnitude that account for the fraction of a measurable phenotype (e.g. the expression level of a gene) inferred to be caused by a system element (e.g. a gene product). The measured phenotype is modeled by multiple influences acting throughout the inferred network. Our method is based on the classical genetic interaction approach of observing how genetic perturbations interact to affect phenotypes, thereby revealing functional relationships such as activation, repression, and pathway order.[19] Genetically "direct" (not necessarily molecularly direct) effects from regulator genes on the phenotype are separated from the genetically "indirect" effects that involve genetic interactions between regulator genes. The model correctly implicated novel genes in a cell differentiation process and proved accurate in predicting phenotypes for novel combinations of genetic knockouts.

A key aspect of the model is that it infers the effective biological activity of each regulator, rather than assuming a correspondence between the regulator's gene and protein expression level.

This modeling strategy was based on the fact that the net contribution of a gene to a phenotype is the result of many steps including transcription, translation, and various post-translational modifications like phosphorylation. In our initial study, a wide range of activity levels were inferred for a set of regulators in response to variations in genetic background.[18] These inferred activities were indirectly validated when the model was used to successfully predict the outcome of novel pair-wise combinations of gene knockouts. This flexibility in inferred gene activity makes this modeling approach amenable to studying more diverse allele forms, such as multiple copies of a gene.

In this work, we assess the capacity of our modeling approach to predict the effects of a multi-copy genetic perturbation, and how that perturbation interacts with deletions of other network genes. Furthermore, we show that our models can predict the effects of triple-mutant combinations. Following our previous methods, we first inferred a model from gene expression data for single and pair-wise gene deletions. We then extended our methods to predict how multiple copies of one of the network genes modify the activity of all regulator genes in the network (including its own) and, in turn, the downstream effects on genome-wide gene expression patterns. These quantitative predictions were directly tested in the laboratory by constructing the required strains and collecting additional microarray data. The model accurately predicted complex patterns of gene expression.

2. Network Model Inference

A central result of our previous work was the mapping and verification of a molecular network that regulates the filamentous growth response in yeast.[18] Many of the genes and pathways in this network were previously implicated in filamentous growth, and most of the newly identified genes exhibited strong knockout phenotypes. We chose four genes from that network for further study: *TEC1*, *CUP9*, *CIN5*, and *YAP6*. *TEC1* and *CUP9* were two of the five transcription factor genes chosen in our previous study. The transcription factors Yap6 and Cin5 were implicated through that study as occupying central positions in the regulatory network and candidate transmitters of genetic influences from *TEC1* and *CUP9*. We chose *CIN5* as our gain-of-function gene to further explore how alternate mutations interact with other network genes. Furthermore, there was evidence that *YAP6* and *CIN5* were involved in feedback in the network, since *TEC1* and *CUP9* were mapped in the network as both upstream regulators and downstream targets of transcript regulation. Therefore the subnetwork comprising the genes *TEC1*, *CUP9*, *YAP6*, and *CIN5* is a suitable test case for modeling the effects of genetic copy-number perturbation in a highly interactive genetic network.

2.1.1. *Yeast Gene Expression Profiling*

We collected expression profiles in triplicates[20] of wild type and mutant strains grown under filamentous-form conditions for 10 hours, as previously described.[21] Target labeling with the Affymetrix GeneChip® One-Cycle Target Labeling kit and hybridization to Yeast Genome 2.0

Arrays was done according to the manufacturer's protocols. Microarray data were collected for yeast diploid strains of 15 genotypes including wild type, deletion mutants for each of the four transcription factors, all six double-deletion combinations, and five strains transformed with a plasmid bearing *CIN5* and its native promoter (Supplemental Material and Table S2). Microarray data were normalized using robust multi-array averaging (RMA)[22] as implemented in the BioConductor software package.[23] Each gene expression data point was taken as the mean of the three corresponding biological replicates. We verified that the mean variation between biological replicates was less than the mean variation between different strains. Expression intensities for each gene were transformed into Log2 ratios relative to yeast-form wild-type expression. We restricted subsequent analysis to a set of 1267 differentially expressed genes, defined as having a factor of two difference between their lowest and highest expression intensities.

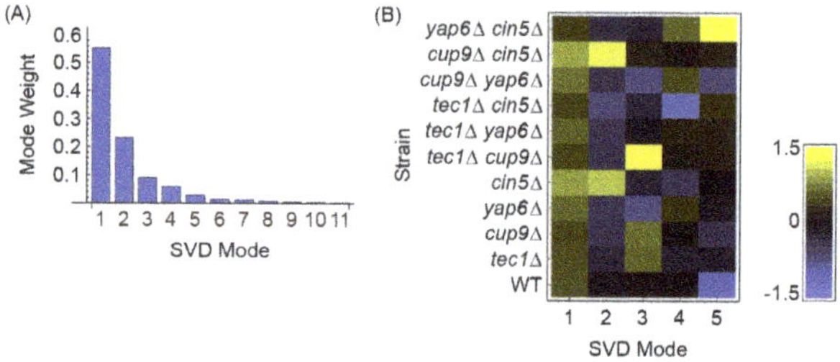

Figure 1. SVD eigenvalues and eigengenes matrix. (A) Bar chart of eigenvalues and (B) raster plot of eigengenes matrix are shown for the first five SVD modes.

2.1.2. *Singular Value Decomposition Analysis*

To identify the global expression patterns we performed singular value decomposition (SVD) on the matrix of differentially regulated genes (Figure 1).[24] SVD is an unsupervised algebraic method that mathematically separates a data matrix into a set of 'modes' determined by quantitative patterns within the data. Each mode is manifest in the data as a global expression-pattern component, or eigengene, that contributes to the expression of each gene to a degree varying from negligible to predominant. We performed SVD on data for wild-type, single-deletion, and double-deletion strains (11 in total). Of the 11 resulting SVD modes, the first five modes account for 96% of the information in the data set. This provides support for the suitability of a linear model, because it is consistent with the dimensional reduction of 11 experimental conditions to five linearly independent modes (the four seed genes plus the collective remainder of the genome), plus a small noise component. We followed the procedures in previous publications[18,25] to identify gene sets positively and negatively associated with each mode and to query those sets for enriched Gene Ontology annotations and transcription factor targets.

2.1.3. *Genetic Influences Decomposition*

We derived a network model of genetic influences using genetic influences decomposition, a technique designed to dissect the complexities of genetic interactions.[18] We considered the data set of all single and double deletions of genes *CIN5*, *CUP9*, *TEC1*, and *YAP6*. We refer to these four genes hereafter as "seed genes" in the analysis. This procedure reformulates an expression data matrix D as the product of two matrices: (i) an influence matrix, X, of coefficients for the genotype-independent influences of the seed genes on target genes; and (ii) a genotype matrix, G, of inferred activity levels for the seed genes in each genotype. Thus, the genetically "direct" (not necessarily molecularly direct) influences from the seed genes to target genes are separated quantitatively from the genetically "indirect" effects that involve a second seed gene and a genetic interaction. In the genotype matrix, G, we define the wild-type activities to be equal to one ($g_A^{WT} = g_B^{WT} = \ldots = 1$); activity levels of null alleles are fixed at zero. Note that other allele types can be accommodated readily with a measured or inferred level of activity relative to the wild type. Other genotype matrix elements (capturing genetic interactions) are unknown a priori, but they can be calculated as activity changes relative to wild type under perturbations of other seed genes ($g_A^{B\Delta}, g_B^{A\Delta}$, $g_C^{A\Delta B\Delta}$, etc.). We performed a least-squares best-fit solution for the decomposition in terms of X and G. The form of matrix G guarantees the existence a unique best-fit solution due to the strict arrangement of ones and zeros required by the genotypes (i.e., the rows of matrix G are linearly independent). We used SVD to aid in finding the best-fit solution. Genetic-influences decomposition was performed on the first five SVD modes. The full influences matrix X was determined by multiplying the results by the SVD eigenarrays. Finding the best-fit solution was a tractable problem using commercial software (Mathematica) on a PC. In matrix notation, this procedure is summarized:

$$D = u.v.w^T \cong u.v.x.G = X.G \tag{1}$$

where the symbol $\cong$ denotes a best-fit solution. The matrices u, v, and w^T are the singular value matrices for the first six modes. The eigengene matrix w^T is further decomposed into $x.G$ with a least-squares best fit. The 5 x 5 square matrix x encodes the expression influences for the first five eigengenes. The 1267 x 5 matrix X contains the expression influences for each gene.

To model how the seed genes influence one another, we inferred and quantified the influences the seed genes exert on each others' activity level. We stress that these parameters correspond to influences on inferred regulatory activity, rather than mRNA levels. Inferring interactions between seed genes is a further dimensional reduction of G into a matrix of gene-gene interactions, M, and a matrix of basal activities, G_0, with diagonal elements { G_0^{TEC1}, G_0^{CUP9}, G_0^{YAP6}, G_0^{CIN5}} and all off-diagonal elements equal to zero. Diagonal elements of M were set to zero because self-interactions cannot be mathematically distinguished from basal activity. Gene deletions are computed by taking the limit of the deleted genes basal activity to zero, which effectively removes that gene from the network. The equations used to calculate these quantities are written for each strain background (i.e. each column in G omitting the first row value of 1):

$$
\begin{aligned}
&Wild\ Type: && (G)^{WT} = [(G_0)^{-1} - M]^{-1}.1 \\
&Single\ Deletion: && (G)^{A\Delta} = lim_{G_0^A \to 0}[(G_0)^{-1} - M]^{-1}.1 \\
&Double\ Deletion: && (G)^{A\Delta B\Delta} = lim_{G_0^B \to 0}\, lim_{G_0^A \to 0}[(G_0)^{-1} - M]^{-1}.1
\end{aligned}
\tag{2}
$$

The vector 1 is defined as {1, 1, ..., 1} with length equal to the number of seed genes (four in this case). For the derivation of these equations see our previous work.[18] With the fit for G found above, we found least-squares best fit solutions for $G_0$and M as shown in Figure 2. We found strong negative influences from *CIN5* on the activities of *TEC1*, *YAP6*, and *CUP9*, with positive influences from *TEC1* and *YAP6* back to *CIN5*. The gene pairs *TEC1* and *YAP6* and *TEC1* and *CUP9* are both mutually influential with positive values. A major consequence of this network complexity is the possibility of non-linear behavior generated by the sum of linear influences.

To assess the goodness of fit for the genetic influences decomposition, we compared fitted double-mutant expression values with those predicted by an additive control model, in which the expression of gene X in a double-mutant background is:

$$X^{A\Delta B\Delta} - X^{WT} = (X^{A\Delta} - X^{WT}) - (X^{B\Delta} - X^{WT}) \tag{3}$$

This estimates the expression of every double mutant as the sum of effects for the two single mutants. To identify the most significant influences, we performed a series of bootstrap cross-validations by repeatedly re-analyzing half of the 1267 genes.[18] We obtained approximately 18000 solutions and defined significant influences as those with a mean more than four standard deviations from zero, which correspond to an empirical significance of $p = 6 \times 10^{-6}$. These influences are shown in Figure 2. Although these were the dominant influences, all influences (Figure S2 and Table S3) were used in calculations.

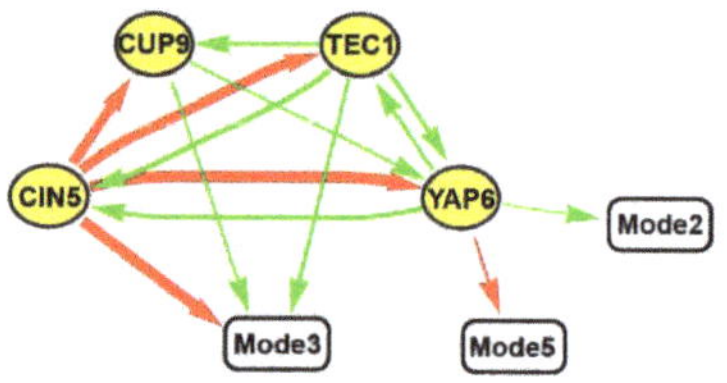

Figure 2. Network of significant positive (green) and negative (red) gene-to-gene and gene-to-expression influences. Yellow nodes are regulator genes, white nodes are expression patterns shared by multiple genes. Each edge corresponds to a parameter inferred in the genetic model with width proportional to influence magnitude.

We compared our model fit with the additive control (Eq. 3). For the expression of each gene, we computed the Pearson correlation between the experimental data and (i) the model fit and (ii) the fit to the additive control. The median correlation for the model fit was 0.88, compared to 0.56 for the additive control. Since the additive control is designed to fit the single mutants, the difference in fit quality between model and control is much greater for the six double-mutants. For this subset of strains, the median correlations were 0.96 for the model and 0.36 for the additive control. We also assessed the fit for the expression of each gene in a given double-knockout strain relative to its expression in the wild-type strain, and summary statistics in terms of expression fold-change are shown in Figure 3. These results demonstrate that the genetic interactions account for much of the goodness of fit.

We identified the genes that significantly exhibit each SVD pattern of gene expression, allowing us to associate a positive-valued and negative-valued set of genes with each mode. Each set of genes was then queried for statistical enrichment of Gene Ontology annotations[26] and transcription-factor binding targets[27-30] (Supplementary Table 1). This allowed us to characterize the gene sets by the enriched annotations and regulators. Of particular interest, we identified the set of 197 genes that exhibit the Mode 3 expression pattern, defined as having eigenarray coefficients greater than one standard deviation above the mean[25]. This set of genes is enriched in carbohydrate metabolism ($p = 5.5 \times 10^{-7}$) and targets of the following nine transcription factors: Phd1, Sok2, Sut1, Msn2, Rox1, Flo8, Mga1, Ume6, and Skn7 (Set 3-Positive, Table S1). No other set had this many enriched regulators. Four of these transcription factors (Phd1, Sok2, Rox1, and Skn7) were found to have enriched targets in filamentation-related genes in our earlier study, and three (Phd1, Sok2, and Rox1) were previously mapped downstream of the four seed genes.[18] Mga1 and Ume6 were implicated in the larger network that controls filamentation in that study. The remaining three (Sut1, Msn2, and Flo8) are known regulators of filamentation.[7,31] Given the number and functional coherence of transcription factors that regulate the Mode 3 gene set, we identified this as the most functionally relevant gene set for our filamentous growth study. While this is not the quantitatively largest expression pattern in the data, the greater expression patterns are signatures of more general processes such as cell growth (Mode 2) which generate more generic expression changes. Among the four seed genes, the Mode 3 expression pattern is positively influenced by *TEC1* and *CUP9* and negatively influenced by *CIN5* (Figure 2). *YAP6* indirectly influences this set via its influences on the other seed genes.

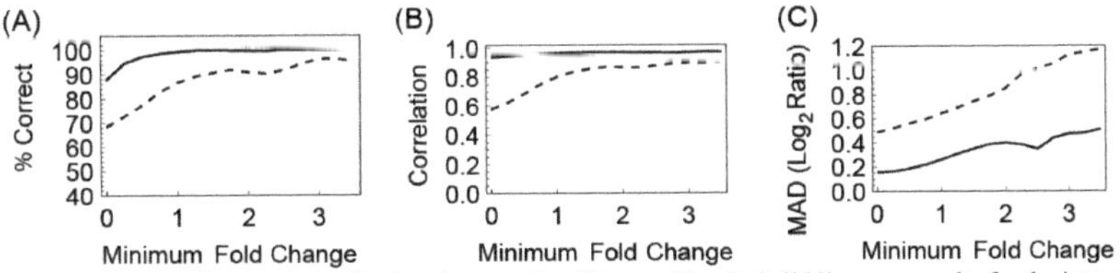

Figure 3. Fit accuracy for the six double-knockout strains (data combined). Solid lines are results for the interactive model; dashed lines are results for the additive control. Plots show (A) percentage of correct up or down-expression, (B) overall correlation of fit and data, and (C) median absolute deviation (MAD) for fit versus measured expression. All quantities are plotted as a function of fold-change, where the x-axis denotes the subset of expression ratios with the x-value or greater fold change.

2.2. *Predictions and Validation for a Multicopy Perturbation*

2.2.1. *Prediction for Multi-Copy Strains*

In previous work we limited our predictions to gene deletion strains. Here, we extend the method to address a multicopy allele. To predict the effects of a *CIN5* gain-of-function (*CIN5**) we replaced the basal level of *CIN5* activity, which is represented by the parameter G_0^{CIN5}. Network interactions adjust this value to produce the *CIN5* activities in the G matrix (Eq. 1). The precise increase in the value of G_0^{CIN5*} is difficult to fix for many reasons, including: the number of extra copies of *CIN5* borne by a plasmid with the 2μ replication origin is variable; microarray assays are performed on large cell populations and thus represent population-averaged signals; and the relationship between number of gene copies and regulatory activity is uncertain in our model. We find, however, that the predictions from our model are insensitive to increases in G_0^{CIN5*} for all values above 5 times the original value G_0^{CIN5} (which is equal to 0.58). This is demonstrated for an example gene in Figure 4. Therefore to predict global expression for *CIN5* multi-copy strains we artificially increased the basal activity value for *CIN5* as $G_0^{CIN5} \rightarrow G_0^{CIN5*} = 50G_0^{CIN5}$ with the factor 50 chosen as an asymptotic value. We computed new columns for the G matrix corresponding to these strains, following Eq. 2 with zero limit replaced by the multi-copy limit. Thus multiple copies of Gene A with a deletion of Gene B forms a column with 1 followed by:

$$Deletion\ and\ Multicopy: \quad (G)^{A\Delta B*} = lim_{G_0^B \to 50G_0^B}\, lim_{G_0^A \to 0}[(G_0)^{-1} - M]^{-1}.1 \qquad (4)$$

For instance, the net activity level predicted for *CIN5*, g_{CIN5}^{CIN5*}, is 0.40. We then multiplied this new column in G by the previously-derived matrix X to produce a column of 1267 predictions for the expression of each gene for each of the four new strains.

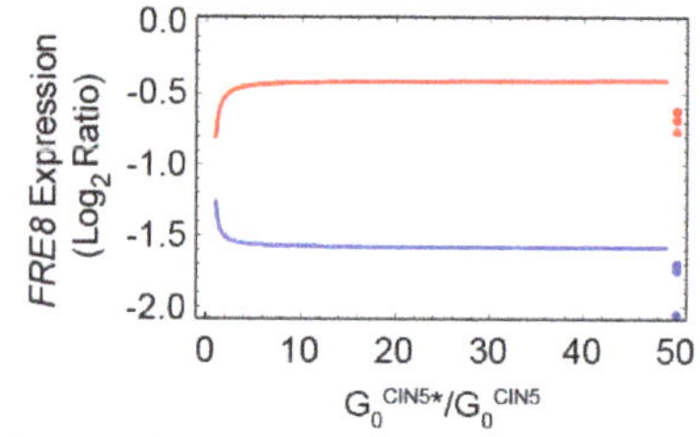

Figure 4. Predicted expression of an example gene (*FRE8*) is insensitive to basal *CIN5* activity beyond small values. The red line is predicted expression for *CIN5** strain background, and the blue line is for *yap6Δ CIN5** double mutant. Both lines begin at best-fit values for wild-type *CIN5* (multiplier of 1). Red and blue points are experimental results (in triplicate) for the corresponding strains.

That gene expression predictions rapidly approach asymptotic values that are stable for arbitrarily high values of G_0^{CIN5*} is a direct result of the self-regulating interactions between the seed genes. Although we have not derived a dynamic model, the steady-state solution from our system of equations is sufficient to limit the effect of an arbitrarily high activity for the *CIN5* gene.

Although the model is composed of linear influences, the network of interactions predicts a highly nonlinear effect on gene expression for *CIN5**. The network model derived above (Figure 2) predicts that an attempt to exogenously increase the activity of *CIN5* will suppress the activities of *TEC1* and *YAP6*, which will in turn repress an increased activity of *CIN5*. Moreover, rather than predicting an increase in *CIN5* activity, the model predicts an effective reduction relative to the wild type. Our model predicts an activity level for *CIN5* in the *CIN5** strain to be 0.40 relative to the wild-type value fixed at 1. This behavior is the result of the genetic interactions inferred between the seed genes (Figure 2).

2.2.2. *Experimental Test of Predictions*

To test our predictions, we constructed a set of *CIN5* gain-of-function (*CIN5**) strains by transforming the wild-type, *tec1*Δ, *yap6*Δ, *cup9*Δ, and *tec1*Δ *cup9*Δ strains with a 2μ plasmid bearing the *CIN5* gene. The *CIN5* gene was cloned with its native promoter to amplify *CIN5* expression without modifying possible genetic interactions mediated by transcription. We collected microarray data as specified in Section 2.1.1. Our prediction of nonlinear behavior in *CIN5* activity was verified by gene expression for the *CIN5** strains, as our predictions matched measured levels very well. These predictions are summarized in Figure 5, which shows statistics for the expression of each gene in a given multi-copy strain relative to its expression in the wild-type strain (formatted as in Figure 3). Prediction accuracy increased with the magnitude of predicted effects (Figure 5), suggesting that predictive power was greatest above experimental noise. For the 777 expression predictions of over one-fold change relative to the wild-type expression, we found a Pearson correlation between experiment and data of 0.60 and agreement between up/down differential expression 81% of the time. For the 147 predictions of two-fold or greater change, the Pearson correlation was 0.71 and 90% of the predictions were directionally correct. These predictions can be contrasted with the additive model, which fares increasingly poorly as the fold-change increases in terms of correlation and directional accuracy (Figure 5). Most substantially, the additive model does not include regularizing genetic interactions and thus the numerical predictions are off by a factor proportional to the estimated number of extra gene copies (here set to 50) (Figure 5C). The model correctly predicted genome-wide expression *CIN5* gain-of-function effects and how these effects interact with deletions of *TEC1*, *CUP9*, and *YAP6*. In the context of our network model, the explanation for this effect is the inferred negative feedback from Cin5 to the three other seed genes.

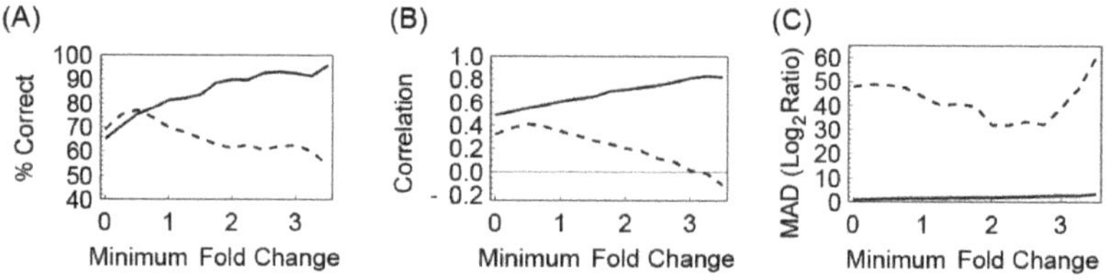

Figure 5. Prediction accuracy for *CIN5* gain-of-function in wild-type, *tec1Δ*, *cup9Δ*, *yap6Δ*, and *tec1Δcup9Δ* backgrounds (data combined). Solid lines are results for the interactive model; dashed lines are results for the additive control. Plots show (A) percentage of correct up or down-expression, (B) overall correlation of predictions and data, and (C) median absolute deviation (MAD) for predicted versus measured expression. All quantities are plotted as a function of fold-change, where the x-axis denotes the subset of expression ratios with the x-value or greater fold change. Predictive power generally increases with differential expression.

3. Discussion and Conclusions

The model we derived agrees well with the template of molecular interactions we mapped in our previous study.[18] In that study, we mapped genetic influences between *TEC1*, *CUP9*, and three other genes. *YAP6* and *CIN5* were implicated through the integration of molecular interaction data as candidate intermediaries of influences that traveled from *TEC1* and *CUP9* to downstream transcriptional targets. A number of the implications of that model were borne out in this study. For example, we mapped *CUP9* influence passing through Yap6 and Cin5 to regulate a set of filamentation-related genes. Indeed, our model shows *CUP9* influencing *CIN5* via *YAP6*. The seed gene with the greatest direct influence on the filamentation-related genes (Mode 3) in this study is *CIN5*, consistent with our earlier model that placed Cin5 downstream of the other three seed genes. Furthermore, our previous network exhibited strong hints of regulatory feedback involving the four genes studied in this work. The network derived in this study captures this type of feedback and it was experimentally validated by the feedback-dependent predictions.

The importance of an interactive network model was underscored by the fact that fit and prediction accuracy increased with the number of genetic perturbations involved. For the knockout data, the model fit generally predicted double-deletion expression values better than those of single deletions. This was also true for predictions of multi-copy *CIN5*. Our least successful predictions were for the *CIN5** strain, particularly for genes with minor expression changes from wild-type values. Overall, the effects of *CIN5** were small compared to *cin5*Δ effects. Therefore a possible reason for the relatively poor predictions is that the model was derived from the very strong effects of the *CIN5* deletion, which might have over-predicted the effects of multiple *CIN5* copies by assigning large values in the X matrix. However, when combined with one or two additional perturbations, the model was able to predict expression patterns with high accuracy. The *tec1*Δ *cup9*Δ *CIN5** triple-perturbation predictions were the most accurate and correctly captured the moderation of the double-knockout effects by the *CIN5* gain-of-function (Supplementary Figure S1). For genes with one or greater fold change, 98% of predictions were correct compared to 81% for all strains (Figure 5A) and the correlation between prediction and experiment was 0.88 compared to 0.60 for all strains (Figure 5B). Predicting such complex interactions necessarily requires a network model.

Although our focus in this study was forward prediction, the modeling methods could be used to infer the relative activity levels of uncharacterized alleles in novel data. This would be a straightforward application of the influences decomposition defined in Eq. 1. For example, if we treat our validation data as an independent data set and directly fit the activity level for *CIN5* in a *CIN5** strain (*i.e.*, the parameter G_0^{CIN5*}) the result is 0.75. This is an inferred reduction from the wild-type baseline of 1, and similar to our forward predicted value of 0.40. Thus for alleles with uncertain activity, our approach has the flexibility to make either reverse inferences on experimental data or forward predictions based on training data (knockout data in this instance). Reverse inference may prove particularly useful when addressing populations with natural genetic variation, because in many cases the trait-associated genes will harbor partial loss or gain of function alleles rather than complete gene deletions or multiplications.

We note that while our linear modeling technique accurately predicted general trends in gene expression, it did not always accurately predict precise effects. Inaccuracies in prediction were primarily due to errors in magnitude rather than misprediction of overall trends of up and down-regulation. The model generally over-predicted the effects of novel genetic perturbations, suggesting that there are additional genes which attenuate the perturbations we introduced. This is not surprising given that we have only analyzed four genes embedded in a highly complex system. However it is also certain that the linear assumption of our influence decomposition over-simplifies the complex biochemical relationships between gene products. Thus we view the resulting genetic influences model (Figure 2) as a template for more detailed biochemical models that require additional parameters (*e.g.* kinetic rate constants) but provide more precise predictions.

Predicting the effects of multifactorial genetic variation will require understanding how the variant genes operate in a network of genetic interactions. In this work we were able to demonstrate the power of a network inferred from gene knockouts to predict combinations of both hypermorphic and hypomorphic genetic variants. Prediction accuracy depended on genetic interactions being included in the model. This predictive power was a direct result of the feedback interactions in the inferred model. As the potential activity of the *CIN5* gene grows, interactions with the other network genes quickly stabilize its actual activity and, therefore, downstream phenotypes like global gene expression. We stress that this behavior was predicted in a model inferred from genetic deletions of *CIN5*, and the model was capable of predicting the outcomes of a genetic perturbation with a priori unknown effects on the levels of activity of the gene product. This is potentially an important capability when addressing genetically diverse populations with uncharacterized alleles. Indeed, genetic studies are increasingly revealing the role of diverse genetic variants, such as copy-number variation and promoter polymorphisms, in human disease. The capability to model and predict how these variants interact with null mutations will be of critical value in genomic medicine.

4. Supplementary Material

Supplementary Material located at http://www.jax.org/research/faculty/carter/supplement.html contains yeast strain information, supplementary tables, and a supplementary figure.

5. Acknowledgments

We are grateful to Susanne Prinz for helpful advice and comments on the manuscript.

References

1. M. M. Iles, *PLoS Genet* **4**, e33, (2008).
2. P. M. Visscher & G. W. Montgomery, *JAMA* **302**, 2028, (2009).
3. T. A. Manolio, F. S. Collins, N. J. Cox *et al.*, *Nature* **461**, 747, (2009).
4. R. E. Mills, K. Walter, C. Stewart *et al.*, *Nature* **470**, 59, (2011).
5. A. M. Dudley, D. M. Janse, A. Tanay, R. Shamir & G. M. Church, *Mol Syst Biol* **1**, 2005 0001, (2005).
6. J. Zhu, B. Zhang, E. N. Smith *et al.*, *Nat Genet* **40**, 854, (2008).
7. B. L. Drees, V. Thorsson, G. W. Carter *et al.*, *Genome Biol* **6**, R38, (2005).
8. R. P. St Onge, R. Mani, J. Oh *et al.*, *Nat Genet* **39**, 199, (2007).
9. A. H. Tong, G. Lesage, G. D. Bader *et al.*, *Science* **303**, 808, (2004).
10. N. Van Driessche, J. Demsar, E. O. Booth *et al.*, *Nat Genet* **37**, 471, (2005).
11. M. Costanzo, A. Baryshnikova, J. Bellay *et al.*, *Science* **327**, 425, (2010).
12. B. Lehner, *J Exp Biol* **210**, 1559, (2007).
13. S. L. Ooi, X. Pan, B. D. Peyser *et al.*, *Trends Genet* **22**, 56, (2006).
14. X. Pan, P. Ye, D. S. Yuan *et al.*, *Cell* **124**, 1069, (2006).
15. M. Schuldiner, S. R. Collins, N. J. Thompson *et al.*, *Cell* **123**, 507, (2005).
16. R. B. Brem & L. Kruglyak, *Proc Natl Acad Sci U S A* **102**, 1572, (2005).
17. S. Prabhakar, A. Visel, J. A. Akiyama *et al.*, *Science* **321**, 1346, (2008).
18. G. W. Carter, S. Prinz, C. Neou *et al.*, *Molecular systems biology* **3**, 96, (2007).
19. L. Avery & S. Wasserman, *Trends Genet* **8**, 312, (1992).
20. K. L. Thompson, B. A. Rosenzweig, R. Honchel *et al.*, *Mol Carcinog* **32**, 176, (2001).
21. S. Prinz, I. Avila-Campillo, C. Aldridge *et al.*, *Genome research* **14**, 380, (2004).
22. R. A. Irizarry, B. Hobbs, F. Collin *et al.*, *Biostatistics* **4**, 249, (2003).
23. R. C. Gentleman, V. J. Carey, D. M. Bates *et al.*, *Genome Biol* **5**, R80, (2004).
24. O. Alter, P. O. Brown & D. Botstein, *Proc Natl Acad Sci U S A* **97**, 10101, (2000).
25. G. W. Carter, S. Rupp, G. R. Fink & T. Galitski, *Genome Res* **16**, 520, (2006).
26. M. Ashburner, C. A. Ball, J. A. Blake *et al.*, *Nat Genet* **25**, 25, (2000).
27. A. R. Borneman, J. A. Leigh-Bell, H. Yu *et al.*, *Genes Dev* **20**, 435, (2006).
28. C. T. Harbison, D. B. Gordon, T. I. Lee *et al.*, *Nature* **431**, 99, (2004).
29. K. D. MacIsaac, T. Wang, D. B. Gordon *et al.*, *BMC Bioinformatics* **7**, 113, (2006).
30. J. Zeitlinger, I. Simon, C. T. Harbison *et al.*, *Cell* **113**, 395, (2003).
31. H. Liu, C. A. Styles & G. R. Fink, *Genetics* **144**, 967, (1996).

INTEGRATIVE NETWORK ANALYSIS TO IDENTIFY ABERRANT PATHWAY NETWORKS IN OVARIAN CANCER

LI CHEN[1,2], JIANHUA XUAN[1,*], JINGHUA GU[1], YUE WANG[1],
ZHEN ZHANG[2], TIAN-LI WANG[2], IE-MING SHIH[2]

[1]*The Bradley Department of Electrical and Computer Engineering, Virginia Tech*
900 N. Glebe Road, Arlington, VA 22203, USA
Email: {lchen06, xuan, gujh, yuewang}@vt.edu

[2]*Department of Pathology, Johns Hopkins Medical Institutions*
1550 Orleans Street, Baltimore, MD 21231, USA
Email: {lchen93, zzhang7}@jhmi.edu, shihie@yahoo.com, tlw@welch.jhu.edu

Ovarian cancer is often called the 'silent killer' since it is difficult to have early detection and prognosis. Understanding the biological mechanism related to ovarian cancer becomes extremely important for the purpose of treatment. We propose an integrative framework to identify pathway related networks based on large-scale TCGA copy number data and gene expression profiles. The integrative approach first detects highly conserved copy number altered genes and regards them as seed genes, and then applies a network-based method to identify subnetworks that can differentiate gene expression patterns between different phenotypes of ovarian cancer patients. The identified subnetworks are further validated on an independent gene expression data set using a network-based classification method. The experimental results show that our approach can not only achieve good prediction performance across different data sets, but also identify biological meaningful subnetworks involved in many signaling pathways related to ovarian cancer.

1. Introduction

Ovarian cancer is the fifth leading cause of cancer-related deaths among women in the United States [1]. It has been estimated that 21,990 women will be newly diagnosed and 15,460 women will die of ovarian cancer in 2011 [1]. Most ovarian cancers are serous ovarian carcinomas and only less than 20% of them can be early detected. Prognosis for high grade serous carcinoma patients remains unsatisfactory because most patients develop resistance to chemotherapy after surgery and eventually die [2]. Therefore, chemoresistance has been a critical clinical problem and it is important to understand the biological mechanism of ovarian cancer to overcome the resistance to chemotherapy. Many research topics, such as gene mutation analysis [3], biomarker identification [4], etc., have been carried out to study the chemotherapy resistance. Among them, identification of pathway networks in ovarian cancer becomes an important topic in study.

---

* To whom correspondence should be addressed.

New technologies have generated large amounts of high-throughput genomic and proteomic data related to ovarian cancer, which make it possible to conduct a comprehensive study from the computational point of view to better examine the cancer genome. The Cancer Genome Atlas (TCGA) project is the one of the studies that collects various biological data for ovarian cancer using genome analysis techniques. It provides opportunities and challenges to develop computational methods to study cancers based on multiple biological data, which can reveal different aspects and levels of biological system function. Traditional computational or statistical approaches, mainly focusing on one type of data source, cannot provide a system view of complex biological system. Current and future needs require sophisticated integration of diverse sets of data, aiming to better understand the main features (e.g., components and their interactions) of biological processes or systems [5, 6]. Many integrative approaches have been proposed to study glioblastoma based on TCGA data portal [7-9]. A recent study explored mRNA expression, microRNA expression, promoter methylation and DNA copy number data on TCGA ovarian cancer samples and provided biologically meaningful results successfully [10]. However, the paper [10] has not conducted a sophisticated integrative analysis across different data sources. Here, we propose a new integrative framework for pathway network identification in ovarian cancer.

Our integrative approach is based on the hypothesis that the cancer phenotype can be reflected by gene expression profiles, which are driven by genomic changes at the copy number level. It is also based on the hypothesis that the highly conserved copy number altered genes might not be differentially expressed at the gene expression level. Therefore, our approach first detects the consensus regions in the DNA copy number data in high-grade ovarian cancer patients and regards the genes with conserved copy number altered as the seed genes. Then a network identification method, based on gene expression profiles and protein-protein interaction network [11], is used to identify significant subnetworks from seed genes that could differentiate two different phenotypes among patients according to the overall survival time. Finally, the identified subnetworks are cross validated on a public gene expression data set using a network-based prediction model. Our results show that the proposed integrative approach can achieve good prediction performance with a high reproducibility across different data sets. Moreover, it also identifies several important pathway related networks, such as ErbB signaling pathway and Notch signaling pathway, which are likely associated with the development of ovarian cancer.

2. Materials and method

2.1. *Integrative framework*

Fig.1 illustrates an integrative framework for copy number analysis, subnetwork identification and prediction by integrating DNA copy number data, mRNA gene expression profiles, protein-protein interaction network and clinical information. From DNA copy number data, we detect significant consensus amplified and deleted regions using Genomic Identification of Significant Targets in

Cancer (GISTIC) algorithm [12] and then extract the genes located in these regions. We consider these genes are highly related to the ovarian cancer mechanism and may function as 'drivers' to form different phenotypes. We also curate ovarian cancer related genes from literature [2] and combine them with the consensus genes from copy number data as seed genes for subnetwork identification. Then we identify subnetworks based on the seed genes using bootstrapping Markov random field-based method (BMRF), which integrates protein-protein interaction networks and the gene expression profiles. For the purpose of evaluation, we adopt a public gene expression data set in the study. We train a classifier on the TCGA gene expression data set on the identified significant subnetworks and then test on the public data set using network-constrained support vector machines (netSVM). Finally we measure the prediction performance and conduct survival analysis on these two gene expression profiles.

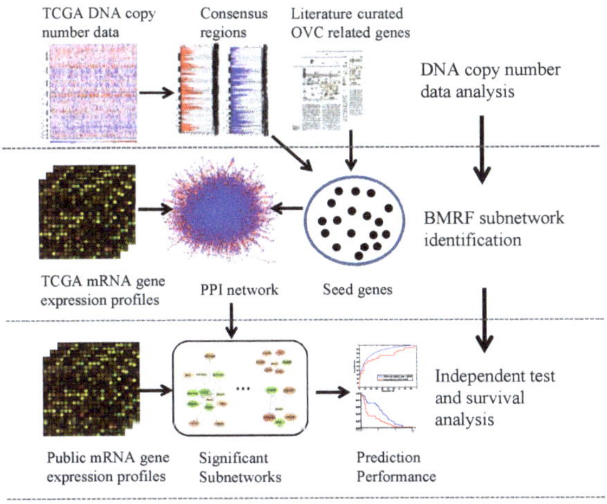

Fig. 1. Illustration of an integrative framework for TCGA ovarian cancer data analysis.

2.2. *Data description*

DNA copy number data and mRNA gene expression data for ovarian serous cystadenocarcinoma cases are obtained from TCGA data portal. 157 patients have DNA copy number expression, mRNA gene expression, as well as the clinical survival information. Among them, 58 patients have overall survival time less than 2 years and 45 patients have overall survival time larger than 4 years. We categorize these two groups as 'high risk' and 'low risk', respectively. An independent public mRNA gene expression data set (GSE3149) for ovarian cancer cases is obtained from Bild *et al.* [13]. Accordingly, 45 patients are grouped in 'high risk' and 49 patients are grouped in 'low risk'. The copy number data are arrayed by Agilent Human Genome CGH microarray 244A chips and both mRNA gene expression profiles are arrayed by Affymetrix HG U133A chips. The copy number data are normalized using dChip software [14] and mRNA gene expression data are normalized by Plier and quantile normalization methods [15]. Subnetworks are identified from protein-protein interaction (PPI) network obtained from the HPRD database [16], which contains about 9,000 genes and 35,000 interactions. We convert probe set IDs used in gene expression data to Entrez gene IDs. The probe set ID with the largest variance across patients' samples is used where multiple probe set IDs are linked to one Entrez gene ID. By mapping the PPI network and two data sets we obtain 7,249 genes in 27,885 interactions to be investigated.

2.3. *DNA copy number consensus region detection*

The segmentations on the DNA copy number data are detected by Circular Binary Segmentation (CBS) method [17]. The CBS is a modified binary segmentation method that splits the chromosome into regions of equal copy number to reduce noises and estimates parameters through permutation distribution. The significantly amplified or deleted genomic regions across the ovarian cancer samples are detected by GISTIC algorithm [12] on the segmented DNA copy number data. The GISTIC algorithm takes segmented copy number data and identifies regions of the genome that are significantly amplified or deleted across a set of samples. A G-score is assigned to each aberration that considers the amplitude of the aberration as well as the frequency of its occurrence across all samples. False Discovery Rate q-values are then calculated for the aberrant regions, and regions with q-values below a user-defined threshold are considered significant. Here we set both amplification threshold and deletion threshold as 1 (4 copies for amplification and 1 copy for deletion), and the false discovery rate q-value threshold as 0.01.

2.4. *Network identification by bootstrapping MRF (BMRF)*

We apply a bootstrapping Markov random file (BMRF) method to the TCGA gene expression data by integrating protein-protein interaction network to identify significant subnetworks that could distinguish the expression patterns between 'high risk' and 'low risk'. BMRF method follows a maximum a posteriori (MAP) principle to form a novel network score that explicitly considers pairwise gene interactions in PPI networks.

Let's first define a random variable vector $\mathbf{f} = \{f_1, \cdots, f_m\}$ to represent a set of discriminative scores of m genes (or proteins) between two phenotypes. In the context of a PPI network, Let S represent a gene set of m genes in a network and N_i represent connected neighbors of gene i. We define a pairwise clique C_2 on N_i and S as $C_2 = \{\{i, i'\} \mid i' \in N_i, i \in S\}$.

The random variable vector $\mathbf{f}$ is said to form a Markov random field on S with respect to N_i and subject to the following conditions:

$$P(\mathbf{f}) > 0, \ \forall \mathbf{f} \in \mathbf{F}$$
$$P(f_i \mid f_{S-\{i\}}) = P(f_i \mid f_{N_i}) \quad (1)$$

The second criterion in Equation (1) is the Markov property of a random field, which states that the probability of a certain configuration at gene i is statistically independent of the configurations of all other genes ($j \in S$) given configuration N_i.

The possible configuration **f** of a set of random variable vector **F** obeys a Gibbs distribution if the joint distribution takes the following form:

$$P(\mathbf{f}) = \frac{1}{Z} \times e^{-\frac{1}{T}U(\mathbf{f})}, \quad (2)$$

where Z is a normalizing constant given by $Z = \sum_{\mathbf{f} \in \mathbf{F}} e^{-\frac{1}{T}U(\mathbf{f})}$ and U is given by $U(\mathbf{f}) = \sum_{c \in C} V_c(\mathbf{f})$.

U is an energy function that is determined by a sum of clique potentials $V_c(\mathbf{f})$ over all cliques. Clique potentials allow the modeling of knowledge (*a priori*) about the contextual interactions between genes at neighboring sites. For simplicity, we usually assign 0 potential to all cliques of size greater than 2. The energy $U(\mathbf{f})$ corresponds to the probability of that configuration. From Equation (2), we can see that lower energies correspond to more likely configurations. The parameter T is often referred to as 'temperature' that controls the sharpness of the distribution. Z is a normalization constant and does not need to be calculated.

Denote the observed discriminative scores of genes between two phenotypes as $\mathbf{z} = \{z_1, \cdots, z_m\}$. Here, we define z_i as the z-score of its corresponding p-value p_i using $z_i = \Phi^{-1}(1 - p_i)$, where Φ^{-1} is the inverse normal cumulative density function (CDF) [18]. We assume that the observed discriminative score is a result of the addition of independent zero mean Gaussian noise to the underlying discriminative score; $\mathbf{z} = \mathbf{f} + \mathbf{e}$, $\mathbf{e} \sim N(0,1)$. One possible estimate of the underlying discriminative score **f** is the MAP estimate $\hat{\mathbf{f}}$ that maximizes the likelihood of posterior probability ($\log P(\mathbf{f} \mid \mathbf{z})$); with the help of Bayes' rules and Gibbs distribution, it is equivalent to state that the MAP estimate $\hat{\mathbf{f}}$ minimizes the following posterior potential function: $\hat{\mathbf{f}} = \arg\min_{\mathbf{f}}(U(\mathbf{f}) + U(\mathbf{z} \mid \mathbf{f}))$. The first term in the posterior potential function is the prior potential given by:

$$U(\mathbf{f}) = \sum_{i \in S} V_1(f_i) + \sum_{i \in S} \sum_{i' \in N_i} V_2(f_i, f_{i'}) = \frac{-1}{m} \sum_{i \in S} f_i + \frac{\lambda}{k} \sum_{(i,i') \in E} \left(\frac{f_i}{\sqrt{d_i}} - \frac{f_{i'}}{\sqrt{d_{i'}}} \right)^2, \quad (3)$$

where d_i is the degree of gene i in the PPI network, k is the number of interactions (or edges), and λ is a trade-off parameter. The first term in Equation (3) is the average discriminative score in a subnetwork; the second term in Equation (3) imposes the smoothness across the subnetwork, while putting more weights on the genes with large degrees. Note that the posterior potential function is normalized by the number of genes and the number of edges in the subnetwork, hence, independent of the subnetwork size.

The second term in the posterior potential function is the likelihood potential given by:

$$U(\mathbf{z} \mid \mathbf{f}) = \frac{\gamma}{m} \sum_{i \in S} (z_i - f_i)^2 / 2, \tag{4}$$

where γ is a trade-off parameter. The likelihood potential gives the average square of difference between observed and underlying discriminative scores, given the assumption of a Gaussian distribution of the noise signal with 0 mean and 1 standard deviation.

Thus, we can define the subnetwork score as the negative posterior potential function that takes into account the dependency among the genes of a subnetwork, which, in the form of estimated discriminative scores, can be defined as follows:

$$NetScore(G) = -U(\hat{\mathbf{f}} \mid \mathbf{Z}) = \frac{1}{m} \sum_{i \in S} \hat{f}_i - \frac{\lambda}{k} \sum_{(i,i') \in E} \left(\frac{\hat{f}_i}{\sqrt{d_i}} - \frac{\hat{f}_{i'}}{\sqrt{d_{i'}}} \right)^2 - \frac{\gamma}{m} \sum_{i \in S} (z_i - \hat{f}_i)^2 / 2. \tag{5}$$

Once we define the MRF-based network score, a modified simulated annealing search algorithm is then developed to efficiently find optimal or suboptimal subnetworks with maximal network scores. Finally, to improve their reproducibility across data sets, a bootstrapping scheme is implemented to statistically select confident subnetworks. BMRF method has the advantage in identifying hub genes that usually express little changes among different phenotypes and are hard to detect, therefore improves the mechanism study of ovarian cancer. In this experiment, we determine the significant subnetworks according to network size and network score. A network is considered as significant if its size is greater than 5 and the network score is larger than 1.65 ($p \leq 0.05$; normal distribution).

2.5. *Network constrained support vector machines (NetSVM)*

Given a training sample set ($\mathbf{x}_1$, y_1), ..., ($\mathbf{x}_l$, y_l) with p features and l samples, where $\mathbf{x}_i \in R^p$ and $y_i \in \{-1,+1\}$, the SVM learning algorithm aims to find a linear function of the form $f(\mathbf{x}) = \boldsymbol{\beta} \cdot \mathbf{x} + b$, with $\boldsymbol{\beta} \in R^p$ and $b \in R$ such that a data point $\mathbf{x}$ is assigned to a label +1 if $f(\mathbf{x}) > 0$, and a label -1 otherwise. Consider a gene network that is represented by a graph $G = (V, E, W)$, where V is a set of vertices that correspond to p genes, $E = \{u \sim v\}$ is a set of edges indicating that gene u and v are linked on the network and W is the weights of the edges. The degree of a vertex v is defined as $d_v = \sum_u w(u,v)$, where $w(u, v)$ indicates the weight of edge $u \sim v$. For this application, the weights could represent the probabilities of having edges between two vertices. Following Chung *et al.* [19], we define the Laplacian matrix $\mathbf{L}$ of G with the uv^{th} element to be:

$$\mathbf{L}(u,v) = \begin{cases} 1 - w(u,v)/d_u & \text{if } u = v \text{ and } d_u \neq 0 \\ -w(u,v)/\sqrt{d_u d_v} & \text{if } u \text{ and } v \text{ are adjacent} \\ 0 & \text{otherwise} \end{cases}. \tag{6}$$

This matrix is symmetric and non-negative definite and its corresponding eigenvalues or spectra reflect many properties of the graph as detailed in [19].

We define the network-constrained SVM given non-negative parameter η as follows:

$$\min_{\beta,b,\xi} \frac{1}{2}\boldsymbol{\beta}^T\boldsymbol{\beta} + \eta\boldsymbol{\beta}^T L\boldsymbol{\beta} + C\sum_{i=1}^{l}\xi_i \quad s.t.\ y_i\left(\boldsymbol{\beta}\cdot\mathbf{x}_i + b\right) \geq 1-\xi_i, \xi_i \geq 0. \tag{7}$$

Note that **L** can be written as $\mathbf{L} = \mathbf{SS}^T$, where **S** is the matrix whose rows are indexed by the vertices and whose columns are indexed by the edges of G such that each column (corresponding to an edge e = $\{u, v\}$) has an entry $\sqrt{w(u,v)}/\sqrt{d_u}$ in the row corresponding to u, an entry $-\sqrt{w(u,v)}/\sqrt{d_u}$ in the row corresponding to v, and zero entries elsewhere. Therefore we can see that $\boldsymbol{\beta}^T\mathbf{L}\boldsymbol{\beta}$ can be re-written as

$$\boldsymbol{\beta}^T\mathbf{L}\boldsymbol{\beta} = \sum_{u\sim v}\left(\frac{\beta_u}{\sqrt{d_u}} - \frac{\beta_v}{\sqrt{d_v}}\right)^2 w(u,v). \tag{8}$$

From this representation we can understand that the added regularization term $\eta\boldsymbol{\beta}^T\mathbf{L}\boldsymbol{\beta}$ imposes the smoothness of parameters (coefficients) $\boldsymbol{\beta}$ over the network via penalizing the weighted sum of squares of the scaled difference of coefficients between neighboring vertices in the network.

The solution of Equation (7) could be obtained by reducing it to a conventional SVM optimization problem based on the property of **L** that is symmetric and semi-positive definite. We set equal weight 1 for all connections in PPI database in our experiment.

2.6. *Classification performance merits and survival analysis*

For the identified subnetworks, we conduct three-fold cross validation on TCGA data set and independent test on the public data set. During the cross validation iteration, each time we leave one fold as validation set and the others as training set. Note that the folds are stratified so that they contain the approximately same proportions of labels as the original data. The three-fold cross validation procedure is repeated 100 times in order to get more reliable performance estimation by different randomizations. The average validation performance is reported.

We evaluate the prediction performance through several statistical analyses. Given the true labels of samples and prediction results, we use the Receiver Operating Characteristic (ROC) curve [20] and the area under the curve (AUC) to measure the prediction accuracy of the classifier. ROC curve is a graphical plot of true positive rate (TPR) vs. false positive rate (FPR). AUC is an important performance measure that provides an overall measure of accuracy for the prediction. Furthermore, accuracy, sensitivity and specificity are calculated as well.

Also, given the sample survival time information, we conduct the Kaplan-Meier survival analysis [21] for the prediction results to generate plots for overall survival time. To compare the difference of two survival curves, p-value and hazard ratio are reported. P-value is calculated by using the log-rank test and hazard is calculated using the Cox proportional model.

3. Results and discussion

Fig. 2 shows the heatmap of the TCGA copy number data and detected consensus regions for both amplification and deletion. There are 869 amplification regions and 979 deletion regions that are significantly consensus across all the high grade ovarian cancer tumor samples. 751 and 816 genes are located in the amplification regions and deletion regions, respectively. Some of these genes are functionally related to kinase, transcription factor, oncogenes, etc. For example, CCNE1 and MYC are located in the focal amplification regions and they are oncogenes. SMAD4 is located in the focal deletion region and it is a tumor suppressor gene. These findings are biologically interesting; however, many of them do not have clear functional annotation because of the limited knowledge. Most of genes are isolated in the context of PPI network; and it is difficult to understand the underlying biological mechanism even though they are important to sever as the potential 'drivers' in ovarian cancer. Therefore we treat these genes as the seed genes in the subnetwork identification methods.

We also collect 74 genes associated with ovarian cancer pathways from literatures [2], which are functionally related to tumor suppressor, oncogenes, signaling and tumor biology. Finally we obtain 511 genes as the seed genes from DNA copy number data after mapping the genes to PPI network.

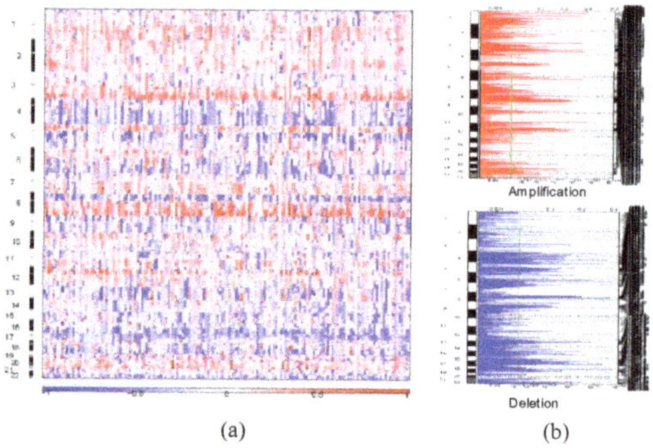

Fig. 2. (a) Heatmap of TCGA copy number data and (b) detected significant consensus regions.

We identified subnetworks from TCGA gene expression data and the PPI network, using BMRF method for all seed genes. Based on the network score and network size, 36 subnetworks are significantly ($p < 0.05$) showing different expression patterns between 'high risk' and 'low risk' groups on the TCGA data set. We trained a classifier using netSVM based on these subnetworks with cross validation. The accuracy (standard deviation) of three-fold cross validation is 79.53% (0.0313) with 76.02% (0.0584) sensitivity and 82.26% (0.0312) specificity. The classifier is further validated on an independent gene expression data set [13] and achieves 74.47% accuracy with 62.22% sensitivity and 85.71% specificity. The ROC curves for cross validation and independent test are shown in Fig. 3. Kaplan-Meier analysis of independent test in Fig. 4 also shows significant different ($p = 0.0003$) in overall survival between two groups predicted as 'high risk' and 'low risk'. We have also performed prediction using traditional gene selection method T-test (220 genes with $p < 0.01$) and conventional SVM and achieved 86.92% accuracy for cross validation and 69.15% accuracy for independent test. As a comparison, our proposed method achieves better reproducibility across different data sets. Moreover the identified genes by the proposed method are more biologically meaningful than the ones selected by T-test in terms of GO functional annotation, where many signaling pathway related genes are identified by the proposed method (see below), while no significant pathway is enriched in the genes selected by T-test.

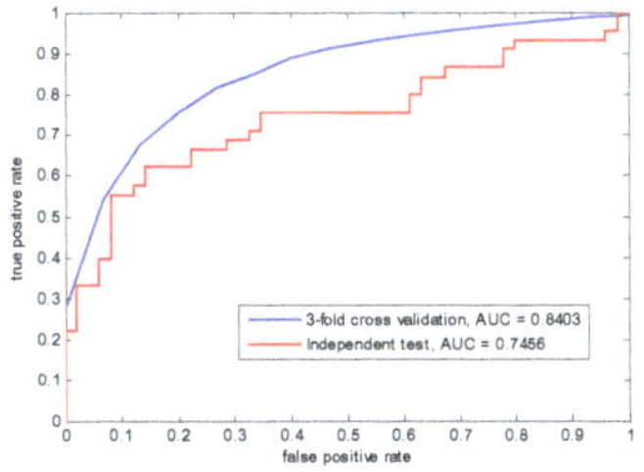

Fig. 3. ROC curves of three-fold cross validation on the TCGA data set and independent test on the public data set.

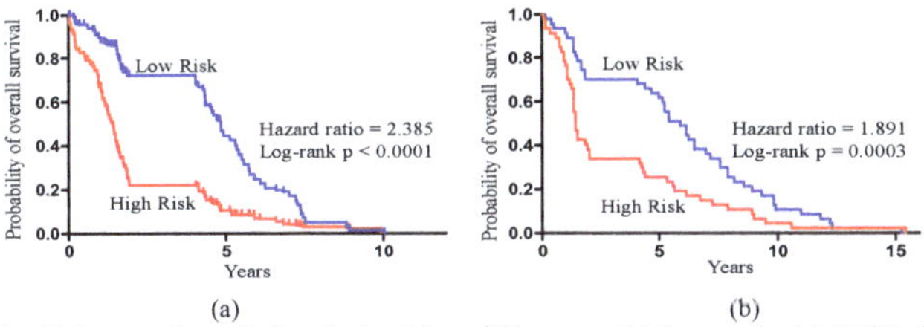

Fig. 4. Kaplan-Meier overall survival analysis of three-fold cross validation on the (a) TCGA data set and (b) independent test on the public data set.

There are in total 224 genes in the identified subnetworks, grown from 36 seed genes. Among these seed genes, 29 genes are from copy number altered genes and 11 genes are from literature collection (four genes: CCNE1, JAK2, SMAD4 and MYC are overlapped). In terms of gene family of identified seed genes, seven are oncogenes (EGFR, JAK2, JUN, LPP, MYC, NOTCH2 and RAF1); two are tumor suppressors: BRCA1 and SMAD4; and other genes are belonging to transcription factors, cell differentiation markers, protein kinases, etc. Note that CCN1, MYC, BRCA1 are also reported in the study of [10], which indicates that our proposed method could identify biological meaningful genes.

We then conducted functional annotation and pathway analysis using MsigDB database [22] for the identified subnetworks. The functions of 'Cell cycle', 'Apoptosis', 'Nucleus' and 'DNA repair' are significantly enriched in some of the subnetworks, shown in Fig. 5. These findings are consistent with our understanding of the cancer development.

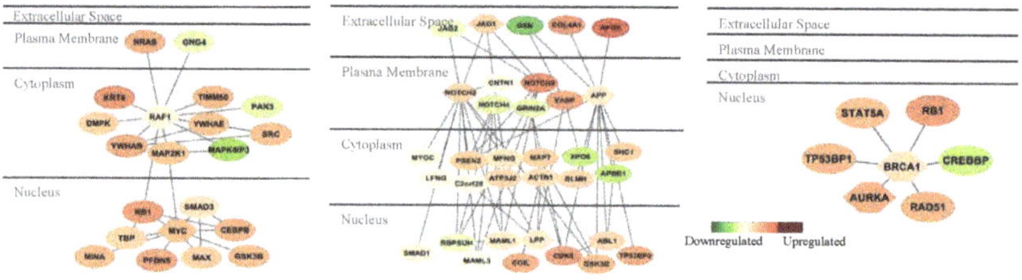

(a) (b) (c)

Fig. 5. Subnetworks identified from TCGA ovarian cancer gene expression data set. Subnetworks are merged if more than 2 genes are common. Node shape indicates the seed gene (hexagon) or non-seed gene (ellipse). Node color indicates the fold change between 'high risk' and 'low risk' groups. Red represents over-expressed in 'high risk' group and green reflects over-expressed in 'low risk' group. Enriched pathways and GO functional annotations are: (a) ErbB signaling pathway: p=2.43e-10; Cell cycle: p=1.47e-07. (b) Notch signaling pathway p=1.11e-16; Apoptosis p=3.76e-04. (c) Genes involved in Homologous Recombination Repair p=8.56e-08; Nucleus p=6.61e-04.

Interestingly, many signaling pathways are also significantly enriched in several subnetworks, for example, ErbB signaling pathway (Fig. 5(a), Fig. 6(a)), Notch signaling pathway (Fig. 5(b), Fig. 6(b)), NFκB signaling pathway (Fig. 7(a)), and TGF beta signaling pathway (Fig. 7(b)). Signaling pathway is more complicated and diverse. Epidermal growth factor receptor (EGFR) and ERBB2/HER-2 are members of the ErbB family of tyrosine kinase receptors. The studies have shown that the aberrant activity of EGFR and ERBB2 are important in tumor growth and development. Moreover, the overexpression of EGFR and ERBB2 and their downstream targets is associated with resistance to ovarian cancer chemotherapy [23]. Notch signaling pathway has been studied in many papers showing that it is active in ovarian cancer [24-27]. It is suggested that the inhibition of Notch signaling may be a therapeutic strategy for ovarian cancer [27]. NFκB transcription factors are key regulators of cell proliferation and apoptosis [28]. It is believed that changes in the upstream pathways will deregulate NFκB activation in cancer. Notice that many studies have focused on gene RSF-1 in NFκB network (Fig.7 (a)) and shown that it is involved in paclitaxel resistance in ovarian cancer [29, 30]. Transforming growth factor-beta (TGF-beta) is a tumor suppressor, which is involved in many types of human cancer, including ovarian cancer [31]. Recent study also shows that the activated TGF-b signaling pathway in omental metastases of ovarian cancer is a potential therapeutic target [32].

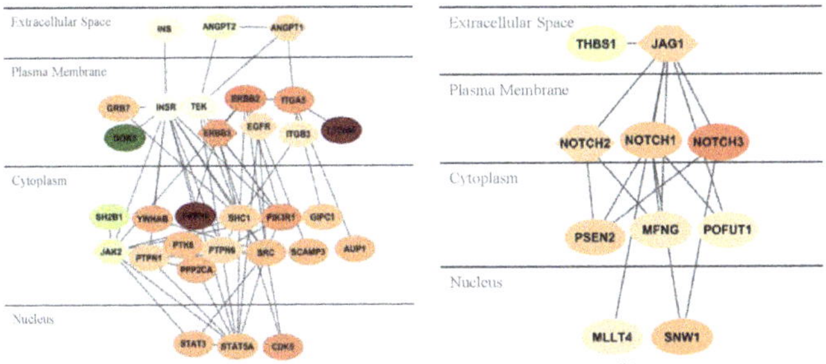

(a) (b)

Fig. 6. Enriched pathways and GO functional annotations are: (a) ErbB signaling pathway: p=9.10e-13; Signal transduction p=2.33e-07. (b) Notch signaling pathway p=2.86e-15. Figure legends are same as the ones in Fig. 5.

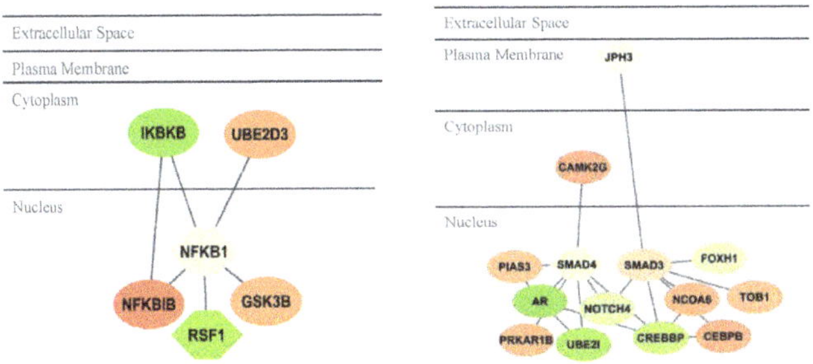

(a) (b)

Fig 7. Enriched pathways and GO functional annotations are: (a) NFκB signaling pathway p=7.03e-05. (b) TGF beta signaling pathway p=1.88e-06. Figure legends are same as the ones in Fig. 5.

4. Conclusion

We have proposed an integrative framework to analyze TCGA ovarian cancer data by integrating DNA copy number data, microarray data, protein-protein interaction data and patient clinical information. After detecting highly conserved copy number altered genes, we have applied network-based methods to identify pathway networks that can differentiate different expression patterns between different phenotypes. The experimental results have shown that our identified subnetworks could achieve good prediction performance and generalizability on independent data set using a network-constrained classifier. The results also show that our integrative approach can identify biological meaningful subnetworks that related to the development of ovarian cancer and drug resistance.

For the future work, the proposed method will be further enhanced through optimal parameter selection, statistical assessment and multiple hypothesis testing. Moreover, more ovarian cancer samples from TCGA portal and public data sets will be investigated.

5. Acknowledgments

This research was supported in part by NIH Grants (CA139246, CA149653, CA149147, and NS29525-18A1) and a DoD/CDMRP grant (BC030280).

References

1. Jemal, A., F. Bray, *et al.*,CA Cancer J Clin, 61(2): p. 69-90,(2011).
2. Bast, R.C., Jr., B. Hennessy, and G.B. Mills,Nat Rev Cancer, 9(6): p. 415-28,(2009).
3. Norquist, B., K.A. Wurz, *et al.*,J Clin Oncol,(2011).
4. Tcherkassova, J., C. Abramovich, *et al.*,Tumour Biol,(2011).
5. Hanash, S.,Nat Rev Cancer, 4(8): p. 638-44,(2004).
6. Taylor, I.W., R. Linding, *et al.*,Nat Biotechnol, 27(2): p. 199-204,(2009).
7. Cooper, L.A., J. Kong, *et al.*,IEEE Trans Biomed Eng, 57(10): p. 2617-21,(2010).
8. Ovaska, K., M. Laakso, *et al.*,Genome Med, 2(9): p. 65,(2010).
9. Gundem, G., C. Perez-Llamas, *et al.*,Nat Methods, 7(2): p. 92-3,(2010).
10. TCGA,Nature, 474(7353): p. 609-615,(2011).
11. Clarke, R., N. Brunner, *et al.*,Proc Natl Acad Sci, 86: p. 3649-3653,(1989).
12. Beroukhim, R., G. Getz, *et al.*,Proc Natl Acad Sci U S A, 104(50): p. 20007-12,(2007).
13. Bild, A.H., G. Yao, *et al.*,Nature, 439(7074): p. 353-7,(2006).
14. Zhao, X., C. Li, *et al.*,Cancer Res, 64(9): p. 3060-71,(2004).
15. Affymetrix,Edited by Affymetrix I. Santa Clara, CA,(2005).
16. Mishra, G.R., M. Suresh, *et al.*,Nucleic Acids Res, 34(Database issue): p. D411-4,(2006).
17. Olshen, A.B., E.S. Venkatraman, *et al.*,Biostatistics, 5(4): p. 557-72,(2004).
18. Ideker, T., O. Ozier, *et al.*,Bioinformatics, 18 Suppl 1: p. S233-40,(2002).
19. Chung, F., *Spectral Graph Theory*. CBMS Reginal Conferences Series. Vol. 92. 1997: American Mathematical Society, Providence.
20. Witten I. and Frank E., *Data Mining: Practical Machine Learning Tools and Techniques with Java Implementations*. 2000: Morgan Kaufmann.
21. Kaplan, E. and P. Maier,J AM Stat Assoc, 53: p. 457-481,(1958).
22. Subramanian, A., P. Tamayo, *et al.*,Proc Natl Acad Sci U S A, 102(43): p. 15545-50,(2005).
23. de Graeff, P., A.P. Crijns, *et al.*,Br J Cancer, 99(2): p. 341-9,(2008).
24. Rose, S.L.,Int J Gynecol Cancer, 19(4): p. 564-6,(2009).
25. Choi, J.H., J.T. Park, *et al.*,Cancer Res, 68(14): p. 5716-23,(2008).
26. Park, J.T., X. Chen, *et al.*,Am J Pathol, 177(3): p. 1087-94,(2010).
27. Shih, I.M. and T.L. Wang,Cancer Res, 67(5): p. 1879-82,(2007).
28. Rayet, B. and C. Gelinas,Oncogene, 18(49): p. 6938-47,(1999).
29. Sheu, J.J., B. Guan, *et al.*,J Biol Chem, 285(49): p. 38260-9,(2010).
30. Choi, J.H., J.J. Sheu, *et al.*,Cancer Res, 69(4): p. 1407-15,(2009).
31. Zhang, Y.Y., X. Li, *et al.*,Yi Chuan Xue Bao, 31(8): p. 759-65,(2004).
32. Yamamura, S., N. Matsumura, *et al.*,Int J Cancer,(2011).

ROLE OF SYNTHETIC GENETIC INTERACTIONS IN UNDERSTANDING FUNCTIONAL INTERACTIONS AMONG PATHWAYS

SHAHIN MOHAMMADI*, GIORGOS KOLLIAS, AND ANANTH GRAMA

Department of Computer Science, Purdue University, West Lafayette, IN 47906, USA
**E-mail: mohammas@purdue.edu*

Synthetic genetic interactions reveal buffering mechanisms in the cell against genetic perturbations. These interactions have been widely used by researchers to predict functional similarity of gene pairs. In this paper, we perform a comprehensive evaluation of various methods for predicting co-pathway membership of genes based on their neighborhood similarity in the genetic network. We clearly delineate the scope of these methods and use it to motivate a rigorous statistical framework for quantifying the contribution of each pathway to the functional similarity of gene pairs. We then use our model to infer interdependencies among KEGG pathways. The resulting KEGG crosstalk map yields significant insights into the high-level organization of the genetic network and is used to explain the effective scope of genetic interactions for predicting co-pathway membership of gene pairs. A direct byproduct of this effort is that we are able to identify subsets of genes in each pathway that act as 'ports' for interaction across pathways.

Keywords: Synthetic genetic interactions, Modified congruence score, Neighorhood overlap graph, KEGG pathways, Functional dependencies

1. Introduction

Systematic deletion of genes in *Saccharomyces cerevisiae* (yeast) reveals that only about 18% of its nearly 6000 genes are essential for cell growth on rich glucose media.[1] This observation suggests a strong underlying compensatory mechanism in the cell against genetic perturbations. Gene duplications and the existence of parallel (or alternative) pathways are well-known buffering mechanisms in the cell that are responsible for its robustness.[2–4] Gene duplications have been studied extensively. However, it is now believed that distributed compensatory mechanisms, such as parallel pathways, contribute the most towards cell robustness.[2,5] The underlying structure of these parallel pathways is determined by genetic interactions among genes having dependent, but not identical functions.[6,7] This is the rationale behind synthetic genetic interactions, in which double-mutant phenotypes significantly deviate from the expected (or typical) phenotype for gene pairs.

High-throughput technologies based on diploid synthetic lethal analysis by microarrays (dSLAM)[8,9] and/or synthetic genetic arrays (SGA) and its variants,[10–13] provide us with large datasets of genetic interactions in Saccharomyces cerevisiae. These interactions may be broadly classified into two main groups – negative (or aggravating/ synergistic) interactions, which describe double mutants that exhibit a more severe phenotype than expected; and positive (or alleviating/ epistatic) interactions, in which double mutants exhibit a less severe phenotype than expected. Synthetic lethal (SL) interactions correspond to a specific subset of negative interactions, in which the double-mutant is fatal while the individual mutations are vital. Analyzing these rapidly growing datasets of genetic interactions in order to understand the functional machinery of the cell is a challenging task. Three main models have been proposed for the higher order organization of genetic interactions – the *between-pathway model* among redundant functions, the *within-pathway model* among genes with additive effect, and the *indirect effects model*.[14] These models have been used both to predict functional similarity of gene pairs, implicitly, and to shed light on the higher level organization of pathways.

The rationale behind function prediction methods is that gene pairs that share "many" neighbors are more likely to be functionally similar. This thesis has been used to construct a complementary network, orthogonal to the initial genetic network, in which edges code functional similarity of gene pairs. Tong et al.[11] use hierarchical clustering based on the similarity of genetic interaction patterns and show that sets of genes within the same pathway tend to cluster together. Schuldiner et al.[12] also use genetic interaction patterns, which they refer to as *phenotypic signatures*, on an E-MAP dataset that consists of both quantitative alleviating and aggravating interactions in early secretory pathway (ESP). They confirm that pairs of genes with intermediate levels of correlated phenotypic signatures have a higher probability of negatively interacting using aggravating links, while highly correlated pairs typically show no aggravating interactions, and in many cases, they are connected using alleviating interactions. Costanzo et al.[13] construct a complementary weighted network, called the 'functional map' of the cell, in which every edge weight encodes the strength of the Pearson correlation among interacting partners of gene pairs. After applying a force-directed layout on the constructed functional map, they find that genes participating in same pathway tend to be in close proximity to each other. Ye et al.[15] construct a complementary network based on the statistical significance of shared neighbors, called the congruence network. They show that there is a strong correlation between GO annotations and the congruence score of gene pairs. Another class of studies uses models of genetic interactions to identify the high-level organization of pathways, and provides insight into interactions among them. These methods either use a bottom-up approach, for e.g., Segre et al.[16] and Bandyopadhyay et al.,[17] to group genes into functional modules and then analyze functional enrichment of modules, or a top-down technique, for e.g., Zhang et al.,[18] to infer interactions among known pathways directly.

In spite of these research efforts, there remain significant unresolved questions: How accurate are genetic interactions in predicting different functional pathways? What is the scope of these predictions (in terms of functions)? What is the specific structure around each pathway that allows us to capture the relationships among its member genes? In order to answer these questions, we start with a comprehensive evaluation of existing methods for predicting functional similarity of gene pairs. We demonstrate that, surprisingly, all of the methods considered have limited scope of effectiveness. To study this shortcoming in terms of the specific structure around each pathway, we developed a novel statistical framework (along with efficient associated algorithms) to quantify the contribution of various pathways to the functional similarity of each gene pair. We use this framework to construct the neighborhood overlap graph (NOG) for each pair of pathways and assess their functional dependence. These dependencies result in the KEGG crosstalk map, which not only explains the performance limitations of genetic interactions in predicting functional similarity of genes, but also uncovers the high-level organization of the cellular machinery. An important consequence of our method is that it enables us to identify individual genes within pathways that act as 'ports' of interaction across pathways. These genes can further be used to interpret the effective scope of functional similarity prediction methods.

2. Methods

2.1. *Notations*

Aggravating genetic interaction network, or *genetic interaction network (GIN)*, is a genetic map of the cell, represented as a graph $G = (V, E)$. Vertices V in the graph correspond to genes and edges encode aggravating interaction among gene pairs. A GIN of size $n = |V|$ can be represented by its *adjacency matrix*, A, which is an $n \times n$ matrix. The (i, j)th element in this matrix is 1 if gene v_i interacts with v_j and 0 otherwise. We denote the set of genes having aggravating interaction with a specific gene, $v_i \in V$, as $N(v_i)$. Given the set of *core genes*, that are annotated with at least one KEGG annotation, aggravating interactions are classified as *internal*, where both genes reside in the core set, *crossing*, in which one end of interaction is in the core set and the other is not, and *external*, in which both ends of the interaction reside outside of the core set.

Each KEGG pathway is represented as a subgraph of the GIN, represented by $P_A = (V_A, E_A)$, in which the set V_A consists of all genes in the pathway, and edge set E_A consists of all interactions incident on these genes. For each pathway pair $< P_A, P_B >$, we define the *neighborhood overlap graph (NOG)*, represented by $H_{A \to B}$, as the graph defined over same vertices as V_A, with edges encoding the significance of common neighbors of gene pairs in P_A with respect to P_B. An *interaction port* is defined as a functionally coherent subset of genes in pathway P_A that have a (statistically) significant number of shared neighbors in P_B, which can be equivalently defined as a dense subgraph of $H_{A \to B}$.

We rely on standard statistical definitions – the binomial coefficient, $C(r, s) = \frac{r!}{s!(r-s)!}$, is the number of groups of s items selected from a population of r items when order does not matter. Hypergeometric distribution, represented by $HG(k|N, s, t)$, is the probability of having k successes in a sequence of s random draws without replacement, from a size N, where t items are classified as successes, and $N - t$ items are classified as failures. The tail of hypergeometric distribution, denoted by $HGT(k|N, s, t)$, is the probability that a random variable $X \sim HG(k|N, s, t)$ is greater than or equal to k.

2.2. *Performance of local methods for predicting functional similarity of gene pairs*

Functional similarity prediction of gene pairs is assessed by their co-pathway membership, as determined by the KEGG pathways. Each of the methods considered generates a symmetric matrix that codes predictions for co-pathway membership of genes. In order to evaluate these methods, we rank gene pairs based on the computed similarity matrix and use the *minimum HyperGeometric* (mHG)[19] score to evaluate each ranked-list. This method has been previously used to evaluate the enriched GO terms in a ranked list of genes.[20] Here, we extend this method to assess the significance of top-ranked *gene pairs*, as opposed to genes, that correctly predict co-pathway membership of genes with respect to a given pathway.

Given a ranked list of gene pairs and a cut-off value that partitions this list into *top-ranked* (target set) and *low-ranked* (background set) gene pairs, we are interested in evaluating relative density of functionally similar gene pairs in the target set versus background set. Denote the total number of gene-pairs by $N = C(n, 2)$, total number of gene-pairs in a given pathway P_A by N_A, and the size of target set as l. Also, let the random variable X denote the number of gene pairs in the target set that participate in P_A. X has hypergeometric distribution and its p-value can computed as follows:

$$
\begin{aligned}
P - value(X = k) &= Prob(k \leq X) = HGT(k|N, N_A, l) \\
&= \sum_{x=k}^{min(N_A,l)} \frac{C(l,x)C(N-l, N_A - x)}{C(N, N_A)}
\end{aligned}
\tag{1}
$$

Without any prior information about the optimal cutoff, we need to automatically find l by finding the partition that minimizes the p-value. We encode the co-membership of gene-pairs in P_A using a binary vector $\lambda = \lambda_1, \lambda_2, ...\lambda_N \in \{0,1\}^N$, having N_A ones and $N - N_A$ zeros. Here λ_i is 1 if both of the genes in the ith ranked gene pair are members of pathway P_A, and 0 otherwise. The minimum HyperGeometric (mHG) score is then defined as:

$$
mHG(\lambda) = min_{1 \leq l \leq N} HGT(b_l(\lambda); N, N_A, l), \tag{2}
$$

where $b_l(\lambda) = \sum_{i=1}^{l} \lambda_i$. After correcting mHG scores for multiple comparison using the Bonferroni method, $-log(mHG)$ is used to quantify the performance of different methods in recapturing the co-pathway membership of different KEGG pathways.

2.3. *Constructing the neighborhood overlap graph (NOG)*

The neighborhood overlap graph (NOG) of a pathway P_A with respect to pathway P_B is an unweighted, undirected graph defined over the same vertices as P_A. In this graph, there is an edge between vertices v_i and v_j if their shared neighborhood, with respect to P_B, is statistically significant. Denote the degree of nodes v_i and v_j by D_i and D_j, respectively. Furthermore, denote the number of neighbors of nodes v_i and v_j, and their common neighbors that reside in a pathway P_B of size n_B nodes, using d_i^B, d_j^B, and k_{ij}^B, respectively. Figure 1 illustrates these parameters in a sample case.

Let the random variable X denote the number of common neighbors of v_i and v_j that reside in P_B, if we were choosing their neighbors from P_B at random. Let the random variable Y_i denote the number of node v_i's neighbors that reside in P_B if we were randomly selecting D_i neighbors of v_i from KEGG core genes (Y_j is defined similarly). It can be shown that Y_i and Y_j are independent and that all three variables follow the hypergeometric distribution. The significance of the neighborhood structure of v_i and v_j with respect to P_B, called the *modified congruence score (MCS)*, is defined as $Prob(Y_i = d_i, Y_j = d_j, k_{ij}^B \leq X | n, n_B, D_i, D_j)$, which can be computed using Bayes rule as follows:

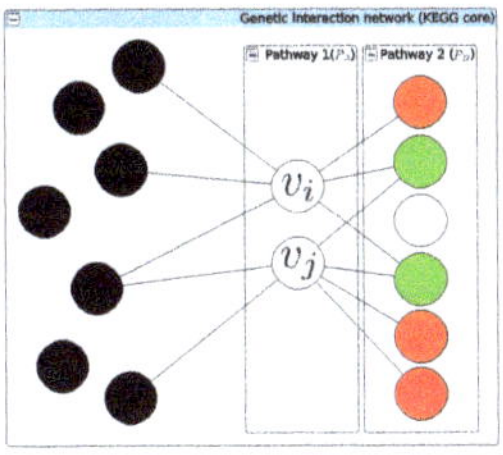

Fig. 1. A sample neighborhood configuration for v_i and v_j. Here $n = 15, D_i = 6, D_j = 6, n_B = 6, d_i^B = 3, d_j^B = 4$, and $k_{ij}^B = 2$

$$\begin{aligned}
MCS_B(i,j) &= Prob(Y_i = d_i, Y_j = d_j, k_{ij}^B \leq X | n, n_B, D_i, D_j) \\
&= Prob(k_{ij}^B \leq X | n_B, Y_i = d_i, Y_j = d_j) * Prob(Y_i = d_i, Y_j = d_j | n, n_B, D_i, D_j) \\
&= HGT(k_{ij}^B \leq X | n_B, Y_i = d_i, Y_j = d_j) \\
&* HG(Y_i = d_i | n, n_B, D_i) * HG(Y_j = d_j | n, n_B, D_j). \qquad (3)
\end{aligned}$$

For any given vertex pair $< v_i, v_j > \in V_H * V_H$, there will be an edge in E_H if their MCS is less than or equal to a predefined threshold. Algorithm 2.1 computes the neighborhood overlap graph of a given pathway pair $< P_A, P_B >$ and threshold level α. It may be noted here that under null model of random neighbor selection for gene pairs in V_A, the NOG follows the Erdos-Renyi random graph model with parameters $|V_A|$ and α, $ER(|V_A|, \alpha)$. We use this property later to assess the density significance of each subgraph in NOG, as explained in the Section 2.4.

Algorithm 2.1.

Input: *$P_A = (V_A, E_A)$: Source pathway, $P_B = (V_B, E_B)$: Destination pathway, n: Network size ($|V|$), α: Significance threshold*

Output: *$H = (V_H, E_H)$: Neighborhood overlap graph*

- ***Initialization:*** $n_B = |V_B|, V_H = V_A, E_H = \{\}$
- ***for each*** *vertex pair* $< v_i, v_j > \in V_A * V_A$ ***do***
  - $D_i = |N(v_i)|, D_j = |N(v_j)|$
  - $d_i^B = |N(v_i) \cap V_B|, d_j^B = |N(v_j) \cap V_B|$
  - $k_{ij}^B = |N(v_i) \cap N(v_j) \cap V_B|$
  - $MCS_B(i,j) = HGT(k_{ij}^B | n_B, d_i^B, d_j^B) * HG(d_i^B | n, n_B, D_i) * HG(d_j^B | n, n_B, D_i)$.
  - ***if*** $MCS_B(i,j) \leq \alpha)$ ***then***
    - $E_H = E_H \cup < i,j >$
  - ***end if***
- ***end for***
- ***return*** H

2.4. *Identifying interaction ports and inferring cross-pathway dependencies*

If we decompose the neighborhood of gene pair $(v_i, v_j) \in P_A$ with respect to different pathways, we can identify specific pathways (say P_B) that contribute the most to the functional similarity of genes v_i and v_j, as predicted by local neighborhood similarity methods. In this case, we say that gene pair (v_i, v_j) depends on P_B. Similarly, we can identify a functionally coherent subset of genes in P_A that depends on P_B. This subset of genes in P_A is referred to as an *interaction port* between P_A and P_B. As defined in Section 2.3, the neighborhood overlap graph (NOG) represents the significance of the common neighbors of each gene pair $< v_i, v_j > \in P_A$ with respect to pathway P_B. This suggests that interaction ports can equivalently be defined in terms of dense subgraphs in the NOG.

In order to find interaction port(s), we use the concept of graph cores.[21] The k-core of a given graph G is its maximal subgraph (not necessarily connected) such that every vertex is connected to at least k other vertices in the subgraph. We use iterative peeling to find a cohesive subset of vertices in NOG and record connected components at each core. As mentioned in Section 2.3, the Erdos-Renyi (ER) random graph provides a good null model for comparison with NOG. Using this random graph model, we process connected components of each core separately based on their density and size, as proposed by Koyuturk et al.,[22] and assign p-values to each one of them. During this iterative process, we can find multiple ports and as we go to the inner cores, we find more refined ports with higher densities, but possibly fewer vertices (specialization of ports). We infer that pathway P_A depends on pathway P_B if its most significant interaction port has a smaller p-value than a predefined threshold.

3. Results

3.1. *Datasets*

3.1.1. *Genetic interaction network*

We obtain the genome-scale quantitative genetic interactions from a recent SGA experiment in budding yeast, *Saccharomyces cerevisiae*.[13] This dataset consists of 1712 query genes – 1378 null alleles of non-essential genes and 334 hypomorphic or conditional alleles of essential genes (214 temperature-sensitive and 120 DAmP alleles), crossed over 3885 array strains, which spans a total of 5.4 million gene pairs that cover different biological processes. The aggravating genetic interaction network is constructed from the SGA dataset after applying the lenient cutoff.[13] Different hypomorphic alleles of each essential gene are merged together and represented by a single node, such that each node represents a unique gene in the SGA experiment. The final network consists of 363078 aggravating interactions among 4417 genes.

3.1.2. *Functional annotations*

We downloaded KEGG[23] annotations for budding yeast from the KEGG FTP site on January 4th, 2011. This catalogue organizes genes in the epistatic map into 97 different pathways, ranging from metabolic pathways to genetic information processing pathways. In order to filter out pathways that are too general or too specific, we eliminate pathways with more than 100 genes (*Metabolic pathways* and *Biosynthesis of secondary metabolites*, which are members of the Global Map) or less than 5 genes (15 pathways). The final dataset consists of 80 pathways, covering 1026 genes in the genetic interaction network.

3.1.3. *Availability*

All datasets and scripts used in this study can be downloaded from `http://www.cs.purdue.edu/homes/mohammas/PSB`.

3.2. *Similarity of genetic neighborhood as a predictor of functional similarity*

We evaluate three commonly used methods for predicting co-pathway membership of genes – sum of length-2 paths (or the number of shared neighbors), congruence score,[15] and the Pearson correlation of interaction profiles of gene pairs. After computing *mHG* score for each of these methods, we apply significance threshold $\alpha = 0.01$ to identify the enriched pathways. Each pathway is considered enriched, if it is enriched in at least one of the tested methods. Figure 2 illustrates the performance of these methods in identifying co-pathway membership of genes in different KEGG pathways. Enriched pathways are sorted based on the overall performance of all three methods, and are labelled accordingly.

As depicted in the figure, there are only 42 pathways among 80 ($\sim 50\%$) that are enriched. Further analysis of these enriched pathways indicates that most KEGG pathways that are involved in *genetic information processing* are highly enriched, while the bulk of the pathways that are *not* enriched are annotated as metabolic pathways. Among enriched pathways, *ribosome* had zero mHG score to machine precision (in spite of normalization before plotting log scores), which indicates high coverage of its member gene pairs among top ranked scores. One aspect of analysis that is not reflected in the figure is the coverage of each method at the point that minimizes mHG score, as well as the rank that caused the mHG to minimize. A detailed analysis of different methods indicates that congruence score and Pearson correlation methods have better precision with lower recall. This can be observed from the fact that the cut that minimizes mHG for these methods usually resides in top 1% of scores, while the same cut for number of shared neighbors method has better coverage of member genes, at the cost of reduced precision.

This observation suggests a specific structure around enriched pathways that boosts their similarity scores and allows us to recapture their co-pathway membership. In order to see how much of this signal comes from other KEGG pathways, we prune all crossing and external interactions in the genetic interaction network and recompute the enrichment scores only using KEGG core genes. Figure 3 illustrates the performance of local methods on the vertex induced subgraph of GIN using KEGG core genes. Most enriched pathways remain highly enriched even if we only use the internal interactions among KEGG pathways to compute functional similarity scores. This internal signal can be interpreted using the between-pathway model of genetic interactions and suggests a bi-cliquish structure around enriched pathways.

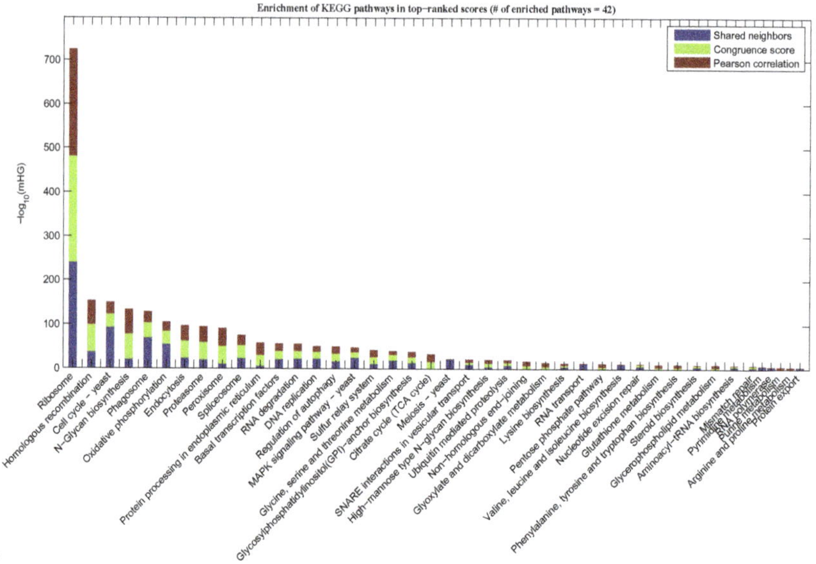

Fig. 2. Performance of local neighborhood methods in predicting co-pathway membership of gene pairs in different KEGG pathways. Enriched pathways are sorted based on their overall enrichment as computed by different methods, and are labelled appropriately (rest of pathways are used for spacing among enriched pathways).

3.3. *Constructing KEGG crosstalk map*

We used the induced subgraph of genetic interaction network defined by internal interactions to decompose the neighborhood of each pathway and to construct the interaction map among KEGG pathways. Figure 4 shows the density of the constructed map as a function of thresholding dependency scores. The final density drop in the plot is related to p-values that were normalized, since they were zero (to the machine precision). As can be seen from the figure, less than 3% of pathway pairs depend on each other. There is also a significant density drop to 1.5% as we start increasing the threshold. We extract all interactions after this specific threshold and construct the KEGG crosstalk map. Figure 5 includes approximately 50% of captured dependencies among all non-isolated pathways. Isolated nodes (pathways) that do not have any dependencies are eliminated from the figure. Referring to Figure 3, we observe that most isolated pathways in KEGG crosstalk map have not been enriched using functional prediction methods either. This suggests a high correlation between the specific structure around pathways, which is manifested as pathway dependencies, and the performance of functional similarity methods in predicting co-pathway membership of their genes. Figure 6 illustrates an example port of interaction between protein processing in ER and proteasome pathways.

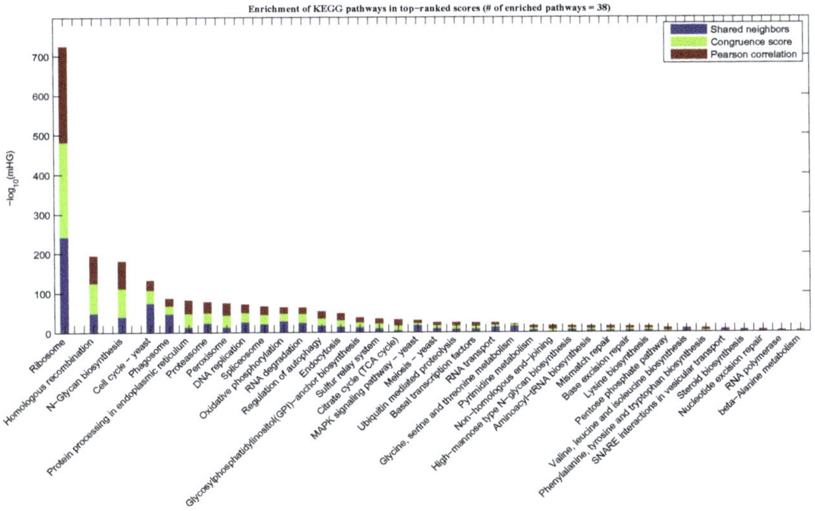

Fig. 3. Performance of local neighborhood methods in predicting co-pathway membership of gene pairs in different KEGG pathways, after pruning the crossing and external interactions.

Port genes are colored in red and are overlaid on the diagram of protein processing in ER (pathway diagram is adopted from KEGG).

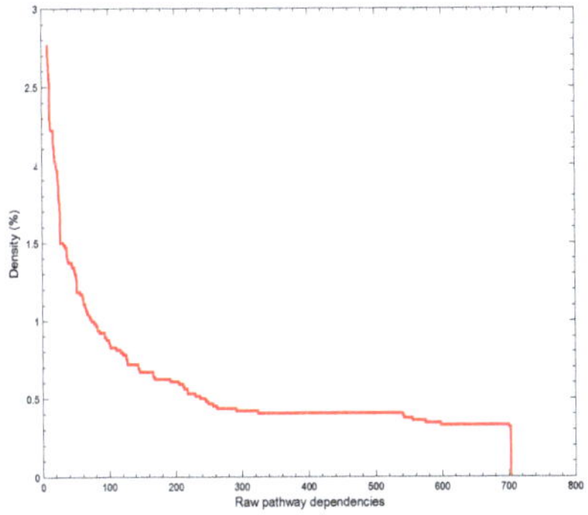

Fig. 4. Density of KEGG crosstalk map as a function of thresholding.

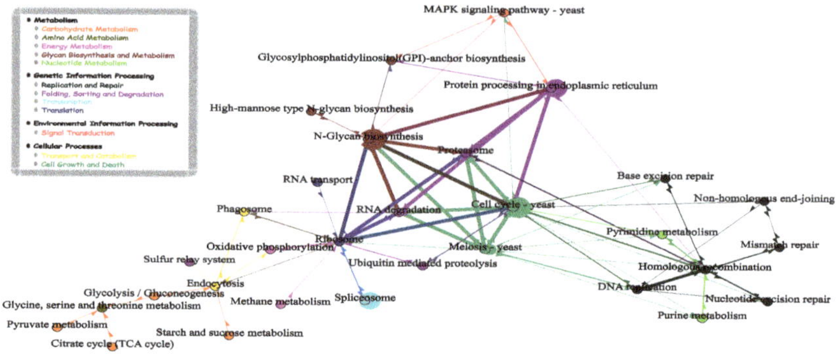

Fig. 5. Computed KEGG crosstalk map.

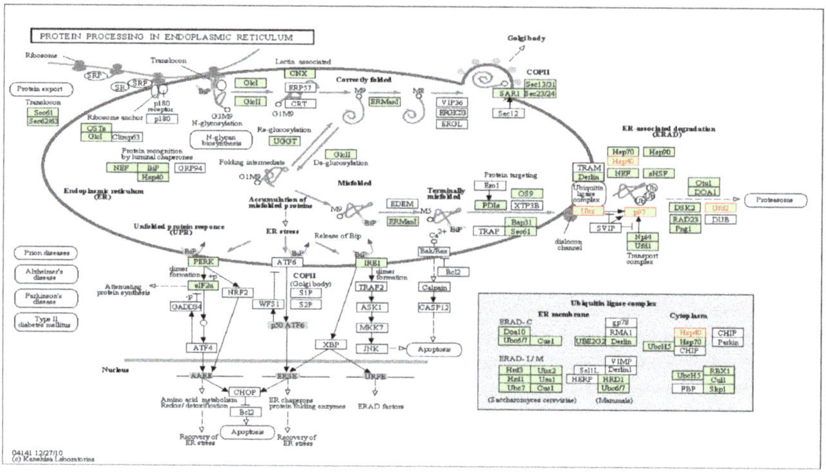

Fig. 6. Protein processing in ER, port genes interacting with proteasome pathway are colored in red (pathway diagram is adopted from KEGG website).

4. Discussion

Figure 5 illustrates a strong internal buffering mechanism in *N-Glycan biosynthesis*, *cell cycle*, *protein processing in endoplasmic reticulum* and *spliceosome* pathways, which can be interpreted using the within-pathway model of genetic interactions.This buffering against genetic perturbations, preserved through evolution, hints at the important role of these pathways in maintaining the functional stability of the cell. Referring to Figure 3, we see that all of these pathways are highly enriched, thus the internal redundancy of these pathways is a significant source of support for the functional similarity of their member genes.

Further analysis of the dependency map suggests that dependencies among functionally related pathways cluster them into coherent modules. The ribosome pathway resides at the heart of the map, having dependencies with both pre- and post-translational pathways. The former machinery can protect the cell against defective RNA constructed during transcription, through RNA degradation, and mis-spliced pre-mRNA constructed through Spliceosome. The latter mechanism regulates the concentration of specific proteins and degrades misfolded proteins by tagging them using ubiquitin mediated proteolysis and targets them for unfolding in proteasomes. This machinery is also known to be essential for many other cellular processes, including the cell cycle, the regulation of gene expression, and responses to oxidative stress. Perturbation of the oxidative phosphorylation pathway can also endanger cell integrity, since RNA molecules are known to be particularly vulnerable to oxidative stress.[24] The varied dependency of ribosome on a number of different pathways can explain its exceptionally high enrichment score, as measured with mHG and shown in Figures 2 and 3.

Most of the synthesized proteins in ribosome can fold themselves spontaneously into their active conformation. However, some of them need further post-translational modification, such as glycosylation, to aid their folding. The endoplasmic reticulum (ER) serves as the gateway for modification of newly synthesised proteins with the help of lumenal chaperones. Terminally mis-folded proteins are degraded through the proteasome using a process called ER-associated degradation (ERAD). The high-level organization of the protein processing in ER is shown in Figure 6, with genes identified as the interaction port marked in red. It can be observed that the ERAD sub-function is faithfully recovered as the interaction port between protein processing in ER and the proteasome pathway. These functional dependencies can be also observed in the highly significant triangular dependencies between protein processing in ER, proteasome, and N-glycan biosynthesis in Figure 5.

Another observation relates to the densely connected module of DNA repair, clustered in bottom right part of the map. This module consists of the DNA replication pathway, together with both single strand repair pathways; including the base excision repair, mismatch repair, and nucleotide excision repair pathways; and double-strand repair pathways, namely the non-homologous end-joining and the homologous recombination pathways. The Purine and Pyrimidine metabolic pathways, both of which help in metabolising nucleotide, are also parts of this module. Referring back to the result of functional similarity prediction methods, we note that all of these pathways are highly enriched.

In this study, we demonstrated that synthetic genetic interactions provide powerful means for predicting functional similarity of gene pairs. However, their scope of effectiveness is limited to a specific subset of pathways. This subset includes most pathways involved in genetic information processing and cellular processes, while missing many of the metabolic pathways. To further study the specific structure around these pathways that boosts their functional similarity, we decomposed the neighborhood around each pathway and identified the dependencies among them, inferred from the observed topological patterns. We discovered that the strength of these dependencies, manifested as the significance of the interaction ports between pathways, is highly correlated with the performance of functional prediction methods, and can be used to interpret their effective scope. In addition, observed dependencies reflect our previous understanding of higher level organization of pathways, which suggests that the KEGG crosstalk map by itself can be used to uncover the cellular machinery.

5. Acknowledgments

This work supported by the Center for Science of Information (CSoI), an NSF Science and Technology Center, under grant agreement CCF-0939370, and by NSF grants DBI 0835677 and 0641037.

References

1. G. Giaever *et al.*, *Nature* **418**, 387 (July 2002).
2. A. Wagner, *Nature genetics* **24**, 355 (April 2000).
3. Z. Gu *et al.*, *Nature* **421**, 63 (January 2003).
4. K. Hannay, E. M. Marcotte and C. Vogel, *BMC genomics* **9**, p. 609 (December 2008).
5. A. Wagner, *BioEssays : news and reviews in molecular, cellular and developmental biology* **27**, 176 (February 2005).
6. Y. Qi *et al.*, *Genome Research* **18**, 1991 (December 2008).
7. S. J. Dixon *et al.*, *Annual review of genetics* **43**, 601 (August 2009).
8. X. Pan *et al.*, *Molecular cell* **16**, 487 (November 2004).
9. S. L. Ooi, D. D. Shoemaker and J. D. Boeke, *Nat Genet* **35**, 277 (October 2003).
10. A. H. Tong *et al.*, *Science* **294**, 2364 (December 2001).
11. A. H. Tong *et al.*, *Science (New York, N.Y.)* **303**, 808 (February 2004).
12. M. Schuldiner *et al.*, *Cell* **123**, 507 (November 2005).
13. M. Costanzo *et al.*, *Science* **327**, 425 (January 2010).
14. C. L. Tucker and S. Fields, *Nature Genetics* **35**, 204 (November 2003).
15. P. Ye *et al.*, *Molecular Systems Biology* **1**, msb4100034 (November 2005).
16. D. Segrè, A. Deluna, G. M. Church and R. Kishony, *Nature genetics* **37**, 77 (January 2005).
17. S. Bandyopadhyay, R. Kelley, N. J. Krogan and T. Ideker, *PLoS Comput Biol* **4**, p. e1000065 (April 2008).
18. L. V. Zhang, O. D. King, S. L. Wong, D. S. Goldberg, A. H. Tong, G. Lesage, B. Andrews, H. Bussey, C. Boone and F. P. Roth, *Journal of biology* **4**, 6+ (2005).
19. E. Eden, D. Lipson, S. Yogev and Z. Yakhini, *PLoS Comput Biol* **3**, p. e39 (March 2007).
20. E. Eden *et al.*, *BMC Bioinformatics* **10** (Feb 2009).
21. S. B. Seidman, *Social Networks* **5**, 269 (1983).
22. M. Koyutürk, W. Szpankowski and A. Grama, *Journal of Computational Biology* **14**, 747 (August 2007).
23. M. Kanehisa and S. Goto, *Nucl. Acids Res.* **28**, 27 (January 2000).
24. Q. Kong and C. L. Lin, *Cell Mol Life Sci* **67**, 1817 (June 2010).

DISCOVERY OF MUTATED SUBNETWORKS ASSOCIATED WITH CLINICAL DATA IN CANCER

FABIO VANDIN, PATRICK CLAY, ELI UPFAL, BENJAMIN J. RAPHAEL

Department of Computer Science, and Center for Computational Molecular Biology, Brown University, Providence, RI 02912, U.S.A.

E-mail: {vandinfa,pclay,eli,braphael}@cs.brown.edu

A major goal of cancer sequencing projects is to identify genetic alterations that determine clinical phenotypes, such as survival time or drug response. Somatic mutations in cancer are typically very diverse, and are found in different sets of genes in different patients. This mutational heterogeneity complicates the discovery of associations between individual mutations and a clinical phenotype. This mutational heterogeneity is explained in part by the fact that driver mutations, the somatic mutations that drive cancer development, target genes in cellular pathways, and only a subset of pathway genes is mutated in a given patient. Thus, pathway-based analysis of associations between mutations and phenotype are warranted. Here, we introduce an algorithm to find groups of genes, or pathways, whose mutational status is associated to a clinical phenotype *without* prior definition of the pathways. Rather, we find subnetworks of genes in an gene interaction network with the property that the mutational status of the genes in the subnetwork are significantly associated with a clinical phenotype. This new algorithm is built upon HotNet, an algorithm that finds groups of mutated genes using a heat diffusion model and a two-stage statistical test. We focus here on discovery of statistically significant correlations between mutated subnetworks and patient survival data. A similar approach can be used for correlations with other types of clinical data, through use of an appropriate statistical test. We apply our method to simulated data as well as to mutation and survival data from ovarian cancer samples from The Cancer Genome Atlas. In the TCGA data, we discover nine subnetworks containing genes whose mutational status is correlated with survival. Genes in four of these subnetworks overlap known pathways, including the focal adhesion and cell adhesion pathways, while other subnetworks are novel.

1. Introduction

A major goal of cancer sequencing projects such as The Cancer Genome Atlas is to identify genetic and epigenetic alterations that determine clinical phenotypes, such as survival time or drug response. There are a number of reports of genes whose mutational status is associated with survival such as KRAS mutations,[1] EGFR amplifications,[2] and PTEN mutations.[3] Despite the rapid increase in catalogs of somatic mutations in cancer genomes,[4,5] progress in determining which mutations, or mutations in which genes, determine clinical phenotypes remains slow. The difficulty is due in part to the extensive *mutational heterogeneity* exhibited by cancer genomes, where the somatic mutations or mutated genes vary widely across patients. This mutational heterogeneity is a consequence of two features of the somatic mutation process in cancer. First, the somatic mutations in each cancer genome are a mixture of functional *driver* mutations responsible for cancer, and random *passenger* mutations that accumulate during tumor progression but are inconsequential for cancer. Second, driver mutations target not just single genes, but also groups of genes in signaling or regulatory pathways.[6,7] Thus, different patients may have different subsets of driver mutations in key pathways, and thus driver mutations are distributed over a large number of genomic locations in many different genes. A natural approach for finding clinically relevant mutations is first to determine driver mutations, and then to test each of these for clinical association. However, the problem of distinguishing driver from passenger mutations is itself a challenge. Many cancer genome studies attempt to predict driver mutations (or mutated genes) by finding those with a statistical significant frequency of occurrence in a large cohort of patients. But the power of this approach is reduced by mutational heterogeneity. Moreover, it may be mutations within a pathway, and not a single gene, that determine a clinical phenotype in a heterogeneous cohort of cancer patients.

A more powerful approach to test associations between mutation and phenotype is to test these associations at the pathway level, rather than at the level of single mutation or single genes. Ideally, one would perform such a test using collection of all relevant biological pathways. No such comprehensive collection currently exists, although databases such as KEGG[8] , Reactome,[9,10] and others are important efforts in this direction. An alternative source of information is genome-scale interaction networks that record (binary) interactions between proteins. Examples of such data sources for human are STRING[11] and HPRD.[12] Interaction networks have proven to be a useful source of information for analyzing genomic data, particularly gene expression data. Chuang *et al.*[13] introduced a method to find subnetworks of interaction networks whose gene expression predicts progression to metastasis in breast cancer. This work extends an earlier approach to find differentially expressed subnetworks introduced in Ideker *et al.*[14] and extended in Dittrich *et al.*[15] and in Beisser *et al.*[16] Also, Vaske *et al.*[17] present a method to infer patient-specific genetic activities. The latter method relies on curated pathway interactions among genes and predict the degree of alteration of pathway's activities in the patient using probabilistic inference. More recently, methods to discovery mutated subnetworks have been introduced.[18,19]

Here we introduce an approach to find subnetworks of genes in an interaction network with the property that mutations in the genes in the subnetwork are correlated with a clinical parameter. Specifically, for the clinical parameter of survival time, we identify subnetworks such that the survival time of patients with mutations in the subnetwork is significantly different from patients with no mutation in the subnetwork. To the best of our knowledge this is the first approach to find subnetworks whose mutations are correlated with survival. Previous methods[13–16] utilized gene expression data and incorporate a variety of different scoring methods to identify subnetworks/modules and to compute their statistical significance. We accomplish this goal by extending the HotNet algorithm previously introduced by some of us.[18] Our algorithm represents association scores for individual genes as sources of "heat" on the vertices (genes) of the interaction network, and uses a heat-diffusion model to propagate heat to surrounding vertices. We extract "significantly hot" subnetworks with a statistical test that rigorously bounds the false discovery rate (FDR) on the derived subnetworks. We apply our method to finding mutated subnetworks correlated with survival in simulated data and ovarian cancer data from The Cancer Genome Atlas. In the TCGA data, we discover nine subnetworks containing genes whose mutational status is correlated with survival. Genes in four of these subnetworks overlap known pathways, including the focal adhesion and cell adhesion pathways, while other subnetworks are novel.

2. Methods

2.1. *Generalized HotNet*

Our new method* builds on the HotNet algorithm.[18] A schematic of our method is given in Figure 1. Suppose we are given a gene/protein interaction network $G = (V, E)$ where the vertices represent genes and each edge represents a (binary) interaction between pair of genes, or their corresponding proteins. We are also given a score σ_g for each gene g. Our goal is to find subnetworks of G whose combined scores are statistically significant. There are two challenges that must be addressed. First, it is not feasible to evaluate all possible subnetworks: the large number of subnetworks of a reasonable size (e.g. containing at least 5 genes) implies that testing each incurs large computational burden, and more importantly would require a severe multiple hypotheses testing correction. Second, the topology of the interaction network means that the individual subnetworks cannot be treated as independent hypotheses. We first describe how our method addresses the topology of the network and then describe the statistical methodology.

The subnetworks of interest are determined both by the scores of their genes, and the interactions between the genes. However, human interaction networks are fairly irregular, and typically include "hub genes" with very high degree. Thus, it is important to account for the topology of the interaction network when evaluating subnetworks. For example, consider the following two scenarios. In the first scenario, two genes with high score are connected by a single vertex of degree 2 in the network. In the second scenario, the two genes with high score are connected by a vertex of very high degree. Because there are many such paths through the high degree vertex, it is more likely to see genes with high score connected through a high degree node (the second scenario) than a low degree one (the first scenario). This situation generalizes to more complicated scenarios where neighbors of neighboring vertices may have widely varying degrees. To formalize the intuition behind these scenarios, the HotNet algorithm considers the score of a gene as a quantity of "heat" placed on the gene, and allow heat to diffuse over the edges. Heat placed on a vertex of low degree will diffuse slowly through the graph and thus the neighbors of the vertex will remain hot for long periods of time. In contrast, heat placed on a vertex of high degree will diffuse quickly to the neighbors and none of these vertices will be hot.

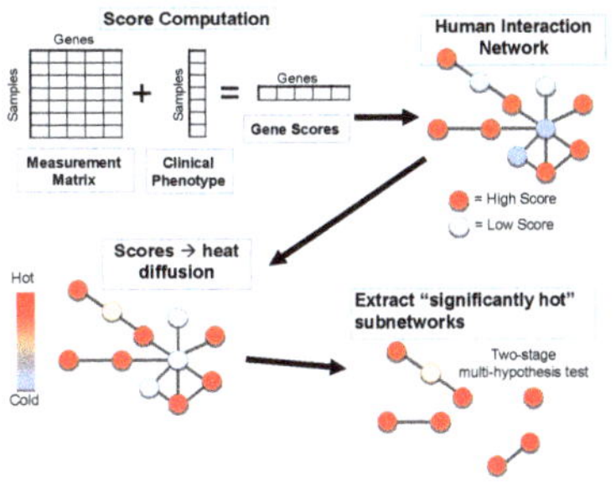

Fig. 1. Generalization of HotNet for clinical data.

HotNet uses a heat diffusion process and a statistical test to derive "significantly hot"

*An implementation of Generalized HotNet is available as a separate tool at: `http://cs.brown.edu/~braphael/software.html`

subnetworks. Significant subnetworks are thus determined by both the score of the genes in the subnetwork and the local topology of the subnetwork. HotNet consists of several steps. First, we use a heat diffusion process to derive a measure of influence between two genes in the interaction network. This measure depends only on the topology of the network. Next, the scores of genes are used to enhance the influence measure, defining the heat exchanged between any two pair of genes. Then, we remove cold edges (i.e., edges with low exchange of heat), dividing the network into subnetworks. Finally, we apply a two-stage statistical test to rigorously bound the false discovery rate (FDR) of the identified subnetworks. We briefly describe each of these steps. Additional details and analyses of the FDR bound are in Vandin *et al.*[18]

Let A be the adjacency matrix of the interaction network G. That is, let $A(i,j) = 1$ if genes g_i, g_j are connected by an edge in G, and let $A(i,j) = 0$ otherwise. Let D be a diagonal matrix with $D(i,j) = 0$ if $i \neq j$, and $D(i,i) = deg(i)$, where $deg(i)$ is the degree of g_i in the graph (i.e., the number of genes that interact with g_i in the network). The matrix $L = -A + D$ is called the Laplacian matrix of G and the matrix $H_t = e^{-Lt}$ is the *heat kernel* of the graph[20,21] for a real number $t \geq 0$. Here e^X denotes the matrix exponential: $e^X = \sum_{k=0}^{\infty} \frac{1}{k!} X^k$. The entry $H_t(i,j)$ represents the heat found on gene i when a unit of heat is placed on gene j and allowed to diffuse for time t. Note that a different diffusion model based on the equilibrium value of a diffusion with loss[22] was employed in the original HotNet algorithm.[18] Fix a value $t > 0$, and let heat diffuse on the interaction network for time t. We define an influence measure $i(u,v)$ between pairs of genes u and v in the interaction network as $i(u,v) = H_t(v,u)$. Thus, the influence $i(u,v)$ of gene u on gene v is the heat observed on v after a length of time t when there is only a unit source of heat on node u at time 0. Since this measure is asymmetric (i.e., for a general diffusion kernel $i(u,v) \neq i(v,u)$), we define a symmetrized influence as $\tilde{i}(u,v) = \min\{i(u,v), i(v,u)\}$. The influence is thus defined only by the topology of the network and the value of the parameter t, and is defined for all pairs of genes based on the heat observed on the vertices.

HotNet uses the score for each gene to enhance the influence measure and then breaks the graph into connected components based on this influence measure. In particular, let $\sigma(g)$ be the score for gene g. Then for each pair (u,v) of genes we define a weight $w(u,v) = \max\{\sigma(u), \sigma(v)\} \times \tilde{i}(u,v)$. Next, given a parameter δ, HotNet removes all the edges (u,v) with $w(u,v) < \delta$. HotNet returns as output the connected components identified after the removal of the edges.

HotNet returns a list of subnetworks, each containing at least s genes, and employs a two-stage statistical test to assess the statistical significance of the returned list of subnetworks. The first stage of the test computes a p-value for the *number* of subnetworks with at least s genes that are returned, for different values of s, under a suitable null hypothesis (see Section 2.2.3 for the description of the null hypothesis used for the results presented in this work). Since the statistic is the number of subnetworks with at least s genes, and the possible values of s (and thus the number of tested hypotheses) is bounded by the number of genes with a score, the number of hypothesis is bounded by a quantity that is much smaller than the number of possible subnetworks of the interaction network. We can thus determine an s such that the number of connected components of size $\geq s$ is significant, with a particular

p-value. This p-value measures the significance of the number of subnetworks of a minimum size s, but does not say which, if any, of the individual subnetworks is significant. The second stage of the test estimates the false discovery rate (FDR) for the subnetworks in the list. (See Ref. 18 for further details of the statistical test.) In summary, HotNet returns: (i) a list of subnetworks, each with at least s genes; (ii) a p-value for the observed number of subnetworks; (iii) an estimated FDR for the list.

2.2. *Adaptation to Clinical Data*

There are three major steps in adapting the HotNet algorithm to applications in clinical data analysis: (i) selection of scoring function $\sigma(g)$ for individual genes; (ii) selection of the parameters t and δ; (iii) determination of the null hypothesis distribution.

2.2.1. *Gene Scores*

The input to HotNet is a score $\sigma(g)$ for each gene g. In order to find associations between mutations in subnetworks and clinical phenotypes, we derive a score for each gene g from the p-value of a statistical test that measures the association between the mutational status of g and a clinical parameter of interest. The particular statistical test depends on the clinical parameter. For example, if the clinical parameter is categorical, (e.g., classes for "response to treatment") the χ^2 test for independence can be used. For survival analysis we use the logrank[23] test to assess the significance of the difference between the survival curve of the set $M(g)$ of patients where g is mutated and the survival curve of the set $\overline{M}(g)$ of patients with no mutations in g.

For a gene g, we define the score $\sigma(g) = -2\log_e p_g$, where p_g is the p-value of the statistical test. Note that with this choice of score, when $t = 0$ the heat on a subnetwork $S = \{g_1, \ldots, g_{|S|}\}$ is equal to $-2\sum_i \log_e p_{g_i}$. This sum correspond to the statistic of Fisher's Method for combining p-values for (independent) statistical tests[24] . When $t > 0$, the total heat on a subnetwork S will depend on the topology of S and the topology of the entire network. For example, if S does not contain any high degree node (e.g., it is a linear path) or if it is a dense subgraph (e.g., a clique), the heat on S will be close to the case $t = 0$, and thus to the statistic of Fisher's Method. However, when vertices in S are connected to many other vertices outside S (e.g. a linear path going through an high degree vertex), the heat on S will diffuse to many other vertices, and result in a reduction in the combined p-value of the genes in S. Note that the gene scores are not a function of the null hypothesis employed by HotNet to assess the significance of the discovered subnetworks, but depends on the nature of the clinical data analyzed.

2.2.2. *Selection of parameters t and δ*

As described in Section 2.1, the execution of HotNet depends on the choice of parameters: t, the length of time that heat diffuses, and δ, the threshold for removing cold edges; i.e. edges of weight less than δ. In this way HotNet divides the network into subnetworks.

The parameter t controls the distance at which the score of a gene will diffuse in the network. If $t = 0$ there is no diffusion, while for $t = +\infty$, the heat distribution reaches equilibrium where all genes have the same heat. Using simulations we studied how the influence $i(u, v)$

changes as the distance between u and v increases for different values of t (data not shown). We fixed $t = 0.1$ since with this choice nodes at different distances receives distinguished amounts of heat with the diffusion process.

We use the following procedure to choose δ. We generate 100 datasets with the distribution of the null hypothesis (see Section 2.2.3), and consider how the number of connected components of size at least s varies as a function of δ for small values of s (i.e., $s = 3, 4, 5$). Note that since we are considering datasets from the null distribution, the subnetworks observed should not be significant. We make a conservative choice, by choosing a first δ that gives the largest (average) number of subnetworks of size at least s. In our experiments we fix $s = 5$, but comparable values of δ are obtained for $s = 3, 4$. Since increasing the value δ corresponds to a more conservative test, we also consider values $\delta' \geq \delta$. In particular, we consider values $\delta' = 1.1\delta, 1.2\delta, \ldots, 2.0\delta$. Note that if we consider δ_1, δ_2 with $\delta_1 < \delta_2$, the subnetworks of size at least s found with δ_2 will all be part of the subnetworks of size at least s found with δ_1. Thus, increasing δ limits the output to subnetworks with larger enhanced influence.

2.2.3. *The Null Hypothesis Distribution*

The accuracy of HotNet depends on the appropriate choice of the null hypothesis distribution. Once this distribution is determined, we generate a large number of instances (1000 in our experiments) to estimate the p-values of the various events.

For survival data we use the following two null hypothesis:

(1) H_0^P: the mutation matrix is fixed and the survival data is permuted across the patients. That is, let c_i be the survival associated to patient i in the observed data, with $i \in \{1, \ldots, m\}$. In a dataset generated under the H_0^P null hypothesis the survival data for patient i is $c_{\pi(i)}$, where $\pi(i)$ is a permutation chosen uniformly at random among all permutations over the set $\{1, \ldots, m\}$.
(2) H_0^M: for each gene g, the set of patients in which g is mutated is chosen uniformly at random and independently of other mutations, preserving the frequency of mutation of the genes. That is, if g is mutated in f_g patients, in a dataset generated under the H_0^M null hypothesis g is mutated in a set of f_g patients chosen uniformly at random, independently of other mutations. (The survival data are not explicitly randomized, since randomizing mutations will already account for randomization of the survival data.)

The main difference between the two models is that H_0^M removes correlations between occurrences of mutations in the genes, which are preserved in H_0^P. Thus, H_0^M removes possible correlations between scores of genes in the network (since genes with correlated mutations have correlated scores). Note that the parameter δ depends on the particular null hypothesis chosen, since it is based on the distribution of the number of subnetworks of size at least s under the null hypothesis.

3. Results

We tested our algorithm on both simulated data and ovarian cancer data from The Cancer Genome Atlas (TCGA). For simulated data, we considered the patients as divided into two classes, thus looking for groups of genes with different mutation status in the two classes. For cancer data we considered a test statistic for survival. Thus, we aim to find groups of genes whose mutation status was associated with patient survival.

3.1. *Simulated data*

We first assessed our method using simulated data. We simulated whole exome sequencing data coming from two classes C_1, C_2 containing 150 patients each. For example, C_1 is a class of patients that respond to a particular treatment, while C_2 is a class of patients that do not respond to the treatment. We assumed that a mutated subnetwork is only one possible cause of a patient being in class C_1. Thus, we *planted* non-random mutations into a subnetwork S consisting of the 4 genes JAG1, NOTCH1, MAML1, and CDK8 into 20% of the patients in C_1. In particular, for a subset P consisting of 20% of the patients of C_1 we mutated one of the genes in S. For all other patients and all other genes we generated mutations according to a background mutation rate of 1.7×10^{-6}, consistent with recent studies on cancer.[25] Note that the patients in P also contain random mutations (in all genes but the ones in S), and that the 4 genes in S were mutated at random in all patients but P.

For each gene g, we built a 2×2 contingency table in which the row variable is the mutation status (i.e., mutated or not) of g and the column variable is the class (i.e., C_1 or C_2) of a patient. We used a χ^2 test for the 2×2 contingency table to obtain the p-value of a gene. None of the four genes turned out to be significant after correction for multi-hypothesis testing. We then used the scores defined in Section 2.2.1 as input for the generalized version of HotNet. The only significant subnetwork reported by our algorithm corresponds to the set of 4 genes JAG1, NOTCH1, MAML1, CDK8 ($p \leq 0.01$, $FDR \leq 0.05$). This shows that our algorithm identifies a subnetwork associated with clinical data when none of the genes in the subnetwork were identified as significant when considered individually, and when the subnetwork is non-randomly mutated in only 20% of the patients of one class.

3.2. *Ovarian TCGA data*

We next considered mutation data from ovarian cancer patients from The Cancer Genome Atlas. For each of these patients, we marked a gene as mutated if a somatic point mutation (or small indel) was present (as measured by exome resequencing) or if a focal copy number aberration (CNA) was present (as measured by array copy number data). Thus, the somatic mutation data was reduced to an $m \times n$ binary mutation matrix measuring the mutation status of $m = 316$ patients for $n = 17301$ genes. Genes not mutated in any patient were removed. Moreover, we removed CNAs for which the sign of the aberration was not the same in at least 90% of the patients with the aberration. This data was the same used in the HotNet analysis of TCGA ovarian publication.[25] The clinical data we used is preliminary data for 266 of these 316 patients, and thus we restricted our attention to these 266 patients. We considered the overall survival (in months) that measures the time from the patient's first surgery to their last followup or death in months, and vital status values (LIVING or DECEASED) that describe if the data is censored or not. For the H_0^P null hypothesis, the overall survival and vital status were treated as combined survival unit.

For each gene g, we considered the set $M(g)$ of patients in which g is mutated and the set $\overline{M}(g)$ of patients in which it is not mutated. We compared the Kaplan-Meier survival curve obtained for patients in $M(g)$ with the survival curve for patients in $\overline{M}(g)$. (We used the R package `survival` to compute the survival curves.) For each gene, we used the *logrank* statistical test to test the hypothesis that there is no difference between the two survival curves. In particular, we used the `survdiff` function of `survival` package in R to compute the p-value for each gene. To focus on genes having a non negligible effect on the set of patients analyzed, we removed genes that were mutated in fewer than five patients. For each p-value we derived a score as described in Section 2.2.1. We also removed outliers: five genes (C2orf65, DOK1, DQX1, LOXL3, and SEMA4F) with p-value less than 10^{-10}. The remaining scores constitute the input to HotNet. Using the procedure of Section 2.2.2 we computed a threshold δ for null hypothesis H_0^P and a threshold δ for null hypothesis H_0^M. In both cases we obtained the threshold $\delta = 0.11$. We then ran our algorithm for thresholds $\delta' = 1.0\delta, 1.1\delta, 1.2\delta, \ldots, 2.0\delta$. The best results were obtained using the threshold $\delta' = 0.22$, and are reported below.

Using this approach HotNet identifies 12 candidate subnetworks containing at least 10 genes: the p-value for observing 12 subnetworks containing at least 10 genes is ≤ 0.05 under H_0^P and < 0.008 under H_0^M, and the FDR for the set of 12 subnetworks is ≤ 0.57 under H_0^P and ≤ 0.43 under H_0^M. The FDR is a conservative estimate of the ratio of false positives among all subnetworks reported by our algorithm, and implies that approximately 5 of the subnetworks reported by HotNet are significant. Since we included CNAs in our analysis, our results may contain potential artifacts resulting from functionally related genes that are both neighbors on the interaction network and close enough on the genome that they are affected by the same CNA. To reduce such artifacts, we applied two heuristics. First, we removed candidate subnetworks returned by HotNet that contain 3 or more genes in the same focal CNA in more than 1% of the patients. Second, for subnetworks with 2 genes g_1, g_2 in the same focal CNA in more than 1% of the patients, we removed the genes that are not found in the subnetwork when alterations in either g_1 or g_2 are removed. Of the 12 subnetworks identified by HotNet, 9 remain after these CNA filtering heuristics. To gain additional support for individual subnetworks and to focus attention on subnetworks with known biological function, we computed the overlap between the genes in candidate subnetworks and known pathways from the KEGG database.[8] For subnetworks that are enriched for at least one KEGG pathway, the best enrichments are reported in Table 1. Those 4 subnetworks are reported in Figs. 2–5. We note that none of the genes in those 4 subnetworks would be flagged as significant using a log rank test on single genes (with FDR ≤ 0.2).

All the subnetworks in Table 1 are enriched for at least one pathway reported in Crijns *et al.*[26] as containing more genes whose expression is correlated with overall survival than expected by chance. One of the subnetworks we found significantly overlaps genes in the focal adhesion pathway, that have been shown to be associated with survival. In particular, increased expression of the focal adhesion kinase (FAK) has been associated with shorter survival[27,28] in ovarian cancer, and the overexpression of ADAM9 has been correlated with brain metastasis in non-small cell lung cancer.[29] Moreover the depletion of MAP1S, part of another subnetwork S_2 we identify, has been associated with reduced survival.[30] To investigate the impact on survival

Table 1. Significant subnetworks identified by our method and significantly overlapping KEGG pathways. For each subnetwork, the genes in the subnetwork, the most significant KEGG pathways given by hypergeometric enrichment and the corresponding p-values are shown.

subnetwork	genes	pathway	enrichment p-val
S_1	ADAM9 ITGAV ITGA6 ITGA3 ITGB5 LIMK1 FGFR2 DLST UMPS PAK4 GATAD2A	Regulation of actin cytoskeleton	1.1×10^{-9}
		Hypertrophic cardiomyopathy (HCM)	2.0×10^{-5}
		Arrhythmogenic right ventricular cardiomyopathy (ARVC)	1.2×10^{-5}
		Dilated cardiomyopathy	2.0×10^{-5}
		Focal adhesion	2.2×10^{-5}
S_2	POLR2I POLR2H POLR2K HELZ MAP1S SMYD3 LRPPRC NFKBIB POLR2B	RNA polymerase	3.1×10^{-8}
		Pyrimidine metabolism	5.2×10^{-6}
		Purine metabolism	3.6×10^{-5}
S_3	MTF1 PVR ATP5J2 CD96 AP1G1 AP1M1 SHPRH LDLR PIGR LPP SCRIB GAK AP1M2 PVRL3 PVRL2 LAMP1	Lysosome	2.3×10^{-4}
		Cell adhesion molecules (CAMs)	4.8×10^{-2}
S_4	RAB8B OPTN RAB3IP TUBB RYR1 BIRC6 RYR2 CACNA1C RABIF GDI2 TUBA4A RAB8A HOMER3 CACNA1S RIMS2	Calcium signaling pathway	7.4×10^{-4}
		Cardiac muscle contraction	3.0×10^{-3}

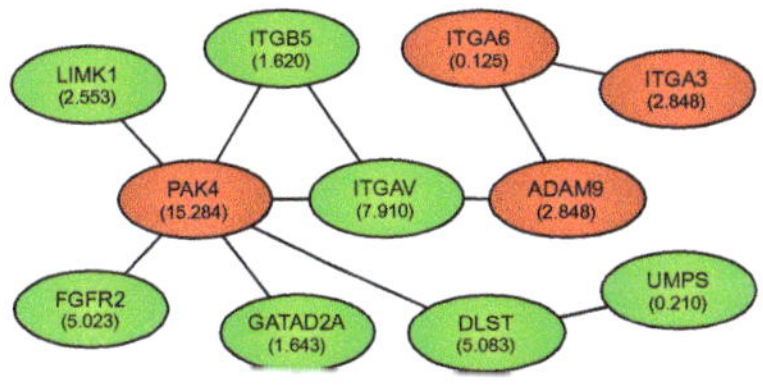

(a)

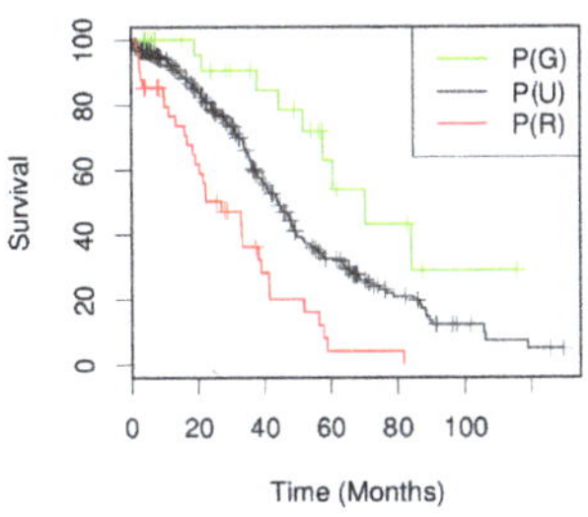

(b)

Fig. 2. Subnetwork S_1 identified by HotNet, significantly overlapping the focal adhesion pathway. (a) Interactions among genes as reported in HPRD. Genes are colored by their effect on the survival of patients: gene g is green if the median survival of patients where g is mutated exceeded that of patients where g is not mutated; gene g is red otherwise. The median survival is determined by the Kaplan-Meier estimator of the R function `survfit`. Numbers are the gene scores $(-2\log_e p_g)$. (b) Survival curves for: patients $P(G)$ with mutations only in green genes (green curve), patients $P(R)$ with mutations only in red genes (red curve), and patients $P(U)$ with mutations in neither red nor green genes (gray curve).

of mutations in the subnetworks identified by our algorithm we compared the survival of patients not having mutations in a particular subnetwork, with the survival of patients having a mutation in the subnetwork. Let $M(g)$ be the set of patients in which a gene g is mutated, and let $\overline{M}(g)$ be the set of patients in which g is not mutated. For a subnetwork S, let $S(G)$ be the set of genes in S for which the the mean survival of the patients in $M(g)$ is larger than the mean survival of the patients in $\overline{M}(g)$. Let $S(R)$ be the subset of S for which the

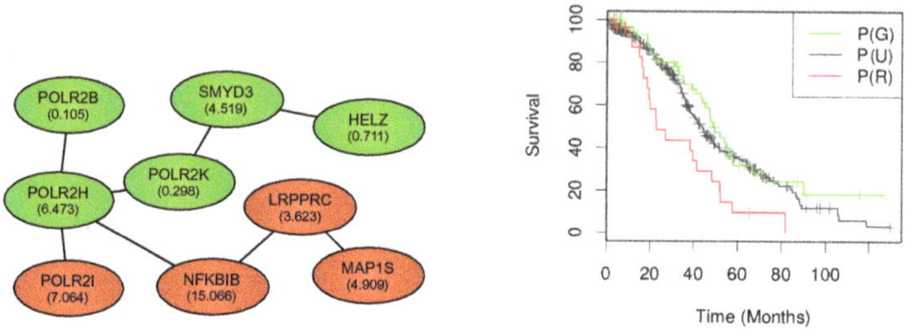

(a) (b)

Fig. 3. Subnetwork S_2 identified by HotNet, significantly overlapping the RNA polymerase pathway. (a) Interactions among genes as reported in HPRD. Genes are colored as in Figure 2 (b) Survival curves for: patients $P(G)$ with mutations only in green genes (green curve), patients $P(R)$ with mutations only in red genes (red curve), and patients $P(U)$ with mutations in neither red nor green genes (gray curve).

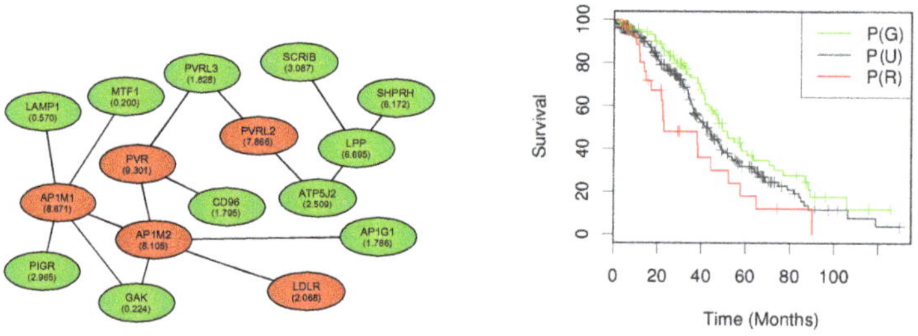

(a) (b)

Fig. 4. Subnetwork S_3 identified by HotNet, significantly overlapping the cell adhesion molecules pathway. (a) Interactions among genes as reported in HPRD. Genes are colored as in Figure 2 (b) Survival curves for: patients $P(G)$ with mutations only in green genes (green curve), patients $P(R)$ with mutations only in red genes (red curve), and patients $P(U)$ with mutations in neither red nor green genes (gray curve).

opposite holds. For each of the four subnetworks identified by our algorithm, we considered (i) patients $P(R)$ with mutations in genes in $S(R)$ but not in genes in $S(G)$; and (ii) patients $P(G)$ with mutations in genes in $S(G)$ but not in genes in $S(R)$. We also considered the set $P(U)$ of patients not having mutations in the subnetwork S. We then compared the survival curves of $P(R)$ and $P(U)$, and the survival curves of $P(G)$ and $P(U)$. The survival curves are reported in Figs 2–5, and the p-values (from logrank) for the tests are reported in Table 2. (In Table 2, A vs B denotes the comparison of survival for patients in A and of survival for patients in B.) For three of the subnetworks we found that at least one of the two tests has p-value $< 6 \times 10^{-3}$, and for the forth subnetwork both p-values are ≈ 0.06. This demonstrates that our method identifies subnetworks whose mutations are associated with survival.

We also used our method separating the analysis for genes with high survival and low

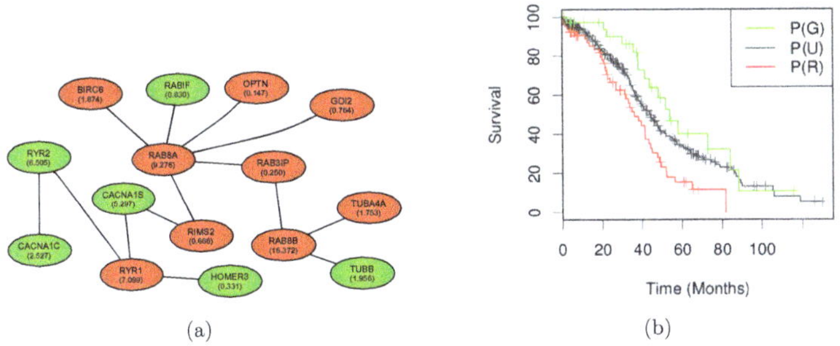

Fig. 5. Subnetwork S_4 identified by HotNet, significantly overlapping the calcium signaling pathway. (a) Interactions among genes as reported in HPRD. Genes are colored as in Figure 2 (b) Survival curves for: patients $P(G)$ with mutations only in green genes (green curve), patients $P(R)$ with mutations only in red genes (red curve), and patients $P(U)$ with mutations in neither red nor green genes (gray curve).

Table 2. p-values from logrank test for the subnetworks of Table 1 enriched in KEGG pathways.

Subnetwork S	$P(R)$ vs. $P(U)$	$P(G)$ vs. $P(U)$
S_1	4.9×10^{-6}	1.5×10^{-2}
S_2	4.3×10^{-3}	2.7×10^{-1}
S_3	5.5×10^{-2}	5.9×10^{-2}
S_4	5.8×10^{-3}	1.9×10^{-1}

survival phenotypes. In particular, we defined a gene to have high survival if the median survival of the patients with a mutation in the gene was higher than the median survival of the patients without mutation in the gene. Otherwise, we defined the gene to have low survival. We then ran HotNet twice: once considering only genes with high survival, and then only genes with low survival, applying the same filtering steps described above. For the high survival genes, our method identified 7 subnetworks of at least 10 genes (p-value ≤ 0.01, FDR ≤ 0.51 under H_0^M, and p-value ≤ 0.3, FDR ≤ 0.8 under H_0^P). Only 3 of those subnetworks remains after the CNA filtering heuristics, and they are not enriched for known pathways. For the low survival genes, our method identifies 5 subnetworks of at least 10 genes (p-value ≤ 0.05, FDR ≤ 0.63 under H_0^M, and p-value ≤ 0.37, FDR ≤ 0.82 under H_0^P). Only 1 of those subnetworks remains after the CNA filtering heuristics, and it is not enriched for known pathways.

4. Discussion

We described an extension of our HotNet algorithm to finding subnetworks of genes whose mutational status is associated with a phenotype of interest. We applied our algorithm to ovarian cancer patients from The Cancer Genome Atlas and find 9 subnetworks associated with survival, 4 of which significantly overlap well-known pathways. Although we presented results using either a χ^2 test or logrank test, the extensions to HotNet described here apply to any gene score, thus allowing for a variety of statistical tests to be used for testing associations to clinical phenotypes.

5. Acknowledgements

This work is supported by NSF grants IIS-1016648 and CCF-1023160. BJR is supported by a Career Award from the Scientific Interface from the Burroughs Wellcome Fund and an Alfred P. Sloan Research Fellowship.

References

1. A. Lievre *et al.*, *Cancer Res.* **66**, 3992 (2006).
2. K. Iida *et al.*, *Br J Cancer* (2011).
3. J. S. Smith *et al.*, *J. Natl. Cancer Inst.* **93**, 1246 (2001).
4. M. Meyerson, S. Gabriel and G. Getz, *Nat. Rev. Genet.* **11**, 685 (2010).
5. L. Ding, M. C. Wendl, D. C. Koboldt and E. R. Mardis, *Hum. Mol. Genet.* **19**, R188 (2010).
6. B. Vogelstein and K. W. Kinzler, *Nat. Med.* **10**, 789 (2004).
7. W. C. Hahn and R. A. Weinberg, *Nat. Rev. Cancer* **2**, 331 (2002).
8. M. Kanehisa and S. Goto, *Nucleic Acids Res.* **28**, 27 (2000).
9. L. Matthews *et al.*, *Nucleic Acids Res.* **37**, D619 (2009).
10. D. Croft *et al.*, *Nucleic Acids Res.* **39**, D691 (2011).
11. D. Szklarczyk *et al.*, *Nucleic Acids Res.* **39**, D561 (2011).
12. T. S. Keshava Prasad *et al.*, *Nucleic Acids Res.* **37**, D767 (2009).
13. H. Y. Chuang, E. Lee, Y. T. Liu, D. Lee and T. Ideker, *Mol. Syst. Biol.* **3**, p. 140 (2007).
14. T. Ideker, O. Ozier, B. Schwikowski and A. F. Siegel, *Bioinformatics* **18 Suppl 1**, S233 (2002).
15. M. T. Dittrich, G. W. Klau, A. Rosenwald, T. Dandekar and T. Muller, *Bioinformatics* **24**, i223 (2008).
16. D. Beisser, G. W. Klau, T. Dandekar, T. Muller and M. T. Dittrich, *Bioinformatics* **26**, 1129 (2010).
17. C. J. Vaske, S. C. Benz, J. Z. Sanborn, D. Earl, C. Szeto, J. Zhu, D. Haussler and J. M. Stuart, *Bioinformatics* **26**, i237 (2010).
18. F. Vandin, E. Upfal and B. J. Raphael, *J. Comput. Biol.* **18**, 507 (2011).
19. E. Cerami, E. Demir, N. Schultz, B. S. Taylor and C. Sander, *PLoS ONE* **5**, p. e8918 (2010).
20. R. I. Kondor and J. Lafferty, Diffusion kernels on graphs and other discrete structures, in *International Conference on Machine Learning*, 2002.
21. F. Chung, *Proceedings of The National Academy of Sciences* **104**, 19735 (2007).
22. Y. Qi, Y. Suhail, Y. Y. Lin, J. D. Boeke and J. S. Bader, *Genome Res.* **18**, 1991 (2008).
23. N. Mantel, *Cancer Chemother Rep* **50**, 163 (1966).
24. R. A. Fisher, *Statistical Methods for Research Workers* (Oliver and Boyd, 1938).
25. The Cancer Genome Atlas Research Network, *Nature* **474**, 609 (2011).
26. A. P. Crijns *et al.*, *PLoS Med.* **6**, p. e24 (2009).
27. A. K. Sood *et al.*, *Am. J. Pathol.* **165**, 1087 (2004).
28. M. K. Siu *et al.*, *Proc. Natl. Acad. Sci. U.S.A.* **107**, 18622 (2010).
29. Y. Shintani, S. Higashiyama, M. Ohta, H. Hirabayashi, S. Yamamoto, T. Yoshimasu, H. Matsuda and N. Matsuura, *Cancer Res.* **64**, 4190 (2004).
30. R. Xie, S. Nguyen, K. McKeehan, F. Wang, W. L. McKeehan and L. Liu, *J. Biol. Chem.* **286**, 10367 (2011).

INTRINSICALLY DISORDERED PROTEINS: ANALYSIS, PREDICTION, SIMULATION, AND BIOLOGY

JIANHAN CHEN
Department of Biochemistry
Kansas State University, Manhattan, KS 66506, USA
jianhanc@ksu.edu

JIANLIN CHENG
Department of Computer Science
Informatics Institute
University of Missouri, Columbia, MO 65211, USA
chengji@missouri.edu

A. KEITH DUNKER
Center for Computational Biology and Bioinformatics
School of Informatics
Indiana University, Indianapolis, IN 46202-5122, USA
kedunker@iupui.edu

1. Introduction

Intrinsically disordered proteins (IDPs) are an important newly recognized class of proteins that rely on the lack of stable structures, both in vitro and in vivo, for function. Before the computational studies of the late 1990s, the tendency was to believe that IDPs were artifacts, resulting for example from denaturation during isolation or from the loss of critical factors such as metal ions or small molecules that were needed for the formation or maintenance of structure. However, the bioinformatics work of the late 1990s showed that most of the IDPs have amino acid compositions distinctly different from the compositions of structured proteins. These studies suggested that, in addition to the concept of "sequence → structure → function" there is another concept of "sequence → disordered ensemble → to function."

Furthermore, the functions carried out by structured proteins and disordered ensembles are very different. The main functions for structured proteins are catalysis, membrane transport, and reversible binding of small molecules, whereas the main functions of IDPs are reversible binding of large molecules, assembly, regulation, and control. Finally, from the amino acid compositional differences between structured and disordered proteins, predictors of disorder were developed. Application of these predictors to various proteomes shows that eukaryotic cells contain a much higher fraction of IDPs compared to bacteria and archaea. For instance, recent predictions suggest that 35% – 45% of human proteins contain long regions that have IDP features, and that, overall, about 45% of the amino acids in humans are located in regions with IDP features.

Thanks to the important and perhaps pivotal role played by computation in the development of the IDP field, it is now recognized by many but not yet all that IDPs play fundamental roles in crucial biological functions and are extensively involved in human diseases. The study of IDPs is rapidly evolving into a field of growing prominence. Nonetheless, the dynamic and heterogeneous nature of IDPs has remained a significant challenge. There are still many open questions in the field, including how to further improve the accuracy of computational IDP predictions, how to integrate IDP predictions with the study and prediction of other protein structural features, how to better predict IDP sequence-to-function relationships, how to better compute and simulate different states of IDPs, what is the most effective means for integrating structural data from simulation and experiment, how to computationally probe mechanistic aspects of coupled binding and folding and to understand how IDP recognition is regulated, and how to develop rational strategy for modulating IDP function in human diseases. Thus, there is a great demand for further development of computational IDP approaches that are more efficient and more accurate, and at the same time, can improve our understanding of biology from the molecular to the system level. This represents exciting opportunities for computational approaches to make crucial contributions, from aspects ranging from prediction and analysis to simulation.

One of the main goals of this Session is thus to introduce and discuss: (1) important advances in all frontiers of computational "IDPology", (2) available computational capabilities in prediction, analysis and simulation of IDPs, and more importantly, (3) outstanding challenges, further directions and key biological questions to be addressed. The other goal is to promote communication and collaboration between scientists working in different areas of IDP computation and between experiment and computation.

2. Papers in this Session

This session includes 11 high-quality papers addressing a wide range of important problems in prediction, simulation and analysis of IDPs.

Analysis of IDPs' function and evolution

Elucidating biological processes and pathways that IDPs take part in and the consequent evolutionary pressures on IDPs is an important topic. **Huang et al.** developed a method to classify proteins into four categories - structured, mixed, disordered and rare, based on charge-hydropathy and cumulative distribution function. Their investigation of functions of proteins in different categories discovered that disordered class is highly active in mitosis-related processes and the mixed class is highly associated with signaling pathways. **Patil et al.** reported a method and a web tool to group intrinsically disordered domains using a similarity metric based on amino acid composition. The groups identified from a large number of disordered domains by their method were enriched with specific Gene Ontology (GO) and Pfam function annotations, suggesting that this method may be applied to functional annotation of IDPs. **Hsu et al.** designed an experiment to study the sequence conservation of binding regions of multiple partner proteins that interact with the same disordered segment of another protein. The experiment showed that residues interacting with disordered segments were substantially more conserved that other buried or exposed residues, suggesting that these disordered segments may play important function roles (e.g. signaling) in cells. **Jeong and Kim** provided a comprehensive study of co-evolution of intrinsically disordered regions. The study revealed that the degree of residue co-evolution significantly decreased in disordered regions and differd in function categories.

Studying how disordered regions modulate molecular binding and recognition of IDPs is an active research topic. **Vuzman et al.** discovered an interesting mechanism through which disordered tails in DNA-binding proteins were used by post-translational modification to modulate DNA-protein binding affinity. The study demonstrated that post-translational modifications such as acetylation and phosphorylation, which often took place at disordered sites, could modulate the interactions of proteins with DNA by changing the local and global properties of the disordered tails. The binding affinity may be tuned by adjusting the number of modifications and the cross-links between them. **Guo et al.** used a combination of bioinformatics and statistical methods to study the distribution of disordered residues in human transcription factors (TFs) and non TFs. It was found that the regions flanking DNA Binding Domains (DBD) in TFs were significantly enriched with disordered residues. This specific identification of the locations of enriched disordered regions in TFs may shed light on the functions of these disordered regions. **Gao and Xu** studied the correlation between correlation between posttranslational modification (PTM) and intrinsic disorder in proteins using both functional and structural data. Their investigation showed that most PTMs (e.g. phosphor-serine/-threonine/-tyrosine) preferentially occurred in disordered regions, while a few PTMS occurred in ordered regions more often. The work also suggests that disorder-order transitions may be introduced by PTMs.

Simulation of IDPs' conformation

Generating an ensemble of realistic conformations for a disordered protein is critical for studying its structural and functional properties. **Fisher et al.** introduced a computationally efficient algorithm called Variational Bayesian Weighting with Structure Selection (VBWSS). VBWSS aims to construct an ensemble of an IDP using a minimal number of conformations and provides estimates for the uncertainties in various properties of the ensemble. The authors successfully validated the algorithm against reference ensembles and applied it to construct an ensemble for the 140-residue IDP, α-synuclein. **Burger et al.** carried out long timescale explicit solvent molecular dyanamics (MD) simulations to quantify the ligand-free state of nuclear receptor co-activator binding domain (NCBD) of CREB binding protein. While NCBD has been shown to be a molten globule, the intrinsic disordered nature has prevented detailed structural characterization using traditional methods. Their simulation and analysis was able to capture certain rare states at the tails of the conformational distribution, which, intriguingly, included three of four ligand bound states that were mutually accessible via a complex network of pathways. **Mitrea et al.** employed a series of structure-based and sequence-based methods to elucidate the mechanism of how the N-terminal oligomerization domain of nucleophosmin protein (Npm-N) was transformed from ordered states into disordered states. Their study led to an interesting hypothesis that Npm-N had evolved energetic switches within its structure to enable transformation to a disordered state and the transformation was triggered by sequential phosphorylation of solvent exposed hot spots.

Prediction of IDPs

Predicting disordered regions from protein sequence serves as the basis for many kinds of computational IDP studies. Developing more accurate disorder prediction methods or integrating existing base predictors is important for structural and functional analysis of IDPs. **Peng et al.** designed a new regression-based model to quantify quality of the majority-vote consensus of a given triplet of disorder predictors based on their individual performance and their complementarity measured at the disorder residue and segment levels. The method can effectively integrate individual disorder prediction methods to improve prediction accuracy.

Acknowledgements

We would like to thank all the authors who submitted manuscripts to this session and all the anonymous reviewers who rigorously and constructively reviewed the manuscripts. We greatly appreciate the IDP community's enthusiastic support for this session.

QUASI-ANHARMONIC ANALYSIS REVEALS INTERMEDIATE STATES IN THE NUCLEAR CO-ACTIVATOR RECEPTOR BINDING DOMAIN ENSEMBLE

VIRGINIA M. BURGER[1,4*], ARVIND RAMANATHAN[2,*], ANDREJ J. SAVOL[1,4], CHRISTOPHER B. STANLEY[3], PRATUL K. AGARWAL[2] and CHAKRA S. CHENNUBHOTLA[4,†]

[1]*Joint Carnegie Mellon University-University of Pittsburgh Ph.D. Program in Computational Biology,* [2]*Computational Biology Institute and Computer Science and Mathematics Division,* [3]*Neutron Scattering Science Division, Oak Ridge National Laboratory, Oak Ridge, Tennessee 37831, USA.* [4]*Department of Computational and Systems Biology, University of Pittsburgh, Pennsylvania 15260, USA.*
[†]*E-mail: chakracs@pitt.edu*

The molten globule nuclear receptor co-activator binding domain (NCBD) of CREB binding protein (CBP) selectively recruits transcription co-activators (TCAs) during the formation of the transcription pre-initiation complex. NCBD:TCA interactions have been implicated in several cancers, however, the mechanisms of NCBD:TCA recognition remain uncharacterized. NCBD:TCA intermolecular recognition has challenged traditional investigation as both NCBD and several of its corresponding TCAs are intrinsically disordered. Using $40\mu s$ of explicit solvent molecular dynamics simulations, we relate the conformational diversity of ligand-free NCBD to its bound configurations. We introduce two novel techniques to quantify the conformational heterogeneity of ligand-free NCBD, dihedral quasi-anharmonic analysis (dQAA) and hierarchical graph-based diffusive clustering. With this integrated approach we find that three of four ligand-bound states are natively accessible to the ligand-free NCBD simulations with root-mean squared deviation (RMSD) less than 2Å. These conformations are accessible via diverse pathways while a rate-limiting barrier must be crossed in order to access the fourth bound state.

Keywords: intrinsically disordered proteins; NCBD; quasi-anharmonic analysis; graph-theoretic spectral clustering

1. Introduction

Intrinsically disordered proteins (IDPs) play a vital role in regulating cellular processes in eukaryotic cells.[1,2] Structural studies have revealed that unlike well-folded globular proteins, IDPs exist as highly dynamic ensembles even under equilibrium conditions, with diverse and constantly fluctuating secondary/tertiary structure.[3] The ability of IDPs to adapt their binding surface to recognize various binding partners provides a novel means of regulating various cellular activities.[4] Given the abundance of IDPs in the human genome and their involvement in neurodegenerative, cardiovascular, and amyloid-related diseases,[5,6] there is tremendous interest in understanding the basic molecular mechanisms by which IDPs recognize their binding partners and facilitate their specific functions. For example, some IDPs possess the remarkable ability to undergo synergistic folding upon recognizing their binding partners.[7] The contrasting ability of IDPs to achieve a high degree of structural plasticity while retaining binding specificity presents a serious challenge in characterizing their sequence-structure-function relationships.

The intrinsically disordered nuclear co-activator binding domain (NCBD) of the CREB binding protein (CBP) interacts with numerous transcription co-activator proteins (TCA), including the steroid receptor co-activators (SRC),[8] p53,[9] p73,[10] interferon regulatory factors (IRF)[11] and the vi-

*These authors contributed equally to this work

ral protein Tax.[12] As NCBD aids recruitment of the transcriptional machinery, its dysfunction (and that of its binding partners) is implicated in several forms of leukemia[13] and lung cancer.[14] Circular dichroism (CD) and ultra-violet (UV) spectroscopic studies reveal that native NCBD adopts a compact structure with a high degree of helicity but lacks the sigmoid unfolding curve characteristic of folded proteins.[15] Structural studies using nuclear magnetic resonance (NMR) and X-ray crystallography indicate that NCBD adopts unique conformations when complexed with specific partners[15,16] and that synergistic folding facilitates the interdigitation of three helices, a feature common in NCBD's bound topology (identified by $\alpha_1 - \alpha_3$; see Fig. 1).[17,18] Increasingly, the specific orientations of these three α-helices are thought to confer the specificity inherent to NCBD:TCA intermolecular recognition.[15–18]

While a number of studies point to the behavior and structure of NCBD in its bound state,[15–18] the conformational heterogeneity of *apo*-form NCBD has been challenging to characterize. Emerging evidence from NMR experiments[16] suggest that native NCBD can adopt conformations that largely resemble the SRC/ACTR-bound conformation. However, that study also revealed that ligand-free NCBD does not sample states that resemble the IRF-bound conformations. Moreover, Fraenkel et al.[19] have determined the *apo*-form of NCBD to be quite different from Poulsen et al.[16] Based on the current insights gained from experimental studies, the biophysical mechanisms underlying NCBD:TCA recognition process remain unclear. Likewise, a quantitative description of *disorder-to-order* transitions between the ligand-free or ligand-bound NCBD ensembles is lacking.

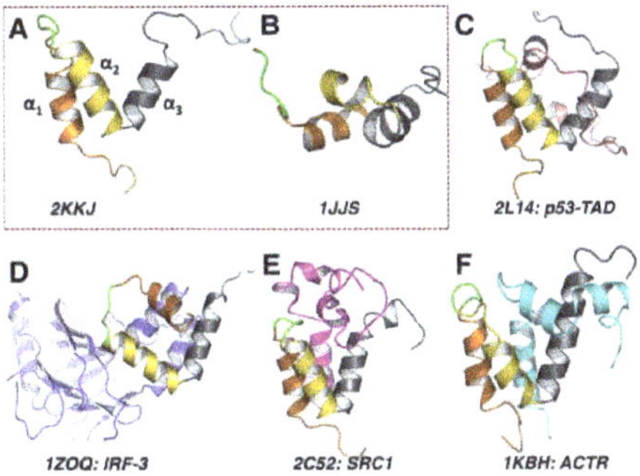

Fig. 1: **Bound and unbound forms of NCBD.** NMR ensembles of the ligand-free structures: 2KKJ (A) and 1JJS (B); NCBD in complex with (C) p53 trans-activation domain (TAD) (2L14: TAD in pink); (D) interferon regulatory factor 3 (IRF3) (1ZOQ: IRF3 in pale blue); (E) steroid receptor coactivator 1 (SRC1) (2C52: SRC1 in magenta); (F) interaction domain of activator for thyroid hormone and retinoid receptors (ACTR) (1KBH: ACTR in cyan). In all panels, the three helix bundle of NCBD is highlighted in orange (α_1), yellow (α_2) and gray (α_3), while the specificity loop (PSSP) is in green.

In this paper, we address the aforementioned issues and outline an integrated experimental and computational strategy to analyze disorder-to-order transitions in NCBD's conformational landscape. Our aims are to: (a) obtain insights into the nature of intrinsic fluctuations accessible to ligand-free NCBD, (b) identify regions within NCBD that are implicated in its disorder-to-order transitions and (c) elucidate whether ligand-free NCBD can access conformations that resemble the ligand-bound conformations. To this end, we exploit recent advances in molecular simulation technologies to extensively sample ligand-free NCBD. Using graphics processing units (GPUs), we accelerate conventional all-atom explicit solvent molecular dynamics simulations to microsecond time-scales. The aggregate dataset constitutes $40\mu s$ of MD simulation and required approximately two months of total clock-time.

Long time-scale simulations challenge conventional trajectory analysis methods. In particular, biophysically relevant events within such trajectories are often difficult to detect.[20,21] Likewise, experimental techniques also present modeling challenges; results from small-angle neutron scattering (SANS) experiments on NCBD suggest a distinctly long-tail (or anharmonic) behavior in the distributions of radius of gyration and end-to-end distance values.[22] This long-tailed behavior implies that atomic fluctuations in NCBD involve significant higher-order correlations, which are commonly overlooked with typical trajectory analysis tools.[23] Recently, we introduced quasi-anharmonic analysis (QAA) as an effective computational model to quantify these higher-order correlations which emerge prominently within long simulations.[24] QAA provides insights into the inherent *anharmonicity* in atomic fluctuations and is thus ideal for quantifying the disorder-to-order transitions in NCBD observed from both experiments and simulations. Furthermore, QAA organizes the conformational heterogeneity in NCBD fluctuations into a small set of conformational sub-states that share structural and energetic homogeneity.

Markov state models (MSMs) and their variants also provide organizational principles for molecular simulations. These methods exploit the kinetic connectivities[27] or structural similarities[26] between conformational sub-states and have been useful for determining transition pathways between conformational sub-states.[20,25] As a comparison to QAA, MSMs discretize conformation space into a network or graph of sub-states rather than projecting it into a low-dimensional, continuous representation. A central contribution of the work here is an approach which exploits both the dimensionality reduction (and visual interpretability) of QAA and rigorous graph theoretic methods to determine a hierarchy of transitions between sub-states. With this integrated approach, we determined that ligand-free NCBD can indeed access conformations representative of the ligand-bound form. Within our simulations, NCBD's α_1 and α_2 helices in the ligand-free and ligand-bound conformations are largely similar; α_3 however, can exhibit a wide degree of flexibility and does not generally sample conformations that are similar to the ligand-bound state.

2. Approach

We performed ten $4\mu s$ all-atom explicit solvent MD simulations of apo-NCBD (Section 3). To identify biophysically relevant motions within these simulations, we developed a novel, broadly extensible, dimensionality reduction framework based on quasi-anharmonic analysis in the dihedral angle space, called *dihedral QAA* or *dQAA* (Section 4). To validate our simulations we used two order parameters: radius of gyration (R_g) and helicity (H; defined here as the percentage of NCBD that adopts α-helical structure as assigned by STRIDE[28]), which can be measured experimentally via SANS[22] and circular dichroism[16] experiments respectively.

To determine meta-stable conformational states, we invoke a multi-scale Markov diffusion approach (Section 5) to group similar conformations in the dQAA space. Iterative diffusion-based clustering in the dQAA space results in a hierarchical description of the NCBD conformational landscape. Each level of the hierarchy provides a set of increasingly broad (or inclusive) meta-stable states, allowing the conformational landscape of NCBD to be viewed as a collection of nested sub-states. As we demonstrate, dQAA coordinates provide a natural framework for organizing the conformational heterogeniety of the apo-NCBD ensemble and help identify disordered or compact conformational states. In addition, the Markov diffusion approach captures meta-stable states that provide insight into the nature of structural changes that NCBD must undergo in order to sample conformations close to the ligand-bound state (Section 6).

3. Molecular Simulations for NCBD

A total of six NMR and X-ray NCBD structures are available in ligand-free and ligand-bound form. Fig. 1 shows the variation in the orientation of the three α-helices between these structures. While NCBD adopts very similar helical orientations when binding ACTR, SRC1 and p53, the interfaces and helical turns of NCBD when complexed to each ligand are quite different. Furthermore, NCBD adopts a radically different orientation for interacting with IRF3; α_3 twists and rests on a very different axis from that in the ACTR interaction.

In the interest of sampling the large conformational space of ligand-free NCBD, we initiated a 4μs long simulation for each of the 10 conformations in the NMR ensemble (2KKJ) that is representative of the ligand-free state. We used the AMBER suite of tools[29] and the ff99SB[30] force-field to model the proteins. Each of the ten conformations was immersed in a cubic box of SPC water molecules such that the solvent box boundary was never less than 10Å from the protein. Counterions consisting of 10 Cl^- were added to ensure system neutrality. The box sizes were approximately $90 \times 90 \times 90$ Å^3 (with slight variations for each of the ten simulations). Using the protocol highlighted in our previous work,[31] each of the simulation systems was subjected to energy minimization and equilibration. A final MD equilibration of 1.0ns duration was run to ensure the systems reached a stable conformation. All the simulations were carried out at 300K using the NVE ensemble. Each of the ten systems had between 9,000 and 12,000 water molecules, resulting in system sizes varying between 18,000 and 22,000 atoms.

Production runs were carried out using the recently developed ACEMD (accelerated MD) code specifically for graphics processing unit (GPU) systems.[32] In order to accelerate the MD simulations to reach microsecond time-scales, the systems were simulated using a time-step of 4fs using a hydrogen mass-partitioning scheme.[33] The alteration to the dynamics due to the mass-partitioning scheme is minimal since individual atom masses do not appear explicitly in the equilibrium distribution.[32] Ten production runs sampling 4μs per simulation were performed. Coordinates were saved every 200 ps, resulting in about 20,000 conformations per simulation or an aggregate total of 200,000 conformations for all simulations (40μs total).

Comparison with NMR: To compare our production runs with NMR data, we used SPARTA[34] to predict the 1H, 13C, and 15N chemical shifts for the ensembles generated from MD simulations. SPARTA uses backbone ϕ and ψ torsion angles, side-chain χ_1 angles, and sequence information to predict backbone chemical shifts of protein structures.[34] We found that the simulations show reasonable agreement with the chemical shifts from the experimental ensembles (2L14, 1KBH and 2KKJ). In particular, the correlation coefficients between the mean MD and the experimental 15N shifts are 0.74, 0.78, and 0.88, respectively, for the 2L14, 1KBH and 2KKJ data. We note that computed 1H and 13C chemical shifts are less consistent with respective experiments presumably due to force-field inaccuracies and the 4 fs MD integration time-step.[35] While the agreement between experiments and computations is a cursory check on the quality of data obtained, we must also note that the chemical shifts from the experimental ensembles may not be fully representative of the conformational heterogeneity of apo-NCBD.

Comparison with SANS: We next compare simulation results with experimentally derived R_g values from small-angle neutron scattering (SAS) experiments. The distribution of R_g values from MD simulations is observed to be more constrained than that obtained from SANS, possibly due to MD sampling deficits (Fig. 2a panel B, blue: aggregate simulations; red dash: single simulation; red: SANS data). This is in part because MD trajectories are strongly biased by the chosen starting pose, which is commonly an energy-minimized X-ray or NMR ensemble structure.[36,37] We note that the range of SANS-derived R_g values suggests that NCBD may undergo disorder-to-order motions on a larger scale than observed in the present simulations.

From a molten globule state to a near ACTR-bound form: To quickly overview significant conformational events in the MD trajectory, we track R_g on-line along a subset of one of the simulation trajectories using two different exponential window smoothing timescales (Fig. 2b). We observe that NCBD changes from a molten-globule form (high R_g) to a near ACTR-bound form (gray cartoon for comparison, shown along with RMSDs). The pathway chosen by this trajectory is highly dynamic, involving several significant rearrangements of the α_1-α_2 (PSSP) loop and α_3. Interestingly, the conformational changes persist across the timescales of the exponential window, confirming the evolution of NCBD from a molten globule state to a near ACTR-bound form. In this particular trajectory, generated from model 2 of the NMR ensemble (2KKJ), NCBD adopts a form that is about 4.27 Å (C^α-RMSD) from the bound form; however, other trajectories adopt conformations that are much closer to the ACTR-bound form (see Section 5).

4. dQAA: Quasi-anharmonic analysis in the dihedral angle space

The conformational heterogeneity we observed in long timescale simulations of NCBD motivated us to eliminate the sensitivity to Cartesian alignment by analyzing the NCBD ensemble in the dihedral angle space. For a N residue protein there are a total of $2N$ backbone ϕ and ψ angles, $\phi = \{\phi_i\}_{1,\ldots,N}$, $\psi = \{\psi_i\}_{1,\ldots,N}$. Each backbone dihedral angle pair (ϕ_i, ψ_i) can be converted into a Euclidean representation by $x_{i-3} = \cos(\phi_i)$; $x_{i-2} = \sin(\phi_i)$; $x_{i-1} = \cos(\psi_i)$; $x_i = \sin(\psi_i)$, yielding a $4N$ vector x. We first considered dihedral PCA (dPCA), where a covariance matrix is generated from this data and is diagonalized to obtain a low-dimensional representation of the conformational ensemble.[45–47] We observed that NCBD conformers projected into low-dimensional dPCA space lacked coherency (or homogeneity) with respect to the R_g values, indicating that dPCA is unable to fully describe the disorder-to-order motions of NCBD (data not shown).

Protein motions are anharmonic; therefore, capturing the conformational diversity of protein fluctuations requires effective models that quantify anharmonic motional signatures.[23,38–42] Anharmonicity is best summarized by higher-order statistics.[38,39] Our previously developed framework, quasi-anharmonic analysis (QAA), exploits these higher-order statistical signatures of protein motions.[24] When applied to μs time-scale simulation data of proteins involved in molecular recognition and enzyme catalysis, QAA revealed (i) functionally relevant, hierarchically-organized conformational sub-states and (ii) a set of on-pathway intermediates between these sub-states. This result is consistent with the understanding that proteins sample from a hierarchical, multilevel energy landscape with minima and maxima separated by energy barriers.[43,44] We observed that the sub-states determined with QAA were energetically coherent, indicating that our low-dimensional representation appropriately depicts energetically-related conformers as neighbors. We emphasize, however, that the resultant energy coherence within observed sub-states is an emergent property of QAA,

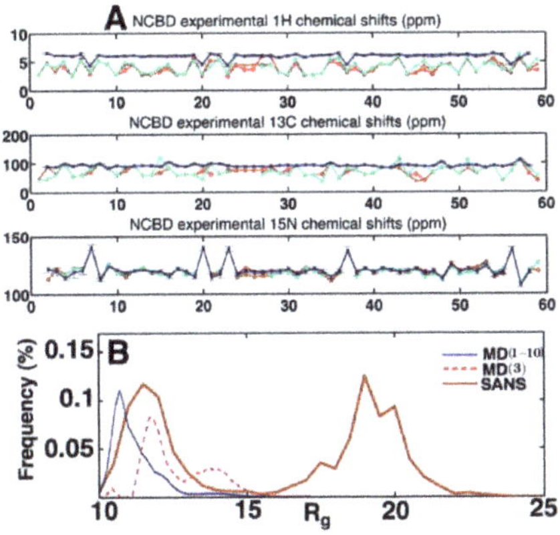

(a) Comparing simulations with experiments

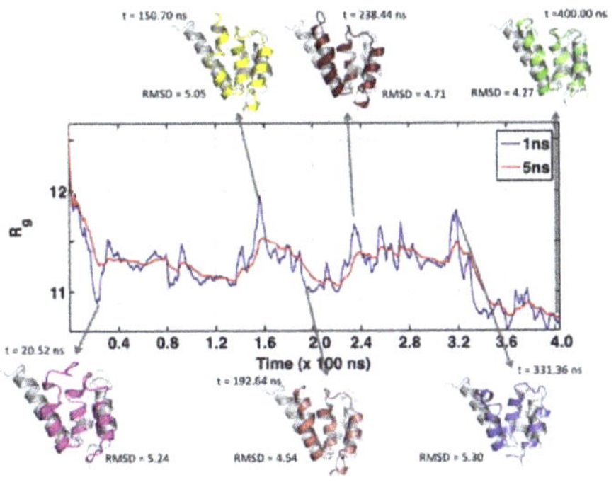

(b) Online tracking of R_g

Fig. 2: **Disorder-to-order transitions in NCBD ligand-free ensemble** (a) A comparison of simulated NCBD ensembles with NMR (A) and SAS (B) experimental data, illustrating qualitative agreement. Chemical shift data is taken from three ensembles, 2KKJ (16363cat.bmrb, red), 2L14 (17071cat.bmrb, brown), 1KBH (5228cat.bmrb, cyan), and compared to computed mean chemical shifts from the simulations. (B) R_g is shown for SANS data (tan, solid), aggregated MD data (blue, normalized), and a single MD trajectory (2KKJ, model 3)(dashed red, normalized). Not all of the conformational landscape is sampled by MD, as is evident from the second SANS peak. (b) R_g during first 400ns of a single MD trajectory (2KKJ, model 2), with 1ns (blue) and 5ns (red) exponential smoothing showing disorder-to-order transitions. Conformations at six timepoints are aligned to crystal structure 1KBH.

indicating that our higher-order statistical approach selects meaningful reaction coordinates.

With the intention of capturing anharmonic disorder-to-order motions, we pursued anharmonicity as an informative statistic in the form of dihedral QAA (dQAA), basing our technique on the diagonalization of a tensor of fourth-order statistics in the dihedral angle space. This tensor describes dihedral angle fluctuations and their couplings and can be efficiently diagonalized with a technique called joint-diagonalization of cumulant matrices (JADE), a well known machine learning algorithm for analyzing multi-variate data.[48] To begin with, second-order correlations are removed from the dihedral angle fluctuation data. Next, a fourth order cumulant tensor $\mathcal{K}$ is computed consisting of both auto- and cross-cumulants. The cumulant tensor will have a total $4N \times (4N + 1)/2$ matrices each of size $4N \times 4N$ accounting for auto- and cross-cumulant terms. Finally, the fourth order dependencies denoted by the sum of the cross-cumulant terms are minimized, a procedure equivalent to diagonalizing $\mathcal{K}$. No closed form solution exists for diagonalizing a tensor, however an approximate solution can be found using efficient algebraic techniques such as Jacobi rotations.[49] Just as an eigenbasis diagonalizes a covariance matrix, a matrix U is found to approximately diagonalize the cumulant tensor. The basis matrix U represents anharmonic modes of motion derived by minimizing the fourth-order dependencies in dihedral angle fluctuations, in addition to eliminating the second-order correlations as is the case with dPCA. Unlike in dPCA, the column vectors of U (sorted

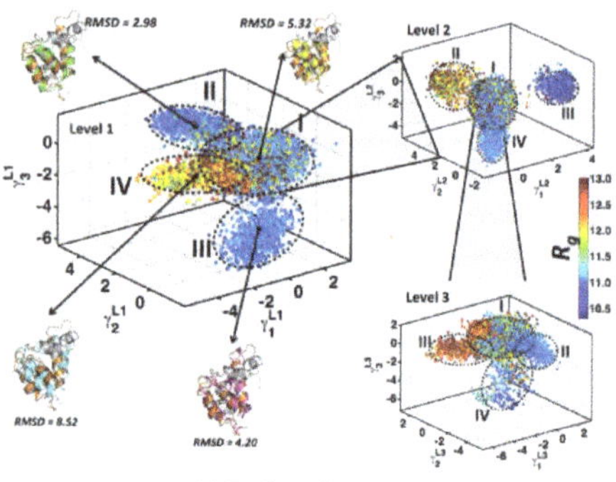

(a) Radius of gyration

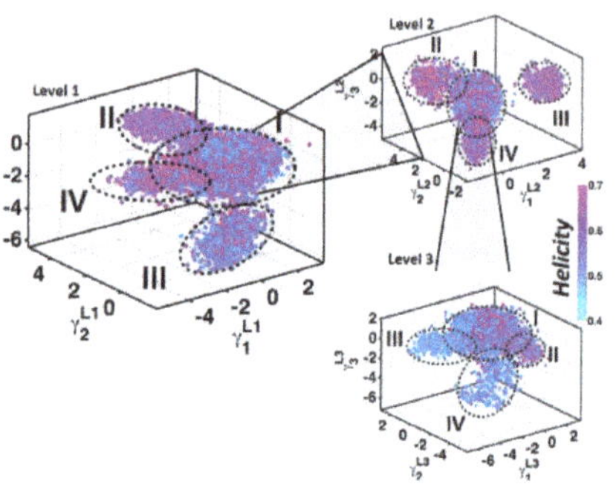

(b) Helicity percentage

Fig. 3: **dQAA identifies a hierarchy of disorder-order promoting motions and homogeneous clusters in 2KKJ μs timescale ensemble**. MD trajectory frames are projected along the top three dQAA modes and colored by (a) R_g and (b) Helicity. (a) Level 1 of the dQAA hierarchy reveals two compact, low R_g clusters (II and III). Cluster IV has high R_g values (red) indicating a more open conformation. Mean conformers in each cluster (I: yellow, II: green, III: maroon, IV: blue) are superimposed on the bound conformer of NCBD-ACTR (orange) and the respective RMSDs are given. Successive application of the dQAA analysis to heterogenous clusters (Level 2 and 3) highlight a rich conformational diversity when painted with R_g values values. (b) In level 1, dQAA clusters I and III are predominantly low in helicity (blue) and dQAA clusters II and IV are predominantly high in helicity (pink). The ability to separate ordered (high helicity) from disordered (low helicity) conformers improves as dQAA is applied recursively to subsets of conformers.

decreasingly by amplitude ($\|U_i\|$) can be non-orthogonal and hence intrinsically coupled.

Results: Using 40 μs simulations of NCBD, we performed dQAA to reduce 232-dimensional input data (from 58 dihedral angles in each conformer) to a 50-dimensional subspace. For visualization, we projected the conformers along the top three QAA modes as shown in Fig. 3. To assess if the projected conformers share any structural similarities, we colored the conformations using two biophysically relevant order parameters: (a) R_g and (b) H (helicity). The dQAA space colored with R_g revealed two compact (homogeneous) clusters with low R_g values, one open conformation cluster with high R_g and one heterogeneous cluster. Thus, dQAA modes can reveal disorder-to-order motions, an ability that can be further tested by recursively applying dQAA on the heterogeneous cluster. The results from a recursive decomposition highlight the rich conformational diversity present in the simulated NCBD ensemble and illustrate the ability of dQAA to capture meaningful conformational transitions. Although dQAA cannot directly compensate for the deficiencies of MD sampling, the determined anharmonic modes suggest functionally relevant disorder-to-order transitions. Similar results can be seen by coloring the dQAA space with helicity values, showing that the sub-states involve transitions in NCBD from a more extended form to a more helically compact form. This emergent homogeneity in dQAA space suggests a new strategy to identify metastable states in the MD trajectory, which we discuss next.

5. Hierarchical clustering in the dQAA-space to identify meta-stable states

Observing that neighboring conformers in dQAA-space have similar R_g and H values, and noting that this coherence is an emergent property of dQAA representation, we hypothesize that nearest neighbors in dQAA-space are dynamically and kinetically related. We use the conformational coordinates returned by dQAA to build long-lived metastable states using graph-theoretic spectral clustering approaches. To this end, we consider each frame in the trajectory as a node in an undirected graph and connect each node to 10 of its nearest Euclidean neighbors in the three-dimensional dQAA space. The edges are assigned weights inversely proportional to the difference in their radius of gyration values, thus merging both the dynamic and emergent properties of the dQAA space into the edge weights. We then cluster this network using a hierarchical Markov diffusion framework.[50] This approach is an adaptation of our earlier work developing spectral graph partitioning algorithms for segmenting natural images,[50] understanding protein dynamics and allosteric propagation,[51] relating signal propagation on a protein structure to its equilibrium dynamics,[52] and finally discovering metastable states in MD trajectories.[53]

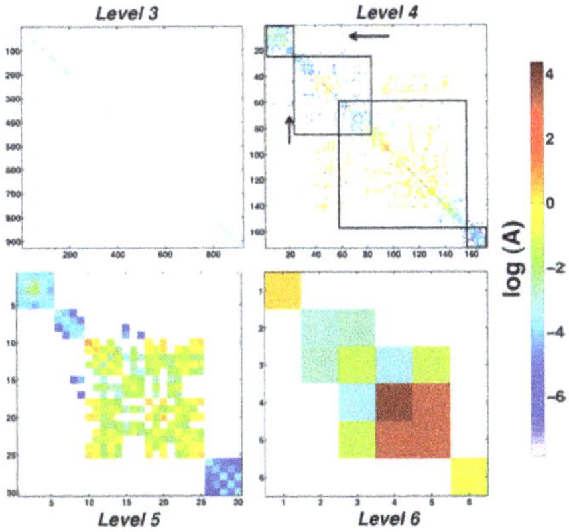

Fig. 4: **A hierarchy of conformational sub-states in the disorder-to-order transitions of NCBD conformational landscape.** A total of 6 levels are found by the hierarchical clustering. For hierarchy levels 3-6, the log of the affinity between each sub-state pair is shown.

We begin hierarchical clustering by constructing a Markov transition matrix using an affinity matrix of edge weights between conformer pairs in the dQAA space. We then initiate a Markov chain (or random walk) on the weighted undirected network. As Markov transition probabilities homogenize through diffusion, an implicit clustering emerges from the network. First, a set of nodes representing the putative clusters are identified. The number of clusters chosen is determined by the algorithm so that every node in the network has some Markov probability of transitioning into at least one of the clusters. Then, a Markov transition matrix is newly constructed using this reduced representation. The important principle behind this construction is that upon reaching a stationary distribution at the coarsest hierarchy level, the Markov chain has also converged at finer (more local) network levels. This consistency regulates the overall topology of the network and helps build a multi-resolution representation of metastable states.

We expect that fine-grained hierarchy levels will produce many small clusters containing close neighbors in the QAA space; that is, within each such cluster most members will be drawn from the same, narrow time-window. As Markov diffusion progresses (fine-grained to coarse-grained), conformers that are more distant neighbors will be connected by edges in the diffused network and will therefore be assigned to the same cluster. Thus, the hierarchical clustering can highlight dynamical connections between conformers at different timescales.

Results: The affinity matrix hierarchy derived by the clustering algorithm is shown in Fig. 4. The affinity matrices show several regions of high cross-talk at lower levels of the hierarchy. Iterative diffusion of the Markov chain derived from the initial affinity matrix (200000 × 200000), results in six hierarchy levels (Table 1). The mean C^{α}-RMSD to cluster center at the bottom hierarchy level is 3.2Å, indicating that clustering in dQAA-space also captures structural similarity between trajectory frames in Cartesian-space. Clusters with low mean RMSDs to the four experimental bound conformations and the two experimental unbound conformations occur at each hierarchy level. At the finest level of the hierarchy, the clusters representing the bound conformations are very small, but as the hierarchy progresses, they are found in more dominant sub-states, indicating that the bound conformations are energetically accessible. As seen in Table 1, the alignment to 1ZOQ is poor. However, if only helices α_2 and α_3 are considered, the RMSD is very low (data not shown). In contrast, for the three other ligand bound states, α_1 and α_2 align well to the simulations. Thus, a barrier involving the repositioning of this helix may need to be crossed in order to access the IRF-3 bound state.

PDB	ligand-free 1JJS	ACTR 1KBH	IRF3 1ZOQ	SRC1 2C52	ligand-free 2KKJ	p53 2L14	
Level	rank/ RMSD(Å)	rank/ RMSD(Å)	rank/ RMSD(Å)	rank/ RMSD(Å)	rank/ RMSD(Å)	rank/ RMSD(Å)	Total number of clusters
3	895/5.3	928/1.8	313/7.3	928/1.9	928/1.4	910/5.2	928
4	49/6	110/1.9	122/7.3	168/2.0	81/1.5	132/5.2	172
5	10/6.3	30/1.9	25/7.4	30/2.1	30/1.5	30/5.3	30
6	1/6.4	3/2.0	5/7.4	3/2.2	3/1.6	3/5.3	6

Table 1: Conformational similarity between determined sub-states and extant structural models. Sub-states are ranked according to membership, 1 being the largest. For the coarsest hierarchy levels, sub-state rank and RMSD from sub-state center to experimental conformation is given for the sub-state with lowest RMSD to the experimental conformation.

6. Intermediate states of ligand-free NCBD access ligand-bound conformations

The organization of the ligand-free NCBD ensemble indicates the presence of six large conformational sub-states that interconvert between each other. One can visualize the six sub-states from the coarsest hierarchy level as illustrated in Fig. 5(a). Of the six sub-states, sub-states 4 and 5 constitute over 88% of the entire ligand-free ensemble, consisting of 98,143 and 79,672 conformers respectively. The remaining sub-states (1, 2, 3 and 6) represent rare transitions in the landscape. It is interesting to observe that sub-states 1 and 6 are somewhat isolated from the conformational states, however a sizable population of conformations exist in each state (see affinity map in 5(a)). Although one may attribute the isolation to the MD sampling protocol, it is important to note that descending through the various levels of the hierarchy (Level 5 through Level 2) indicates that both sub-states 1 and 6 are connected via extremely lowly populated states (see Fig. 4), indicating that multiple paths exist through which states 1 and 6 can be reached. We also note that while certain pairs of sub-states (such as [2,3] and [4,5]) freely interconvert between each other, sub-state 3 alone can access conformations that are similar to that of sub-state 5. Therefore, sub-state 3 acts as an intermediate state from which conformations in sub-states 2, 4 and 5 interconvert.

Sub-state 1 (rank 3) represents the state closest to the bound conformations observed experimentally (Table 1). As illustrated in Fig. 5(b), a representative structure from sub-state 1 is compared with two ligand-bound structures, namely 1KBH (panel A) and 2C52 (panel B). Sub-state 1 represents the third least populated state of all sub-states (9,488 or 4.7% of conformers). However, when compared with the bound structures, on an average, it exhibits smaller RMSD values to the bound 1KBH (RMSD: 2.0 Å) and 2C52 (RMSD: 2.2 Å) conformers. This observation indicates that the ligand-free state of NCBD can access sub-states resembling the bound state.

It may be tempting to conclude that sub-state 1 is isolated from other conformational sub-states. However, as noted above, closer examination of the cluster hierarchy (Fig. 4, Level 4) reveals that concerted structural changes along a complex pathway are required for NCBD to adopt a binding competent conformation. By descending through the hierarchy, one can observe from Level 4 that a small subset of states (indicated by arrows on Fig. 4) closely resemble conformations in sub-state 1. This conformational state arises out of a rare state mostly consisting of conformers similar to sub-states 2 and 3 in level 6 of the hierarchy. Note that sub-state 2 in level 6 of the hierarchy consists of just 938 (or less than 0.05%) of the overall conformers, representing a rare transition. In this sub-state, the α_3 helix adopts a conformation that is more extended and hence represents an intermediate state that mediates a transition from sub-states 4 and 5 to the bound sub-state 1.

The observed clusters and conformational changes also provide a hypothesis for interconversions necessary for facilitating NCBD-ligand binding. For one, if NCBD is relatively compact, as in sub-states 4 and 5, then α_3 must initially undergo partial unfolding, seen in sub-states 2 and 3, to allow for the ligand to bind. Only then can α_3 adapt itself to form a full α-helix, as seen from experimental ensembles. Since we have not performed a comparison of our simulations with the ligand-bound state of either 1KBH or 2C52, we cannot provide a quantitative picture about the nature of changes that are required. However, based on the structural information available from experiments, such a partial unfolding-refolding pathway may indeed be responsible for facilitating NCBD's recognition of its binding partners. A similar scenario can also be proposed for α_1, which twists when binding with IRF3 (seen in Fig. 1D), although these experiments will be pursued in the future.

7. Conclusions and Future Work

As part of pursuing further work in the area, we propose to incorporate simulations from a second NMR ensemble (1JJS) as well as several ligand-bound conformations to map out the conformational landscape of NCBD. Furthermore, by extending the Markov diffusion framework, we will elucidate the kinetic rates of significant conformational transitions.

The methodologies we have put forward yield the following insights: (a) ligand-free NCBD can indeed access conformations representative of the ligand-bound form and (b) structural changes required for ligand-free NCBD to access states that resemble ligand-bound conformations require concerted changes throughout the protein. We show that within our simulations, ligand-free α_1 and α_2 orientations largely resemble those of ligand-bound conformations; α_3 however, can exhibit a

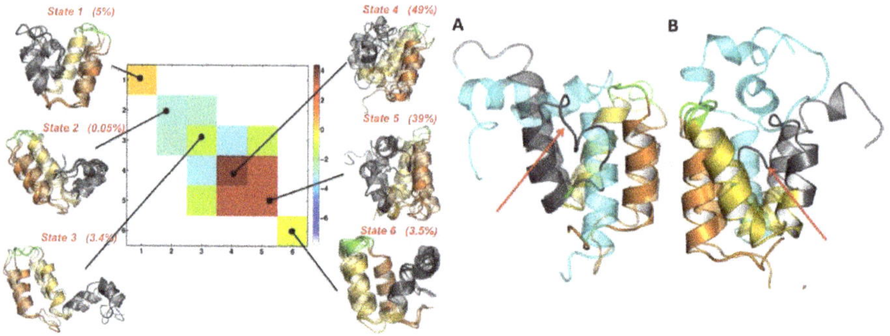

(a) Sub-states at level 6 (b) Rate limiting structural changes

Fig. 5: **Intermediate states of ligand-free NCBD enable access to ligand-bound conformations** Intermediate states of ligand-free NCBD enable access to ligand-bound conformations (a) Log affinities between sub-states at hierarchy level 6 are shown. For each of the 6 clusters, an ensemble of random conformers within that cluster are shown, and the percent of total frames within the cluster is given. High affinity (red) between two clusters indicate that those clusters are similar in dQAA space. Low affinity (blue - white) indicates that clusters have low similarity in dQAA space. (b) Comparing NCBD ensembles with the bound ligands (A) ACTR (1KBH; cyan) and (B) SRC1 (2C52; cyan) showing the orientations of α_3 indicated by red arrows.

wide degree of flexibility and does not generally sample conformations that are similar to ligand-bound states.

References

1. M. Fuxreiter, P. Tompa, I. Simon, V. N. Uversky, J. C. Hansen and F. J. Asturias, *Nat. Chem. Biol.* **4**, 728 (2008).
2. M. M. Babu, R. van der Lee, N. S. de Groot and J. Gsponer, *Curr. Opi. Struct. Biol.* **21**, 432 (2011).
3. H. J. Dyson and P. E. Wright, *Nat. Rev. Mol. Cell Biol.* **6**, 197 (2005).
4. A. K. Dunker, I. Silman, V. N. Uversky and J. L. Sussman, *Curr. Opi. Struct. Bio.* **18**, 756 (2008).
5. M. Stefani, *Biochimica et Biophysica Acta (BBA) - Molecular Basis of Disease* **1739**, 5 (2004).
6. Q. C. Zhang, T. l. Yeh, A. Leyva, L. G. Frank, J. Miller, Y. E. Kim, R. Langen, S. Finkbeiner, M. L. Amzel, C. A. Ross and M. A. Poirier, *J. Biol. Chem.* **286**, 8188 (2011).
7. M. L. Acevedo and W. L. Kraus, *Mol. Cell. Biol.* **23**, 335 (2003).
8. H. Li and J. D. Chen, *J. Biol. Chem.* **273**, 5948 (1998).
9. W. Gu, X.-L. Shi and R. G. Roeder, *Nature* **387**, 819 (1997).
10. S. Burge, D. P. Teufel, F. M. Townsley, S. M. V. Freund, M. Bycroft and A. R. Fersht, *Proc. Natl. Acad. Sci. USA* **106**, 3142 (2009).
11. R. Lin, C. Heylbroeck, P. M. Pitha and J. Hiscott, *Mol. Cell. Biol.* **18**, 2986 (1998).
12. K. E. S. Scoggin, A. Ulloa and J. K. Nyborg, *Mol. Cell. Biol.* **21**, 5520 (2001).
13. C. G. Mullighan, J. Zhang, L. H. Kasper, S. Lerach, D. Payne-Turner, L. A. Phillips, S. L. Heatley, L. Holmfeldt, J. R. Collins-Underwood, J. Ma, K. H. Buetow, C.-H. Pui, S. D. Baker, P. K. Brindle and J. R. Downing, *Nature* **471**, 235 (2011).

14. M. I. Torres-Arzayus, J. Font de Mora, J. Yuan, F. Vazquez, R. Bronson, M. Rue, W. R. Sellers and M. Brown, *Cancer Cell* **6**, 263 (2004).
15. C. W. Lee, M. A. Martinez-Yamout, H. J. Dyson and P. E. Wright, *Biochemistry* **49**, 9964 (2010).
16. M. Kjaergaard, K. Teilum and F. M. Poulsen, *Proc. Natl. Acad. Sci. USA* **107**, 12535 (2010).
17. S. J. Demarest, M. Martinez-Yamout, J. Chung, H. Chen, W. Xu, H. J. Dyson, R. M. Evans and P. E. Wright, *Nature* **415**, 549 (2002).
18. S. J. Demarest, S. Deechongkit, H. J. Dyson, R. M. Evans and P. E. Wright, *Protein Sci.* **13**, 203 (2004).
19. C. H. Lin, B. J. Hare, G. Wagner, S. C. Harrison, T. Maniatis and E. Fraenkel, *Mol. Cell* **8**, 581 (2001).
20. G. Bowman, K. Beauchamp, G. Boxer and V. S. Pande, *J. Chem. Phys.* **131**, p. 124101 (2009).
21. M. A. Balsera, W. Wriggers, Y. Oono and K. Schulten, *J. Phys. Chem.* **100**, 2567 (1996).
22. G. Zaccai and B. Jacrot, *Annu. Rev. Biophysics. Bioeng.* **12**, 139 (1983).
23. M. Kurylowicz, C.-H. Yu and R. Pomès, *Biophys. J.* **98**, 386 (2010).
24. A. Ramanathan, A. Savol, C. Langmead, P. Agarwal and C. Chennubhotla, *PLoS ONE* **6**, p. e15827 (2011).
25. G. R. Bowman G. and V. S. Pande, *Proc. Natl. Acad. Sci. USA* **107**, 10890 (2010).
26. F. Morcos, S. Chatterjee, C. L. McClendon, P. R. Brenner, R. López-Rendón, J. Zintsmaster, M. Ercsey-Ravasz, C. R. Sweet, M. P. Jacobson, J. W. Peng and J. A. Izaguirre, *PLoS Comput. Biol.* **6**, p. e1001015 (2010).
27. N. Deng, W. Zheng, E. Gallicchio and R. M. Levy, *J. Am. Chem. Soc.* **133**, 9387 (2011).
28. M. Heinig and D. Frishman, *Nucl. Acids Res.* **32**, W500 (2004).
29. D. A. Case, T. E. Cheatham, T. Darden, H. Gohlke, R. Luo, K. M. Merz, A. Onufriev, C. Simmerling, B. Wang and R. J. Woods, *J. Comp. Chem.* **26**, 1668 (2005).
30. V. Hornak, R. Abel, A. Okur, B. Strockbine, A. Roitberg and C. Simmerling, *Proteins: Struct. Func. Bioinfo.* **65**, 712 (2006).
31. A. Ramanathan and P. K. Agarwal, *J. Phys. Chem. B* **113**, 16669 (2009).
32. M. J. Harvey, G. Giupponi and G. D. Fabritiis, *J. Chem. Theory Comput.* **5**, 1632 (2009).
33. K. A. Feenstra, B. Hess and H. J. C. Berendsen, *J. Comp. Chem.* **20**, 786 (1999).
34. Y. Shen and A. Bax, *J. Biomol. NMR* **38**, 289 (2007).
35. J. Mittal and R. B. Best, *Biophys. J.* **99**, L26 (2010).
36. L. S. Caves, J. Evanseck and M. Karplus, *Protein Sci.* **7**, 649 (1998).
37. A. Grossfield, S. E. Feller and M. C. Pitman, *Proteins: Struc. Func. Bio.* **67**, 31 (2007).
38. B. Mao, M. R. Pear, J. A. McCammon and S. H. Northrup, *Biopolymers* **21**, 1979 (1982).
39. T. Ichiye and M. Karplus, *Biochemistry* **27**, 3487 (1988).
40. F. Pontiggia, G. Colombo, C. Micheletti and H. Orland, *Phys. Rev. Lett.* **98**, p. 048102 (2007).
41. K. N. Woods, *Phys. Rev. E* **81**, p. 031915 (2010).
42. G. Schiró, C. Caronna, F. Natali and A. Cupane, *J. Am. Chem. Soc.* **132**, 1371 (2010).
43. H. Frauenfelder, F. Parak and R. D. Young, *Ann. Rev. Biophys. Biophys. Chem.* **17**, 451 (1988).
44. M. A. Miller and D. J. Wales, *J. Chem. Phys.* **111** (1999).
45. A. Altis, P. Nguyen, R. Hegger and G. Stock, *J. Chem. Phys.* **126**, p. 244111 (2007).
46. Y. Mu, P. Nguyen and G. Stock, *Proteins: Struct. Func. Bio.* **58**, 45 (2004).
47. G. G. Maisuradze and D. M. Leitner, *Proteins: Struct. Func. Bio.* **67**, 569 (2007).
48. J.-F. Cardoso, *Neural Computation* **11**, 157 (1999).
49. G. H. Golub and C. F. Van Loan, *Matrix Computations* (Johns Hopkins University Press, 1996).
50. C. Chennubhotla and A. Jepson, *Neural Info. Proc. Sys.* **17**, 273 (2005).
51. C. Chennubhotla and I. Bahar, *Mol. Sys. Biol.* **2**, p. 36 (2006).
52. C. Chennubhotla and I. Bahar, *PLoS Comput Biol* **3**, 1716 (2007).
53. A. Savol, V. Burger, P. Agarwal, A. Ramanathan and C. Chennubhotla, *Bioinformatics* **27**, i52 (2011).

EFFICIENT CONSTRUCTION OF DISORDERED PROTEIN ENSEMBLES IN A BAYESIAN FRAMEWORK WITH OPTIMAL SELECTION OF CONFORMATIONS

CHARLES K. FISHER

Committee on Higher Degrees in Biophysics, Harvard University
Cambridge, Massachusetts 02139-4307, United States
Email: ckfisher@fas.harvard.edu

ORLY ULLMAN

Department of Chemistry, Massachusetts Institute of Technology
Cambridge, Massachusetts 02139-4307, United States
Email: orly@mit.edu

COLLIN M. STULTZ*

Harvard-MIT Division of Health Sciences and Technology, Department of Electrical Engineering and Computer Science, and the Research laboratory of Electronics, Massachusetts Institute of Technology
Cambridge, Massachusetts 02139-4307, United States
**Corresponding Author Email: cmstultz@mit.edu*

Constructing an accurate model for the thermally accessible states of an Intrinsically Disordered Protein (IDP) is a fundamental problem in structural biology. This problem requires one to consider a large number of conformations in order to ensure that the model adequately represents the range of structures that the protein can adopt. Typically, one samples a wide range of structures in an attempt to obtain an ensemble that agrees with some pre-specified set of experimental data. However, models that contain more structures than the available experimental restraints are problematic as the large number of degrees of freedom in the ensemble leads to considerable uncertainty in the final model. We introduce a computationally efficient algorithm called Variational Bayesian Weighting with Structure Selection (VBWSS) for constructing a model for the ensemble of an IDP that contains a minimal number of conformations and, simultaneously, provides estimates for the uncertainty in properties calculated from the model. The algorithm is validated using reference ensembles and applied to construct an ensemble for the 140-residue IDP, monomeric α-synuclein.

1. Introduction

Intrinsically Disordered Proteins (IDPs) are a class of polypeptides that populate diverse ensembles of conformations under physiological conditions.[1, 2] It is believed that a number of IDPs play a critical role in the development of neurodegenerative disorders including Alzheimer's and Parkinson's – diseases that affect millions of people each year.[3, 4] As a result, gaining an understanding of the conformational properties of these proteins is an important task, which could pave the way for the discovery of new therapeutics through structure based drug design.[5]

A model for the ensemble of an IDP consists of a set of structures $S = \{s_1, \ldots, s_n\}$ and a set of weights $\vec{w} = \{w_1, \ldots, w_n\}$, where w_i corresponds to the equilibrium probability of conformation s_i. Typically, these structures and weights are chosen so that averages calculated from the ensemble agree with experimental observations;[2, 6, 7] for example, so that the radius of gyration calculated from the ensemble is similar to its experimentally determined value. Previous studies have shown that agreement with experimental observations is not sufficient to ensure that an ensemble is accurate, because there may be many different ways of choosing the structures and weights to achieve a good fit to the experimental data.[8, 9] Therefore, it is important to develop methods that can quantify the amount of uncertainty associated with a model of an IDP ensemble.

We previously developed an algorithm, called Bayesian Weighting (BW), that uses Bayesian inference to construct an ensemble for an IDP that agrees with experimental observations, while simultaneously estimating the uncertainty associated with this model.[8] The BW method calculates a 'posterior' probability distribution over all ways of weighting the structures in a pre-specified conformational library. Point estimates and error bars for various properties of the ensemble can be computed by calculating an average over this probability distribution. An important feature of the algorithm is that it provides a built-in error check in the form of an uncertainty parameter, or posterior uncertainty, which is related to the error in the estimated population weights. Our previous study suggests that this uncertainty parameter is a metric that assesses model correctness.[8] When the uncertainty parameter is 0, one can be relatively sure that the model is correct. By contrast, a value of 1 suggests that the ensemble is inaccurate and values calculated from the ensemble will be associated with very large confidence intervals.

Of course, the quality of the structural library will also affect the accuracy of the resulting ensemble. The structural library must be diverse enough to capture the states populated by the IDP, but if it is too large then the problem will be under-restrained, leading to a large posterior uncertainty. One way to overcome this problem is to use variable selection techniques to identify an optimal subset of conformations from within a larger structural library.[10] Such an algorithm might begin by estimating the population weights using a large conformational library and iteratively discarding lowly weighted conformations to improve the ensemble. In practice, performing this type of structure selection algorithm within a fully Bayesian framework would be computationally intractable because BW uses Monte Carlo methods to estimate the weights, and these calculations can take a long time to converge; therefore, we introduce an approximate algorithm called Variational Bayesian Weighting and Structure Selection (VBWSS) that can perform the calculations quite rapidly – providing a decrease in computational time of roughly 4 orders of magnitude compared to BW. In this work we describe the VBWSS method and validate the approach using 'reference' ensembles. Lastly we use the method to characterize the ensemble of the intrinsically disordered protein, α-synuclein (αS).

2. Theory

2.1. *Optimal Structure Selection*

When constructing an ensemble for an IDP, it is necessary to use a structural library that is diverse enough to cover the entire range of accessible conformations. However, increasing the size of this library adds degrees of freedom to the model making the problem more underdetermined. In our prior work, we described a method for calculating a posterior distribution that assigns a probability to each possible choice of weights as a way of quantifying uncertainty in the ensemble.[8] The posterior probability density function (PDF) is calculated using Bayes' theorem:[8]

$$f_{\vec{W}|\vec{M},S}\left(\vec{w}|\vec{m},S\right)=\frac{f_{\vec{M}|\vec{W},S}\left(\vec{m}|\vec{w},S\right)f_{\vec{W}|S}\left(\vec{w}|S\right)}{f_{\vec{M}|S}\left(\vec{m}|S\right)} \tag{1}$$

where $S=\{s_i\}_{i=1}^{n}$ denotes the set of structures, $\vec{w}=\{w_i\}_{i=1}^{n}$ denotes the set of population weights, $\vec{m}=\{m_i\}_{i=1}^{k}$ denotes the set of k experimental measurements. To calculate eq. (1) we must specify a likelihood function, $f_{\vec{M}|\vec{W},S}\left(\vec{m}|\vec{w},S\right)$, and a prior distribution, $f_{\vec{W}|S}\left(\vec{w}|S\right)$. The normalizing constant, or marginal likelihood (ML), is given by $f_{\vec{M}|S}\left(\vec{m}|S\right)=\int f_{\vec{M}|\vec{W},S}\left(\vec{m}|\vec{w},S\right)f_{\vec{W}|S}\left(\vec{w}|S\right)d\vec{w}$. The specific forms for each of these terms will be given in section 2.2. Using eq. (1), the Bayesian estimate for the weight of structure s_i is:

$$w_i^B \equiv \int w_i f_{\vec{W}|\vec{M},S}\left(\vec{w}|\vec{m},S\right)d\vec{w} \tag{2}$$

It may be possible to obtain a diverse ensemble that has a low uncertainty by considering different subsets, $S\subseteq\mathbb{Z}$, of a heterogeneous structural library, $\mathbb{Z}$, with n conformations.

In order to use Bayesian variable selection techniques to identify an optimal set of conformations, $S^*\subseteq\mathbb{Z}$, we have to specify an a priori probability for each subset of structures. If we do not have a priori knowledge to guide this choice, it is reasonable to assume that every possible subset is equally probable; i.e., $f_S\left(S\right)\propto 1$, where $f_S\left(S\right)$ is the probability of subset S. In this case, the posterior probability for a subset of conformations is:

$$f_{S|\vec{M}}\left(S|\vec{m}\right)\propto f_{\vec{M}|S}\left(\vec{m}|S\right) \tag{3}$$

where $\vec{w}=\{w_i\}_{i=1}^{l}$ is the vector of weights for the subset $S\subseteq\mathbb{Z}$ that contains $l\leq n$ structures. Therefore, the optimal subset of structures is obtained by maximizing the ML, $f_{\vec{M}|S}\left(\vec{m}|S\right)$. Note that it is usually not tractable to search through all subsets of structures, because the number of possibilities may very large. Instead, we begin with the entire structural library, where the weights are estimated using eq. (2) and the ML is calculated. Then, the lowest weighted structure is thrown out if its estimated weight is below a cutoff, w_{cut}. The weights of the remaining structures are then recalculated along with the new value of the ML and, again, the lowest weighted structure is discarded if its weight is below w_{cut}. The algorithm repeats this process until either the ML converges or all of the weights are greater than w_{cut}. The set of structures that had the largest value of the ML is chosen for the final ensemble. In what follows, we develop a variational BW, or VBW, method to facilitate efficient calculation of the weights at each step.

2.2. *Variational Bayesian Weighting*

At each iteration of the algorithm, after a set of structures has been chosen, an approximation to the ML can be calculated efficiently using 'variational' Bayesian inference.[11] As stated above, the ML is proportional to the criterion for selecting a set of structures. In addition, by maximizing an approximate form of the ML we can arrive at an optimal set of weights for the structural subset under consideration. In variational Bayesian inference, the posterior PDF given by eq. (1) is approximated with a simpler PDF that allows one to easily calculate quantities of interest.[11] Since we are interested in vectors of weights that are positive and sum to one, we choose our simple distribution to be a Dirichlet distribution with PDF:[12]

$$g(\vec{w}|\vec{\alpha},S)=\frac{\Gamma(\alpha_0)}{\sum_{i=1}^{n}\Gamma(\alpha_i)}\prod_{i=1}^{l}w_i^{\alpha_i-1} \tag{4}$$

where $\Gamma(\cdot)$ is the gamma function, $\{\alpha_i>0\}_{i=1}^{l}$ are parameters of the distribution and $\alpha_0=\sum_{i=1}^{l}\alpha_i$.[13] Using this distribution, we can calculate a lower bound on the logarithm of the ML as follows:

$$\log f_{\vec{M}}(\vec{m}|S)\geq\int g(\vec{w}|\vec{\alpha},S)\log\left\{\frac{f_{\vec{M}|\vec{W}}(\vec{m}|\vec{w},S)f_{\vec{W}|S}(\vec{w}|S)}{g(\vec{w}|\vec{\alpha},S)}\right\}d\vec{w}\equiv-L(\vec{\alpha}|S) \tag{5}$$

We note that our overall goal is to maximize the ML; however, since equation (5) provides a lower bound for the ML, we instead maximize $-L(\vec{\alpha}|S)$, or equivalently minimize $L(\vec{\alpha}|S)$. Furthermore, it is easy to show that $L(\vec{\alpha}|S)$ is equal (up to an additive constant) to the Kullback-Leibler (KL) distance[14] between $g(\vec{w}|\vec{\alpha},S)$ and $f_{\vec{W}|\vec{M},S}(\vec{w}|\vec{m},S)$. Therefore, by finding the set of parameters that minimizes $L(\vec{\alpha}|S)$ we also obtain an approximation to $f_{\vec{W}|\vec{M},S}(\vec{w}|\vec{m},S)$.

The feature that makes the VBW algorithm computationally efficient is that the objective function can be obtained in closed-form for a suitable choice of prior distribution. Thus, one can apply a standard minimization protocol to find the set of parameters that minimize $L(\vec{\alpha}|S)$. Before sketching the derivation, we need to define all of the terms in eq. (1).

We denote the current set of structures under consideration as $S=\{s_i\}_{i=1}^{l}$ and presume that we have a set of experimental measurements, $\vec{m}=\{m_i\}_{i=1}^{k}$, which have experimental errors $\{\varepsilon_i^{\exp}\}_{i=1}^{k}$. We assume that it is possible to calculate the ith experimental observable in the jth conformation, denoted as $\hat{m}_{ij}$, with an accuracy ε_i^{pre}. For example, if the Cα chemical shift for the first structure is calculated using SHIFTX, then ε_1^{pre} denotes the error reported for the calculation of a Cα chemical shift using SHIFTX, which is approximately 0.98 ppm.[15] The total uncertainty that results from experimental error and inaccurate prediction algorithms is defined as $\varepsilon_i = \sqrt{\left(\varepsilon_i^{\exp}\right)^2 + \left(\varepsilon_i^{pre}\right)^2}$. We assume that the likelihood function for each experimental measurement follows a Gaussian distribution, such that the total likelihood function (assuming that the measurements are independent) can be written as follows:

$$f_{\vec{M}|\vec{W},S}\left(\vec{m}|\vec{w},S\right) = \prod_{i=1}^{k}\left(\frac{1}{\varepsilon_i\sqrt{2\pi}}\right)\exp\left\{-\sum_{i=1}^{k}\frac{1}{2\varepsilon_i^2}\left(m_i - \sum_{j=1}^{l}\hat{m}_{ij}w_j\right)^2\right\} \tag{6}$$

Finally, for the prior distribution, $f_{\vec{W}|S}\left(\vec{w}|S\right)$, we choose a Dirichlet distribution with PDF:

$$f_{\vec{W}|S}\left(\vec{w}|S\right) = \frac{\Gamma(n/2)}{n\Gamma(1/2)}\prod_{i=1}^{l} w_i^{-1/2} \tag{7}$$

This choice of prior is a type of Jeffreys' distribution – a class of non-informative prior distributions that are widely used in Bayesian inference.[16]

Beginning with the definition of $L\left(\vec{\alpha}|S\right)$ in eq. (5), we can rewrite the objective function as:

$$L\left(\vec{\alpha}|S\right) = \int g\left(\vec{w}|\vec{\alpha},S\right)\log\frac{g\left(\vec{w}|\vec{\alpha},S\right)}{f_{\vec{W}|S}\left(\vec{w}|S\right)}d\vec{w} - \int g\left(\vec{w}|\vec{\alpha},S\right)\log f_{\vec{M}|\vec{W},S}\left(\vec{m}|\vec{w},S\right)d\vec{w} \tag{8}$$

The first term is the KL distance between two Dirichlet distributions, i.e. the variational and the prior distributions, and has been previously reported.[17] The second term is given by:

$$-\int g\left(\vec{w}|\vec{\alpha},S\right)\log f_{\vec{M}|\vec{W},S}\left(\vec{m}|\vec{w},S\right)d\vec{w} = \sum_{i=1}^{k}\frac{1}{2\varepsilon_i^2}\int g\left(\vec{w}|\vec{\alpha},S\right)\left(m_i - \sum_{j=1}^{l}\hat{m}_{ij}w_j\right)^2 d\vec{w} + C \tag{9}$$

where C is a constant that depends on the experimental errors. The integral in eq. (9) can be calculated exactly giving an analytical expression for eq. (8):

$$\begin{aligned} L\left(\vec{\alpha}|S\right) &= \log\frac{\Gamma(\alpha_0)}{\Gamma(l/2)} + \sum_{i=1}^{l}\log\frac{\Gamma(1/2)}{\Gamma(\alpha_i)} + \sum_{i=1}^{l}(\alpha_i - 1/2)\left\{\psi(\alpha_i) - \psi(\alpha_0)\right\} \\ &+ \frac{1}{2}\sum_{i=1}^{k}\varepsilon_i^{-2}\left(m_i - \frac{1}{\alpha_0}\sum_{j=1}^{l}\hat{m}_{ij}\alpha_j\right)^2 + \frac{1}{2}\sum_{i=1}^{l}\sum_{j=1}^{l}\left(\sum_{l=1}^{k}\frac{\hat{m}_{li}\hat{m}_{lj}}{\varepsilon_l^2}\right)\frac{\alpha_i(\alpha_0 - \alpha_i)\delta_{ij} - \alpha_i\alpha_j(1-\delta_{ij})}{\alpha_0^2(\alpha_0+1)} \end{aligned} \tag{10}$$

where δ_{ij} is the Kronecker delta function, $\psi(\cdot)$ is the digamma function and the constant from eq. (9) has been neglected because constant terms do not play a role in function minimization.

Finally, suppose that we have used an optimization algorithm to identify the set of parameters, $\{\hat{\alpha}_i\}_{i=1}^{l}$, that minimizes $L(\vec{\alpha}|S)$ for the current set of conformations. The estimates for the population weights are given by the simple formula $w_i^B = \hat{\alpha}_i / \sum_{j=1}^{l} \hat{\alpha}_j$.

2.3. *Variational Bayes with Structure Selection*

The Variational Bayesian Weighting with Structure Selection (VBWSS) algorithm is:

(1) Initialize the set of conformations to the entire structural library, i.e. set $S = \mathbb{Z}$.

(2) Use simulated annealing to find the set of parameters that minimizes $L(\vec{\alpha}|S)$.

(3) Remove the lowest weighted structure if it has $w_i^B < w_{cut}$ and go to step 4, else go to step 5.

(4) If $L(\vec{\alpha}|S)$ has not improved for 10 iterations go to step 5, else return to step 2.

(5) Exit the algorithm and return the set of structures and parameters that produced the smallest value of $L(\vec{\alpha}|S)$.

The simulated annealing algorithm in step 2 used a Gaussian cooling schedule $T(t) = T(0)\exp\left(-(5t)^2\right)$, where t is the fraction of steps completed. For the first iteration with the entire structural library, the number of steps was set to $100 \cdot n$ and the initial temperature was set to $T(0) = 2$. Since each iteration of the structure selection algorithm involves throwing out a lowly weighted structure, we reasoned that the set of parameters identified in iteration *j* should be reasonable guess for the parameters for iteration *j+1*. Thus, each iteration after the first was initialized using the parameters identified in the previous iteration and run for $50 \cdot l$ steps beginning from a temperature of $T(0) = 1$. The step size was optimized during a short equilibration period at the start of each iteration to target a 50% acceptance ratio at $T = 1$.

2.3. *Approximate Confidence Intervals*

The variational approximation to the posterior PDF can be used to calculate confidence intervals (CI) for parameters of the ensemble using a simple analytical approximation. Here, we suppose that our final set of conformations is $S = \{s_i\}_{i=1}^{l}$ and the corresponding set of variational parameters is $\{\hat{\alpha}_i\}_{i=1}^{l}$. Again, we define $\hat{\alpha}_0 = \sum_{i=1}^{l} \hat{\alpha}_i$. As an example, suppose that we are interested in estimating the ensemble average distance between two atoms in the protein. The distance between the two atoms in the ith structure is D_i. The Bayesian point estimate for the ensemble averaged distance is $\langle D \rangle = \hat{\alpha}_0^{-1} \sum_{i=1}^{l} D_i \hat{\alpha}_i$ and the variance in the ensemble averaged distance is $\mathrm{var}[D] = \hat{\alpha}_0^{-2} (\hat{\alpha}_0 + 1)^{-1} \sum_{i=1}^{l} \sum_{j=1}^{l} D_i D_j \left(\hat{\alpha}_i \hat{\alpha}_0 \delta_{ij} - \hat{\alpha}_i \hat{\alpha}_j \right)$, which comes from the formula for the variance of a linear combination and the covariance matrix of a Dirichlet distribution.[12] Thus, an approximate 95% CI can be calculated using $\langle D \rangle \pm 1.96\sqrt{\mathrm{var}[D]}$; the

1.96 is the number of standard deviations of a Gaussian distribution that corresponds to a 95% CI.[12]

3. Results and Discussion

3.1. *Validation with Reference Ensembles*

The ultimate goal of the VBWSS algorithm is to construct an accurate, parsimonious representation for the conformational ensemble of an IDP using a minimum amount of computational effort while, simultaneously, estimating the uncertainty in the resulting model. Thus, there are a number of criteria by which the algorithm must be judged. First, we will address the 2 most important criteria: that the algorithm provides a means for accurately modeling conformational ensembles and for estimating their uncertainties. To address these questions, and to choose the weight threshold for structure selection, we used the method of reference ensembles.

A reference ensemble is a pre-defined 'truth' for which both the set of conformations and their weights are known. The same 20 reference ensembles that were used to validate the previously reported BW algorithm were also used in this study to facilitate a comparison between BW and the new VBWSS algorithm.[8] To review, each of the reference ensembles consisted of a set of 95 conformations for the small peptide, met-enkephalin, and a set of weights. The different sets of weights were chosen so that the reference ensembles had different amounts of entropy. Backbone NMR chemical shifts were calculated for each reference ensemble using SHIFTX and were randomly perturbed by 0.1 ppm to model experimental uncertainty.[15]

The ensembles were compared by measuring the distances between the vectors of weights. For the measure of distance, we used the Jensen-Shannon divergence (JSD), which ranges from 0 to 1 for identical and maximally different vectors of weights, respectively, and is given by:[18, 19]

$$\Omega^2\left(\vec{w}^{(1)},\vec{w}^{(2)}\right)=\frac{1}{2}\sum_{i=1}^{n} w_i^{(1)}\log_2\left(\frac{2w_i^{(1)}}{w_i^{(1)}+w_i^{(2)}}\right)+\frac{1}{2}\sum_{i=1}^{n} w_i^{(2)}\log_2\left(\frac{2w_i^{(2)}}{w_i^{(1)}+w_i^{(2)}}\right) \tag{11}$$

For VBWSS, if a structure was not included in the ensemble then its weight was set equal to zero. The square root of the JSD can also be used to quantify the uncertainty in an ensemble by calculating the expected distance from the point estimate for the weights.[8] It was shown previously that such an estimate, $\sigma_{\vec{w}^B}=\sqrt{\int\Omega^2\left(\vec{w},\vec{w}^B\right)f_{\vec{W}|\vec{M},S}\left(\vec{w}|\vec{m},S\right)d\vec{w}}$, was strongly correlated to the distance between the 'true' weights of a reference ensemble and the estimated weights, $\vec{w}^B$. This is an important feature of BW because it provides a built-in error check on the accuracy of the ensemble. To ensure that this feature was preserved in VBWSS, the weight cutoff for structure selection was chosen to maximize the correlation between $\sigma_{\vec{w}^B}$ and $\Omega\left(\vec{w}^T,\vec{w}^B\right)$ for the 20 reference ensembles, yielding a value of $w_{cut} = 0.004$; this value of w_{cut} was used throughout the rest of the analysis.

A comparison of the accuracy of the VBWSS and BW algorithms on the 20 reference ensembles is shown in Fig. 1. As shown in Fig. 1A, the estimates for the weights obtained from the VBWSS algorithm are similar in accuracy to those obtained from BW. In addition, the correlation between the uncertainty estimate, $\sigma_{\vec{w}^B}$, and the actual error in the estimated weights, $\Omega\left(\vec{w}^T, \vec{w}^B\right)$, obtained from VBWSS is R = 0.9. The error in the weights is related to $\sigma_{\vec{w}^B}$ by $\Omega\left(\vec{w}^T, \vec{w}^B\right) \approx 1.54 \cdot \sigma_{\vec{w}^B}$. Thus, VBWSS obtained a similar level of accuracy as BW and maintained the ability to quantify the uncertainty in the ensemble.

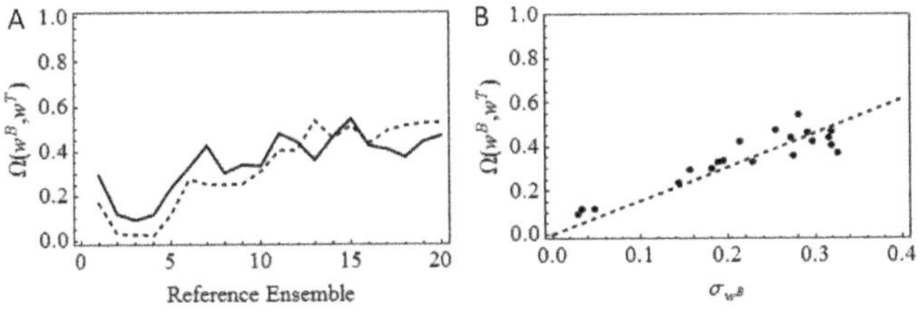

Figure 1. (A) The error between the true and estimated weights for the 20 reference ensembles. The solid and dashed lines indicate VBWSS and BW, respectively. The ensembles are ordered by increasing entropy. (B) The correlation (R = 0.9) between $\sigma_{\vec{w}^B}$ and $\Omega\left(\vec{w}^T, \vec{w}^B\right)$ obtained from VBWSS. The best fit, $y = 1.54x$, is shown as a dashed line.

It is also important to ensure that CIs calculated from VBWSS provide reasonable estimates of the errors in parameters of the ensemble. We assessed this feature of the algorithm by comparing ensemble averaged inter-atomic distances calculated from the VBWSS ensembles to their corresponding values in the reference ensembles. The approximate formula for a CI defined in section 2.4 should be fairly accurate if, on average, $\left(\langle D\rangle - D_{ref}\right)^2 \approx \mathrm{var}[D]$; here, D_{ref} is the ensemble average value of the distance calculated from the reference ensemble. In practice, we found that the standard deviation, calculated using VBWSS was too small, which is a problem that is known to affect variational approximations in Bayesian inference. To correct for the bias in the estimation of the standard deviations from VBWSS, we determined an empirical equation, $\langle D\rangle \pm 1.96 \cdot \lambda \cdot \sqrt{\mathrm{var}[D]}$, that gave the CIs for the inter-atomic distances approximately 95% coverage frequency of their reference ensemble values. The best fit value was $\lambda \approx 1.54$, which was used for subsequent calculations of CIs.

3.2. *α-Synuclein Ensemble*

αS is one of the most studied members of the IDP family. This 140 residue protein has been implicated in the pathology of a family of diseases known as synuclepathies.[20] The most common among these is Parkinson's disease, a neurodegenerative disorder characterized by intracellular aggregates known as Lewy bodies and the loss of dopaminergic neurons.[21] While the exact relation between the aggregates and neuronal death remains a subject of debate, understanding the nature of the unfolded protein and its ability to aggregate may prove crucial to the design of new therapies. Furthermore, although it has been suggested that αS adopts a tetrameric structure with considerable helical content in the native cell environment[22], the formation of aggregates likely involves the dissociation of these tetramers into disordered monomers that can form aggregates rich in cross-beta structure[22]. Consequently, there is still a need to understand the structure of the unfolded form of the disordered monomeric form of αS because such data may provide insights into the aggregation process.

We applied the VBWSS algorithm to a previously generated structural library of αS that was created by pruning a large sample of $\sim 10^5$ structures to a representative library of 299 conformations. The exact same set of experimental measurements which was used to generate the BW ensemble was used here to generate the VBWSS ensemble. More specifically, we used C, Cα, Cβ and N chemical shifts,[23] N-H RDCs[24] and the radius of gyration from SAXS experiments[25] along with the calculated values of chemical shifts using SHIFTX,[15] RDC using PALES[26] and radius of gyration through CHARMM.[27] RDCs predicted from PALES are frequently scaled to account for uncertainty in predicting the magnitude of alignment.[8] Here, the predicted RDCs were scaled by 0.25, which was found using a simple grid search to minimize the VBW objective function.

We found that the VBW algorithm was extremely efficient compared to the BW algorithm. The VBW algorithm (without structure selection) took less than 30 seconds, compared to approximately 2 weeks for BW, running in parallel on eight 2.4 GHz Intel Xeon processors – a decrease in computational time of roughly 4 orders of magnitude. The increased computational efficiency of the variational approximation more than made up for the increase in computational effort due to structure selection, with the total VBWSS algorithm taking less than 30 minutes.

Structure selection reduced the ensemble size from 299 to 78 non-zero weighted conformations, as shown in Fig 2A. The VBWSS ensemble also fits the experimental data well (Figs. 2B-C). In addition, the calculated ensemble average value of the radius of gyration was 40.4Å with a 95% CI of [39.2-41.6]Å compared to the experimental value of 40±2Å.[25]

We cross validated our ensemble by comparing inter-residue distances obtained from a recently published FRET experiment described in Grupi et al[28] (Table 1). Our calculated 95% CIs for the ensemble average distances, measured between the Cα atoms, are in good agreement with the corresponding experimentally determined values. It is important to note, that we made no explicit use of any inter-residue distance data in the ensemble generation procedure.

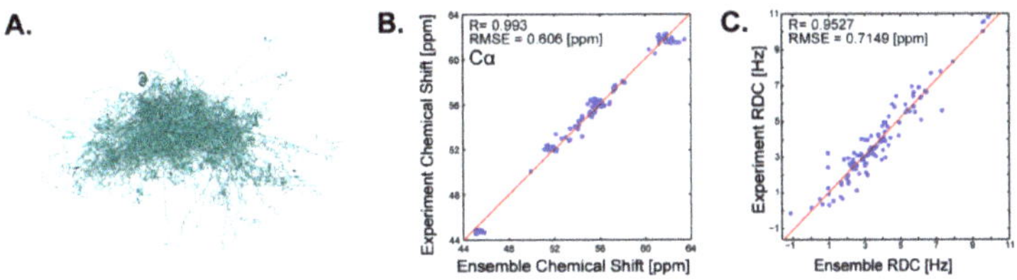

Figure 2. αS VBWSS algorithm. (A) Alignment of non-zero weighted structures. (B) Agreement of calculated and experimental Cα chemical shifts. (C) Agreement of calculated and experimental RDCs.

Table1. Cross validation through inter-residue distances

Probe pair	Experiment [Å]	Ensemble [Å]
18,26	15.1	16.5-17.8
26,39	21.8	20.2-22.7
4,18	16.9	17.9-19.9
18,39	29.4	30.0-33.6
4,26	33.6	27.9-31.2
66,90	29.6	36.5-40.1
39,66	40.1	40.2-44.5
4,39	43.0	39.5-44.3

One of the most interesting features of αS is its ability to form aggregates that contain cross-beta structure under the right set of experimental conditions.[29] Because the formation of αS aggregates was shown to be involved in the pathology of Parkinson's disease, it is of great interest to gain further knowledge about aggregation prone conformations within the disordered state. It was previously found that the minimal toxic aggregating segment is located in the NAC region of the protein, residues 68-78 or in reference to the NAC, NAC(8-18).[30] Therefore, we focused on assessing regions with aggregation propensity within the ensemble. Because structures that place the segment in an extended and solvent exposed orientation may be aggregation-prone, we calculated the percentage solvent accessible surface area (%SASA) of the NAC(8-18) segment using CHARMM[27] and the secondary structure content of each structure in the ensemble. Percentage solvent accessible surface area was calculated by dividing the SASA values for N-C-Cα atoms by the SASA values of these atoms in a fully extended conformation. Results of this analysis are shown in Fig 4. Similar analysis can be found in Ullman & Stultz.[31]

We found that approximately 12% (8-15% is the 95% CI) of the structures have the NAC(8-18) segment in an extended (more than 3 extended residues) and solvent exposed (%SASA>40) orientation. These results are similar to those found using the BW algorithm[31] and suggest that the ensemble of αS contains conformations that can readily form toxic, beta-sheet rich aggregates.

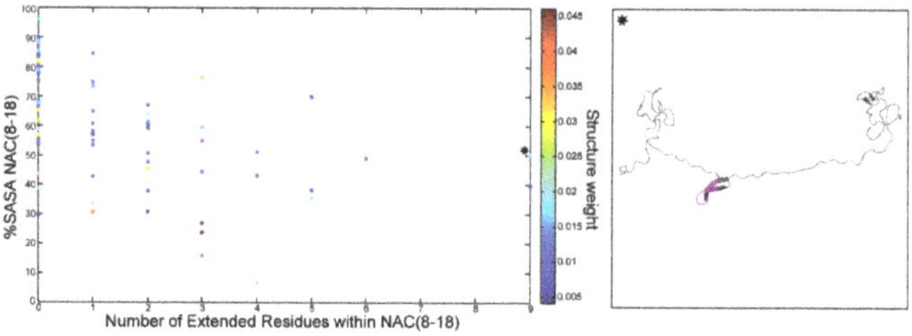

Figure 4. Aggregation propensity in the ensemble. Colors represent the weight of each of structure. The star denotes a structure, shown to the right, with a relatively high weight.

4. Conclusions

Constructing an ensemble for an IDP is a difficult task that requires reliable methods for estimating parameters from the ensemble and their associated uncertainties. Here, we introduced an algorithm for selecting an optimal set of conformations and implemented a variational approximation to the BW algorithm that provides a decrease in computational time of roughly 4 orders of magnitude. The two methods were combined in the VBWSS algorithm, which provides a computationally efficient method for constructing IDP ensembles. The VBWSS algorithm was validated using reference ensembles, and was found to produce a similar level of accuracy as the BW algorithm. In addition, accurate estimates for the uncertainties in characteristics of the ensemble can be calculated from VBWSS using simple formulas.

In general, proteins with a larger amount of disorder result in VBWSS ensembles with a larger amount of uncertainty, all other things being equal. Nevertheless, certain characteristics of the ensemble may be well defined even when the other aspects are highly uncertain. This highlights the importance of using confidence intervals to make inferences about ensemble characteristics.

We applied the VBWSS algorithm to construct an ensemble for αS and found that the ensemble agrees very well with experimental data. In addition to a dramatic decrease in computational time over BW, the VBWSS resulted in an improved fit to some experimental data. The BW algorithm obtains optimal structure weights for a given set of conformations while VBWSS obtains an optimal set of weights as well as an optimal set of conformations that fit the existing set of experimental data. Given this extra degree of freedom, it is not surprising that, at least in the case of α-synuclein, VBWSS obtains slightly better fits to the experimental data. In addition, we were able to reproduce experimentally measured inter-residue distances that were not included as restraints in the algorithm. The ensemble suggests that αS populates aggregation prone conformations in the disordered state. These results illustrate that the VBWSS algorithm provides an efficient, and accurate, method for constructing models of IDP ensembles.

5. Acknowledgements

This work was supported by NIH Grant 5R21NS063185-02

References

1. A. Huang and C. M. Stultz, Future Medicinal Chemistry **1** (3), 467-482 (2009).
2. C. K. Fisher and C. M. Stultz, Curr Opin Struct Biol **21** (3), 426-431 (2011).
3. A. J. Lees, J. Hardy and T. Revesz, Lancet **373** (9680), 2055-2066 (2009).
4. K. Blennow, M. J. de Leon and H. Zetterberg, Lancet **368** (9533), 387-403 (2006).
5. S. J. Metallo, Curr Opin Chem Biol **14** (4), 481-488 (2010).
6. L. Salmon, G. Nodet, V. Ozenne, G. Yin, M. R. Jensen, M. Zweckstetter and M. Blackledge, J Am Chem Soc **132** (24), 8407-8418 (2010).
7. T. Mittag, J. Marsh, A. Grishaev, S. Orlicky, H. Lin, F. Sicheri, M. Tyers and J. D. Forman-Kay, Structure **18** (4), 494-506 (2010).
8. C. K. Fisher, A. Huang and C. M. Stultz, J Am Chem Soc **132** (42), 14919-14927 (2010).
9. D. Ganguly and J. Chen, J Mol Biol **390** (3), 467-477 (2009).
10. L. Wasserman, Journal of Mathematical Psychology **44** (1), 92-107 (2000).
11. J. T. Ormerod and M. P. Wand, American Statistician **64** (2), 140-153 (2010).
12. J. A. Rice, *Mathematical Statistics and Data Analysis: Third Edition*. (Thomson Higher Education, Belmont, CA, 2007).
13. J. Jensen, Acta Mathematica **30** (1), 175-193 (1906).
14. S. Kullback and R. A. Leibler, Annals of Mathematical Statistics **22** (1), 79-86 (1951).
15. S. Neal, A. M. Nip, H. Zhang and D. S. Wishart, Journal of Biomolecular NMR **26** (3), 215-240 (2003).
16. H. Jeffreys, Proceedings of the Royal Society of London Series a-Mathematical and Physical Sciences **186** (1007), 453-461 (1946).
17. D. M. Blei, K. Franks, M. I. Jordan and I. S. Mian, Bmc Bioinformatics **7**, - (2006).
18. D. M. Endres and J. E. Schindelin, IEEE Transactions on Information Theory **49** (7), 1858-1860 (2003).
19. J. Lin, IEEE Transactions on Information Theory **37**, 145-151 (1991).
20. J. E. Galvin, V. M. Y. Lee and J. Q. Trojanowski, Archives of neurology **58** (2), 186 (2001).
21. M. G. Spillantini and M. Goedert, Annals of the New York Academy of Sciences **920** (THE MOLECULAR BASIS OF DEMENTIA), 16-27 (2000).
22. T. Bartels, J. G. Choi and D. J. Selkoe, Nature **477** (7362), 107-110 (2011).
23. J. N. Rao, Y. E. Kim, L. S. Park and T. S. Ulmer, Journal of molecular biology **390** (3), 516-529 (2009).
24. C. W. Bertoncini, C. O. Fernandez, C. Griesinger, T. M. Jovin and M. Zweckstetter, Journal of Biological Chemistry **280** (35), 30649-30652 (2005).
25. A. Binolfi, R. M. Rasia, C. W. Bertoncini, M. Ceolin, M. Zweckstetter, C. Griesinger, T. M. Jovin and C. O. Fernandez, Journal of the American Chemical Society **128** (30), 9893-9901 (2006).
26. M. Zweckstetter and A. Bax, J. Am. Chem. Soc **122** (15), 3791-3792 (2000).
27. B. R. Brooks, R. E. Bruccoleri, B. D. Olafson, D. J. States, S. Swaminathan and M. Karplus, Journal of Computational Chemistry **4** (2), 187-217 (1983).
28. A. Grupi and E. Haas, Journal Of Molecular Biology **405** (5), 1267-1283 (2011).
29. L. C. Serpell, J. Berriman, R. Jakes, M. Goedert and R. A. Crowther, P Natl Acad Sci USA **97** (9), 4897 (2000).
30. O. M. A. El-Agnaf and G. B. Irvine, Biochemical Society Transactions **30**, 559-565 (2002).
31. O. Ullman and C. M. Stultz, Submitted (2011).

CORRELATION BETWEEN POSTTRANSLATIONAL MODIFICATION AND INTRINSIC DISORDER IN PROTEIN

JIANJIONG GAO* and DONG XU†

Department of Computer Science, C.S. Bond Life Sciences Center, University of Missouri Columbia, Missouri 65211, USA

Protein intrinsic disorder has been shown to play an important role in some posttranslational modifications (PTM). In this paper, we systematically investigated the correlation between protein disorder and dozens of PTMs using data from UniProt/Swiss-Prot and 3-D structures solved by NMR from Protein Data Bank. We observed that many PTMs have a preference for occurrence in disordered regions, including phospho-serine/-threonine/-tyrosine, hydroxylation, sulfotyrosine, S-geranylgeranyl cysteine, deamidated glutamine, 4-carboxyglutamate, 6'-bromotryptophan and most of methylation; while a few PTMs have a preference for occurrence in ordered regions, including 4-aspartylphosphate, S-nitrosocysteine, tele-methylhistidine, FMN conjugation, 4,5-dihydroxylysine, 3-methylthioaspartic acid, most of ADP-ribosylation, and most of FAD attachment. It is also noted that acetyllysine does not show any significant preference for occurrence in either disordered or ordered regions. Further analysis of NMR structures suggested disorder-to-order transitions might be introduced by modifications of phospho-serine/-threonine, mono-/di-/tri-methyllysine, sulfotyrosine, 4-carboxyglutamate, and potentially 4-hydroxyproline. This study sheds light on the functions and mechanisms of various PTMs.

1. Background

Almost all proteins undergo certain chemical modifications on their side chains, called posttranslational modifications (PTM) at some cellular state. Many PTM sites have been shown to occur in disordered regions. For example, it has been reported that phosphorylation was overrepresented in disordered regions;[1,2] the regions containing acetylated and methylated lysines in histone proteins was shown to be disordered;[3] methylated arginine was observed to be enriched in disordered regions;[4] various aspects of ubiquitination process were reported to occur predominately in disordered regions,[5] and protein disorder was suggested to facilitate hydroxylation of proline residues.[6] Large-scale studies on the relationship between protein disorder and PTM were also carried out previously. Pang et al. investigated the correlation between 44 types of PTMs and surface accessibility/disorder.[7] Xie et al. studied the correlations between predicted disorder and PTMs annotated by Swiss-Prot functional keywords, and they reported significant associations between PTMs and predicted protein disorder.[8] We noted that these two large-scale studies were both based on computationally predicted disorder. While most disorder prediction tools achieve high accuracies when predicting long intrinsic disordered

* Current address: Computational Biology Program, Memorial Sloan-Kettering Cancer Center, New York, NY 10065, USA.

† Corresponding author: xudong@missouri.edu.

regions, they typically do not target prediction of short disordered regions. Besides, the current way to define disorder regions (residues with atoms missing in X-ray structures) for training these tools may not be reliable. Therefore, correlations between PTM and disorder implied from these prediction tools may have some bias.

In this study, we systematically investigated the correlations between various PTMs and protein disorder/flexibility. We utilized large-scale PTM annotations from UniProt/Swiss-Prot too but with more specific types and on a finer level than noted in previous studies (e.g. Xie et al). We excluded N/C-terminal modifications (e.g. N-linked acetylation and N-linked methylation, and C-linked amidation) for analysis, since terminal residues in general are inherently more flexible/disordered. In addition, to overcome the limitations of predicted disorder, we explored for the first time NMR 3-D structures in Protein Data Bank (PDB)[9] to study the relationship between PTM and disorder/flexibility. In this regard, we do not treat protein regions with binary states, i.e. order and intrinsic disorder. Instead, we characterize protein order/disorder using a continuous and quantitative measure, i.e. flexibility defined by the distribution of multiple structural models in the same PDB file. By combining results of predicted disorder and NMR structures, PTMs were categorized according to their correlations with protein disorder more reliably.

We also compared the disorder/order state before and after modifications occur. Previous studies suggested disorder-to-order transitions after modifications such as phosphorylation.[10] Here, we further studied the NMR data innovatively approaching this issue in a more systematical way thereby observing modification-induced disorder-to-order transitions for several PTMs.

2. Results

2.1. *Correlation of PTM sites and their predicted disorder scores*

Protein sequences and annotations of known PTMs were retrieved from UniProt/Swiss-Prot (release 2010_09)[11], using a bioinformatics tool, Musite.[2] Disordered regions were predicted for the retrieved proteins by applying a widely used protein disorder prediction tool VSL2B.[12] The disorder prediction scores of PTM sites were then extracted and compared with those of non-PTM sites, as shown in Table 1 with major findings summarized below:

- For phosphorylation, the average predicted disorder scores of phosphoserines, phosphothreonines and phosphotyrosines are significantly greater than those of their unmodified counterparts. However, phosphohistidine and 4-aspartylphosphate have significant lower mean disorder scores than their unmodified counterparts.
- We investigated 17 subtypes of methylation/dimethylation/trimethylation. Ten of them (including all 4 subtypes of dimethylation) have significantly greater mean predicted disorder scores, including cysteine methyl ester, asymmetric dimethylarginine, N6-methyllysine, symmetric dimethylarginine, N6,N6-dimethyllysine, glutamate methyl ester (Glu), omega-N-methylarginine, Omega-N-methylated arginine, leucine methyl ester and N6-methylated lysine. Tele-methylhistine and S-methylcysteine have significantly lower mean disorder scores. N6,N6,N6-trimethyllysine does not have any significant difference in mean disorder score between PTM and non-PTM lysine residues.

Table 1. Comparison of predicted disorder scores between PTM sites and non-PTM sites. Predicted disorder score rages from 0 to 1. Higher score mean larger possibility of being in a disordered region. The p-value for the significance of the disorder score difference is based on two-tailed, two-sample student t-test.

PTM	UniProt keyword	# of PTM sites	Amino acid	Mean disorder score PTM sites		Mean disorder score Non-PTM sites	p-value
Phosphorylation	Phosphoserine	53080	S	0.81	>	0.59	0
Phosphorylation	Phosphothreonine	13477	T	0.69	>	0.49	0
Phosphorylation	Phosphotyrosine	8075	Y	0.46	>	0.35	7.58E-287
Phosphorylation	Phosphohistidine	357	H	0.40	<	0.47	2.28E-14
Phosphorylation	4-aspartylphosphate	529	D	0.20	<	0.51	1.74E-240
Methylation	Cysteine methyl ester	491	C	0.78	>	0.35	6.82E-199
Methylation	N6-methyllysine	499	K	0.78	>	0.56	2.22E-64
Methylation	Glutamate methyl ester	40	E	0.84	>	0.58	8.63E-21
Methylation	Omega-N-methylarginine	98	R	0.81	>	0.52	1.01E-20
Methylation	Leucine methyl ester	14	L	0.69	>	0.41	5.59E-11
Methylation	N5-methylglutamine	10	Q	0.57	>	0.55	9.22E-02
Methylation	Pros-methylhistidine	10	H	0.37	<	0.47	1.07E-01
Methylation	N4-methylasparagine	65	N	0.46	<	0.49	6.14E-02
Methylation	5-methylarginine	11	R	0.46	<	0.53	5.30E-02
Methylation	S-methylcysteine	12	C	0.26	<	0.35	3.14E-02
Methylation	Tele-methylhistidine	47	H	0.12	<	0.47	7.77E-167
Dimethylation	Asymmetric dimethylarginine	385	R	0.90	>	0.52	3.67E-171
Dimethylation	Symmetric dimethylarginine	57	R	0.91	>	0.52	3.76E-25
Dimethylation	N6,N6-dimethyllysine	340	K	0.72	>	0.56	8.13E-25
Dimethylation	Omega-N-methylated arginine	247	R	0.72	>	0.52	1.04E-18
Trimethylation	N6-methylated lysine	374	K	0.69	>	0.56	1.36E-18
Trimethylation	N6,N6,N6-trimethyllysine	696	K	0.57	>	0.56	1.92E-01
Hydroxylation	4-hydroxyproline	900	P	0.89	>	0.60	2.27E-236
Hydroxylation	5-hydroxylysine	179	K	0.97	>	0.56	8.65E-136
Hydroxylation	3-hydroxyproline	11	P	1.00	>	0.60	6.28E-128
Acetylation	N2-acetylarginine	32	R	0.67	>	0.52	3.09E-05
Acetylation	N6-acetyllysine	11436	K	0.54	<	0.56	6.83E-09
ADP-ribosylation	ADP-ribosylcysteine	12	C	0.30	<	0.35	3.79E-01
ADP-ribosylation	ADP-ribosylasparagine	8	N	0.15	<	0.49	1.71E-06
ADP-ribosylation	ADP-ribosylserine	7	S	0.46	<	0.59	4.80E-10
ADP-ribosylation	ADP-ribosylarginine	87	R	0.18	<	0.52	3.26E-78
Deamidation	Deamidated glutamine	43	Q	0.77	>	0.55	4.86E-20
Deamidation	Deamidated asparagine	74	N	0.42	<	0.49	3.80E-02
FAD attachment	Pros-8alpha-FAD histidine	21	H	0.54	>	0.47	4.39E-02
FAD attachment	Tele-8alpha-FAD histidine	36	H	0.33	<	0.47	2.04E-11
FAD attachment	S-8alpha-FAD cysteine	52	C	0.21	<	0.35	2.40E-23
FMN conjugation	FMN phosphoryl threonine	33	T	0.37	<	0.49	3.01E-10
FMN conjugation	S-4a-FMN cysteine	21	C	0.20	<	0.35	6.46E-13
Carboxylation	4-carboxyglutamate	647	E	0.65	>	0.58	1.66E-19
Geranyl-geranylation	S-geranylgeranyl cysteine	29	C	0.81	>	0.35	3.20E-12
Palmitoylation	S-palmitoyl cysteine	155	C	0.29	<	0.35	1.03E-02
S-diacylglycerol cysteine	S-diacylglycerol cysteine	37	C	0.39	>	0.35	2.36E-01
S-Nitrosylation	S-nitrosocysteine	56	C	0.27	<	0.35	8.63E-06
Sulfation	Sulfotyrosine	561	Y	0.58	>	0.35	2.87E-77
Dihydroxylation	4,5-dihydroxylysine	8	K	0.32	<	0.56	3.93E-05
Bromination	6'-bromotryptophan	24	W	0.46	>	0.30	1.21E-03
Beta-methylthiolation	3-methylthioaspartic acid	503	D	0.38	<	0.51	0

- All 3 subtypes of hydroxylation (4-hydroxyproline, 3-hydroxyproline and 5-hydroxylysine) have significantly greater mean disorder scores.
- For acetylation, N2-acetylarginines have much greater mean scores than non-PTM arginine residues. N6-acetyllysine has lower mean scores than non-PTM lysine residues. Although the difference is statistically significant, the absolute value of difference is very small. Therefore, we assume that N6-acetyllysine does not have any significant preference for occurrence in either disordered or ordered regions.
- The residues with all 4 subtypes of ADP-ribosylation have lower mean disorder scores than their unmodified counterparts, but the difference for ADP-ribosylcysteine is not significant.
- For deamidation, deamidated glutamine has a significantly greater mean disorder score, while diamidated asparagine has a significantly lower one.
- For FAD attachment, tele-8alpha-FAD histidine and S-8alpha-FAD cysteine have significantly lower disorder scores.
- For FMN conjugation, both FMN phosphoryl threonine and S-4a-FMN cysteine have significantly lower mean disorder scores.
- 4-carboxyglutamate, S-geranylgeranyl cysteine, 6'-bromotryptophan and sulfotyrosine all have significantly greater mean disorder scores than their unmodified counterparts.
- 4,5-dihydroxylysine and S-nitrosocysteine both have significantly lower mean disorder scores than their unmodified counterparts.
- S-palmitoyl cysteine has a lower mean disorder score although the difference is not large.
- S-diacylglycerol cysteine does not have a significantly different mean disorder score.
- 3-methylthioaspartic acid has a significantly lower mean disorder score.

2.2. *Correlation of PTM sites and their spatial fluctuations in NMR 3-D structures*

We also investigated the flexibility of spatial fluctuations of PTM sites in protein 3-D structures determined by NMR spectroscopy. 7,714 NMR-based protein structures were retrieved from PDB (as of May 4th, 2011). Since the number of modified residues in 3-D structures is limited (as shown in Section 2.3), PTM annotations from UniProt/Swiss-Prot were mapped and aligned onto known protein structures, based on the mapping downloaded from European Bioinformatics Institute (ftp://ftp.ebi.ac.uk/pub/databases/msd/sifts/text/pdb_chain_uniprot.lst). We excluded PTM sites that had any mismatch (i.e. with any different types of amino acids) between UniProt and PDB.

The spatial fluctuation score of a residue among multiple NMR models of the same protein was calculated based on the following equation:

$$F = \sqrt{\frac{\sum_{i=1}^{n}(X_i - \overline{X})^T(X_i - \overline{X})}{n}} \qquad (1)$$

Table 2. Comparison of spatial fluctuations in 3-D structure between PTM sites and non-PTM sites. PTM annotations in UniProt/Swiss-Prot were mapped onto NMR-based 3-D structures. PTMs with one or more sites are included. The p-value for the significance of the fluctuation difference is based on two-tailed, two-sample student t-test.

PTM	UniProt keyword	# of PTM sites	Amino acid	Mean fluctuation (Å) PTM sites		Non-PTM sites	p-value
Phosphorylation	Phosphoserine	585	S	1.88	>	1.16	1.3E-21
Phosphorylation	Phosphothreonine	131	T	1.62	>	1.05	7.0E-5
Phosphorylation	Phosphotyrosine	164	Y	1.10	>	0.91	0.037
Phosphorylation	Phosphohistidine	4	H	2.87	>	1.01	0.224
Phosphorylation	4-aspartylphosphate	6	D	0.57	<	1.10	4.0E-5
Methylation	Omega-N-methylarginine	4	R	3.20	>	1.07	0.021
Methylation	N6-methyllysine	28	K	1.51	>	1.07	0.034
Dimethylation	Symmetric dimethylarginine	3	R	2.30	>	1.07	0.008
Dimethylation	N6,N6-dimethyllysine	13	K	2.11	>	1.07	0.011
Dimethylation	Asymmetric dimethylarginine	4	R	2.16	>	1.07	0.163
Trimethyllysine	N6,N6,N6-trimethyllysine	32	K	1.00	<	1.07	0.689
Hydroxylation	4-hydroxyproline	68	P	0.73	<	1.20	6.4E-13
Acetylation	N6-acetyllysine	395	K	1.02	<	1.07	0.348
Bromination	6'-bromotryptophan	5	W	1.50	>	0.86	0.150
Carboxylation	4-carboxyglutamate	62	E	2.13	>	1.07	7.3E-8
Palmitoylation	S-palmitoyl cysteine	9	C	2.24	>	0.87	0.004
S-Nitrosylation	S-nitrosocysteine	10	C	0.56	<	0.87	1.4E-4
Sulfation	Sulfotyrosine	6	Y	1.02	>	0.91	0.102

where n is the number of models in the PDB file, X_i is a 3-element vector representing the 3-D coordinates of the C-alpha atom for the ith model, and $\overline{X}$ is the mean vector of X_i's. Structures with less than 10 models (n<10) were excluded from the analysis. A residue with a larger spatial fluctuation typically results from sparse spatial restraints derived from nuclear Overhauser effects (NOEs) in NMR spectroscopy. Thus, the residue is more flexible and has higher tendency to be disordered. Conventionally, short mobile regions of proteins may not be considered as intrinsic disorder, which often refers to long protein fragments (>40 residues) that cannot be observed in X-ray crystallography. However, we believe the protein flexibility has a continuous spectrum, from highly rigid, to mobile and then to completely disordered. Therefore, we use the parameter F to characterize protein order/disorder, which can quantify the relationship between PTM and protein disorder continuously. The spatial fluctuations of PTM sites and non-PTM sites were then compared using student's t-test as shown in Table 2. Details are explained below:

- For phosphorylation, phospho-serine, -threonine and -tyrosine have significantly greater mean fluctuation scores than corresponding non-PTM residues; and 4-aspartylphosphate has a significantly lower one. These results are consistent with Table 1.

- The result for phosphohistinine is not consistent with the corresponding comparison in Table 1. The mean fluctuation score of phosphohistidine is greater than non-PTM histidine, mainly because the residue H243 in the structure with PDB accession of 1JOY (corresponding to the residue H243 of Swiss-Prot entry P0AEJ4) has very high fluctuation scores of 4.85 Å (chain A) and 5.11 Å (chain B). The fluctuation scores of residues corresponding to the other two phosphohistidines (H842 in PDB:1SR2 and H58 in PDB:1Y6D) are actually low (0.58 Å and 0.96 Å). Since the two high fluctuation scores may be outliers (which could explain the insignificant *t*-test p-value), one cannot make any inference until more data are available.
- The results for the subtypes of methylation are consistent with Table 1, except that asymmetric dimethylarginine does not have a significant p-value. Omega-N-methylarginine, N6-methyllysine, symmetric dimethylarginine and N6,N6-dimethyllysine have significantly greater mean fluctuation scores than corresponding non-PTM residues. Similar to Table 1, N6,N6,N6-trimethyllysine has a mean fluctuation score very close to non-PTM lysine.
- Again, N6-acetyllysine has a mean fluctuation score that is almost the same as non-PTM lysine, providing more evidence that N6-acetyllysine may not have preference on either disordered or ordered regions.
- 6'-bromotryptophan and 4-carboxyglutamate have significantly greater mean fluctuation scores, which are consistent with Table 1.
- S-nitrosocysteine has a significantly lower mean fluctuation score, which is consistent with Table 1.
- Sulfotysine has a greater mean fluctuation score but not statistically significant. More data are needed for a significance test.
- S-palmitoyl cysteine has a significantly higher mean fluctuation score than non-PTM cysteine, which is inconsistent with Table 1. We noted that the disorder scores for both S-palmitoyl cysteine and corresponding non-PTM sites are relatively low in Table 1 (0.27, and 0.35, respectively). We observed from the structures (e.g. C422 in PDB:1Q68 and C5 in PDB:1SPF) that S-palmitoyl cysteine tends to be in short, highly mobile regions (less than 15 amino acids), which are not considered as disordered regions by protein disorder prediction tools.
- The inconstancy of results for 4-hydroxyproline between Table 1 and Table 2 will be explained in Section 2.3.

2.3. *Spatial fluctuation changes in 3-D structure due to PTM*

Although sparse, there are some modified residues characterized in the NMR-based 3-D structures. To study the possibility of conformational changes after modifications, we separated PTM sites used in Section 2.2 into two groups: one group containing those PTM sites that are actually modified in structures; and the other group containing PTM sites (mapped from UniProt/Swiss-Prot) that are in pre-modified apo-forms in structures. For each type of PTM, we then compared the spatial fluctuations between the two groups, if there were cases available in

Table 3. Comparison of spatial fluctuations of PTM sites in 3-D structure before and after modifications. PTMs with one or more modified residue in NMR-based 3-D structure are included. The p-value for the significance of the fluctuation difference is based on two-tailed, two-sample student t-test.

			Modified in structure		Pre-modified in structure		
PTM	UniProt keyword	Amino acid	# of sites	Mean fluctuation (Å)	# of sites	Mean fluctuation (Å)	p-value
Phosphorylation	Phosphoserine	S	9	0.98	576	1.90	8.2E-4
Phosphorylation	Phosphothreonine	T	15	1.07	116	1.69	0.022
Phosphorylation	Phosphotyrosine	Y	3	0.93	161	1.10	0.447
Methylation	N6-methyllysine	K	10	0.94	18	1.84	0.006
Dimethylation	N6,N6-dimethyllysine	K	5	0.84	8	2.90	2.6E-04
Trimethylation	N6,N6,N6-trimethyllysine	K	12	0.71	20	1.18	0.073
Hydroxylation	4-hydroxyproline	P	65	0.74	3	0.71	0.917
Acetylation	N6-acetyllysine	K	7	0.95	388	1.02	0.442
Sulfation	Sulfotyrosine	Y	4	0.93	2	1.19	0.001
Carboxylation	4-carboxyglutamate	E	49	1.76	13	3.52	0.004

both groups, as shown in Table 3. Interestingly, for all groups except 4-hydroxyproline, the modified residues have lower mean spatial fluctuations than unmodified residues. The differences for phosphoserine, phosphothreonine, N6-methyllysine, N6,N6-dimethyllysine, N6,N6,N6-trimethyllysine, sulfotyrosine and 4-carboxyglutamate are significant. This finding could indicate disorder-to-order transitions triggered by those PTMs. For 4-hydroxyproline, most of the PTM sites (65 out of 68) are modified in the structures (Table 3). The mean spatial fluctuation of the three residues in apo-form is lower than the 65 residues in modified form mainly because one of the three unmodified residues, i.e. P6 in PDB:2H8S, has a very low fluctuation score (0.32 Å). If we assume this residue is an outlier, 4-hydroxyproline could also follow the same disorder-to-order transition, which is also supported by the data in Tables 1 and 2. The disorder scores in Table 1 for pre-modified residues suggest 4-hydroxyprolines are very likely to occur on disordered proline residues, and the spatial fluctuations in Table 2 for post-modified residues suggest that 4-hydroxyprolines' transition are more ordered.

As an example, coagulation factor IX (UniProt accession of P00740) undergoes disorder-to-order transition in 3-D structure after gamma-carboxylation on glutamic acid, as shown in Figure 1. PDB entries 1CFI[13] and 1CFH[14] are two NMR solved structures for residues 1-47 of coagulaiton factor IX. 1CFI is heavily carboxylated, containing 12 4-carboxyglutamates (Figure 1(A)), while none of glutamic acid residues in 1CFH are carboxylated (Figure 1(D)). From the secondary structures (Figures 1(B) and 1(E)), it is obvious that 1CFI is substantially more ordered with increased helical content than 1CFH. From Figures 1(C) and 1(F), the structure fluctuation among NMR models for 1CFI is much lower than that of 1CFH.

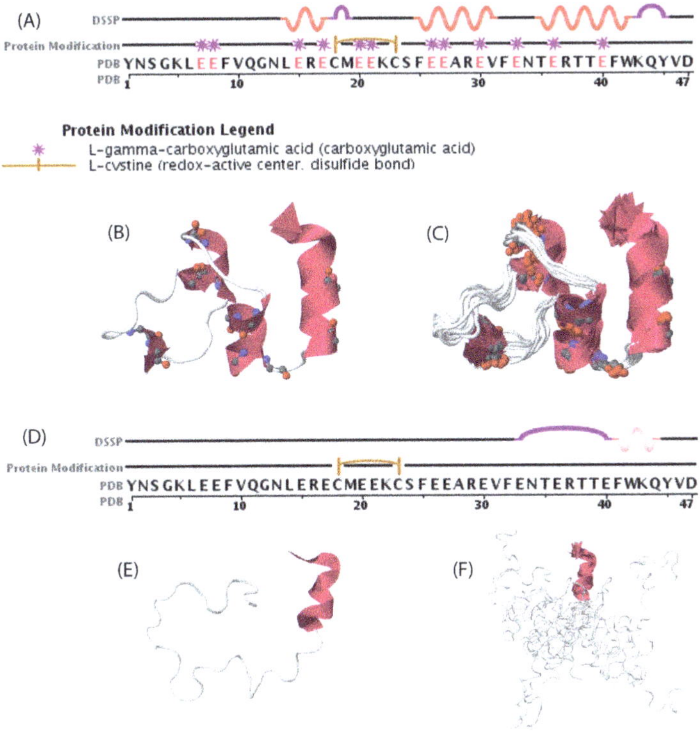

Fig. 1. Disorder-to-order transition after gamma-carboxylation on coagulation factor IX (UniProt accession P00740). (A) Sequence and secondary structures of PDB 1CFI (heavily carboxylated); (B) 3-D visualization of the first model in 1CFI; (C) 3-D visualization of all 17 models overlaid in 1CFI; (D) Sequence and secondary structures of PDB 1CFH (with no carboxylation); (F) 3-D visualization of the first model in 1CFH; (F) 3-D visualization of all 16 models overlaid in 1CFH.

3. Discussion

Based on observed correlations between PTMs and predicted disorder and spatial fluctuation in 3-D structures, we can divide PTMs into three categories: (1) PTMs that have preferences for occurrence in disordered regions; (2) PTMs that have preferences for occurrence in ordered regions; and (3) PTMs that have no obvious preferences for occurrence in either disordered or ordered regions. Some PTMs have positive correlations to both predicted disorder and high spatial fluctuation in NMR structures. Therefore, they have strong preferences for occurrence in disordered regions including phosphoserine, phosphothreonine, phosphotyrosine, omega-N-methylarginine, N6-methyllysine, symmetric dimethylarginine, N6,N6-dimethylarginine, 6'-bromotryptophan, and 4-carboxyglutamate. Some PTMs were observed to be positively correlated to both predicted order and low spatial fluctuation, including 4-aspartylphosphate and S-nitrosocysteine, and thus they are highly overrepresented in ordered regions. Many PTMs have significant positive correlations to predicted disorder only, but with no corresponding NMR structures available including cysteine methyl ester, glutamate methyl easter, N6-methylated lysine, leucine methyl easter, asymmetric dimethylarginine, omega-N-methylated arginine, 5-hydroxylysine, 3-hydroxyproline, N2-acetylarginine, deamidated glutamine, S-geranylgeranyl cysteine, and sulfotyrosine. Some PTMs have positive correlations to predicted order only but are also with no corresponding NMR structures available including tele-methylhistidine, ADP-ribosylasparagine, ADP-ribosylserine, ADP-ribosylarginine, tele-8alpha-FAD histidine, S-8alpha-FAD cysteine, 4,5-dihydroxylysine and 3-methylthioaspartic acid. N6-acetyllysine has no significant correlation to either predicted disorder or spatial fluctuation in 3-D structures. Further analysis of NMR structures also provided evidences of disorder-to-order transitions after modifications of phospho-serine/-threonine, mono-/di-/tri-methyllysine, sulfotyrosine, 4-carboxyglutamate, and potentially 4-hydroxyproline. Disorder-to-order transition could be a general mechanism that many PTMs use to control the functions of proteins. The 4-hydroxyproline residues have high mean predicted disorder but low mean spatial fluctuation. This could be due to disorder-to-order transition after hydroxylation and, therefore, 4-hydroxyproline may still target proline residues predominately in disordered regions.

It is noted that both data of disorder prediction and NMR structures have limitations: predicted disorder may have a certain level of inaccuracy depending on the training data and algorithm, and NMR structures may have bias since the models of some NMR structures may have been selected in an *ad hoc* way by the experimentalists. By combining both data, we hope to better assess the results and hence gain more credibility with consistent results.

Most of the results in this paper are novel findings. It is worth mentioning the differences between this study and previous ones by Pang et al.[7] and Xie et al.[8] These two studies correlated PTMs with predicted disorder/order, while this study departed from tradition to take advantage of NMR structures to verify the correlations and to investigate PTM-induced disorder-to-order transitions. Xie et al. only reported results for general types of PTMs (e.g. methylation and phosphorylation). In contrast, we investigated PTMs with many specific types and subtypes. We found that subtypes of PTMs (e.g. N6-methyllysine and tele-methylhistidine) in the same general type (e.g. methylation) could have different correlations to disorder regions. All these findings provided useful insight into the mechanisms of various PTMs and may facilitate further investigations into the structural and functional implications of these PTMs.

Acknowledgments

We like to thank Dr. A. Keith Dunker for helpful discussions. This work was supported by the funding from the National Institute of Health (grant number R21/R33 GM078601) and National Science Foundation (grant number DBI-0604439).

References

1. L.M. Iakoucheva, P. Radivojac, C.J. Brown, T.R. O'Connor, J.G. Sikes, Z. Obradovic and A.K. Dunker, *Nucleic Acids Res.* **32(3)**, 1037 (2004).
2. J. Gao, J.J. Thelen, A.K. Dunker and D. Xu, *Mol. Cell. Proteomics.* **9(12)**, 2586 (2010).
3. J.C. Hansen, X. Lu, E.D. Ross and R.W. Woody, *J. Biol. Chem.* **281(4)**, 1853 (2006).
4. K.M Daily, P. Radivojac, and A.K. Dunker. *IEEE Symposium on Computational Intelligence in Bioinformatics and Computational Biology.* 475 (2005).
5. P. Radivojac, V. Vacic, C. Haynes, R.R. Cocklin, A. Mohan, J.W. Heyen, M.G. Goebl and L.M Iakoucheva. *Proteins.* **78(2)**, 365 (2010).
6. W.C. Hon, M.I. Wilson, K. Harlos, T.D.W. Claridge, C.J. Schofiel, C.W. Pug, P.H. Maxwell, P.J. Ratcliffe, D.I. Stuart and E.Y. Jones, *Nature.* **417(6892)**, 975 (2002).
7. C.N.I Pang, A. Hayen, and M.R. Wilkins, *J Proteome Res.* **6(5)**, 1833 (2007).
8. H. Xie, S. Vucetic, L.M. Iakoucheva, C.J. Oldfield, A.K. Dunker, Z. Obradovic and V.N. Uversky, *J Proteome Res.* **6(5)**, 1917 (2007).
9. P.W. Rose, B. Beran, W.F. Bluhm, D. Dimitropoulos, D.S. Goodsell, A. Prlic, M. Quesada, G.B. Qulinn, J.D. WestBrook, J. Young, B. Yukich, C. Zardecki, H.M. Berman and P.E. Bourne, *Nucleic Acids Res.* **39**, D392 (2011).
10. M.I. Lilia, P. Radivojac, C.J. Brown, T.R. O'Connor, J.G. Sikes, Z. Obradovic, and A.K. Dunker, *Nucleic Acids Res.*, **32(3)**, 1037 (2004).
11. N. Farriol-Mathis, J.S. Garavelli, B. Boeckmann, S. Duvaud, E. Gasteiger, A. Gateau, A.L. Veuthey and A. Bairoch, *Proteomics.* **4(6)**, 1537 (2004).
12. Z. Obradovic, K. Peng, S. Vucetic, P. Radivojac and A.K. Dunker, *Proteins.* **61(S7)**, 176 (2005).
13. S.J. Freedman, B.C. Furie, B. Furie and J.D. Baleja, *J. Biol. Chem.* **270**, 7980 (1995).
14. S.J. Freedman, B.C. Furie, B. Furie and J.D. Baleja, *Biochemistry.* **34**, 12126 (1995).

INTRINSIC DISORDER WITHIN AND FLANKING THE DNA-BINDING DOMAINS OF HUMAN TRANSCRIPTION FACTORS

XIN GUO[1], MARTHA L. BULYK[2†], and ALEXANDER J. HARTEMINK[1†]

[1]*Department of Computer Science, Duke University, Box 90129, Durham, NC 27708-0129, USA*
E-mail: {xinguo,amink}@cs.duke.edu

[2]*Division of Genetics, Department of Medicine; Department of Pathology; Brigham and Women's Hospital and Harvard Medical School, Boston, MA 02115, USA*
Harvard-MIT Division of Health Sciences and Technology, Harvard Medical School, Boston, MA 02115, USA
E-mail: mlbulyk@receptor.med.harvard.edu

†*Correspondence should be addressed to M.L.B. and A.J.H.*

While the term 'protein structure' is commonplace, it is increasingly appreciated that proteins may not possess a single, well-defined structure: some regions of proteins are intrinsically disordered. The role these intrinsically disordered regions (IDRs) play in protein function is an area of significant interest. In particular, because proteins containing IDRs are largely involved in processes related to molecular recognition, the question arises whether IDRs are important in these recognition events. It has been observed that IDRs are enriched in transcription factors (TFs) in comparison with other proteins, and we sought to explore this enrichment more precisely, with an eye toward functional dissection of the prevalence and locations of IDRs in different classes of TFs. Specifically, we considered the occurrences of 76 classes of DNA-binding domains (DBDs) within a comprehensive set of 1,747 human, sequence-specific TFs. For each DBD class, we analyzed whether a significant level of disorder was present within the DBD itself, the N-terminal or C-terminal sequence flanking the DBD, or both flanking sequences. We found that although the DBDs themselves exhibit significant order, the regions flanking the DBDs exhibit significant disorder, which suggests a functional role for such IDRs in TF DNA binding. These results may have important implications for studies of TFs not just in human but across all eukaryotes, and suggest future studies focused on testing the roles of N- and C-terminal flanking regions in determining or modulating the DNA binding affinity and/or specificity of the associated TFs.

Keywords: intrinsically disordered region, DNA-binding domain, transcription factor

1. Introduction

The function of a protein is encoded in its amino acid sequence (*i.e.*, primary structure). However, protein activity typically depends on the protein being folded properly into its component secondary structure elements (*e.g.*, alpha helices, beta sheets) and the overall, global conformation of the protein (*i.e.*, tertiary structure). Protein structure can be determined experimentally at high resolution either by X-ray crystallography or by nuclear magnetic resonance (NMR). X-ray crystallography is widely used, but cannot provide information on the conformation of regions that are either highly dynamic or unstructured in the crystal. In contrast, while NMR can provide information about flexibility and dynamics in proteins, it is currently limited to smaller proteins.

Through a combination of structural and biochemical studies, it has become increasingly appreciated that a protein may not adopt a single, well-defined 'structure', a term connoting a measure of rigidity. Rather, a protein may sample an ensemble of global conformations; parts of the protein may be largely constantly structured across this ensemble, while other parts may be quite variable or flexible across the ensemble. These latter regions are sometimes termed 'intrinsically disordered regions' (IDRs), though they may adopt a more structured conformation upon interaction with another molecule, whether a protein, DNA, or other ligand [1].

Proteins are largely involved in processes related to molecular recognition (*e.g.*, binding, signaling, complex formation, enzymatic catalysis), and IDRs may enable these recognition events either directly (*e.g.*, serving as the recognition domain of a protein) or indirectly (*e.g.*, serving as a hinge that allows two ordered regions of a protein to come together to effect recognition). For this reason, IDRs have been studied rather extensively over the past decade, and a large number of computational methods have been developed for the prediction of IDRs on the basis of amino acid sequence, though this remains an imperfect art (see [2] for a review).

In this study, we were interested in exploring the role(s) that IDRs might play in the recognition tasks of transcription factors (TFs) in particular. Computational explorations have found that IDRs are generally more prevalent in TFs than would be expected by chance, especially in eukaryotes [3–5]. As a specific example, careful molecular studies have shown that a region of fifteen amino acids within the DNA-binding domain (DBD) of the estrogen receptor (ER) is disordered in solution, and makes contacts with DNA (and with another ER DBD monomer), as shown in a co-crystal structure of the ER DBD bound to DNA [6]. Moreover, IDRs outside the homeodomain DBD have also been found to impact the DNA-binding affinity of the *Drosophila* TF Ubx [7]. In addition, the region N-terminal to the proximal accessory region of the *Saccharomyces cerevisiae* C2H2 zinc finger TF Adr1 is disordered in solution (even after binding DNA) and increases the affinity for non-specific DNA, mainly by increasing the DNA association rate; increased affinity for non-specific DNA might allow a protein to find its specific sites more quickly after translocation from non-specific sites that are bound initially [8]. Finally, DBDs often have N- or C-terminal extensions, referred to as 'arms' or 'tails', that bind DNA but are disordered when free in solution [9]. Intrigued by this ensemble of findings pointing to the importance of IDRs in TFs and their interactions with DNA, we sought to explore the connection between IDRs and TF function more precisely and systematically. We were particularly interested in determining whether IDRs were more prevalent in the regions flanking the DBDs that are responsible for the binding of sequence-specific TFs to DNA.

2. Materials and Methods

2.1. *Constructing the TF and non-TF control sets of proteins*

We created two non-redundant datasets of human proteins: a TF set and a non-TF set for use as a control. The procedure for constructing these sets and ensuring their non-redundancy is described below and summarized in Figure 1A.

We assembled the TF set from a published repertoire of human TFs [10]. In their study, Vaquerizas and colleagues manually curated and identified 1,987 TF-coding human genomic loci in the Ensembl database [11]; the list includes 1,960 high-confidence entries and 27 entries

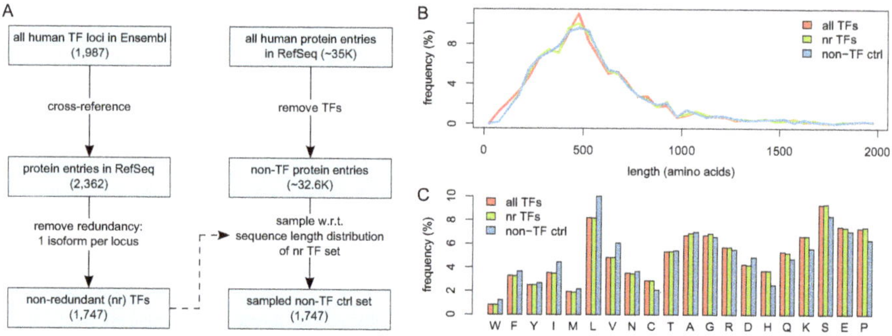

Fig. 1. (A) A schematic of the pipeline for generating the TF set and the non-TF control set. (B) Sequence length distributions of the TF set (nr TFs), the non-TF control set, and the set of all human TFs (with redundancy). (C) The amino acid compositions of the TF set, the non-TF control set, and the set of all human TFs. Amino acids are listed from most order-promoting to most disorder-promoting, according to [13]. It is apparent from the histogram that compared to proteins in general, TFs have fewer order-promoting residues (*e.g.*, W, F, Y, I, M, L, V) and more disorder-promoting residues (*e.g.*, P, E, S, K, Q, H).

curated as probable. We cross-referenced these Ensembl loci against the RefSeq database (release 47) [12] to obtain 2,362 protein isoforms associated with 1,747 genes. To reduce sequence redundancy and thus potential bias, if multiple isoforms were associated with the same gene, we selected only the longest. This resulted in a final total of 1,747 unique TF protein sequences, and in subsequent analysis, we call this our *TF set.*

We assembled our *non-TF control set* by downloading all human proteins from RefSeq, and excluding the 2,362 TF-associated isoforms from above, which yielded a total of 32,567 non-TF proteins. To match the size and sequence length distribution of our TF set, we randomly sampled 1,747 proteins from the 32,567 according to the empirical sequence length distribution of the TF set; to ensure non-redundancy during this process, at each iteration we required that the sampled protein come from a locus not previously sampled. Therefore, the resulting control set contains 1,747 unique non-TF protein sequences.

2.2. *Comparing the TF and non-TF control sets of proteins*

To ensure that the non-TF set represents a well-constructed control for the TF set, we compared various properties of the two sets. First, we compared the sequence length distributions of the TF set and the non-TF control set, in addition to the set of all human TFs (*i.e.*, with redundancy). As shown in Figure 1B, no apparent differences exist between the sequence length distributions in the TF set, the non-TF control set, and the set of all human TFs.

Next, we compared the amino acid compositions of the TF set, the non-TF control set, and the set of all human TFs (Figure 1C). The amino acid composition of sequences in IDRs has been shown to be significantly different from that in ordered regions [14], and IDRs have been shown to have high prevalence in TFs [4], so we might expect compositional differences between the TF sets and the non-TF control set. Indeed, compared to the non-TF control set, both TF sets are enriched in disorder-promoting amino acids (*e.g.*, P, E, S, K, Q, H), and depleted in order-promoting amino acids (*e.g.*, W, F, Y, I, M, L, V) [13, 14], as expected. However, the amino acid compositions of our non-redundant TF set and the set of all human TFs are nearly identical, suggesting that our procedure for removing redundancy introduces no significant compositional bias.

2.3. *Identifying DNA-binding domains (DBDs) and their locations within proteins*

Our goal is to investigate the prevalence and locations of IDRs within human TFs, and in particular, the spatial relationships between IDRs and DBDs in TFs. To identify all sequence-specific DBDs that occur within human TFs, we started with the entire set of human proteins from RefSeq and identified every Pfam domain [15] that was contained in a human protein with a p-value below 0.05. We manually filtered for those domains whose text descriptions in the Pfam or InterPro [16] databases indicated that the domain mediates sequence-specific DNA binding, resulting in 76 domains which we henceforth call *Pfam DBDs*.

Using HMMER [17] with default parameters, we searched for the locations of matches to Pfam DBDs within our TF set. We found 71 of the 76 Pfam DBDs matched to proteins in our TF set, with 32 DBDs appearing more than five times. Of the 1,747 proteins in our TF set, 669 contained only a single DBD, while another 642 contained multiple DBDs; proteins with multiple DBDs are typically those containing multiple zinc fingers, which are annotated as separate domains even if they occur in tandem within a protein. Indeed, the TF with the highest number of DBDs is zinc finger protein 91 (RefSeq: NP_003421), which contains 31 zf-C2H2 (zinc finger, C2H2-type) domains. The zf-C2H2 domain is interesting in its own right as it is by far the most prevalent domain in our TF set, appearing a total of 4,154 times, almost 20 times as often as the next most prevalent domain.

2.4. *Predicting intrinsically disordered regions (IDRs) and their locations within proteins using multiple existing methods*

To perform our analysis, we first needed to predict the ordered and disordered regions within proteins using existing computational tools. Since this remains a bit of an imperfect art, we took care to ensure that our conclusions would not be overly dependent on the predictions of any single choice of method. Consequently, we chose to use three distinct disorder prediction tools, each demonstrated to perform with high accuracy [2]: PONDR VSL2 [18], DISOPRED2 [19], and PreDisorder 1.1 [20].

PONDR VSL2 (also called DisProt VSL2) was evaluated as the top-ranked disorder predictor in CASP7 in 2006 [21], and PreDisorder was ranked among the top methods in disorder prediction during CASP8 in 2008 and CASP9 in 2010. These methods employ a variety of techniques to analyze sequence and structural information for IDR prediction: PONDR VSL2 uses support vector machines (SVMs) to separately address prediction problems in short versus long sequence regions, and then merges the results using a logistic regression model; DISOPRED2 is also based on SVMs, and compared to other prediction methods, the main difference is that it is directly trained on the whole sequence using various combinations of binary-encoded amino acid sequence, secondary structure predictions, and sequence profiles; and PreDisorder 1.1 is based on an *ab initio* prediction method along with a meta-prediction method.

2.5. *Defining disorder features: spatial relationships of IDRs relative to DBDs within TFs*

Given the annotated DBDs and the predicted disorder regions in the TF set and the non-TF control set, we sought to systematically analyze the association between TF DBDs and predicted IDRs by testing for enrichment of IDRs at different locations relative to DBDs. Specifically, we were interested in IDRs within the DBD itself, as well as the regions flanking the DBD, and we developed five distinct 'disorder features': we say that *a DBD is disordered* if at least a fraction f of its residues are predicted to be disordered; we say that *the N-terminal flank of a DBD is disordered* if at least a fraction f of the 30 residues flanking the DBD in the N-terminal direction are predicted to be disordered; analogously, we say that *the C-terminal flank of a DBD is disordered* if at least a fraction f of the 30 residues flanking the DBD in the C-terminal direction are predicted to be disordered; we say that *both flanks of a DBD are disordered* if both the N-terminal and C-terminal flanks are disordered; and finally, we say that *an entire TF is disordered* if at least a fraction f of all of its residues are disordered. We wanted to be fairly stringent in identifying these disorder features, so that we could focus on those with the highest confidence; therefore, we chose the value of 0.8 for f.

2.6. *Calculating statistical significance of disorder features*

To assess whether the prevalence of disorder features within and flanking DBDs was unusually high or low, we needed to determine a suitable measure of significance. Moreover, since different computational tools predict IDRs at different rates (see Section 3.2 below), our significance measure needed to enable the comparison of results across methods, and not be biased by methods that are systematically more or less likely to predict disorder within proteins.

We thus developed two different null models to test for the significance of our disorder features (*e.g.*, disordered DBD, N-terminal flank, or C-terminal flank). The first null model pretended that the location of a DBD occurred uniformly at random within each sequence, and was based on the TF set. The second null model also pretended that the location of a DBD occurred uniformly at random in each sequence, but was based on the non-TF control set. In summary, these two null models—in which the location of a DBD was chosen uniformly at random—were designed to test whether the spatial relationships between IDRs and DBDs were statistically significant or simply occurred by chance.

With each null model providing a baseline expectation for how often a disorder feature might be found by chance, we could then compute a significance measure based on the p-value from a hypergeometric distribution (*i.e.*, Fisher's exact test). For each disorder feature we considered, we computed two separate p-values, one for each null model. Consistency of significance across the two different null models thus gave us some confidence that our results were robust to the specific choice of null model.

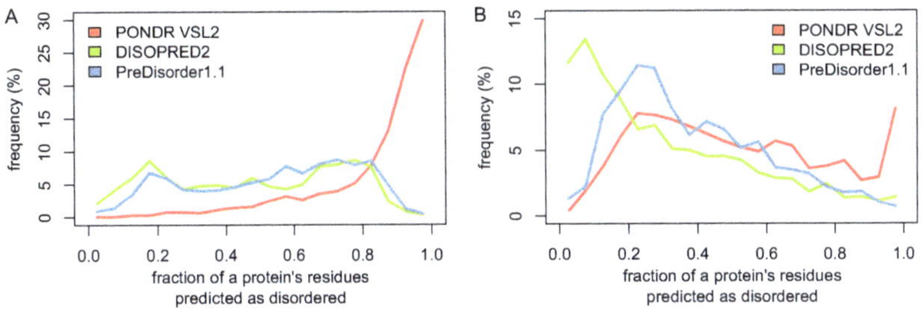

Fig. 2. Distributions of the fraction of each protein's residues predicted as disordered by each method for the proteins in (A) the TF set and (B) the non-TF control set.

Table 1. Statistics summarizing disorder predictions on all the residues of all the proteins in both the TF set and the non-TF control set using three different disorder prediction tools.

	TF set		**non-TF ctrl set**	
	% of residues predicted in IDRs	average length of IDRs	% of residues predicted in IDRs	average length of IDRs
PONDR VSL2	83.2%	106	53.3%	39
DISOPRED2	47.4%	44	34.1%	36
PreDisorder 1.1	50.1%	19	38.3%	18

3. Results

3.1. *Comparing the three methods for predicting IDRs within proteins*

We used three different disorder prediction tools to predict IDRs in both the TF set and the non-TF control set. Though the purpose of this paper is to make use of existing prediction methods and not to evaluate them (which has already been done by others, for example [2, 21]), it is important to have at least an overall sense of how each method is performing on our protein sets. A summary of the results of the three methods is shown in Figure 2 and listed in Table 1. In Figure 2, we compare the fraction of each protein's residues predicted as disordered by each method. In Table 1, we calculate the total percentage of protein residues predicted as disordered by each method, along with the average length of each predicted IDR. The figure and table reveal that all three methods consistently predict proteins in the TF set to have a greater fraction of disordered residues, more disordered residues, and longer IDRs than proteins in the non-TF control set, confirming earlier findings that IDRs are enriched in TFs.

As an aside, it is apparent that PONDR VSL2 is far more likely than the other two methods to call a residue as disordered, in both the TF set and the non-TF control set, suggesting that the method is probably operating at a different point on its receiver operating characteristic (ROC) curve, with high sensitivity but also perhaps a relatively high false positive rate [21]. In addition, the average length of IDRs predicted by PONDR VSL2 is higher than the other two methods, which may be related to the previous point, but may also be because the method uses different SVMs to predict IDRs in short and long sequences separately.

3.2. *Assessing significance of order or disorder within and flanking human TF DBDs*

To systematically study the associations between IDRs and DBDs, for each occurrence of a DBD class within a human TF, we calculated 30 different p-values: the significance under two different null models (based on the TF set and the non-TF control set) of five different kinds of disorder features (DBD, N-terminal flank, C-terminal flank, both flanks, and entire TF) as computed by three different prediction methods (PONDR VSL2, DISOPRED2, and PreDisorder 1.1). For each combination of null model and feature, we say that *the feature exhibits significant disorder under that null model* if at least two of the three prediction methods predict disorder at p-value ≤ 0.005; on the other hand, we say that *the feature exhibits significant order under that null model* if at least two of the three prediction methods predict disorder at p-value ≥ 0.995. Note that it is certainly possible for a feature to be neither significantly ordered nor significantly disordered under a particular null model.

Although we computed whether features exhibited significant order or disorder across all Pfam DBDs occurring in our TF set, to avoid artifacts due to small sample size, we restricted our subsequent analysis to the 32 DBD classes with at least five occurrences in the TF set. Many of the most frequent DBD classes, including the 10 most prevalent ones, are structurally similar and can be roughly classified into two groups: (1) those containing zinc fingers, and (2) those containing a basic helix-turn-helix type of domain, domains in which helices are separated by loops (*e.g.*, Homeobox, HLH, Fork_head, Ets). The enrichment analysis results for these 32 DBD classes are listed in Table 2; at the bottom of the table, we also included the Pfam domains Basic, AT_hook, and P53 (Basic and AT_hook are included because we mention them below in comparison to another study; P53 is a well-studied DBD included for general interest).

The top 10 most frequently occurring DBD classes in human TFs all exhibit significant order within the DBD itself, suggesting that structural flexibility within these domains is rather limited. Strikingly, our results indicate that although the DBDs themselves exhibit significant order, the regions flanking the DBDs are likely to exhibit significant disorder. Only in the case of zf-C2H2 do the flanking regions exhibit significant order (this will be discussed further in the next section). In contrast, 26 of the other 31 DBDs exhibit significant disorder in either the N-terminal flank, the C-terminal flank, or both; and none of the other 31 DBDs exhibit significant order in either flank under either null model. This is consistent with prior studies in which it was found that DBDs are often separated by flexible linker regions, allowing TFs to bind DNA with fine control over DNA binding affinity [22, 23].

3.3. *Investigating detailed spatial relationships of IDRs relative to DBDs within TFs*

To further investigate the detailed spatial relationships of the IDR predictions of the three different methods to protein DBDs, we generated a meta-plot of the average predicted order/disorder in the vicinity of each Pfam DBD according to each prediction method. To do this, we first identified all occurrences of a Pfam DBD in the TF set, and then across all those occurrences, calculated the average (mean) order/disorder score predicted by each method at

Table 2. Enrichment analysis of significantly occurring ordered and disordered regions within and flanking human TF DBDs.

					DBD		N-terminal flank		C-terminal flank		both flanks		whole TF sequence	
No.	**DBD (Pfam)**	**TF family**	**average DBD length (res.)**	**number of DBDs in TFs**	TF set	non-TF ctrl set	TF set	non-TF ctrl set	TF set	non-TF ctrl set	TF set	non-TF ctrl set	TF set	non-TF ctrl set
1	PF00096	zf-C2H2	23.1	4154	**OR**	**OR**	**OR**	**OR**	**OR**	**OR**	**OR**	**OR**	**OR**	**OR**
2	PF00046	Homeobox	56.3	216	**OR**	**OR**	**ID**	**ID**	**ID**	**ID**	**ID**	**ID**	**ID**	**ID**
3	PF00010	HLH	53.3	100	**OR**	**OR**	**ID**	**ID**	**ID**	**ID**	**ID**	**ID**	–	**ID**
4	PF00505	HMG_box	68.0	56	**OR**	**OR**	**ID**	**ID**	**ID**	**ID**	**ID**	**ID**	–	**ID**
5	PF00250	Fork_head	98.2	47	**OR**	**OR**	**ID**	**ID**	**ID**	**ID**	**ID**	**ID**	**ID**	**ID**
6	PF00105	zf-C4	70.2	45	**OR**	**OR**	**ID**	**ID**	**ID**	**ID**	**ID**	**ID**	**OR**	–
7	PF00249	Myb_DNA-binding	47.2	43	**OR**	**OR**	–	–	–	**ID**	–	–	–	–
8	PF00170	bZIP_1	64.2	34	**OR**	**OR**	**ID**	**ID**	–	–	**ID**	**ID**	**ID**	**ID**
9	PF00178	Ets	85.0	27	**OR**	**OR**	**ID**	**ID**	**ID**	**ID**	**ID**	**ID**	–	–
10	PF00320	GATA	35.1	20	–	**OR**	**ID**	**ID**	**ID**	**ID**	**ID**	**ID**	–	**ID**
11	PF00907	T-box	187.6	18	–	–	**ID**	**ID**	**ID**	**ID**	**ID**	**ID**	–	–
12	PF01530	zf-C2HC	31.0	14	**ID**	**ID**	**ID**	**ID**	**ID**	**ID**	**ID**	**ID**	**ID**	**ID**
13	PF02319	E2F_TDP	73.8	13	–	–	**ID**	**ID**	–	–	–	**ID**	–	–
14	PF00313	CSD	68.6	12	–	–	–	–	–	–	–	–	–	–
15	PF05485	THAP	89.2	12	–	–	**ID**	**ID**	**ID**	**ID**	**ID**	**ID**	–	–
16	PF01422	zf-NF-X1	21.5	11	–	–	–	–	–	–	–	–	–	–
17	PF03165	MH1	109.9	11	–	–	**ID**	**ID**	**ID**	**ID**	**ID**	**ID**	–	–
18	PF07716	bZIP_2	54.0	10	–	–	**ID**	**ID**	–	–	**ID**	**ID**	–	–
19	PF00292	PAX	125.6	9	–	–	**ID**	**ID**	**ID**	**ID**	**ID**	**ID**	–	–
20	PF00098	zf-CCHC	17.9	8	–	–	–	**ID**	–	**ID**	–	**ID**	–	–
21	PF00808	CBFD_NFYB_HMF	63.1	8	–	–	–	**ID**	–	–	–	–	–	–
22	PF04218	CENP-B_N	52.5	8	–	–	–	–	–	–	–	–	–	–
23	PF00751	DM	47.0	7	–	–	**ID**	**ID**	**ID**	**ID**	**ID**	**ID**	–	**ID**
24	PF01342	SAND	79.0	7	–	–	**ID**	**ID**	**ID**	**ID**	**ID**	**ID**	–	–
25	PF02257	RFX_DNA_binding	72.7	7	–	–	**ID**	**ID**	–	–	–	–	–	–
26	PF02864	STAT_bind	251.9	7	–	–	–	–	–	–	–	–	–	–
27	PF02892	zf-BED	50.1	7	–	–	**ID**	**ID**	**ID**	**ID**	**ID**	**ID**	–	–
28	PF10401	IRF-3	174.0	7	–	–	–	–	–	–	–	**ID**	–	–
29	PF00447	HSF_DNA-bind	104.2	6	–	–	–	–	**ID**	**ID**	**ID**	**ID**	–	–
30	PF04516	CP2	227.2	6	–	–	–	–	–	–	–	–	–	–
31	PF03299	TF_AP-2	208.2	5	–	–	–	**ID**	–	–	**ID**	**ID**	–	–
32	PF05044	Prox1	224.0	5	–	–	**ID**	**ID**	–	–	**ID**	**ID**	–	–
	PF01586	Basic	91.0	4	**ID**	**ID**	**ID**	**ID**	**ID**	**ID**	**ID**	**ID**	**ID**	**ID**
	PF00870	P53	196.3	3	–	–	**ID**	**ID**	**ID**	**ID**	**ID**	**ID**	–	–
	PF02178	AT_hook	109.0	1	**ID**	**ID**	**ID**	**ID**	**ID**	**ID**	**ID**	**ID**	**ID**	**ID**

Notes:
The DBDs with at least 5 occurrences in the TF set are listed in the table, together with Basic, P53, and AT_hook.
ID indicates significant disordered DBDs, DBD flanks, or TFs (in at least two of three methods, p-value ≤ 0.005).
OR indicates significant ordered DBDs, DBD flanks, or TFs (in at least two of three methods, p-value ≤ 0.005).
A dash (–) indicates entries that are neither significantly ordered nor significantly disordered.
The DBDs with fewer than 5 occurrences in the TF set include: Runt, TEA, Basic, HALZ, z-alpha, FYRN, FYRC, P53, ARID, DMA, AKAP95, GATA-N, P53_tetramer, Homez, XPA_N, zf-DHHC, GCM, CG-1, Vert_HS_TF, SIM_C, Rad51, HAND, Beta-trefoil, LAG1-DNAbind, PWI, zf-MYND, SAP, GCR, Oest_recep, Prog_receptor, zf-TRAF, zf-CHY, Vert_IL3-reg_TF, HSA, Rio2_N, BrkDBD, zf-RAG1, AT_hook, and TMF_DNA_bd.

each residue within the DBD match and both of its flanks (up to 30 amino acids). In cases where a TF contained only a partial DBD match and not a full domain according to the HMMER alignment, we considered only the aligned region in our calculations. We normalized the resulting scores for the purpose of comparison across methods, and for uniformity in scale across plots for different DBD classes (Figure 3).

Figure 3 displays meta-plots for five of the ten Pfam DBDs most prevalent in human TFs. Results from DISOPRED2 and PreDisorder 1.1 are fairly consistent across all five domain classes. Moreover, all three methods are in good agreement in zf-C2HC and demonstrate similar prediction trends in zf-C4, Homeobox, and HLH. Extended to all the DBDs listed in Table 2, over 67.2% of the DBD classes that are found to exhibit either significant disorder or significant order are identified as such by all three methods.

Nevertheless, some discrepancies in the results from the different methods are evident, such as zf-C2H2. The C2H2-type zinc finger domain is the most prevalent DBD class found in metazoan TFs, including in human [24]. It is also one of the most highly ordered DBDs; however, the linker regions between these C2H2 zinc finger domains are often disordered [25]. As shown in Figure 3A, PONDR VSL2 reports that the C2H2 domain occurrences in human TFs exhibit significant disorder in both the C2H2 domain itself and the adjacent N- and C-terminal flanks; however, DISOPRED2 and PreDisorder both report the opposite, namely that zf-C2H2 and its flanks exhibit significant order. Liu and colleagues carefully analyzed the difficulties of predicting intrinsic disorder in the zf-C2H2 domains and their linker regions [4]. They concluded that because many linker regions between C2H2 zinc fingers are quite short, the windowing procedures employed by some IDR prediction algorithms prevent them from being detected as disordered; the result is an artifact in which linker regions between C2H2 zinc fingers are over-predicted as being ordered.

3.4. *Analyzing spatial relationships for some DBD classes prevalent in human TFs*

3.4.1. *Zinc fingers*

Zinc fingers are small structural motifs whose folds are stabilized by coordination of one or more zinc ions. Zinc fingers can be classified according to their zinc-coordinating residues and folds. In Figures 3A-C, we show our IDR prediction results for the three major zinc finger domain classes found in human TFs: zf-C2H2 (the most prevalent DBD class in human TFs), zf-C4 (also referred to as nuclear receptors), and zf-C2HC. Although all three classes contain zinc fingers, we find variability in their regions of order and disorder. As discussed above, the C2H2 zinc finger domain is itself believed to be highly ordered, with individual ordered zinc fingers separated by highly flexible linker regions [25]. We find that the C4 domain exhibits significant order within the DBD itself, but significant disorder in flanking regions. In contrast, we find that the C2HC domain exhibits significant disorder in both the DBD and flanking regions.

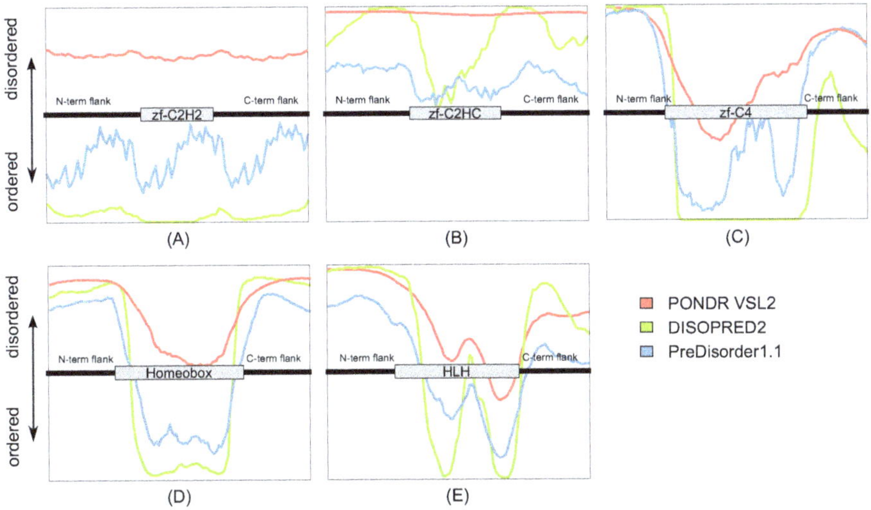

Fig. 3. Shown are meta-plots for five prevalent DBDs in human TFs. (A) zinc-finger C2H2-type (length: ~23 amino acids), (B) zinc-finger C2HC-type (length: ~31 amino acids), (C) zinc-finger C4-type (length: ~70 amino acids), (D) homeodomain fold (length: ~58 amino acids), and (E) helix-loop-helix (length: ~53 amino acids).

3.4.2. *Homeobox*

Homeobox (homeodomain fold) is the second-most abundant DBD class within human TFs. The homeodomain fold consists of an approximately 60 amino acid helix-turn-helix structure in which three alpha helices are connected by short loop regions. Our results (Figure 3D) extend the results of a prior study [7] that found multiple intrinsically disordered sequences located outside the homeodomain DBD of the *Drosophila* TF Ubx, that allow Hox family members (*i.e.*, a subclass of TFs with Homeobox DBDs) to bind DNA with high affinity but relatively low specificity [26, 27].

3.4.3. *HLH*

HLH (basic helix-loop-helix) is the third-most abundant DBD class within human TFs, and is characterized by two α-helices connected by a loop. TFs that have this domain typically bind DNA as either homo- or hetero-dimers, with each monomer contacting DNA through a helix containing basic residues that facilitate DNA binding [28]. As shown in Figure 3E, all three methods report that HLH exhibits significant order within the domain itself, but significant disorder in both the N- and C-terminal flanking regions. Our results also indicate that a short but highly disordered region may frequently occur in the middle of the HLH domain, consistent with prior observations that the linker regions and the loop region of HLH proteins are of higher flexibility, allowing dimerization by folding and packing one smaller helix against the other one [28].

4. Discussion

In this study, we used three different computational disorder prediction methods to investigate the prevalence of IDRs within DBDs and in their flanking regions across essentially the entire repertoire of human, sequence-specific TFs and their associated Pfam DBDs. Our choice of multiple prediction methods was motivated by a desire to be able to draw robust conclusions that were not dependent on any one particular method.

Previously it was found that TFs are enriched for IDRs [3, 4]. At the same time, DBDs responsible for TF binding did not seem themselves to be particularly enriched for IDRs. For example, of the 25 DBDs studied in [4], only the Basic and AT_hook domains exhibited high amounts of disorder; however, those domains are not particularly prevalent in human TFs, occurring in our TF set just four times and one time, respectively.[a] We were intrigued by the possibility that the enrichment of IDRs observed in TFs might be at least partly due to disorder in the regions flanking DBDs; under such a hypothesis, DBDs can be thought of as islands of order flanked by regions of disorder.

Our results support exactly such a hypothesis: the most prevalent DBDs in human TFs exhibit significant order, but the flanking regions of these DBDs generally exhibit significant disorder. Similarly, although DBDs of intermediate prevalence (occurring between 5 and 20 times in our TF set) do not appear often enough to exhibit either significant order or disorder within the domains themselves, most of them still exhibit significant disorder in one or both flanking regions.

The functional role played by the significant prevalence of disorder in the regions flanking DBDs of human TFs is unclear. However, we can speculate that the increased flexibility afforded by these flanking IDRs might contribute to the ability of TFs to 1) recognize target sequences in the DNA appropriately, 2) bind to a wider diversity of DNA target sequences, 3) be anchored with higher affinity to the DNA after recognizing target sequences, 4) bind to other factors and complexes positioned on the DNA or involved in transcriptional regulation, or 5) present activation domains to downstream transcriptional regulatory machinery. It should be emphasized that these possibilities are speculative; however, the results of this study suggest numerous testable hypotheses regarding the roles of N- and C-terminal regions flanking DBDs for many frequently occurring DBDs in hundreds of human TFs. For example, the importance of the predicted disorder in these flanking regions in determining or modulating the DNA binding affinity and/or specificity of the associated TFs could be investigated with protein binding microarrays (PBMs) [29, 30]. PBMs could assay the affinity and/or specificity of proteins representing the DBDs with their flanking regions, as compared to either the DBDs alone or the DBDs with mutant flanking regions predicted not to be significantly disordered. If found to contribute to the DNA binding affinity and/or specificity of TFs, IDRs that flank DBDs would broaden the scope of functional domains to be considered when evaluating the

[a]Though they do not occur often, where they do occur, they exhibit significant disorder in our results as well, corroborating the results in [4]; see Table 2.

potential impact of mutations or natural polymorphisms within exomes, such as in medical sequencing projects.

This study was focused on human TFs; however, since these DBD classes are the predominant DBD classes not just in human TFs but throughout eukaryotes, the results of this study may have important implications for studies of TFs across all eukaryotes.

5. Acknowledgments

We are grateful to Lizzy Rossin for creating the initial list of Pfam DBDs in human TFs. This work was funded by grants from NIH (R01 HG003985) to M.L.B., and from NIH (P50 GM081883-01) and DARPA (HR0011-09-1-0040) to A.J.H.

References

[1] D. Eliezer, *et al.*, *Curr. Opin. Struct. Biol.* **19**, 23 (Feb 2009).
[2] B. He, *et al.*, *Cell Res.* **19**, 929 (Aug 2009).
[3] Y. Minezaki, *et al.*, *J. Mol. Biol.* **359**, 1137 (Jun 2006).
[4] J. Liu, *et al.*, *Biochemistry* **45**, 6873 (Jun 2006).
[5] M. Fuxreiter, *et al.*, *Trends Biochem. Sci.* **36**, 415 (Aug 2011).
[6] J. W. Schwabe, *et al.*, *Structure* **1**, 187 (Nov 1993).
[7] Y. Liu, *et al.*, *J. Biol. Chem.* **283**, 20874 (Jul 2008).
[8] L. E. Schaufler, *et al.*, *J. Mol. Biol.* **329**, 931 (Jun 2003).
[9] C. Crane-Robinson, *et al.*, *Trends Biochem. Sci.* **31**, 547 (Oct 2006).
[10] J. M. Vaquerizas, *et al.*, *Nat. Rev. Genet.* **10**, 252 (Apr 2009).
[11] P. Flicek, *et al.*, *Nucleic Acids Res.* **39**, D800 (Jan 2011).
[12] K. D. Pruitt, *et al.*, *Nucleic Acids Res.* **37**, D32 (Jan 2009).
[13] A. Campen, *et al.*, *Protein Pept. Lett.* **15**, 956 (2008).
[14] A. K. Dunker, *et al.*, *J. Mol. Graph. Model.* **19**, 26 (2001).
[15] R. D. Finn, *et al.*, *Nucleic Acids Res.* **38**, D211 (Jan 2010).
[16] S. Hunter, *et al.*, *Nucleic Acids Res.* **37**, D211 (Jan 2009).
[17] S. R. Eddy, *et al.*, *Genome Inform* **23**, 205 (Oct 2009).
[18] K. Peng, *et al.*, *BMC Bioinformatics* **7**, 208 (2006).
[19] J. J. Ward, *et al.*, *J. Mol. Biol.* **337**, 635 (Mar 2004).
[20] X. Deng, *et al.*, *BMC Bioinformatics* **10**, 436 (2009).
[21] L. Bordoli, *et al.*, *Proteins* **69 Suppl 8**, 129 (2007).
[22] H. X. Zhou, *et al.*, *Biochemistry* **40**, 15069 (Dec 2001).
[23] S. Fukuchi, *et al.*, *J. Mol. Biol.* **355**, 845 (Jan 2006).
[24] R. Tupler, *et al.*, *Nature* **409**, 832 (Feb 2001).
[25] C. O. Pabo, *et al.*, *Annu. Rev. Biochem.* **70**, 313 (2001).
[26] W. J. Gehring, *et al.*, *Cell* **78**, 211 (Jul 1994).
[27] T. Hoey, *et al.*, *Nature* **332**, 858 (Apr 1988).
[28] T. D. Littlewood, *et al.*, *Protein Profile* **2**, 621 (1995).
[29] S. Mukherjee, *et al.*, *Nat. Genet.* **36**, 1331 (Dec 2004).
[30] M. F. Berger, *et al.*, *Cell* **133**, 1266 (Jun 2008).

INTRINSIC PROTEIN DISORDER AND PROTEIN-PROTEIN INTERACTIONS[†]

WEI-LUN HSU, CHRISTOPHER OLDFIELD, JINGWEI MENG, FEI HUANG, BIN XUE[#], VLADIMIR N. UVERSKY[#], PEDRO ROMERO, AND A. KEITH DUNKER[*]

Department of Biochemistry and Molecular Biology, Indiana University School of Medicine
410 W 10th street, Suite 5000
Indianapolis, IN 46202
Emails: {hsu20, cjoldfie, menj, huangfei, promero, or kedunker} @iupui.edu
{binxue or vuversky} @health.usf.edu

Intrinsically disordered proteins often bind to more than one partner. In this study, we focused on 11 sets of complexes in which the same disordered segment becomes bound to two or more distinct partners. For this collection of protein complexes, two or more partners of each disordered segment were selected to have less than 25% amino acid identity at structurally aligned positions. As it turned out that most of the examples so selected had similar 3D structure, the studied set was reduced to just these similar-fold cases. Based on the analyses of the interacting partners, the average sequence identity of the partners' binding regions showed substantially higher conservation as compared to the nonbinding regions: The residue identities, averaged over the 11 sets of partner proteins, were as follows: binding residues, 42 ± 6%; nonbinding residues 20 ± 3%; nonbinding buried residues 26 ± 5%; and nonbinding surface residues 16 ± 3% . The higher sequence identity of the binding residues compared to the other sets of residues provides evidence that these observed interactions are likely to be meaningful biological interactions, not artifacts. Since many of the features of the various interactions indicate that the disordered binding segments were likely to have been disordered before binding, these results also add further weight to the existence and function of intrinsically disordered regions inside cells.

Keywords: Molecular recognition, protein interaction, sequence conservation

1. Introduction

Many proteins execute their biological functions through protein-protein interactions. By binding to interacting partners, proteins can deliver signals to other molecules. For example, hormone neurotransmitters and their receptors trigger various signal transduction pathways following their mutual interaction, antibody recognition of peptide antigens leads to B-cell activation, and the interaction between G-protein coupled receptors and G-proteins leads to the transduction of many signals.

[†] This work is supported by the grants R01 LM007688-01A1, GM071714-01A2 from the National Institute of Health and EF 0849803 from the National Science Foundation

[*] Corresponding author

[#] Present Address: Department of Molecular Medicine, University of South Florida, Tampa, FL 33620

Mapping protein-protein interactions leads to networks that are far from random. A small number of highly connected proteins, called hubs, are observed, while most proteins have only a few connections [1, 2]. We proposed that hubs often use disordered regions to bind to multiple partners (one-to-many signaling), and we also proposed that structured hubs often bind to many partners via their partner's disordered regions (many-to-one signaling) [3]. The common involvement of disorder in hub proteins' interactions has been supported by several subsequent studies [4-6]. Our decision to test whether hub proteins depend on disorder was motivated by prior experiments showing that conformational disorder enabled one particular protein region to bind to multiple partners [7].

These short, partner-binding segments found within disordered regions have been studied under various names, such as eukaryotic linear motifs (ELMs) [8], short linear motifs (SLiMs) [9], ANCHOR-algorithm-indicated binding sites [10], and molecular recognition features (MoRFs) [11-13] among others. ELMs and SLiMs are based on sequence motifs, while ANCHOR-indicated sites and MoRFs both use features from disorder predictors. The motif-based and algorithmic approaches show significant overlap in their identification of their binding sites [14], suggesting that the different approaches associated with the different names are merely emphasizing different aspects of the same types of binding interactions.

The C-terminal region of p53 uses disorder to bind to more than 45 different proteins and to form a tetramer, but only six of these complexes and the tetramer have had their structures deposited in the Protein Data Bank (PDB) [15]. One particular p53 segment "SHLKSKKGQSTSRHKKLMFKTE" (residues 367-388), which is both an ELM and a MoRF and which is located at the C-terminus, morphs into an α-helix when binding with S100ββ, into a β-sheet with sirtuin, into an irregular structure with CREB binding protein (CBP) and into another irregular structure with cyclin A2 as a partner [15].

Very different biological processes are transduced via these four different interactions involving the same segment of p53: The CDK2/cyclin A2 complex regulates progression of S phase of the eukaryote cell cycle by recognizing diverse but structurally constrained target sequences (KXL/RXL motif) from various substrates, including p53 [16]; deacetylase enzymes like the Sir 2 protein, which is a homologue of Sirtuin, can lead to down-regulation of p53-dependent transcription by binding to the acetylated p53 peptide on lysine 382 [17]; the recognition of acetylated lysine 382 in p53 by the conserved bromo-domain of transcriptional coactivator CBP is very specific, leading to the recruitment of p53 acetylation-dependent coactivator following DNA damage and to the activation of cyclin-dependent kinase inhibitor p21 [18]; dimeric S100 calcium binding protein B can sterically block the phosphorylation and acetylation sites of on p53 that are critical for the activation important transcription; finally, the peptide derived from the region of p53 was found to undergo a disorder-to-order conformational change while binding to Ca^{2+} loaded S100ββ [19]. Thus, this same intrinsically disordered segment plays roles in a diverse set of signaling pathways.

Besides p53 other MoRFs that bind to two or more partners and that have structures in PDB have not been systematically compared to understand how disorder can bind to multiple partners. To study more examples, we searched for other instances in PDB in which one disordered segment binds to more than one partner. We found 11 such sets for which the partners have both superimposable 3D structures and pairwise sequence identities lower than 25%. As we show, the flexibility of the disordered regions is important for enabling the same protein segment to bind to alternative partners, thus extending our observations on p53 [15] to 11 other intrinsically disordered protein regions.

2. Results

2.1 Disordered hub dataset

Starting from 4,289 MoRF-containing complexes from PDB, various criteria were applied to extract those examples with the same MoRF sequences bound to collections of different globular partners, such that the partners to a particular MoRF exhibit less than 25% sequence identity. We identified 26 multi-partner sets having a total of 67 complexes. The point of the 25% sequence identity filter was to identify examples in which one MoRF binds to structurally dissimilar partners. Still, for 18 of these 26 examples, the MoRF partners were structurally superimposable even though their sequence identities were quite low and so, only those examples with similar fold were selected for our data set.

The RMSD values and the fraction of the total residues that gave good structural alignments were estimated for the sets of partners for the 18 MoRFs. The 11 sets that had both a relatively low RMSD (2.33 to 5.49 Å) and a relatively high fraction with good structural alignment (0.79 – 0.91), herein called coverage, are given in Table 1. The partners of the DNA repair protein have a reasonable RMSD (3.53) but a low coverage (0.36). This protein was not discarded because one of the partners had a large, non-alignable extra domain that was responsible for the low coverage.

Table 1. Disordered hub examples with structurally similar partners and low sequence identities

		Bound conformation					
Disordered hub examples	Num	Helix	Sheet	Coil	Complex	RMSD	Coverage
Nuclear receptor corepressor 2*	3	2	0	1	0	3.43	0.85
Thyroid receptor associated protein 220 (TRAP 220)*	3	3	0	0	0	3.05	0.91
Nuclear receptor coactivator 1*	2	2	0	0	0	5.49	0.85
Nuclear receptor 0B2 – near N-term*	2	2	0	0	0	3.74	0.86
Troponin I, cardiac muscles	2	0	0	1	1	3.01	0.79
Nuclear receptor 0B2 – near C-term*	2	1	0	1	0	3.88	0.80
Cell death protein GRIM	2	0	2	0	0	2.33	0.79
Beclin-1	2	2	0	0	0	4.10	0.84
BCL-2-like protein 11	2	2	0	0	0	2.72	0.90
Alzheimer's disease amyloid A4 protein homolog	2	0	0	2	0	2.93	0.84
DNA repair protein	2	0	0	2	0	3.53	0.36

*5 of 11 disordered hub examples belonged to the family of coregulatory proteins of nuclear receptor (NR), including thyroid receptor associated protein 220 (TRAP 220)

2.2 Functional consequences of MoRF (or ELM) binding

Interestingly, our dataset indicates that protein disorder is involved in coregulatory proteins of nuclear receptors (NR) such as coactivators and corepressors, making it possible for them to perform one-to-many signaling and to function as disordered hubs. The nuclear receptors (NRs) are a super-family of proteins, associated with other coregulatory proteins involved in the direct mediation and control of the expression of specific gene transcription in response to sensing the presence of hormones and other molecules. Recent data shows that, in addition to direct activation of the basal transcription machinery, nuclear receptors inhibit or enhance transcription by attracting an array of coactivator or corepressor proteins to the transcription complex.

NRs may be classified into two broad categories according to their mechanism of action and subcellular distribution in the absence of ligands. Ligands bind to type I NRs in the cytosol resulting in the dissociation of heat shock proteins, the formation of homo-dimers, translocation from cytoplasm into the cell nucleus, and binding to hormone response elements. Type I NRs include NR subfamily 3, which encompass androgen receptors, estrogen receptors, glucocorticoid receptors and progesterone receptors. Type II NRs, in contrast to type I NRs, are retained in the nucleus and heterodimerize upon binding to DNA in the absence of a ligand, when type II NRs are usually bound to a corepressor. Ligand binding to type II NRs triggers the dissociation of the corepressor, leading to the initiation of transcription by the coactivator. Type II NRs include NR subfamily 1, and receptor molecules such as retinoic acid receptor (RAR), retinoid x receptor (RXR), thyroid hormone receptor (TR) and vitamin D receptor (VDR).

Peroxisome proliferator-activated receptor-binding protein (PBP), also known as thyroid hormone receptor-associated protein 220 (TRAP 220), is an anchor for multi-subunit mediator transcription complex. It functions as a transcription coactivator for nuclear receptors. These coactivator proteins often exhibit histone acetyltransferase (HAT) activity, which weakens the association of the histone to DNA, therefore promoting gene transcription. Three MoRFs in PBP have been found to be involved in multiple interactions with various type II nuclear receptors, such as vitamin D3 receptor (VDR), retinoic acid receptor-beta (RAR-beta) and retinoid x receptor-alpha (RXR-alpha) in our dataset. RAR or VDR, when forming a heterodimer with RXR, can bind to hormone response elements, forming a complex with corepressor protein in the absence of any ligands When a ligand acting as agonist binds to RAR or VDR, it results in the dissociation of the corepressor and recruitment of coactivator which in turn, promotes transcription of downstream target gene.

When gene transcription is repressed by nuclear receptors, it is mediated by interactions with corepressor proteins. This reaction, in turn attracts histone deacetylases (HDACs) to the chromatin, triggering the strong binding of histone to DNA; thus repressing gene transcription. The antagonist further reinforces the binding of corepressor to the nuclear receptor. MoRF mechanisms are also involved in the down regulation of target gene expression when the nuclear receptor corepressor 2 binds to related nuclear receptors such as peroxisome proliferator activated receptor (PPAR), estrogen related receptor gamma (ERR-gamma) and progesterone receptor (PR) [20-22].

In the previous two examples, such coregulatory proteins can interact with various receptors with low sequence identity but high structure similarity using the same MoRF region. The configurations of secondary structures in those MoRFs bound to receptors are also comparable. Further experiments and analyses provided us more detailed and specific explanations regarding how disordered regions facilitate the binding diversity in different complex structures. The analyses of solvent surface area profiles from the examples with structurally different partners, shows different interfaces accommodate a variety of binding partners and those overlapping residues in interfaces bind to different molecules to varied extents. Otherwise, analogous binding profiles were observed within our 11 examples with similar partnerships.

2.3 Binding to multiple partners, conservation at structure-matching sites

In order to answer if having similar binding patterns mean MoRFs tend to bind a specific set of residues on partners, we analyzed the sequence identities of binding regions and nonbinding regions on partner side to determine whether the interacting residues are more selected during evolution (Table 2). Those interacting residues tend to be selected to form connection with same MoRF using similar binding patterns as we expected. The sequence conservation of the binding region is significantly higher than those in other parts of the protein.

Table 2. Sequence identities of binding regions and nonbinding regions in partners of 11 disordered hub examples based on MultiProt structural alignment

Disordered hub examples	B	NB	NB_B	NB_E	Overall
Nuclear receptor corepressor 2	0.36	0.17	0.22	0.15	0.21
Thyroid receptor associated protein 220	0.51	0.23	0.26	0.19	0.24
Nuclear receptor coactivator 1	0.62	0.16	0.22	0.08	0.16
Nuclear receptor 0B2 – near N-term	0.47	0.20	0.30	0.16	0.19
Troponin I, cardiac muscles	0.36	0.34	0.62	0.29	0.25
Nuclear receptor 0B2 – near C-term	0.27	0.16	0.13	0.25	0.12
Cell death protein GRIM	0.33	0.28	0.60	0.09	0.26
Beclin-1	0.45	0.17	0.26	0.06	0.17
BCL-2-like protein 11	0.44	0.16	0.26	0.11	0.21
Alzheimer's disease amyloid A4 protein homolog	0.33	0.19	0.33	0.11	0.15
DNA repair protein	0.43	0.27	0.33	0.11	0.07
	0.42±0.06	0.20±0.03	0.26±0.05	0.16±0.03	0.19±0.04

The values in columns labeled B, NB, NB_B, NB_E, and Overall give the sequence identities of binding, nonbinding, nonbinding buried, nonbinding surface and all residues, respectively.

The overall mean identity and confidence interval for structurally aligned binding and non-binding residues are 42 ± 6% and 20 ± 3%, respectively, as shown in Table 2. These averages were taken over structurally matching residues. In comparison, for a collection of enzymes with ~ 25% sequence identity, the active site residues exhibit sequence identities in the range of 43% to

70% [23]. Interestingly, the binding residues being discussed here for several of the proteins show sequence identities that overlap the lower part of the observed range for enzyme active site residues, which are known to have a high tendency to be conserved. Note also that the binding residues show a much higher conservation for the aligned residues as compared to the nonbinding surface residues (column B versus NB_E) or to the nonbinding buried residues (column B versus NB_B).

Three examples were chosen from our 11 sets in order to assess in more detail how a disordered region uses its conformational flexibility to form interactions with similar but not identical binding pockets. The three examples can be described as an α-MoRF, a β-MoRF and an irregular-MoRF corresponding to the thyroid receptor associated protein 220 (TRAP220), the cell death protein GRIM and the Alzheimer's disease amyloid A4 protein homolog, respectively. Figure 1 shows the interacting residues and binding sites of the three selected cases.

Figure 1.a shows four important residues (M5, L6, L9 and L10) on TRAP220 stretching into the clefts on the surfaces of receptor proteins with small structural variations. On the contrary, those residues on the non-buried side of the α-MoRF have larger conformation fluctuations over the three complexes.

Alanine 1 and tyrosine 4 contribute most of the buried surface areas to the interaction of GRIM and IAP1 (Figure 1.b). The tyrosine side chain makes a huge rotation to fit distinct cavities of IAP1 while the backbone's β-sheet conformation and key interactions didn't change too much (Figure 1.b.1 and 1.b.2).

The MoRF in amyloid A4 protein is mostly coil but has local regions that could be classified as sheet and helix according the DSSP. Its central region (NPTY) adopts coiled conformation and maintains comparably similar structure (colored in black) in both complexes (Figure 1.c). The N-terminus (NGYE) of the green MoRF (Figure 1.c.1) stays in a coiled structure while the N-terminus of the red MoRF in (Figure 1.c.2) turns into β-strand to form an anti-parallel β-sheet with another strand on DAB1 protein. The spatial arrangement of a tyrosine 4 was observed to change substantially, suggesting that this change may facilitate the binding to two different surfaces by the same sequence.

In addition to gathering general evolutionary information for the whole interfaces that MoRFs associate with, some further calculations were carried out on the same three selected examples in order to explore more details at the residue level. We compared the partner residues with which each MoRF residue associates to determine if partner residue variability correlates with overall conformational variation of MoRF itself (Figure 2.a, 2.b and 2.c). Our hypothesis was: the more diverse the amino acids with which a MoRF residue associates, the greater the structural variability of the MoRF backbone. However, the correlation analyses between diversity of partner residues (not shown in Figure 2) and averaged root mean square standard deviation (RMSD) on C-alpha atom (line plots in Figure 2) did not show an obvious and strong correlation. While our particular hypothesis was not supported, Figure 2 nevertheless contains interesting summary information regarding the changes that are observed when one MoRF binds to multiple partners.

Note alternating burial and nonburial of residues (Figure 2.a); this pattern can be traced to the α-helical structure (e.g. an α-MoRF) of the thyroid receptor associated protein 220. Also notice that the buried residues are more hydrophobic, and, except for proline, the nonburied residues are more hydrophilic. While hydrophobic, proline is often found on the surfaces of proteins and, furthermore, frequently occupies positions near the amino-terminal ends of helices.

Several interesting points were found from the sequence and structure analyses of the two residues with the greatest contributions of solvent surface area in the GRIM-IAP1 interaction (Figure 2.b). Although the side chain of alanine 1 is relatively small, a fairly large area becomes buried into the interface. More detailed analysis shows that, for this residue, not only the side chain participates the formation of interaction, but backbone atoms also play a significant role.

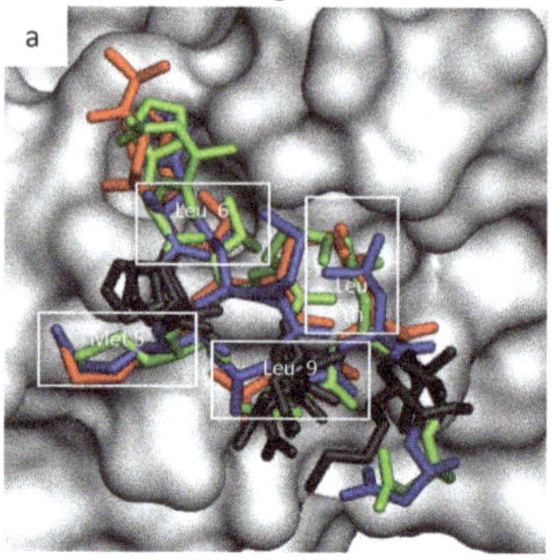

Big rotations on side chains also related to higher backbone structure variations, such as the cases on L9 in the α-MoRF (Figure 2.a), Y4 in β-MoRF (Figure 2.b) and Y4 in irregular-MoRF (Figure 2.c).

The low RMSD of P7 in the irregular-MoRF example (Figure 2.c) may correlate with its capping function in the edge of helix in the X11 protein.

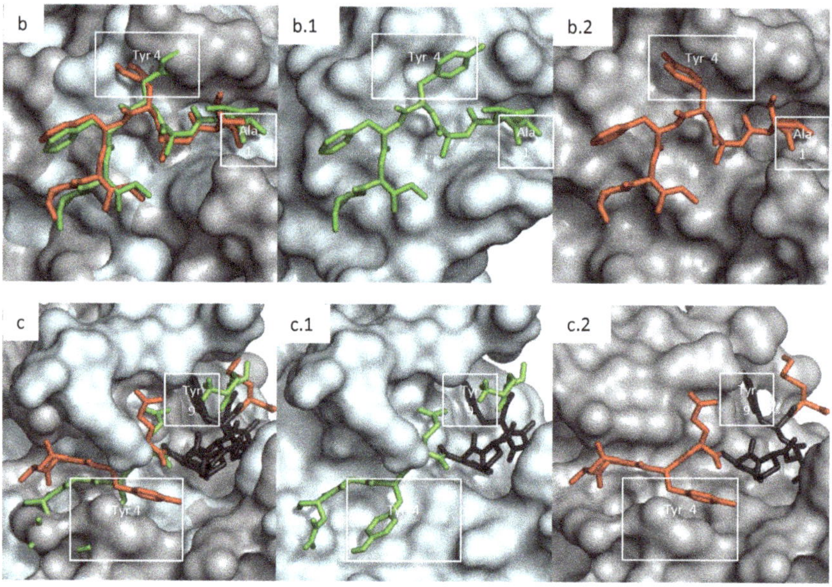

Fig.1. Conformational changes and variations of an identical MoRF binding to its structure-homologous partners from the three selected examples. Different MoRFs are shown in different colors. (a) A fragment from TRAP220 forming α-MoRFs to associate with VDR, RAR-beta and RXR-alpha. Those residues on the exposed side of helixes are colored in black. (b) The binding sites of a β-MoRF from GRIM and apoptosis 1 inhibitor (IAP1). (c) The irregular-MoRF in amyloid A4 protein adapts a highly flexible structure to accommodate the binding pockets of DAB1 and X11. Four structurally constrained residues with lowest RMSD are shown in black.

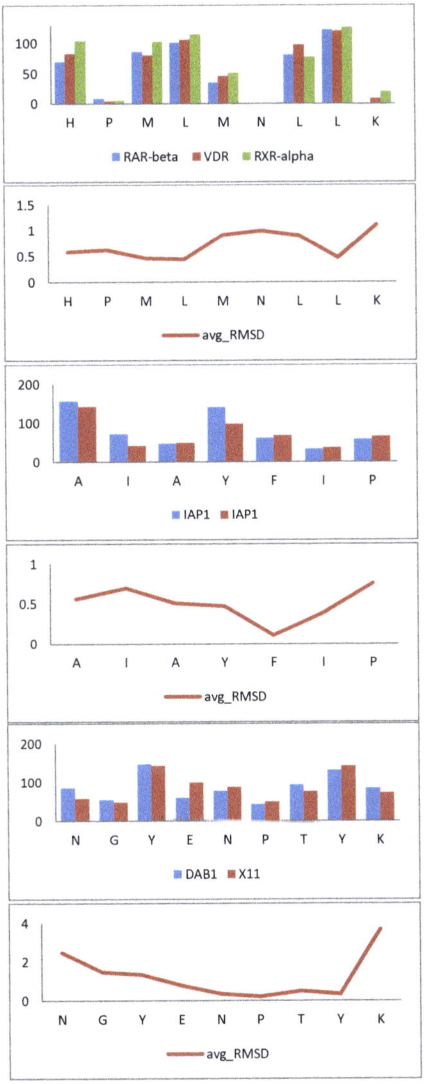

a

		1rkh_A	1xdk_B	1xdk_A
K	1	K		
N	2	E		VEP
H	3	IKEV	IKE	VREEP
P	4	E	E	E
M	5	IPLE	PLE	FVTFE
L	6	IILKLEV	VILLEM	VVLRFEM
M	7	SIE	IIE	LDVE
N	8	K	K	VV
L	9	QIKL	IVKLE	FVEKF
L	10	IKSQIL	VKFIQIL	VKFLQVL
K	11	K	K	VKL
D	12		K	K

b

		1jd5_A	1seo_A
A	1	LMDWNDQWE	IGSWEDERW
I	2	GLMDQWES	EIGSEDEW
A	3	GLMWL	VEIGW
Y	4	TRRGLM	KEIGS
F	5	RCGGGL	KCGVE
I	6	GGWL	VEW
P	7	CGL	CGVWS
D	8		

c

		1oqn_B	1x11_A
Q	1		MQDIG
N	2	DYIAKIR	MYIADQ
G	3	DYIAKR	KMYIAD
Y	4	SYIRFQ	KSYIADAQGF
E	5	RSYIF	TKRSYIF
N	6	VHISYF	LRISYFY
P	7	FQY	SFSY
T	8	VHIFIYK	LRFAYF
Y	9	ARVHEISGEH	KRTISRKQMY
K	10	HGKE	TDGKY
F	11		GKYFL
F	12		RRKRF

Fig.2. Solvent surface area profiles (bar graphs), averaged RMSD (line plots), and interacting residues on various globular partners (right) calculated on each residue from the three selected MoRF sets: (a) α-MoRF: thyroid receptor associated protein 220, (b) β- MoRF: Cell death protein GRIM and (c) irregular-MoRF: Alzheimer's disease amyloid A4 protein homolog. Notice in group (c) the specific interactions between 2 aromatic groups that oriented themselves orthogonally.

3. Discussion

When the idea of hub-based "scale free" protein-protein interaction networks was first proposed [1], a News and Views article pointed out that such multiple interactions were unlike what had been studied up to that time and that an understanding of these multiple interactions would likely require the discovery of new concepts [24]. We immediately tried to suggest that the new principle was likely the use of disordered proteins by means of coupled binding and folding, which had been previously suggested for protein-DNA interactions [25] as well as for one protein binding to several partners [26], but publication of these ideas for hub proteins was delayed somewhat [3]. By now there is strong evidence that, at least for many if not all hub-partner interactions, disorder plays an important role in enabling one protein to bind to multiple partners [3-6].

As we have shown here, one MoRF can bind to multiple partners and, through these sets of interactions, a small region of one protein can play a role in multiple signaling events, thereby affecting several different cellular functions. Limitations on the availability of multiple interactions in PDB led to a small dataset, which, nevertheless, showed consistent results in terms of residue conservation in the binding partners. By restricting sequence identity of binding partners to 25% we were hoping to find MoRFs binding to structurally diverse partners, but it turned out that in most cases the partner's folds were very much conserved, despite the sequence differences. Aromatic residue side chain re-orientation was shown to be important in two and three interactions when the backbone conformation remained constant. Overall, these results paint a more detailed picture of multiple interactions than what has been available to date, and support the notion of intrinsic disorder or flexibility as an important factor in the development of non-random protein interaction networks characterized by multiple binding partners.

Despite the biological and structure-function importance of these disorder-based multiple protein interactions, there have been surprisingly few studies indicating in detail how such multiple interactions are brought about. Investigating more examples in detail is needed to provide a clearer and more general picture of how the one MoRF sequence can bind to two or more different partners.

To broaden our understanding, it makes sense to study different collections of single proteins binding to multiple partners and to study multiple proteins binding to the same partner. Previously we investigated one segment binding to completely unrelated partners (e.g., one-to-many signaling) and a collection of unrelated disordered partners binding to a single binding site on one structured protein (e.g. many-to-one signaling) [15]. Rather than focusing on individual examples as before [15], here we studied a collection of MoRFs involved in binding to more than one partner.

A distinctive feature of this study is the partners to a given MoRF had a sequence identity less than 25% yet displayed the same overall fold. In general, the mutations on one partner are frequently linked to mutations on the other partner, indicating structural compensation or coadaptation across the binding interface [26]. However, the interacting protein pairs we collected here are a special set in that only the partners show amino acid substitutions in their sequences, whereas the MoRFs' sequences are unchanged.

The binding sites on the structured partners may also bind additional disordered sequences that have amino acid substitutions (e.g. many-to-one signaling). If such complexes exist, it would be interesting to determine whether the amino acid changes in the MoRFs compensate for the already observed amino acid changes in the structured binding partners. We have a collection of many-to-one examples (manuscript in preparation), so we can search this set of interactions to determine if any of the structured proteins in the many-to-one collection match any of the structured partners discussed herein. Such a finding would not only provide information about possible mutation compensation across protein-protein interaction interfaces involving disordered protein regions, but would also suggest new concepts with regard to the structural basis of protein-protein interaction networks.

The indispensability of hub proteins is apparent, as they appear to evolve more slowly and are more likely to be vital for survival [1]. MoRF-protein interactions likely have the combination of high specificity coupled with low affinity [12]. The latter property may facilitate the discovery of small drug molecules that block the interactions. This study and others like it have the potential for a new strategy for drug discovery, namely to search for molecules that selectively block certain protein-proteins interactions involving a given protein but not others, by taking advantage of different conformations in the different interactions. This would allow the development of drugs that target specific pathways or even particular pathways in particular tissue types.

Observations have been made that the residues in enzyme active sites tend to evolve more slowly than other parts of the same proteins [23]. We wondered whether the same trend would also be found in the binding sites of the structured partners. By analyzing the 11 sets of interactions that we collected, we found that, like the active-site residues of enzymes, the binding residues of the structured partners exhibited a higher conservation as compared to the non-binding residues.

A recent study showed that protein-protein interactions in which a disordered region binds to a structured partner often involves interactions between two aromatic groups, for which the aromatic residues are frequently not stacked but rather oriented in such a way that a hydrogen of one aromatic ring points towards the centers of the conjugated electron rings of the other [27]. In agreement with this study, many of the protein-protein interactions investigated herein do indeed involve interacting aromatic residues (Figure 2.c), but specific examples without such aromatic-aromatic interactions were also found (Figure 2.a).

Interactions between globular proteins and MoRFs often contain disordered residues as part of the MoRFs [11]. Others have shown that, even though such local regions remain unstructured, they can still affect the overall binding affinity. Such "fuzzy complexes" thus bind to their partners without undergoing complete conversion to structure with the regions that remain disordered still contributing to the energetics of the interaction [28]. From what we have shown here, a search for fuzzy complexes involved in one-to-many and many-to-one signaling could shed new light on these novel and interesting protein-protein "flexible nets".

4. Methods

4.1 Disordered hub dataset

Our disordered hub dataset was extracted from Protein Data Bank (PDB) released on March 28, 2008 by analyzing the complex structures that have short protein fragments bound to globular structured partners. The lengths of MoRFs and their globular structured partners were restricted as 5 to 25 and greater than 70, respectively. Only those interactions whose changes of solvent surface area (ΔASA) upon binding are greater than 400Å2 were considered as biological interactions. Since our goal is to gather all the complexes with MoRF mechanisms which use a specific region to bind multiple structured partners like p53, Fasta program was used to align each MoRF sequence to UniProt database in order to pinpoint its position in its parent sequence (Expectation value was set at 1000). Following that, we only kept those MoRFs which had overlapping regions in their parent sequence mapping and used a cluster algorithm (wherein at least one residue overlapped with the rest of the molecules in the same cluster).

As our research is focused upon those MoRFs from the same disordered region which bind to structurally different partners, we used a blast-cluster program to remove any redundant structured partners in our dataset based on 100% and 25% sequence identity. That meant that those specific MoRFs are in one disordered region, but they may use distinct residues to form bonding with different structured partners. After examination of the entire dataset, several unexpected cases were removed, such as those cases involving one MoRF interacting with more than one partner in a single PDB entry or a partner molecule which may be a part of a subset of the same cluster.

4.2 Sequence and Structure analysis

We previously differentiated the MoRFs into 4 different types (α, β, ι and complex) based on their secondary structure type which has the largest percentage value of the four types mentioned above [11]. If there is no clear preponderance of any one secondary type (which is at least 1% greater than the other 2 types), we called it "a complex-MoRF". Only the residues on the interface were counted. The secondary structure assignment program used here was the DSSP program. The root mean square deviation (RMSD) of pairwise proteins was calculated by CEalign [29]. The coverage of alignable region is calculated by length of aligned regions dividing by average length of all sequences. The transposed coordinates and multiple structure alignments were generated by MultiProt algorithm [30] using the complex structures including both MoRF and partner. Sequence identity calculations are based on the structure alignments.

Both residues in each aligned pair were compared to see if they are both in the binding or nonbinding region. The alignment will be considered identical only when the position in both proteins is assigned in the same class: either binding or nonbinding. For the case with more than 2 partners, we averaged all the identities together. Those aligned residues not consistent with their binding/nonbinding status (one is on binding region, but the other one is not) will be classified into another category that didn't show on table 2. Here, those residues with higher solvent surface changes (greater than 1 $Å^2$) will be considered as interacting residues. Error bars that represent the 95% confidence interval (CI) of a mean are approximated from 3000 random samplings with replacement generated by the bootstrapping method. The molecular images in Figure 1 were generated by PyMol software.

References

1. Jeong, H., et al., *Lethality and centrality in protein networks.* Nature, 2001. **411**(6833): p. 41-2.
2. Barabasi, A.L. and Z.N. Oltvai, *Network biology: understanding the cell's functional organization.* Nat Rev Genet, 2004. **5**(2): p. 101-13.
3. Dunker, A.K., et al., *Flexible nets. The roles of intrinsic disorder in protein interaction networks.* FEBS J, 2005. **272**(20): p. 5129-48.
4. Patil, A., K. Kinoshita, and H. Nakamura, *Hub promiscuity in protein-protein interaction networks.* Int J Mol Sci, 2010. **11**(4): p. 1930-43.
5. Kim, P.M., et al., *The role of disorder in interaction networks: a structural analysis.* Mol Syst Biol, 2008. **4**: p. 179.
6. Haynes, C., et al., *Intrinsic disorder is a common feature of hub proteins from four eukaryotic interactomes.* PLoS Comput Biol, 2006. **2**(8): p. e100.
7. Kriwacki, R.W., et al., *Structural studies of p21Waf1/Cip1/Sdi1 in the free and Cdk2-bound state: conformational disorder mediates binding diversity.* Proc Natl Acad Sci U S A, 1996. **93**(21): p. 11504-9.

8. Gould, C.M., et al., *ELM: the status of the 2010 eukaryotic linear motif resource.* Nucleic Acids Res, 2010. **38**(Database issue): p. D167-80.
9. Edwards, R.J., N.E. Davey, and D.C. Shields, *SLiMFinder: a probabilistic method for identifying over-represented, convergently evolved, short linear motifs in proteins.* PLoS One, 2007. **2**(10): p. e967.
10. Dosztanyi, Z., B. Meszaros, and I. Simon, *ANCHOR: web server for predicting protein binding regions in disordered proteins.* Bioinformatics, 2009. **25**(20): p. 2745-6.
11. Mohan, A., et al., *Analysis of molecular recognition features (MoRFs).* J Mol Biol, 2006. **362**(5): p. 1043-59.
12. Oldfield, C.J., et al., *Coupled folding and binding with alpha-helix-forming molecular recognition elements.* Biochemistry, 2005. **44**(37): p. 12454-70.
13. Cheng, Y., et al., *Mining alpha-helix-forming molecular recognition features with cross species sequence alignments.* Biochemistry, 2007. **46**(47): p. 13468-77.
14. Fuxreiter, M., P. Tompa, and I. Simon, *Local structural disorder imparts plasticity on linear motifs.* Bioinformatics, 2007. **23**(8): p. 950-6.
15. Oldfield, C.J., et al., *Flexible nets: disorder and induced fit in the associations of p53 and 14-3-3 with their partners.* BMC Genomics, 2008. **9 Suppl 1**: p. S1.
16. Lowe, E.D., et al., *Specificity determinants of recruitment peptides bound to phospho-CDK2/cyclin A.* Biochemistry, 2002. **41**(52): p. 15625-34.
17. Avalos, J.L., et al., *Structure of a Sir2 enzyme bound to an acetylated p53 peptide.* Mol Cell, 2002. **10**(3): p. 523-35.
18. Mujtaba, S., et al., *Structural mechanism of the bromodomain of the coactivator CBP in p53 transcriptional activation.* Mol Cell, 2004. **13**(2): p. 251-63.
19. Rustandi, R.R., D.M. Baldisseri, and D.J. Weber, *Structure of the negative regulatory domain of p53 bound to S100B(betabeta).* Nat Struct Biol, 2000. **7**(7): p. 570-4.
20. Wang, L., et al., *X-ray crystal structures of the estrogen-related receptor-gamma ligand binding domain in three functional states reveal the molecular basis of small molecule regulation.* J Biol Chem, 2006. **281**(49): p. 37773-81.
21. Madauss, K.P., et al., *A structural and in vitro characterization of asoprisnil: A selective progesterone receptor modulator.* Molecular Endocrinology, 2007. **21**(5): p. 1066-1081.
22. Xu, H.E., et al., *Structural basis for antagonist-mediated recruitment of nuclear co-repressors by PPAR alpha.* Nature, 2002. **415**(6873): p. 813-817.
23. Grishin, N.V. and M.A. Phillips, *The subunit interfaces of oligomeric enzymes are conserved to a similar extent to the overall protein sequences.* Protein Sci, 1994. **3**(12): p. 2455-8.
24. Hasty, J. and J.J. Collins, *Protein interactions. Unspinning the web.* Nature, 2001. **411**(6833): p. 30-1.
25. Spolar, R.S. and M.T. Record, Jr., *Coupling of local folding to site-specific binding of proteins to DNA.* Science, 1994. **263**(5148): p. 777-84.
26. Fares, M.A., M.X. Ruiz-Gonzalez, and J.P. Labrador, *Protein coadaptation and the design of novel approaches to identify protein-protein interactions.* IUBMB Life, 2011. **63**(4): p. 264-71.
27. Espinoza-Fonseca, L.M., *Aromatic residues link binding and function of intrinsically disordered proteins.* Molecular BioSystems, 2011.
28. Tompa, P. and M. Fuxreiter, *Fuzzy complexes: polymorphism and structural disorder in protein-protein interactions.* Trends Biochem Sci, 2008. **33**(1): p. 2-8.
29. Shindyalov, I.N. and P.E. Bourne, *Protein structure alignment by incremental combinatorial extension (CE) of the optimal path.* Protein Eng, 1998. **11**(9): p. 739-47.
30. Shatsky, M., R. Nussinov, and H.J. Wolfson, *A method for simultaneous alignment of multiple protein structures.* Proteins, 2004. **56**(1): p. 143-56.

SUBCLASSIFYING DISORDERED PROTEINS BY THE CH-CDF PLOT METHOD†

FEI HUANG, CHRISTOPHER OLDFIELD, JINGWEI MENG, WEI-LUN HSU, BIN XUE#, VLADIMIR N. UVERSKY#, PEDRO ROMERO AND A. KEITH DUNKER*

Department of Biochemistry and Molecular Biology, Indiana University School of Medicine

410 W 10th street, Suite 5000,
Indianapolis, IN 46202
Emails: {huangfei, cjoldfie, menj, hsu20, or kedunker} @iupui.edu
promero@compbio.iupui.edu
{binxue or vuversk} @health@usf.edu

Intrinsically disordered proteins (IDPs) are associated with a wide range of functions. We suggest that sequence-based subtypes, which we call flavors, may provide the basis for different biological functions. The problem is to find a method that separates IDPs into different flavor / function groups. Here we discuss one approach, the (Charge-Hydropathy) versus (Cumulative Distribution Function) plot or CH-CDF plot, which is based the combined use of the CH and CDF disorder predictors. These two predictors are based on significantly different inputs and methods. This CH-CDF plot partitions all proteins into 4 groups: structured, mixed, disordered, and rare. Studies of the Protein Data Bank (PDB) entries and homologous show different structural biases for each group classified by the CH-CDF plot. The mixed class has more order-promoting residues and more ordered regions than the disordered class. To test whether this partition accomplishes any functional separation, we performed gene ontology (GO) term analysis on each class. Some functions are indeed found to be related to subtypes of disorder: the disordered class is highly active in mitosis-related processes among others. Meanwhile, the mixed class is highly associated with signaling pathways, where having both ordered and disordered regions could possibly be important.

Keywords: Intrinsically disordered protein, classification, CH plot, CDF prediction

1. Introduction

Unlike structured proteins folded into compact structures, intrinsically disordered proteins (IDPs) exist as flexible ensembles [1]. IDPs are very common. They comprise approximately 25% to 30% of eukaryotic proteomes [2]. Over 50% of eukaryotic proteins and 70% of signaling proteins have long disordered regions [3]. A wide range of biological activities are associated with IDPs, such as providing sites for post-translational modifications, providing sites for binding to partners via short linear motifs, acting as scaffolds by binding to multiple partners , etc. [4-6].

Studies of ordered proteins indicate that homologous proteins typically have conserved 3D structures [7-9]. Thus, structure similarity is used as an important criterion when examining related proteins. Most proteins with similar structure have a common evolutionary origin, and as a consequence their functions are typically closely related [7-9]. Databases such as SCOP [7] and CATH [8] have been constructed using this line of reasoning. These databases serve as a great

---

† This work is supported by the grants R01 LM007688-01A1, GM071714-01A2 from the National Institute of Health and EF 0849803 from the National Science Foundation

* Corresponding author

Current Address: Department of Molecular Medicine, University of South Florida, Tampa, FL 33620

resource for understanding the nature of the various relationships between protein structure and function, and they are widely used in various molecular and biological areas of science [7,8].

Since IDPs lack 3D structure, unique structure can't be used to partition IDPs into subtypes. We previously tried an approach based on disorder prediction to cluster IDP regions into different subtypes, which we called flavors, and some functions showed a weak partitioning among the different flavors [10]. Here our goal is to re-explore the overall idea of partitioning disordered proteins into subtypes, but using a different predictive approach. The previous approach used residue-by-residue order / disorder predictions over IDP regions of proteins [10], but a weakness of that approach was that the disordered regions varied markedly in length, which greatly complicated the interpretation.

Using *mus. musculus* protein data, here we will test an approach in which the order / disorder predictions are binary for the whole protein, indicating that a given protein is more ordered or more disordered overall. The two binary prediction tools are the charge-hydropathy (CH) plot [12,13] and the cumulative distribution function (CDF) [14]. Applying both methods to a protein could have four possible outcomes: both methods predict order, both methods predict disorder, the CH predicts disorder while the CDF predicts order, and vice versa. When both methods predict order, the protein is likely to be predominantly structured and to be found in the Protein Data Bank (PDB) [15]. When both methods predict disorder, the proteins are likely to be IDPs with high net charge and very little structure, and thus are likely to be more extended. If CDF predicts disorder and CH predicts order or vice versa, then these two sets of proteins have both order and disorder tendencies, but with different characteristics for each tendency. Thus, overall, the CH-CDF plot separates proteins into 4 groups with differing order and disorder characteristics.

The CH-CDF plot was previously used to compare the structure-disorder tendencies of the proteomes from several species within the phylum Apicomplexa, which include Plasmodia, Trypanosomes, and Giardia [16]. The CH-CDF plot has also been used to classify the transcription factors associated with the induction of pluripotent stem cells [17] and a collection of plant-specific developmental proteins as well as their distinctive domains [18]. In all three cases, the distributions of the various proteins and domains among the four outcomes provide overviews of similarities and differences between the different sets of proteins [16-18]. According to these prior studies, the CH-CDF plot appears to be useful for identifying overall structure-disorder trends for collections of proteins. Here we apply the CH-CDF plot to the mouse proteome and then investigate whether the four outcomes are associated with differences in structure and function for these proteins.

2. Results

2.1 CH-CDF plot

First, let's illustrate the overall development of the CH-CDF plot. Figure 1A shows the placement of a disordered protein (red) and an ordered protein (blue) onto a CH plot, where the indicated linear discriminant (i.e., a linear classification boundary) was developed from a large training set of proteins [12, 13]. Note that disordered proteins have a higher net charge and lower hydropathy compared to ordered proteins. We use the vertical distances from each protein-representing point to the separation line as the Y-coordinate of that protein in the CH-CDF plot, so when Y is positive, the protein is indicated to be disordered. Figure 1B shows the PONDR VSL2 [19] plots for the same pair of disordered (red) and ordered (blue) proteins. In Figure 1C, the data in 1B are plotted to produce the CDF plot, where the X-axis is the prediction score and the Y axis is the total fraction of sequence loci having that score or lower. Note the different shapes for the CDFs for the ordered (blue) and disordered (red) proteins. An ordered protein's CDF curve occupies the upper part of the graph, while an IDP's CDF curve resides in the lower part of the graph. The optimal separation line, represented as a collection of 7 discrete points, was previously estimated for a large number of structured and disordered proteins [14]. The X-axis for the CH-CDF plot is calculated as the average of the vertical distances from the CDF curve to the seven boundary points. Thus, the ordered proteins are given positive values and disordered proteins are given negative values with respect to the X-axis in the CH-CDF plot.

The entire mouse (*Mus.musculus*) proteome is put onto the CH-CDF plot in Figure 1D. The descriptions of the prediction characteristics for the proteins in each quadrant are included in this plot.

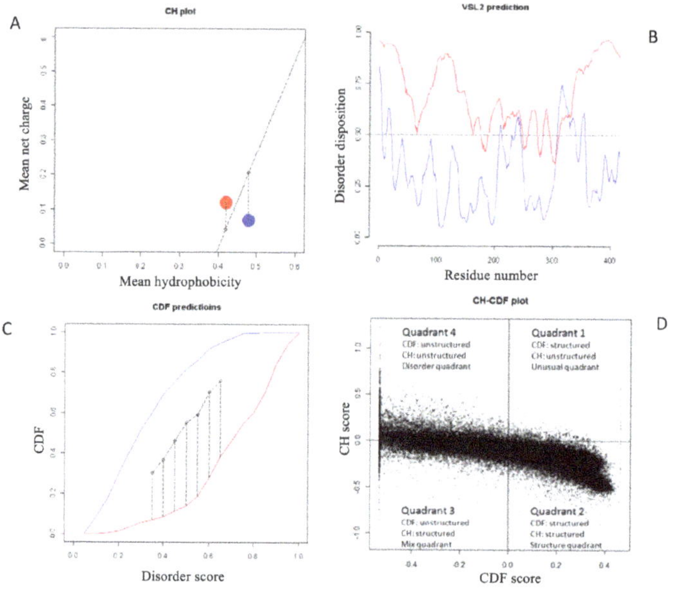

Fig.1. The CH-CDF Plot. A. An example of a CH graph, with a linear classification boundary (y=2.743x-1.109) and a hypothetical IDP and hypothetical structured protein. B. VSL2 prediction curve for an IDP (red) and a structured protein (blue). C. CDF curve of the two proteins in B. Vertical lines are the distance of to calculate CDF score. D. The entire mouse proteome is put onto a CH-CDF plot.

One rationale behind using CH-CDF plot in subclassification of disordered proteins is that CDF examines many more protein attributes than a CH plot, which only uses charge and hydropathy for prediction. Consequently, the CDF curve is more sensitive to disorder than the CH plot [11]. Proteins predicted to be ordered by the CH plot but disordered by CDF (as in Q3) are low in net charge and hydrophobic, but with other features resembling an unstructured protein. Therefore, we propose that such proteins could have both disordered and ordered regions, and we refer them as mixed proteins. Meanwhile, proteins predicted to be unstructured by both methods are referred as disordered (Q4) and proteins predicted to be ordered by both predictors are likely to be structured proteins (Q2). As for proteins in Q1, their number is very small compared to other three quadrants. We do have some assumptions about them, but we have not reached any refined conclusion yet. So here, we refer them as rare proteins.

2.2 PDB coverage

PDB contains protein structures, and thus PDB is biased more towards ordered proteins than disordered. Fig. 2 shows PDB coverage percentages of various proteins vs. their length for each quadrant. By coverage percentage, we mean the percent of a given sequence that forms structure and is observed in PDB. As expected, more of the proteins in Q2 have higher coverage.

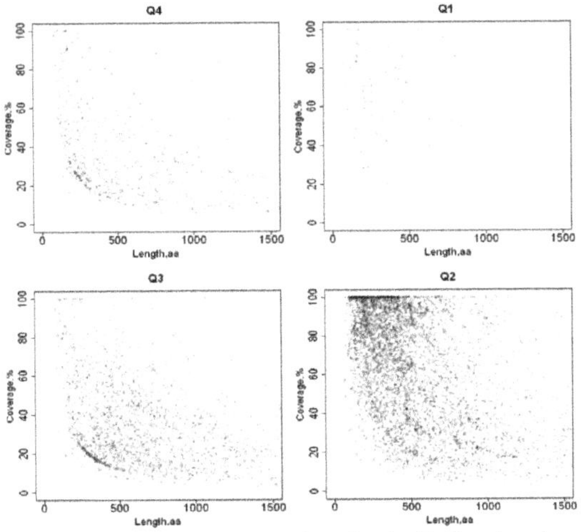

Fig.2. PDB coverage percentages of proteins classified into 4 quadrants.

To quantitate the coverage data of Figure 2, histogram summaries for each quadrant were constructed (Figure 3). When proteins are indicated to be disordered by the CDF (Q3 and Q4), the coverage summaries are similar and mostly show a small fraction of coverage. When proteins are indicated to be structured by CDF (Q1 and Q2), the coverage summaries are similarly biased towards structure. There are other factors to consider as shown below.

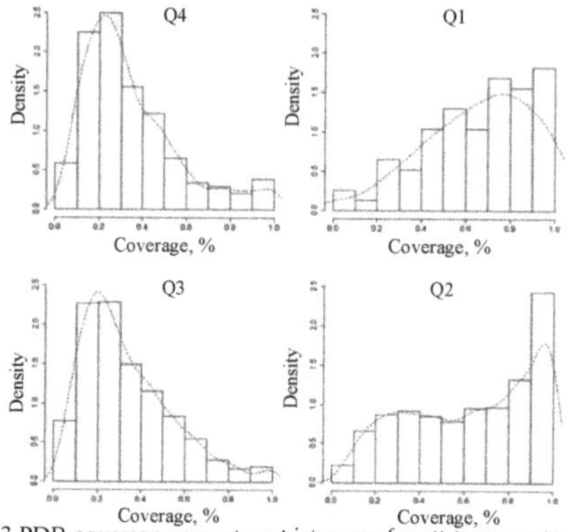

Fig.3 PDB coverage percentage histogram for all four quadrants

Another important consideration is whether a protein has any structure at all in PDB. The structure quadrant (Q2) has the highest fraction of proteins identified with at least one PDB hit, whereas the disorder quadrant (Q4) has the lowest fraction (Figure 4). Note that the mixed quadrant (Q3) actually is the second highest. Its fraction is close to the structure quadrant (Q2), and much higher than the disorder quadrant (Q4). These data suggest that mixed proteins have more structured regions than disordered proteins. Recall Fig. 2 and Fig. 3, which have shown that the coverage percentages for proteins in Q3 are very low, around 20-30% only. Taken together, these mixed proteins are more likely to have structured local regions compared to the disorder quadrant (Q4), so that they have a higher fraction of PDB hits.

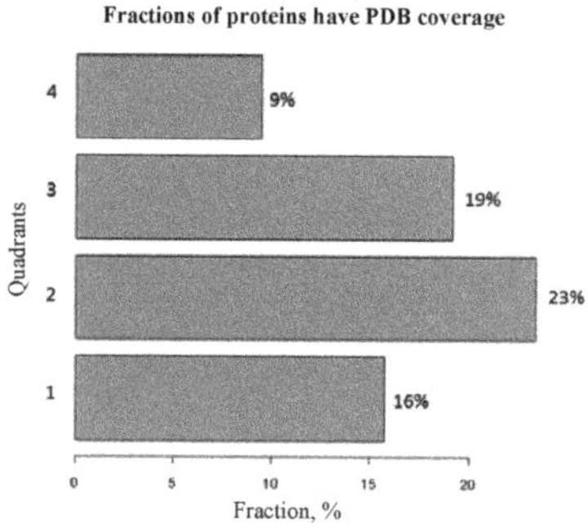

Fig.4. Fraction of protein identified with at least one PDB hit

2.3 Sequence window CH-CDF analysis

To learn more about the Q3 proteins, we dissected each protein sequence into a series of windows of 30 residues and carried out CH-CDF analysis on each segment. Table 1 summarizes results of this analysis. Proteins from the structure quadrant (Q2) have most of the windows in Q2, and the extended disorder quadrant (Q4) protein windows mostly localized in Q4. Interestingly, windows from mixed proteins (Q3) distribute mostly between quadrants Q2 Q3 and Q4, with the most hits in Q4 and slightly less hits in Q2 and Q3, suggesting that mixed proteins very likely contain a balanced distribution of ordered and disordered regions. Proteins from Q1 distribute equally in Q1 and Q2 with slightly less in Q4, again suggesting the presence of disordered regions.

Table 1. Sequence window CH-CDF analysis results

	Window quadrant localization			
	Q1	**Q2**	**Q3**	**Q4**
Q1 sequence windows	35%	35%	4%	26%
Q2 sequence windows	13%	68%	7%	11%
Q3 sequence windows	7%	28%	28%	37%
Q4 sequence windows	7%	13%	16%	64%

2.4 Match PDB coverage to disorder prediction

Since our analysis show that mixed proteins (in Q3) are predicted to have both disordered and ordered regions, here we attempt to verify that these predicted ordered regions are correlated with experimentally determined structures. To this end, we calculated the disorder content of the PDB covered and uncovered regions, respectively. Fig. 5 represents the results of this analysis of disorder content in all 4 quadrants.

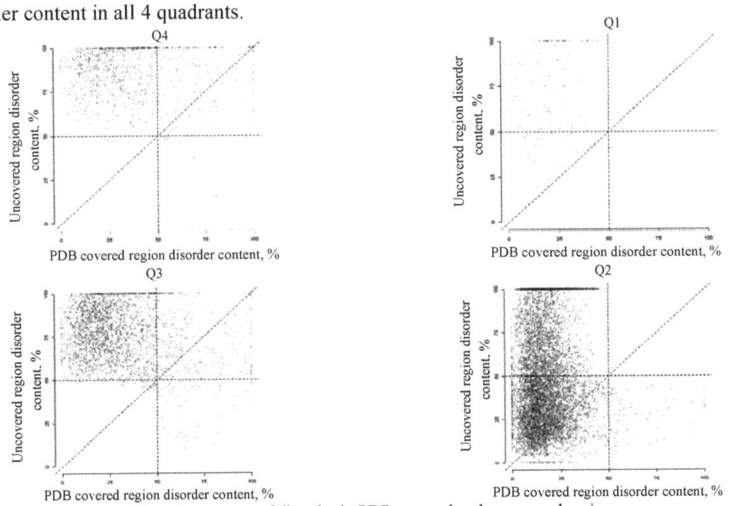

Fig.5. Percentage of disorder in PDB covered and uncovered regions

The X-axis is the disorder content of the PDB covered regions, and the Y-axis is the disorder content on the non-covered region. Disorder higher than 50% means that the corresponding region is largely predicted as disordered, whereas the score of less than 50% suggests that the region is predicted to be structured. In all 4 plots, the majority of the points clustered above the 45 degree diagonal line. We interpret this as that the regions not covered by PDB have more disorder than the covered regions.

The plot for the structure quadrant (Q2) has proteins clustered mainly to the left, both in the upper-left and lower-left corners. Those in the upper-left area could be disordered tails in these structured proteins. Those in the lower left correspond to structured regions that have not yet been crystalized. These segments are expected to be very common because many mouse proteins have multiple structured domains, and given the low percentage of PDB hits (Figure 4), it is likely that many of the structured domains of a given protein fail to make it into PDB.

In contrast to the observations for (Q2), the mixed proteins (Q3) and disordered proteins (Q4) are clustered in the upper-left corner in this plot, meaning that the regions not covered by PDB are predicted to be mostly unstructured, and those regions covered by PDB are predicted to be ordered. This indicates that disorder prediction and PDB coverage are in good agreement. Since we also showed above that mixed proteins are predicted to have both disordered and ordered regions (Table 1), it is likely that the predictions represent the true status of the protein as partially disordered and partially ordered. If this is true, it explains the mixed proteins' somewhat high fraction of PDB hits but low coverage percentages. The ordered regions are aligned to PDB sequences, but they are only a small fraction of the protein.

2.5 Function analysis for each quadrant

Previously we used a complicated prediction scheme to subdivide disordered protein regions into subtypes that we called flavors, and these different disordered flavors showed weak correlations with particular protein functions [10]. Given that the proteins in the four quadrants of the CH-CDF plots have different characteristics, it seemed reasonable to test whether these different structural subtypes exhibit functional separation. Therefore, we analyzed the proteins in each quadrant for their associations with various Gene Ontology (GO) terms.

Table 2 lists those Biological Processes GO terms found to be distinctive for each quadrant. For Q1, four of the five distinctive GO terms deal with RNA. By the CH analysis, these proteins are highly charged, and this feature may be associated with RNA association. For the Q2 structured proteins (Table 2B), most of their GO terms are related to metabolic processes and transporters. These functions are typical for structured proteins. For proteins in Q3, most of these GO terms are related to regulation or developmental pathways, including the Notch and Wnt pathways. As shown above, proteins in Q3 are likely to have both disordered and structured domains. Evidently these functions require both structured and disordered regions in the same proteins. Proteins in disorder quadrant (Q4 and table 1D) are mostly mitosis related.

Table 2. GO term analysis for four quadrants. Number of protein examples found for each GO term is listed on the right side.

Table 2A

Q1 Biological Process
tRNA methylation
tRNA wobble uridine modification
Translational termination
Positive regulation of nitric oxide
rRNA export from nucleus

Table 2B

Q2 Biological Process
Homophilic cell adhesion
Glutamine metabolic process
Phosphorylation
Sterol biosynthetic process
Peptide transport
Isoprenoid biosynthetic process
Calcium ion transport
Nucleotide metabolic process
Proteolysis involved in cellular protein catabolic process

Table 2C

Q3 Biological Process	
Regulation of transcription	Negative regulation of signal transduction
Notch signaling pathway	Pancreas development
Response to heat	Defense response to bacterium
Osteoblast differentiation	Endocytosis
Negative regulation of cell differentiation	Somitogenesis
Regulation of cell proliferation	Actin filament organization
Pituitary gland development	Wnt receptor signaling pathway
Positive regulation of neuron differentiation	Intracellular signaling cascade
Endoderm development	Epithelial cell differentiation
Organ morphogenesis	Transforming growth factor beta receptor signaling pathway

Table 2D

Q4 Biological Process
G1/S transition of mitotic cell cycle
chromosome organization
establishment of cell polarity
response to salt stress
mRNA export from nucleus

3. Discussion

3.1 Overview

IDPs may have subtypes and the different subtypes may have different functions. One previous study indicated that such disordered subtypes may exist [10]. However, the effort to subclassify IDPs such that each class has its own functional features still remains a difficult task.

Instead of relying on the training of existing data to build specific classifiers by machine learning methods, we took an alternative approach based on the hypothesis that different subtypes of IDPs should exhibit different biophysical features. Such features can be readily captured by applying two different prediction tools, CH and CDF, which use different biophysical characteristics for their evaluation of disorder content. We therefore developed a CH-CDF plot for IDP partition.

3.2 Structural Partitioning by the CH-CDF plot

Proteins partitioned by the CH-CDF plot show a very different PDB coverage rate. The structure quadrant (Q2) has many more proteins identified with at least one PDB entry than the disorder quadrant (Q4). The mixed quadrant (Q3) has a fraction of proteins with PDB hits almost comparable to those in the structure quadrant (Q2). However, their coverage rate percentages are typically in the 20%-30% range, while the coverage in the ordered quadrant (Q2) is often as high as 90%-100%. These data suggest that mixed proteins (in Q3) have more crystal-forming ordered regions than those in the disorder quadrant (Q4).

Even though predicted to be structured, the proteins in the structure quadrant (Q2) have a significant fraction of cases with only 20% coverage (Figure 3, Q2). As indicated by the data in Table 1 and Figure 5, this result likely reflects the fact that many mouse proteins have modular structure and contain multiple structured domains. Thus, the entire protein is, overall, predicted to be structured by both the CH and CDF predictors, but if only one of the domains makes it into PDB, then such a protein could have a low coverage.

Some proteins in the mixed quadrant (Q3) and those in the disorder quadrant (Q4) have coverage percentage almost as high as 100%. After examining them individually, some of them are found to bind to ions, DNAs, RNAs, small molecules, etc. Formation of specific complexes could potentially stabilize disordered proteins, and lead to the formation of a crystallizable structure. However, there are indeed cases where such predicted disordered proteins are monomers by themselves. These need further study.

One of our early hypotheses was that proteins with relatively low net charge and high hydropathy, e.g. predicted to be structured by CH, and yet predicted to be disordered by CDF, e.g. located in (Q3), might undergo hydrophobic collapse yet remain lacking stable structure. Such proteins would likely be native molten globules. An alternative hypothesis is that proteins in (Q3) simply contain mixtures of structured and disordered regions.

We first found that Q3 proteins have many more locally ordered sequence windows, indeed far more than the disordered quadrant (Q4), but less than the structure quadrant (Q2) (Table 1). We then showed that the amino acid sequences in proteins are predicted as mostly ordered if a PDB hit is identified for this region (Figure 5). When the sequence region is not matched with a PDB hit, it is most likely predicted to be disordered. Together these observations suggest that the quadrant (Q3) is likely to contain proteins containing relatively balanced contributions of structured and disordered regions. For this reason here we have named the proteins in this quadrant mixed rather than collapsed disorder, a description that may have appeared in previous publications [16-18]. These observations don't rule out the possibility that some of the proteins in (Q3) or even in (Q4) might be native molten globules. Further analysis and experiments are needed to identify native molten globules and determine where they fall on the CH-CDF Plot.

3.2 The rare protein quadrant (Q1)

Proteins in this quadrant are predicted to be unstructured by CH plot, but ordered by CDF. The disordered prediction from CH plot implies that a protein has high charge and is hydrophilic. Such proteins should not be predicted to be structured by CDF, so it is no wonder that the proteins in this quadrant are rare.

The density plot of PDB coverage percentage distribution for the proteins in (Q1) showed a similar pattern when compared to the structure quadrant (Q2) (fig.3). The proteins in (Q1) also have many more proteins identified with a PDB hit than those in the disorder quadrant (Q4) (fig.4). Therefore, one possibility is that these proteins are overall structured, with some high charged or hydrophilic residues, which is just the opposite of proteins in quadrant (Q3). The GO term analysis showed that 4 out of 5 of the significant GO terms are related to nucleotide processing. Further analysis shows that many of the proteins in all four quadrants including (Q1) have net positive charges rather than net negative ones. We are testing whether the positively charged proteins in (Q1) are associated with RNA binding.

3.3 Disorder subtypes and IDP functions

We tested whether the protein compartmentalization by subtypes resulted in function partition as well. For this test, we did an analysis of GO terms to determine if some terms are biased relative to others in the various quadrants (see Methods for details).

Structured proteins exhibited significant biases towards enzymatic processes and transporters. Both of these processes are well known to be associated with structured proteins [4-6]. Meanwhile, the disorder quadrant (Q4) is mainly biased towards GO terms with mitosis-related functions, which again agrees with previous observations [4-6]. On the other hand, the mixed quadrant (Q3) is highly involved in regulation pathways, which are important in development and differentiation. The recent publication on pluripotent stem cell-inducing proteins, which must be heavily involved in gene regulation, showed that these proteins are mostly localized in the mixed quadrant (Q3) [17]. Interestingly, the plant developmental proteins called GRAS straddled the structure and mixed quadrant (Q2, Q3). The N-domains of these proteins localized to the mixed and disordered quadrants (Q4, Q3), while almost all of the C-domains of these proteins localized to the structured quadrant (Q2) [18].

The flexibility provided by disordered regions could be important in such signaling events. The disordered regions could act as linkers connecting function domains. These regions could also directly bind to partners, functioning as Molecular Recognition Features (MoRFs). Such binding is usually accompanied by a disorder → order transition. Because of their flexibility, they might be able to bind to multiple partners, acting as hub proteins in signaling networks. Their flexibility is also capable of fast but short time-span binding, which also may be crucial in signaling events.

4. Methods

4.1 Protein data

The *Mus. musculus* proteome were gathered from Uniprot 15.0 [20]. A total number of 58881 sequences were obtained. Blastclust (http://www.ncbi.nlm.nih.gov/Web/Newsltr/Spring04/ blast lab.html) with default settings were used to reduce redundancy.

4.2 PDB Coverage

PDB monomers data is downloaded from PDBe PISA. Gapped-BLAST algorithm was used to compare query sequences to PDB monomers, with default scoring matrix (BLOSUM 62). A hit was identified only when the hit region is larger than 85% of the PDB monomer sequence, and with more than 30% identity.

4.2 GO term analysis

We downloaded GO terms associated with each protein from GO Database. To reduce protein redundancy, proteins were clustered into protein families by the Blastclust program. If sequence s_i was assigned to cluster $c(s_i)$, and n_i is the total number of proteins assigned to this family, we define a weight $w(s_i)$ for this sequence as $w(s_i) = \frac{1}{n_i}$.

Our 509,214 proteins are in association with 10,703 GO annotations. Protein sequence s_i grouped into a quadrant $k, k = 1,2,3,4$ as group g_k. And for $GO_j, j = 1,2,...10703$, there is a cluster of proteins related to GO_j, as $C_j, j = 1,2,...10703$. Therefore, we calculate $n_{j,k}$, the number of proteins related to GO_j in quadrant k by $n_{j,k} = \sum w(s_i), s_i \in g_k \cap C_j$.

In the next step, we compare $n_{j,1}, , n_{j,2}, , n_{j,3}$ and $n_{j,4}$ for every specific GO term GO_j to examine if GO_j has any bias towards certain quadrant. The expected value of protein frequency in quadrant k is calculated as $E_{j,k} = p_k \cdot \sum_{k=1}^{4} n_{j,k}$, with p_k being the proportion of numbers of proteins in quadrant k. Then we compute X_j^2, sum of expectancy, as $X_j^2 = \sum_{k=1}^{4} (O_{j,k} - E_{j,k})^2 / E_{j,k}$. X_j^2 follows a chi-square distribution with 3 degrees of freedom, $X_j^2 \sim \chi^2(3)$.

Under the null hypothesis that GO_j distributed in 4 quadrants according to expectancy, we can derive p_j as a p-value for GO_j. Since multiple statistic tests are applied, we use the Bonferroni correction to adjust obtained p-value. It reduces the scale of significant results as well. A threshold of 0.05 is chosen, and GO terms with p-values lower than 0.05 are collected.

References

1. Dunker, A. K., C. J. Brown and Z. Obradovic (2002). "Identification and functions of usefully disordered proteins." Adv Protein Chem 62: 25-49.
2. Oldfield, C. J., Y. Cheng, M. S. Cortese, C. J. Brown, V. N. Uversky and A. K. Dunker (2005). "Comparing and combining predictors of mostly disordered proteins." Biochemistry 44(6): 1989-2000.
3. Iakoucheva, L. M., C. J. Brown, J. D. Lawson, Z. Obradovic and A. K. Dunker (2002). "Intrinsic disorder in cell-signaling and cancer-associated proteins." J Mol Biol 323(3): 573-84.
4. Xie, H., S. Vucetic, L. M. Iakoucheva, C. J. Oldfield, A. K. Dunker, V. N. Uversky and Z. Obradovic (2007). "Functional anthology of intrinsic disorder. 1. Biological processes and functions of proteins with long disordered regions." J Proteome Res 6(5): 1882-98.
5. Vucetic, S., H. Xie, L. M. Iakoucheva, C. J. Oldfield, A. K. Dunker, Z. Obradovic and V. N. Uversky (2007). "Functional anthology of intrinsic disorder. 2. Cellular components, domains, technical terms, developmental processes, and coding sequence diversities correlated with long disordered regions." J Proteome Res 6(5): 1899-916.
6. Xie, H., S. Vucetic, L. M. Iakoucheva, C. J. Oldfield, A. K. Dunker, Z. Obradovic and V. N. Uversky (2007). "Functional anthology of intrinsic disorder. 3. Ligands, post-translational modifications, and diseases associated with intrinsically disordered proteins." J Proteome Res 6(5): 1917-32.
7. Murzin, A. G., S. E. Brenner, T. Hubbard and C. Chothia (1995). "SCOP: a structural classification of proteins database for the investigation of sequences and structures." J Mol Biol 247(4): 536-40.
8. Orengo, C. A., A. D. Michie, S. Jones, D. T. Jones, M. B. Swindells and J. M. Thornton (1997). "CATH--a hierarchic classification of protein domain structures." Structure 5(8): 1093-108.
9. Reeck, G. R., C. de Haen, D. C. Teller, R. F. Doolittle, W. M. Fitch, R. E. Dickerson, P. Chambon, A. D. McLachlan, E. Margoliash, T. H. Jukes and et al. (1987). ""Homology" in proteins and nucleic acids: a terminology muddle and a way out of it." Cell 50(5): 667.
10. Vucetic, S., C. J. Brown, A. K. Dunker and Z. Obradovic (2003). "Flavors of protein disorder." Proteins 52(4): 573-84.
11. Uversky, V. N. and A. K. Dunker "Understanding protein non-folding." Biochim Biophys Acta 1804(6): 1231-64.
12. Uversky, V. N., J. R. Gillespie and A. L. Fink (2000). "Why are "natively unfolded" proteins unstructured under physiologic conditions?" Proteins 41(3): 415-27.
13. Oldfield, C.J., Cheng, Y., Cortese, M.S., Brown, C.J., Uversky, V.N., and Dunker, A.K. Comparing and combining predictors of mostly disordered proteins. Biochemistry. 44:1989-2000 (2005).
14. Xue, B., C. J. Oldfield, A. K. Dunker and V. N. Uversky (2009). "CDF it all: consensus prediction of intrinsically disordered proteins based on various cumulative distribution functions." FEBS Lett 583(9): 1469-74.
15. LeGall, T., P.R. Romero, M.S. Cortese, V.N. Uversky, and A.K. Dunker. "Intrinsic disorder in the Protein Data Bank." J Biomol Struct Dyn **24**(4): 325-42 (2007).
16. Mohan, A., Sullivan, Jr., W.J., Radivojac, P., Dunker, A.K., and Uversky, V.N. "Intrinsic disorder in pathogenic and non-pathogenic microbes: discovering and analyzing the unfoldome of early-branching eukaryotes." Molecular Biosystems 4: 328-340 (2008).
17 Xue, B. C.J. Oldfield, Y.Y.Van, A.K. Dunker, and V.N. Uversky. "Protein intrinsic disorder and induced pluripotent stem cells." Mol Biosystems (In Press)
18. Sun, Z., B. Xue, W.T. Jones, E. Rikkerink, A.K. Dunker, and V.N. Uversky. "A functionally required unfoldome from the plant kingdom: intrinsically disordered N-terminal domains of GRAS proteins are involved in molecular recognition during plant development." Plant Mol. Biol. (In Press).
19. Obradovic, Z., K. Peng, S. Vucetic, P. Radivojac and A. K. Dunker (2005). "Exploiting heterogeneous sequence properties improves prediction of protein disorder." Proteins **61 Suppl 7**: 176-82.
20. Magrane, M. and U. Consortium "UniProt Knowledgebase: a hub of integrated protein data." Database (Oxford) **2011**: bar009.

COEVOLVED RESIDUES AND THE FUNCTIONAL ASSOCIATION FOR INTRINSICALLY DISORDERED PROTEINS

CHAN-SEOK JEONG and DONGSUP KIM*

Department of Bio and Brain Engineering, Korea Advanced Institute of Science and Technology (KAIST)
335 Gwahangno, Yuseong-gu, Daejeon 305-701, Republic of Korea
**E-mail: kds@kaist.ac.kr*

The evolution of intrinsically disordered proteins has been studied primarily by focusing on evolutionary changes at an individual position such as substitution and conservation, but the evolutionary association between disordered residues has not been comprehensively investigated. Here, we analyze the distribution of residue-residue coevolution for disordered proteins. We reveal that the degree of coevolved residues significantly decreases in disordered regions regardless of the sequence propensity, and the degree distribution of coevolved and conserved residues exclusively differs in each functional category. Consequently, the coevolution information can be useful for predicting intrinsic disorder and understanding biological functions of a disordered region from the sequence.

1. Introduction

Intrinsically disordered proteins represent proteins that do not have a well-structured fold and undergo dynamic conformational changes. As disordered proteins perform important and complex biological functions such as signaling, regulation, and post-translational modification,[1,2] understanding disordered proteins becomes increasingly important. However, our knowledge on the evolution of disordered proteins and its relationship with functions have been limited.

Because of the absence of well-defined structures, sequence-based analysis plays a crucial role in predicting disordered proteins and the biological functions.[3] Initially, a low sequence complexity has been indicated as a feature of highly disordered proteins.[4] A study that analyzed sequence alignments of disordered proteins has indicated that disordered proteins have different evolutionary patterns of substitutions from ordered proteins.[5] Moreover, a model-based study has revealed that the likelihood of evolutionary mutations and the conservative substitutions for disordered proteins are different from those for ordered proteins.[6] Furthermore, an *in silico* mutagenesis study has compared the conservation of secondary structures such as helices and strands and intrinsic disorders.[7] Although the *in silico* study is based on order-biased evolutionary models,[8] it has shown that long intrinsic disorders are not easily maintained and requires a specific effort for the disorder conservation while the secondary structures are tolerant to random mutations. Therefore, the evolution of disordered proteins exhibits large differences in sequence complexity, amino acid composition, substitution, and conservation.

While the distribution of amino acids at an individual residue has been extensively studied, the evolutionary association between distinct residues, called herein coevolution, has not been explored sufficiently. Coevolution is useful information that complements conservation information, because it reveals inter-relationship between variable residues and differentiate the associated residues from other independently evolved residues. A number of studies have shown that coevolution information is related with structural and functional knowledge.[9–14]

However, the coevolution of disordered protein has not been comprehensively investigated. Although a recent study have indicated that the degree of coevolution in disordered proteins is low because of the rapid evolution rate,[14] the relationship between the coevolution signal and functionality of disordered proteins has not been investigated. The use of coevolution information for disordered proteins can be useful, because conservation information only shows a limited performance in associating the evolutionary information with the functions of disordered proteins. Particularly, according to a recent study where intrinsic disorders are classified with the evolutionary conservation of disorder and sequence, a majority of canonical intrinsic disorders belong to the "flexible disorder" category where the disorder is evolutionarily conserved but the sequences are not.[15] Consequently, the functions frequently occurred with disordered regions are not differentiated solely by the sequence conservation information, and an additional information compensating the sequence conservation is required to understand the functions.

In this study, we investigate the distribution of coevolved residues in disordered proteins experimentally determined. When comparing to non-disordered regions, disordered regions show a low degree of coevolved residues. Additionally, we investigate the relationship between coevolution and biological functions of evolutionary conserved disorders in combination with conservation information, and reveal that each functional category has a unique degree distribution of coevolved and conserved residues.

2. Materials and methods

2.1. *Data set*

Disordered proteins are collected from DisProt v5.7,[16] a manually curated database with experimentally determined structural disorders. To prevent amino acid sequence bias, the disordered proteins are clustered by running CD-HIT[17] with the maximum sequence identity threshold of <60%, and only the representative proteins are taken. Next, the proteins with less than ten homolog sequences in the multiple sequence alignments, and the residues with less than five sequences aligned at the positions are excluded for a reliable estimation of evolutionary information. Consequently, 519 disordered proteins with 57,261 disordered and 163,470 non-disordered residues are considered.

For comparison, ordered protein set consisting of 83 protein with 17,451 residues is used. The proteins have well-determined structures and exclusively represent distinct protein families. The proteins are originally organized and used for evaluating the performance of correlated mutation algorithms that estimating coevolution.[11,18,19]

2.2. *Multiple sequence alignment construction*

For each disordered or ordered protein, the multiple sequence alignment is automatically constructed with the homolog sequences. Homolog sequences are collected by running PSI-BLAST[20] with the options "-e 0.001 -h 0.001 -j 3" against the NCBI NR database clustered with <90% sequence identity, and then the hits with <50% alignment coverage are omitted by running the script program, alignhits.pl in the HHsearch package,[21] with the options "-e 0.01 -cov 50.". Subsequently, the multiple sequence alignment is constructed by running

MUSCLE[22] on the remaining hits with the fastest option "-stable -quiet -maxiters 1 -diags -sv -distance1 kbit20_3."

2.3. *Coevolution estimation*

A coevolution score between two residues is estimated with MIp measurement.[19] The MIp score is derived from the mutual information score defined as

$$\mathrm{MI}(i,j) = \sum_x \sum_y P(x_i, y_j) \log\left(\frac{P(x_i, y_j)}{P(x_i)\,P(y_j)}\right)$$
$$P(x_i) = \sum_y P(x_i, y_j)$$
$$P(y_j) = \sum_x P(x_i, y_j),$$

where $P(x_i, y_j)$ is the joint probability of amino acid x being at position i and amino acid y being at position j. The MIp score is calculated by removing the background noise from the original mutual information score as follows,

$$\mathrm{MIp}(i,j) = \mathrm{MI}(i,j) - \frac{\mathrm{MI}(i,\cdot)\mathrm{MI}(\cdot,j)}{\mathrm{MI}(\cdot,\cdot)},$$

where $\mathrm{MI}(i,\cdot)$ and $\mathrm{MI}(\cdot,j)$ is the mutual information score at position i and position j averaged over all other positions, respectively, and $\mathrm{MI}(\cdot,\cdot)$ is the MI value averaged over all pairs of positions. To prevent an erroneous measurement and achieve an accurate estimation of coevolution signal, we incorporate the pseudocount and marginal probability constraint based on sequence profile. To address the sequence redundancy of multiple sequence alignment, the sequence weight and the effective number of sequences are calculated using the position-based sequence weight[23] and the exponential of negative entropy averaged over all columns,[24] respectively. The above procedure is available via web [a].

The estimated MIp scores are standardized into Z-scores, and the residue pairs whose MIp Z-scores exceed 4.0 are chosen as coevolved residue pairs. Finally, coevolved residues are defined as the residues comprising coevolved residue pairs.

2.4. *Sequence conservation estimation*

Sequence conservation is estimated as Shannon's entropy with the reverse sign.[25] For example, a conservation score at position i is defined as

$$\mathrm{C}(x_i) = \sum_x P(x_i) \log\left(P(x_i)\right),$$

where $P(x_i)$ is the probability of amino acid x being at position i. Similar to the coevolution estimation, the same methods for estimating the sequence weight and the effective number of sequences are also used in the conservation score estimation.

The entropy-based conservation scores are standardized into Z-scores, and the residues whose Z-score exceed 0.8 are chosen as conserved residues.

[a]http://binfolab12.kaist.ac.kr/cmat/

2.5. *Disorder conservation estimation*

The degree of conservation of structural disorder is estimated by the ratio of predicted disordered residues among the residues aligned at the position as following the previous study.[15] For each sequence in multiple sequence alignment, the intrinsic disorders are predicted by running DisEMBL.[26] If a residue is experimentally determined as a disordered residue, and more than 40% of the aligned sequence positions are predicted as disordered residues, the residue is defined as a conserved disordered residue.

2.6. *Functional categories*

To investigate the relationship between the evolutionary information and biological functions of disordered regions, we incorporate the functional categories annotated in DisProt. There exist three kinds of categories, called structural functional type, functional class, and functional subclass, respectively.[2,27–29] The structural functional type represents a structural transition upon performing the function, and the functional class and subclass represent biological functions at different levels of details. As the disorder functions are not categorized exclusively, a disordered region can belong to multiple functional classes and subclasses, and subsequently affect their statistics simultaneously. At functional subclass level, we only consider the functional subclasses assigned at more than 1,000 distinct residues.

3. Results

3.1. *Distribution of coevolved residues for disordered proteins*

We found that the fraction of coevolved residues differs in disordered regions. The fraction of coevolved residues for disordered proteins is 36.02%, similar to that for ordered proteins, 37.12%. However, separating disorder and non-disordered regions in disordered proteins, the fraction of coevolved residues differs as 24.92% and 36.19%, respectively. That is, the degree of coevolved residues significantly drops in disordered regions. As the disordered proteins have different sequence lengths, we also examined the the degree of coevolution with respect to the degree of disorder. Figure 1 shows the frequencies of the different ratios of disordered and coevolved residues for disordered proteins. As the proteins have more disordered residues, the average fractions of coevolved residues decrease. Thus, the degree of coevolution for the disordered proteins is negatively correlated with the degree of intrinsic disorder.

Since the coevolution scores are estimated from multiple sequence alignments, the sequence diversity and the sequence complexity may influence the different degree of coevolution signals between disordered and ordered proteins, and the alignment quality also can. To check the effect of sequence diversity, we first examined the fraction of coevolved residues against disordered and non-disordered residues with the number of aligned sequences and another sequence diversity measure, the effective number of aligned sequences defined as the exponential of negative entropy averaged over all columns of aligned sequences.[24] As shown in Figure 2A, the fraction of coevolved residues for disordered residues is less than that for non-disordered residues. To take account of sequence redundancy, the degrees of coevolution of the disordered and non-disordered residues are compared after binning them according to the effective number of sequences. As shown in Figure 2B, the disordered residues show smaller fractions of coevolved residues, indicating that disordered residues have a lower degree of coevolution signal than non-disordered residues regardless of the diversity of the homolog sequences. Second, we examined the fraction of coevolved residues against disordered and non-disordered residues with various range of sequence complexities as shown in Figure 2C. It has been known that disordered regions exhibit a low sequence complexity, so the different population of coevolved residues may be a result of different sequence complexity. The sequence complexity is calculated by Shannon's entropy measure with a 45-aa window as following the previous study.[4] At every range of sequence complexity, disordered residues have a smaller fraction of coevolved residues than non-disordered residues. Third, we examined the fraction of coevolved residues against disordered and non-disordered residues with various range of sequence conservation score as shown in Figure 2D. As the alignment quality is usually estimated in comparison with structural alignment, the alignment quality for disordered region is hard to estimate. Instead, sequence conservation is used as an indirect measure of alignment quality, because aligned position with high sequence conservation are more likely accurate. The fraction of coevolved residues for disordered residues is consistently less than that for non-disordered residues at every range of sequence conservation score. Consequently, the low degree of coevolution signals in disordered regions can be though as a feature of intrinsic disorder not the bias from sequence pool and alignment error.

We also investigated the degree of coevolved residues regarding to the evolutionary classification of intrinsic disorder introduced by a recent study.[15] According to the study, intrinsic disorder can be classified into three different groups, such as constrained, flexible, and non-conserved disorders. The constrained disorder consists of residues whose disorders and amino acids are evolutionarily conserved, the flexible disorder consists of residues whose disorders are evolutionarily conserved but amino acids are not, and the non-conserved disorder consists of residues whose disorders are not evolutionarily conserved. The two conserved disorders, the constrained and flexible, show the relevant association with biological functions of intrinsic disorder, while the non-conserved disorder does not show a common functional feature. When

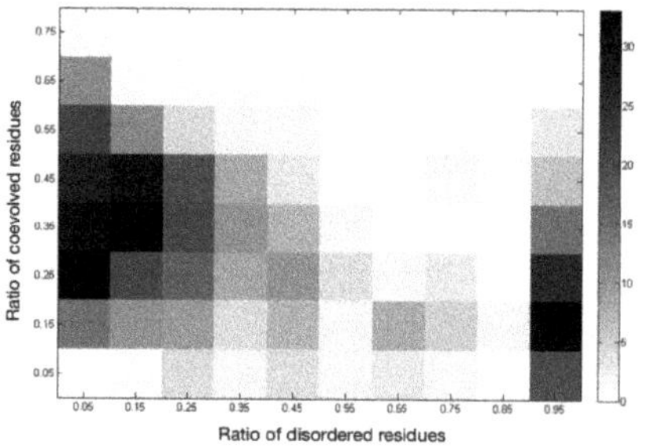

Fig. 1. Densities of disordered proteins with different ratios of disordered and coevolved residues. The intensity represents the frequency of disordered proteins with the ratios of disordered and coevolved residues.

we classified disordered residues according to the the disorder and sequence conservations, the fractions of coevolved residues for the constrained, flexible, and non-conserved disorders appear as 0.2808, 0.2536, and 0.2717, respectively. The flexible disorders show the smaller fraction of coevolved residues than the constrained and non-conserved disorders with the p-values of 2.2×10^{-5} and 6.5×10^{-3}, respectively. We repeated the evolutionary classification by replacing the sequence conservation measure with the maximum frequency of amino acids at aligned sequence positions as following the previous study, but the result does not change much. The fractions of coevolved residues for the constrained, flexible, and non-conserved disorders appear as 0.2799, 0.2457, and 0.2717, respectively, and the p-values of the flexible disorders against the constrained and non-conserved disorders are 4.4×10^{-17} and 1.8×10^{-4}, respectively. The previous study has indicated that flexible and constrained disorders are different in their functional characteristics, as flexible disorder usually presents short linear motifs and performs in signaling,[30] while constrained disorder is associated with the functions requiring tight sequence constraint.[15] Moreover, it has been suggested that contained disorder can often adopt more strict conformations than flexible disorder, so its higher degree of coevolution may come from the same constraint. Another high degree of coevolution observed in non-conserved disorder can be explained with the high rate of ordered protein residue. The aligned protein residues in the non-conserved disorder possibly have a high ratio of structured residues, and thus the background coevolution signal may influence the high degree of coevolved residue in the regions. It is important to note the evolutionarily different disorder groups have distinct functionalities, and in our results, the degrees of coevolved residues differ in those groups.

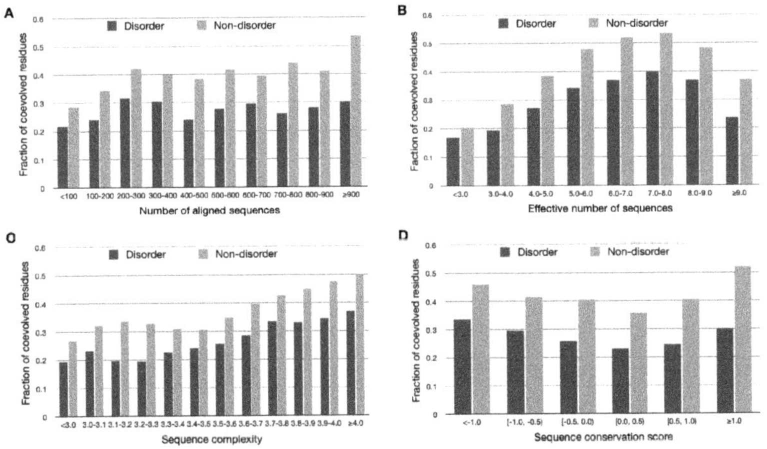

Fig. 2. Degree distribution of coevolved residues in disordered and non-disordered regions. (A) Fraction of coevolved residues to all residues with different ranges of number of aligned sequences. (B) Fraction of coevolved residues to all residues with different ranges of sequence complexity. (C) Fraction of coevolved residues to all residues with different ranges of sequence conservation score.

That is, the coevolution information can be a useful for understanding biological functions of intrinsic disorder.

3.2. *Relationship between coevolution and functions*

Intrinsic disorders have been classified with the evolutionary conservation in the predicted disorder and the sequence, and evolutionary conserved disordered regions have the frequently occurred functions of intrinsic disorders.[15] However, major functions of intrinsic disorder such as signaling and regulation are not differentiated only by the sequence conservation information. Here, we further investigate the relationship between coevolution information and common functions of evolutionary conserved disorders in combination with sequence conservation information.

The content of conserved and coevolved residues significantly differs according to the structural functional type. As shown in Figure 3, two main structural functional types, the disorder that performs the function in disordered state and the disorder-to-order that undergoes structural transition from disordered to ordered state, have distinct frequencies of conserved and coevolved residues. The disorder-to-order type has more conserved and coevolved residues than disorder type, which implies the higher structural constraints upon the structural transition. Additionally, this can support the association between constrained disorder and the structural transition that the previous study has speculated.[15] On the other hand, the disorders whose functions arise within the disordered state are less evolutionarily restrained. As consequence, the structural functional types have distinct proportion of evolutionary conserved and coevolved residues.

Each functional class differs in the ratio of conserved and coevolved residues as shown in Figure 4. The effector class with permanent binding, that recognizes a partner molecule and modifies the activity, shows the highest degree of coevolved residues among the six functional classes. Another class with permanent binding, the assembly class, have relatively more conserved residues than the effector class. The other class with permanent binding, the scavenger class, has the highest ratio of conserved residues. Since the effector and assembly classes have known as related to disorder-to-order transitions,[31–33] their high degrees of conservation and coevolution reflect the functionality. However, the preferred contents are different in combination of both evolutionary information. Specifically, the effector class prefers the coevolved variable residues most. In the similar aspect, the modification site and the chaperone classes share transient binding property,[2,27] but are distinct from each other with the degree of conser-

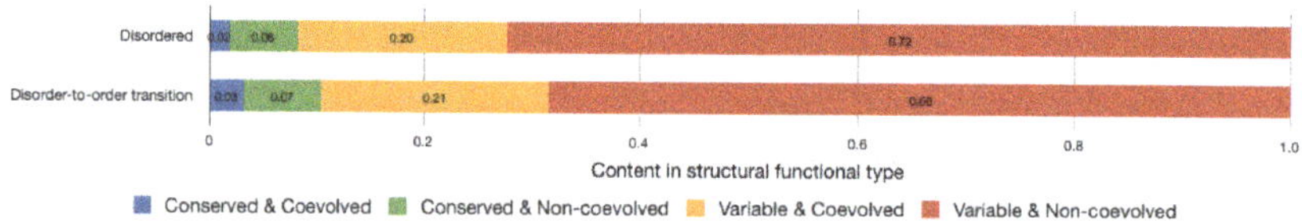

Fig. 3. Relative frequency of conserved and coevolved residues in various structural functional types of disordered residues.

vation and coevolution. The modification site class prefers conservation, while the chaperone class does coevolution of variable residues. The entropic chain class has a large fraction of variable and non-coevolved residues, which reflects the functionality arising from the disordered state.[2] The modification site and the entropic chain classes share a similar content of conservation and coevolution information. Nevertheless, the use of coevolution information reasonably differentiates the effector, the assembly, and the modification site classes, which have almost similar proportions of conserved residues. That is, the degree of coevolution in each functional class can reflect the function and uniquely discriminate the relevant functional class in combination with the degree of conservation.

At the functional subclass level, each functional subclass has a distinct ratio of conserved and coevolved residues as shown in Figure 5. The protein-DNA binding subclass has the highest ratio of conserved residues. The substrate/ligand binding, intra protein interaction, and transactivation subclasses commonly have a high degrees of conserved residues, but significantly differ from each other with the fractions of coevolved residues. In particular, the substrate/ligand binding subclass has a high fraction of coevolved residues, and the other two subclasses have low fractions of coevolved residues. Importantly, various kinds of binding subclasses, such as the protein-protein, protein-DNA, protein-rRNA, substrate/ligand, and metal binding subclasses, are meaningfully separated with the combination of the ratios of coevolved and conserved residues, despite their similarity with permanent binding.[2,27,31–33] The phosphorylation subclass, that comprises a majority of the modification site class, and the autoregulatory subclass, that frequently undergoes some kinds of modifications during playing the role,[2] similarly have a large fraction of variable and non-coevolved residues, but their frequencies conserved and coevolved residues separate the functional subclasses. As at the functional class level, some functional subclasses such as the metal binding, the protein-protein binding, and the phosphorylation subclasses are differentiated by combining coevolution information, even though they have a common enrichment of conserved residues. Therefore, the functional subclasses can be uniquely separated by the degree of coevolution and conservation, likewise the functional classes.

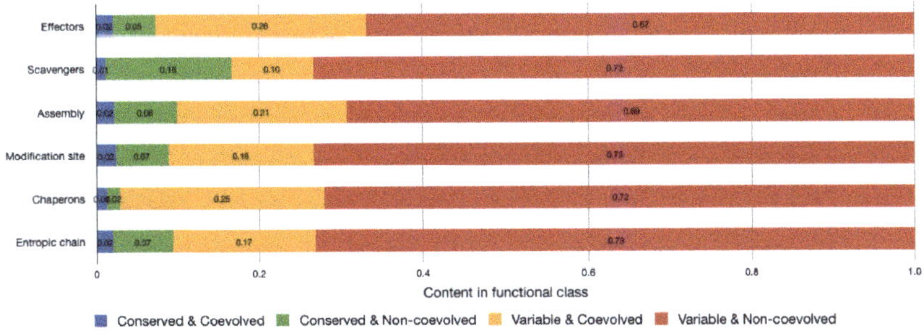

Fig. 4. Relative frequency of conserved and coevolved residues in various functional classes of disordered residues.

4. Discussion

We have demonstrated that coevolved residues are less populated in disordered region. The lack of structural constraints for disordered regions would be the main reason for the degree distribution, because of two reasons. The first reason is that the structural constraints such as intra-molecular contacts are known as one of main factors that develop coevolution signals,[34,35] and the second reason is that the degree of coevolution in disordered regions shows a negative correlation with the degree of disordered residues. That is, unlike ordered residues, disordered residues do not lie in strong structural constraints, lacking a well-structured fold and dynamically changing the conformations, which consequently reduces the degree of evolutionarily coupled residues.

The low degree of coevolved residues in disordered regions can be an effective indicator for disorder prediction. To explore the usefulness of coevolution information for disorder prediction, the enrichment of coevolved residues in neighboring residues within ±7 aa long has been estimated as shown in Figure 6. For disordered residues, 44.4% of disordered residues have zero or one coevolved residues among the neighboring residues. On the other hand, for ordered residues, the number of coevolved residues are evenly distributed regardless of the secondary structure elements, as 5.6–18.0%. Hence, the coevolution information can be useful for disorder prediction. Although some of currently available disorder prediction methods, categorized to as contact-based method, implicitly incorporate coevolution information, those methods do not fully exploit the effectiveness.[36] Some of them just use contact potential without evolutionary information of a target protein.[37,38] Even though Ucons,[39] that directly incorporates contact predictions using evolutionary information,[40] coevolutionary information are not included in the contact prediction procedure. Therefore, the use of coevolutionary information

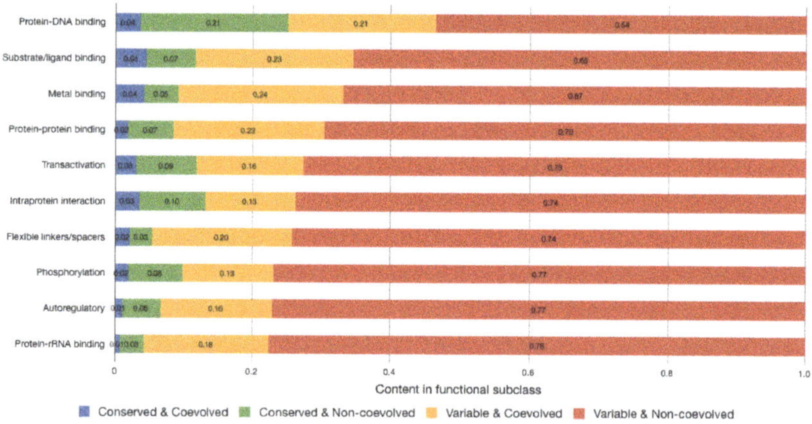

Fig. 5. Relative frequency of conserved and coevolved residues in various functional subclasses of disordered residues.

can effectively improve the current methods for disorder prediction.

Even though disordered residues tend to be less coevolved than ordered residues, a relative ratio of coevolved residues in a disordered region can be an effective indicator to the biological function. When we have examined the proportions of conserved and coevolved residues for various functional annotations at different levels and criteria, the functional annotations have exclusively distinct proportional features. In particularly, many of those functional annotations that share a similar sequence conservation degree can be differently classified in combination with coevolution information. Therefore, the coevolution information can be useful for specifying biological functions of disordered proteins as well as the disorder prediction.

According to Jeon et. al.'s study[14] that explores the relationship between coevolution and conformation changes, the degree of coevolution in flexible regions of ordered proteins encodes the structural transitions. In addition, they have concluded that disordered proteins have a low degree of coevolution, not reflecting the high flexibility. However, in the present study, we have compared the degree distributions of coevolved residues to the overall distribution as background statistics, and revealed the association between coevolution and dynamic behavior such as structural functional types. Accordingly, structural transitions of disordered proteins seem to be encoded as the coevolution signals like those of ordered proteins, but additional studies are needed to clarify the functional association. Because disordered and ordered proteins have different background degrees of coevolution, prior knowledge for the background coevolution upon the structural disorder need to be investigated and incorporated to normalize the coevolution signal, which can be critical to revealing the functional roles of coevolution information.[41]

Coevolution information can be useful for understanding disordered proteins due to three reasons. First, as we have demonstrated, the degree of coevolved residues is related with functional classifications. For example, disordered residues whose function arises from the disordered state show a low degree of coevolution signal, and, on the other hand, disordered residues whose function requires specific recognitions as in effector class show a high degree

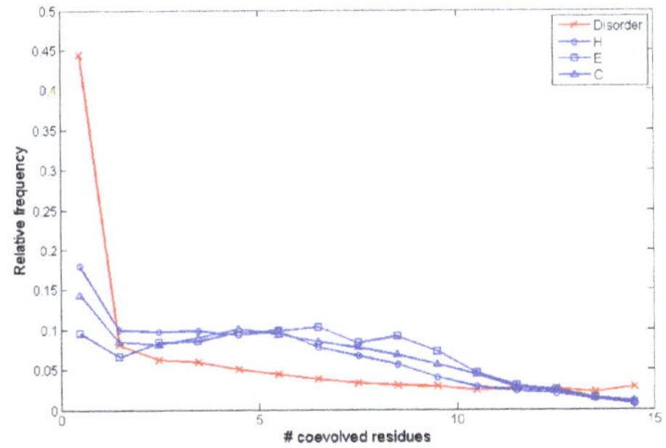

Fig. 6. Distribution of the number of coevolved residues within neighboring positions for disordered and ordered residues. Disorder indicates disordered residues. H, E, and C indicate ordered residues belonging to helix, sheet, and coil by secondary structure, respectively.

of coevolution signal. Second, the coevolution information successfully compensates the conservation information. We have demonstrated a group of functional categories, similar to each other with the degree of conservation but different with the degrees of coevolution signals. Third, coevolution information describes evolutionary association between residues which cannot be represented by conventional sequence analyses. Thus, coevolution analysis would be beneficial to understanding the evolution of disordered proteins.

Although we have incorporated an advanced coevolution estimate with high reliability and robustness, the analysis results are still dependent on multiple sequence alignments. Because the alignment quality is usually estimated in comparison with structural alignment and most of alignment methods are designed for ordered proteins,[42] alignment quality of disordered proteins are difficult to be estimated. Even though, in this study, we have used sequence conservation score as an indirect measure of alignment quality, the effect of alignment quality for disordered proteins needs to be examined more apparently. As a separately calculated substitution matrix for disorder proteins has improved the alignments of them, similarly, a development of multiple sequence alignment methods for disordered proteins would be useful for an accurate coevolution analysis of them.

This study can be extended to a genome-scale analysis by combining disorder prediction methodologies.[43] In the present work, we have chosen only the experimentally determined disordered proteins from DisProt database. Although the sequence redundancy has been removed with sequence identity, the result statistics still depends on the content of the original database. Thus, by incorporating disorder prediction methodologies and then applying the present analysis procedures, more realistic knowledge about coevolution in disordered proteins can be obtained as a lot of prediction-based studies.[15,44–46] Moreover, various kinds of biological information associated with intrinsic disorder can be easily combined and considered in the analysis procedure.

Acknowledgments

We thank all Bioinformatics and Computational Biology Laboratory (BCBL) members for helpful discussion. This work was supported by a grant of the Korea Healthcare technology R&D Project, Ministry for Health, Welfare & Family Affairs, Republic of Korea [A092006]; and the National Research Foundation of Korea (NRF) grant funded by the Korea Government (MEST) [2009-0086964].

References

1. P. Radivojac, L. M. Iakoucheva, C. J. Oldfield, Z. Obradovic, V. N. Uversky and A. K. Dunker, *Biophys. J.* **92**, 1439 (March 2007).
2. A. K. Dunker, C. J. Brown, J. D. Lawson, L. M. Iakoucheva and Z. Obradovic, *Biochemistry* **41**, 6573 (May 2002).
3. Z. Dosztanyi, B. Meszaros and I. Simon, *Brief Bioinformatics* **11**, 225 (March 2010).
4. P. Romero, Z. Obradovic, X. Li, E. C. Garner, C. J. Brown and A. K. Dunker, *Proteins* **42**, 38 (January 2001).
5. P. Radivojac, Z. Obradovic, C. J. Brown and A. K. Dunker, *Pac Symp Biocomput* , 589 (2002).
6. C. J. Brown, A. K. Johnson and G. W. Daughdrill, *Mol Biol Evol* **27**, 609 (March 2010).

7. C. Schaefer, A. Schlessinger and B. Rost, *Bioinformatics* **26**, 625 (March 2010).
8. C. J. Brown, A. K. Johnson, A. K. Dunker and G. W. Daughdrill, *Curr Opin Struct Biol* **21**, 441 (June 2011).
9. S. Chakrabarti and A. R. Panchenko, *PLoS ONE* **5**, p. e8591 (January 2010).
10. A. Kowarsch, A. Fuchs, D. Frishman and P. Pagel, *PLoS Comput Biol* **6**, p. e1000923 (2010).
11. B.-C. Lee and D. Kim, *Bioinformatics* **25**, 2506 (October 2009).
12. N. Halabi, O. Rivoire, S. Leibler and R. Ranganathan, *Cell* **138**, 774 (August 2009).
13. S. Chakrabarti and A. R. Panchenko, *Proteins* **75**, 231 (April 2009).
14. J. Jeon, H.-J. Nam, Y. Choi, J.-S. Yang, J. Hwang and S. Kim, *Mol Biol Evol* (April 2011).
15. J. Bellay, S. Han, M. Michaut, T. Kim, M. Costanzo, B. J. Andrews, C. Boone, G. D. Bader, C. L. Myers and P. M. Kim, *Genome biology* **12**, p. R14 (February 2011).
16. M. Sickmeier, J. A. Hamilton, T. LeGall, V. Vacic, M. S. Cortese, A. Tantos, B. Szabo, P. Tompa, J. Chen, V. N. Uversky, Z. Obradovic and A. K. Dunker, *Nucleic Acids Res* **35**, D786 (January 2007).
17. W. Li and A. Godzik, *Bioinformatics* **22**, 1658 (July 2006).
18. C. M. Buslje, J. Santos, J. M. Delfino and M. Nielsen, *Bioinformatics* **25**, 1125 (May 2009).
19. S. D. Dunn, L. M. Wahl and G. B. Gloor, *Bioinformatics* **24**, 333 (February 2008).
20. S. F. Altschul, T. L. Madden, A. A. Schäffer, J. Zhang, Z. Zhang, W. Miller and D. J. Lipman, *Nucleic Acids Res* **25**, 3389 (September 1997).
21. J. Söding, *Bioinformatics* **21**, 951 (April 2005).
22. R. C. Edgar, *Nucleic Acids Res* **32**, 1792 (January 2004).
23. S. Henikoff and J. G. Henikoff, *J Mol Biol* **243**, 574 (November 1994).
24. A. Biegert and J. Söding, *Proc Natl Acad Sci USA* **106**, 3770 (March 2009).
25. J. Pei and N. V. Grishin, *Bioinformatics* **17**, 700 (August 2001).
26. R. Linding, L. J. Jensen, F. Diella, P. Bork, T. J. Gibson and R. B. Russell, *Structure/Folding and Design* **11**, 1453 (November 2003).
27. P. Tompa, *Trends Biochem. Sci.* **27**, 527 (October 2002).
28. V. N. Uversky, *Protein Sci* **11**, 739 (April 2002).
29. H. J. Dyson and P. E. Wright, *Nat. Rev. Mol. Cell Biol.* **6**, 197 (March 2005).
30. F. Diella, N. Haslam, C. Chica, A. Budd, S. Michael, N. P. Brown, G. Trave and T. J. Gibson, *Front. Biosci.* **13**, 6580 (2008).
31. A. P. Demchenko, *J. Mol. Recognit.* **14**, 42 (2001).
32. H. J. Dyson and P. E. Wright, *Curr Opin Struct Biol* **12**, 54 (February 2002).
33. R. S. Spolar and M. T. Record, *Science* **263**, 777 (February 1994).
34. G. Shackelford and K. Karplus, *Proteins* **69 Suppl 8**, 159 (January 2007).
35. U. Göbel, C. Sander, R. Schneider and A. Valencia, *Proteins* **18**, 309 (April 1994).
36. F. Orosz and J. Ovádi, *Bioinformatics* **27**, 1449 (June 2011).
37. Z. Dosztanyi, V. Csizmok, P. Tompa and I. Simon, *Bioinformatics* **21**, 3433 (August 2005).
38. O. V. Galzitskaya, S. O. Garbuzynskiy and M. Y. Lobanov, *Bioinformatics* **22**, 2948 (December 2006).
39. A. Schlessinger, M. Punta and B. Rost, *Bioinformatics* **23**, 2376 (September 2007).
40. M. Punta and B. Rost, *Bioinformatics* **21**, 2960 (July 2005).
41. A. D. Fernandes and G. B. Gloor, *Bioinformatics* **26**, 1135 (May 2010).
42. J. D. Thompson, B. Linard, O. Lecompte and O. Poch, *PLoS ONE* **6**, p. e18093 (2011).
43. F. Ferron, S. Longhi, B. Canard and D. Karlin, *Proteins* **65**, 1 (October 2006).
44. J. W. Chen, P. Romero, V. N. Uversky and A. K. Dunker, *J. Proteome Res.* **5**, 888 (April 2006).
45. J. W. Chen, P. Romero, V. N. Uversky and A. K. Dunker, *J. Proteome Res.* **5**, 879 (April 2006).
46. J. J. Ward, J. S. Sodhi, L. J. McGuffin, B. F. Buxton and D. T. Jones, *J Mol Biol* **337**, 635 (March 2004).

CRYPTIC DISORDER: AN ORDER-DISORDER TRANSFORMATION REGULATES THE FUNCTION OF NUCLEOPHOSMIN

DIANA M. MITREA

St. Jude Children's Research Hospital, Department of Structural Biology

262 Danny Thomas Place, Memphis TN, 38103, USA

E-mail: Diana.Mitrea@StJude.org

RICHARD W. KRIWACKI

St. Jude Children's Research Hospital, Department of Structural Biology

262 Danny Thomas Place, Memphis TN, 38103, USA

E-mail: Richard.Kriwacki@StJude.org

Abstract

It is now well appreciated that disordered proteins and domains are prevalent in eukaryotic proteomes and that disorder is critically linked with their regulation and functionality. However, our recent observations with the multi-domain protein, nucleophosmin (Npm), suggest that the biological palette of disorder is more diverse than currently understood. The N-terminal oligomerization domain of Npm (Npm-N) can be transformed from a folded, pentameric structure to a monomeric, disordered state through changes in solution ionic strength and, importantly, through physiologically relevant post-translational modifications. Thus, it appears that Npm has been evolutionarily tuned to exist in equilibrium between disordered and ordered states. Results from us and others [1,2] suggest that the function of Npm is regulated through shifts in this equilibrium via post-translational modifications. Interestingly, this polymorphic behavior is not detected using standard secondary structure and disorder prediction algorithms, which show Npm-N to be folded into β-strands, consistent with the structure of the pentameric form. We have used a combination of computational tools, including structure-based analysis, sequence analysis algorithms (NetPhos 1.0, SCRATCH, KinasePhos, GPS2.1, PONDR) and molecular mechanics energy calculations [3,4], to test the hypothesis that the polymorphic behavior of Npm-N can be understood on structural and energetic grounds. This computational strategy has resulted in the identification of unfavorable energetic "hot-spots" within the Npm-N structure which coincide with experimentally observed sites of post-translational modification. Based on these observations, we propose that Npm-N has evolved energetic switches within its structure to enable transformation to a disordered state through phosphorylation. We further propose that the transformation process is triggered by sequential phosphorylation of solvent exposed hot-spots followed by exposure and modification of additional but initially buried sites to completely shift the equilibrium to the disordered state. This regulated, shifting equilibrium is associated with control of Npm localization within the nucleolus, nucleoplasm and cytoplasm, and with its role in regulation of centrosome duplication through interactions with Crm1-Ran [5]. More broadly, we present a general computational strategy to identify transformational hot-spots within proteins and to test the hypothesis that other proteins currently understood to be folded participate in functionally-relevant order-disorder equilibria as we have observed for Npm. The identification of such polymorphic proteins would broaden the palette of protein disorder utilized in biological systems.

1. Biological Function and Structural Features of Npm

The non-ribosomal nucleolar protein nucleophosmin (Npm here; also termed B23 and numatrin) has recently received attention due to its involvement in cell cycle regulation, chromatin remodeling, ribosome biogenesis and most importantly pro- and anti-oncogenic functions, which make it a desirable anticancer drug target [6-8]. Npm primarily resides in the nucleolous, where it regulates, among other systems, the Arf-Hdm2-p53 tumor suppression pathway by stabilizing Arf through nucleolar sequestration [9-11]. It can also shuttle between the nucleus and cytoplasm to fulfill its function in transporting pre-ribosomal particles from their site of synthesis to their site of assembly [7]. Npm exhibits three functional domains (Figure 1A). The 130 N-terminal residues encode the oligomerization domain (Npm-N) [12]; the middle segment encodes a RNA binding and ribonuclease domain; and the C-terminal 54 residues encode a DNA-binding domain [12,13]. Computational analysis of the primary sequence of Npm using the disorder prediction algorithm PONDR [14] indicated that the N- and C-terminal domains are folded and that the middle domain of ~120- residues is disordered (Figure 1B). In agreement with this prediction, the N-terminal domain has been crystallized as a β-sheet-rich pentamer [2,15]. The solution structure for the C-terminal domain has also been elucidated by NMR as being a right-handed 3-helix bundle [13]. To date, there is no structural information on the central region.

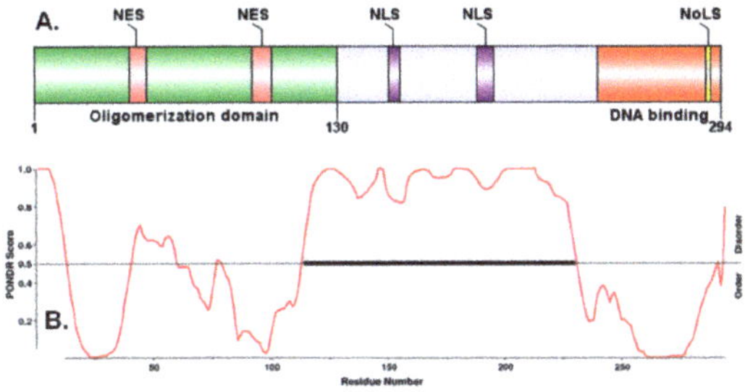

Figure 1. Domain organization of human Npm. (A) Cartoon representation of the three structural domains; oligomerization (green), RNA binding and ribonuclease activity (grey) and DNA binding (orange); (B) Disorder prediction in the secondary structure of full length Npm, using the PONDR algorithm [14].

Mammalian Npm orthologs exhibit a high degree of amino acid sequence conservation with over 69% sequence identity and over 90% sequence similarity observed over the entire protein. The sequence of the oligomerization domain is even more highly conserved, with greater than 90% sequence identity observed across all analyzed mammalian species (Figure 2), suggesting that this domain plays an important, evolutionarily conserved biological role in Npm function. Furthermore, the three dimensional structure of the N-terminal domain is conserved amongst metazoan species [2,15-18]. Initial insights into the importance of Npm oligomerization in biological function emerged from studies in which small molecules or RNA aptamers were used to disrupt its pentameric structure. In these studies, fluorescence microscopy was utilized to show that disruption of oligomers was associated with re-localization of Npm from the nucleolus to the nucleoplasm [19,20]. These results suggest that the pentameric state of Npm is important in the biological function of Npm, with sub-cellular localization being one property that is affected by the presence or absence of folded oligomers.

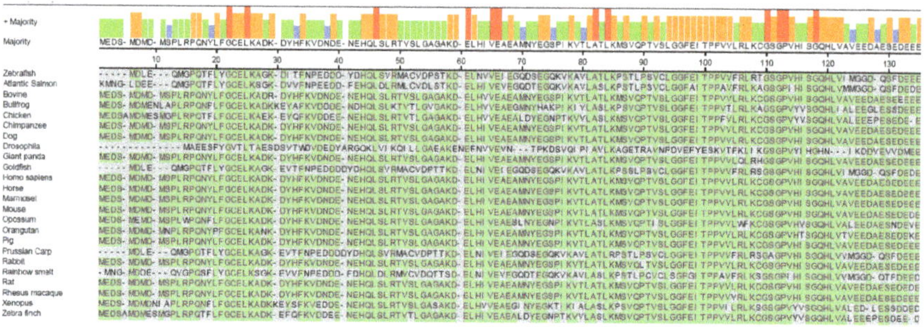

Figure 2. Sequence alignment of Npm-N orthologs using MegAlign (DNASTAR Lasergene 8)

Intriguingly, Olson and co-workers showed in the late 1990s using analytical ultracentrifugation that the oligomerization state of Npm was dependent upon ionic strength *in vitro*. At low and high cation concentrations, Npm sedimented as a monomer and oligomer, respectively [1]. Subsequently, we have shown that the low salt monomeric state of Npm-N is largely disordered, while the high salt state is folded as a pentamer, similar to the crystallographically observed structures [2,15]. Based upon this past report and our recent findings, we initiated investigations into the molecular basis of this apparent ionic strength-dependent, order-disorder transformation.

To gain insight into the intrinsic structural features of the Npm-N amino acid sequence, we performed several different types of sequence analysis. First, secondary structure prediction was performed using the SCRATCH and PredictProtein servers [21,22]. The results of these analyses (Figure 3) were in good agreement with the crystallographic results, with all of the observed β-strands within the pentameric structure accurately predicted by the two servers (only the results from PredictProtein are illustrated in Figure 3). This type of analysis provides little insight, however, into the molecular basis of the ionic strength-dependent structural transformation exhibited by Npm-N. For example, these algorithms do not take into account the influence of ionic strength on secondary and/or higher order structure.

Second, disorder predictions were performed using the PredictProtein and PONDR servers [14,22]. The results were again generally in good agreement with the crystallographic results, with loops that connect β-strands predicted to be disordered (Figures 1A and 3). Again, however, these algorithms have not been developed to allow the influence of ionic strength on disorder or, alternatively, on higher order structure (tertiary and quaternary structure) to be assessed. Consequently, these analyses did not provide insight into the molecular basis of the ionic strength-dependent, order-disorder transformation of Npm-N. A need thus arises for tools that enable us to predict a protein's potential polymorphic behavior.

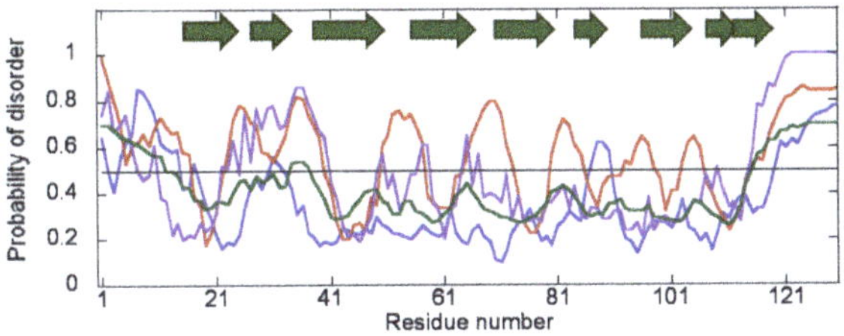

Figure 3. Results of secondary structure and disorder predictions for Npm-N by PredictProtein [22]. Solid traces represent probability for disorder, determined by Blue – NORSnet, Red – PROFbval, Purple - Ucon, Green – MD algorithms; Arrows represent locations of the predicted β-strands.

Next, we discuss our computational strategy to gain insights into the mechanisms responsible for inducing the switch from order to disorder in Npm-N.

2. Alteration of the electrostatic features of Npm-N through phosphorylation

We reasoned that, while variation of ion strength is not a physiologically relevant regulatory mechanism, the order-disorder transformation of Npm-N might be modulated by post-translational modifications of residues within this domain. In particular, we focused attention on phosphorylation as a possible mechanism for perturbation of the Npm-N order-disorder equilibrium. An analysis of the literature, together with results from prediction algorithms, revealed numerous known or putative phosphorylation sites that were evenly distributed through the amino acid sequence (Table 1).

Table 1. Analysis of experimentally characterized and putative phosphorylation sites within Npm-N in the context of the pentameric structure, including the extent of conservation of the modified residue (with the human sequence as reference), solvent accessibility, and the kinase(s) responsible for the modification.

Phosphorylation site	Residue conservation	Solvent accessibility	Modifying kinase
S4	S	Y	PLK2 [23,24]
S10	S/G/N*	Y	CDK [24,25]
Y17	Y/F	Y	TK, PDGFR [24,26]
S43	S	N	PKA, CaM-II, CKI [27]
T46	T	N	CKI [28], CAMK [26 24]
S48	S/T	N	PKA [24,29]
Y67	Y	Y	TK, Abl [24,26]
S70	S/N	Y	CDK [24,25]
T75	T/V*	N	PKA [29,30]
T78	T/S	N	PKC [29]
S82	S	Y	CKI [28]

S88	S	N	PKA, PKG, ATM [27,28,31]
T95	T	Y	PDK [5], cdc2, p38MAPK [29]
S106	S	Y	[32]
S112	S	N	PKA [29]
S125	S	Y	CKII [27,29], CDK2 [30]

* Another aliphatic, hydroxylated phosphorylatable residue (T/S) can be found in close proximity (~ 2-3 residues away from the conserved site).

Results from several large scale proteomic studies demonstrate that numerous residues within Npm-N are phosphorylated in living cells [24,30-32]. Phopshorylation levels are increased in cells undergoing mitosis, when nucleolar localization of Npm is not required [24,30]. Interestingly, while the location and number of Npm's phosphorylatable residues are highly conserved among mammalian species, not all sites shown to be modified are solvent accessible (Figure 4). Based on this observation, we hypothesize that an evolutionarily conserved regulatory mechanism mediates unfolding, or some other type of structural reorganization of pentameric Npm-N to provide kinases access to the buried phosphorylation sites. Our rational was that if phosphorylation at one or more accessible sites destabilized the pentameric protein fold, local or global unfolding may occur, thereby exposing additional sites for modification.

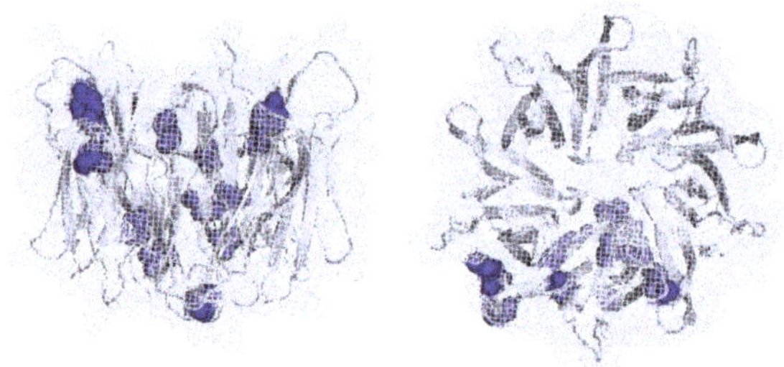

Figure 4. Location of phosphorylation sites (indicated with blue spheres) within the pentameric, β-strand-rich structure of Npm-N. These residues exhibit varying degrees of solvent accessibility (left, side view with modified residues illustrated for one of the five symmetric subunits of the pentamer; right, top view in the same format).

3. *In Silico* site-directed mutagenesis

In order to test the hypothesis discussed above, we employed the empirical force field algorithm FoldX to calculate the energetic effects of phosphorylation at each conserved, potential phosphorylation site of Npm-N. The energy function used by FoldX utilizes the full atomic details of the protein and empirically-derived, weighted physical and statistical energy terms, and is computationally efficient [3,4]. In a recent comparative study, the FoldX algorithm was judged to be among the most accurate protein stability predictors available. Notably, FoldX was one of only two predictors that were able to correctly identify stabilizing mutations [33]. This time and resource efficient approach relies on the availability of a high resolution structural data for the protein under investigation. A reliable force field which accurately estimates the effects of both stabilizing and destabilizing mutations is also critical. For best results, structural models with a resolution of minimum 2.2 Å should be used. Due to the empirical nature of the energy calculations, poorly defined side-chain or backbone orientations in the input model can lead to significant errors [34].

As a starting point for the energy calculations, we utilized the Npm-N pentamer structure refined at 1.8 Å resolution [2]. Due to disorder in the N- and C-termini, residues 1-13 and 119-130 are not visible in the crystal structure, and consequently, we were unable to assess the effect of phosphorylation at Ser4, Ser10 and Ser125 on the stability of Npm-N pentamer. We speculate that, due to the dynamic nature of these regions, their post-translational modification will have only limited impact on the overall fold of the Npm-N oligomer. Modification of these residues, rather, may serve to regulate recognition by various binding partners.

We performed all calculations at the physiological ionic strength of 0.15 M, where experimental evidence places the Npm-N order-disorder equilibrium in favor of the folded pentameric species (ΔG = -149.5 kcal/mol). Reducing the ionic strength to null during FoldX simulations recapitulates the effect observed *in vitro*, namely a considerable reduction in the thermodynamic stability of the pentameric structure ($\Delta\Delta G \sim +75$ kcal/mol). FoldX simulations revealed that phosphorylation of each individual site (within each of the five polypeptide chains present in the pentamer) destabilized the Npm-N structure, with the effects ranging from mild to severe, depending upon the extent of burial on the modified site within the pentameric fold. Due to the empirical nature of the force field, and the rather limited amount of experimental data describing the effect of phosphorylation on protein stability compared to mutagenic studies, we performed calculations of phopsho-mimetic mutations (Ser to Glu and Thr to Asp). The results were in agreement with the phosphorylation results, namely a decrease in the Npm-N stability was observed for all mutated residues. A collection of energetic calculations for the phosphorylated species is given in Table 2. The energy terms within the FoldX force field that contributed (increased in magnitude, corresponding to thermodynamic unfavorability) most significantly to destabilization of the Npm-N pentamer were those for Van de Waals interactions, hydrogen bonds and solvation energy. Counterintuitively, the term for electrostatic interactions was not a major contributor to the change in total stability of the protein, nor did the value of this term for the different phosphorylation mutants correlate with experimental findings on the extent of perturbation of the order-disorder equilibrium destabilization associated with site-specific phosphorylation [2]. Modifications at positions Thr46 or Ser48, shown to be phosphorylated in mitotic cells [24], reduced the stability of Npm-N to extents that were similar to or greater than those associated with reducing the ionic strength parameter to null for the unmodified amino acid sequence. Similarly, modifications at Thr78 and Ser112 resulted in severe energetic penalties (between 60-123 kcal/mol). From a structural and functional perspective, these observations suggest that phosphorylation of these residues causes a shift of the order-disorder equilibrium toward disorder and that this shift may down-regulate Npm function(s) associated specifically with the ordered, pentameric form. These sites, however, are not solvent accessible in the pentameric form of the protein (Table 1 and Figure 4); therefore, it is unlikely that the responsible kinases can directly phosphorylate these sites due to steric inaccessibility. Moderate energetic effects (+10-35 kcal/mol) were observed for *in silico* phosphorylation of Tyr17, Ser43, Thr75, Ser88, a mix of buried and surface exposed residues. The results of our calculations suggest that modifications of surface-exposed sites also induce, albeit to a more limited extent, structural destabilization (e.g., residues Tyr67, Ser82 and Thr95; Table 2). Perhaps not sufficient for inducing complete unfolding of the pentamer, these surface phosphorylation events may shift the population distribution of the ordered pentamer-disordered monomer Npm-N equilibrium, thus providing a mechanism for a buried site to be transiently exposed for phosphorylation.

Table 2. Effects of phosphorylation on the thermodynamic stability of Npm-N pentamer (energies in units of kcal/mol pentamer; $\Delta\Delta G_{total}$ – total free energy; $\Delta\Delta G_{vdw}$ – Van der Waals energy; $\Delta\Delta G_{el}$ – electrostatic energy; $\Delta\Delta G_{solv}$ – solvation energy). Phosphorylation and Alanine substitutions were introduced simultaneously in all five chains of Npm-N; for the AlaScan, the sites were replaced in one polypeptide chain at a time.

Site	Effect of Phosphorylation				Effect of substitution with Alanine				# of chains stabilized by AlaScan
	$\Delta\Delta G_{total}$	$\Delta\Delta G_{vdw}$	$\Delta\Delta G_{el}$	$\Delta\Delta G_{solv}$	$\Delta\Delta G_{total}$	$\Delta\Delta G_{vdw}$	$\Delta\Delta G_{el}$	$\Delta\Delta G_{solv}$	
Y17	14.80	10.69	-0.11	6.38	13.47	10.46	-0.04	5.62	0
S43	18.11	0.62	-3.65	16.49	-5.47	0.81	0.03	-6.99	5
T46	122.80	86.15	6.19	16.72	11.83	3.68	0.00	-2.55	0
S48	66.71	31.02	32.48	-0.70	-2.49	0.74	0.00	-7.78	5
Y67	6.56	7.18	0.35	4.49	4.21	6.81	0.08	4.43	0
S70	-0.30	-1.58	0.67	-1.24	0.15	0.38	0.0	-1.97	2
T75	25.93	8.92	0.85	-0.87	9.12	3.93	-0.89	0.62	0
T78	78.37	44.20	-4.80	19.58	11.58	4.14	0.00	-1.02	0
S82	4.85	-0.85	2.83	-0.31	1.23	0.15	-0.14	-2.44	2
S88	34.93	6.88	2.85	17.99	-1.39	1.33	0.0	-5.24	5
T95	5.33	-1.02	0.83	-3.12	1.66	2.61	0.0	-1.67	2
S112	64.67	35.94	-6.4	30.01	-1.71	1.44	0.0	-8.37	5

4. Probing for structural strain in Npm-N

While the thermodynamic analyses discussed above provide insights into how phosphorylation affects the thermodynamic stability of Npm-N, some questions remain. For example, what are the evolutionary features that enable these types of proteins to switch between an ordered and a disordered species? More importantly, can we identify and understand the energetic hallmarks of this type of structural regulatory mechanism? Can we use the insights gained in our studies of Npm-N to develop general rules to identify other proteins that exhibit similar polymorphic behavior?

To begin to address these questions, we performed systematic *in silico* screens of the effects of structural alteration of each residue in Npm-N on the thermodynamic stability of the pentameric structure. Our rationale was that this strategy would identify residues that experience unfavorable steric or electrostatic clashes within the native pentameric fold. The experimental observations that high solution ionic strength results in stabilization of the ordered, pentameric form of Npm-N [1,2] suggested that unfavorable electrostatic interactions are screened by salt ions and consequently allow the hydrophobic core to more readily collapse and fold. This would not be surprising, considering the acidic loops composed of poly-Glu and -Asp, which connect the β-strands in Npm-N pentamer. As a methodological approach to identify unfavorable side-chain interaction, we utilized the AlaScan function in FoldX. In this procedure, the program replaces every single residue with Ala, on each one of the five polypeptide chains, one at a time, and calculates the resulting change in free energy. The observation of increased stability upon Alanine substitution, which essentially corresponds to removal of the side-chain, suggests that the substituted residue experienced unfavorable interactions with one or more of its neighbors. As depicted in Table 2, replacement of either of the four highly conserved, buried Ser residues (43, 48, 88 and 112) with Alanine, in individual subunits of the Npm-N pentamer, resulted in enhanced stability of the protein. This suggests that removal of the hydroxyl group of the Ser side-chain releases some form of unfavorable "conformational strain". We further simulated the effect of side-chain removal by uniformly mutating each conserved Ser residue within all five protomers within the pentamer. Again, the four Ser residues are notable as introducing unfavorable conformational strain within the pentameric structure, as shown in the selected energetic terms highlighted in Table 2. In accord with the hydrophobic environment of these amino acids, the electrostatic contribution to the increase in stability is negligible. Van der Waals clashes are reduced, while the solvation energy becomes more favorable. These results support the hypothesis that conformational strain within the wild-type Npm-N pentamer provides a mechanism for thermodynamic destabilization of this ordered state. The energy landscape created enables structural polymorphism through an equilibrium with an unfolded, disordered state.

5. Mechanistic insights on Npm's order-disorder polymorphism

The concept that the normal function of proteins involves conformational fluctuations between multiple states is well established. There are several examples wherein the structural and dynamic features associated with transitions amongst alternative conformations have been characterized in detail (*e.g.* ATP synthase[35], calmodulin [36]). However, the function of most proteins is understood in terms of static, folded structures. Our results for Npm-N illustrate that, in contrast to the examples noted above, that conformational fluctuations can occur between two dramatically different, functionally relevant conformations: a folded, pentameric state and a highly disordered, monomeric state. Due to the bias in secondary structure prediction algorithms which are based on the structural data present in the Protein Data Bank, and that of the disorder prediction algorithms, which are based on the current understanding of intrinsically disordered proteins, only the folded state of Npm-N was evidenced in these types of analyses. By using a computational approach to analyze the thermodynamics of conformational strain within the wild-type Npm-N pentamer, we identified residues we describe as structural hot-spots which appear to have dual functions: *(1)* as sites of thermodynamic destabilization and *(2)* as sites for post-translational modification (phosphorylation) to augment thermodynamic destabilization. Furthermore, our results suggest a regulatory mechanism, in which sequential phosphorylations regulate Npm-N structure and function (Figure 5). Modification of solvent exposed residues weakens the pentameric structure of Npm-N, shifting the order-disorder equilibrium toward the disordered state. This enhances the accessibility of buried sites for phosphorylation, leading to further destabilization of the ordered state and consolidation of the disordered state. The first step of this proposed mechanism is supported by a report from Bañuelos and co-workers, who have shown that phosphomimetic mutations at the surface exposed sites in the nucleoplasmin core results in structural destabilization, with no evident changes in the overall packing in the crystal structure. Interestingly, the enhanced dynamics in the protein correlated well with an increase in functional activity in decondensing sperm chromatin [37]. Phosphorylation events have been previously linked to increase in structural flexibility in proteins, often times in order to facilitate binding of a particular partner. In several instances, local unfolding occurs as part of the function regulatory process. The order-to-disorder transitions are restricted, however, to small structural elements (*e.g.* unfolding of an α-helix) or occur between two energetically proximal states [38-41].

It is well understood that post-translational modifications occur on a significant fraction of eukaryotic proteins and that they play a crucial role in modulating their function. Due to technical difficulties, however, more often than not, the structural characterization of mammalian proteins is performed in the absence of these modifications. We propose a perspective from which to view newly characterized proteins and a re-visiting of the previously solved structures of, particularly, multi-functional proteins, in search of alternative, potentially biologically relevant conformations. An energetic assessment of the protein, in the presence of simulated post-translational modifications, can provide insights into the existence of destabilizing hot-spot positions and/or the energy penalty or benefit associated with any particular modification. Taking this strategy a step further, one could compare the location of the hot-spot residues, not only to the positions for post-translational modifications, but also those corresponding to ligand binding or partner-protein binding, which could become accessible for interaction only after a certain set of already exposed sites have been modified. This type of mechanism seems to be quite efficient for regulation of multi-domain, multi-functional proteins, where one segment of the polypeptide chain is unfolded in response to phosphorylation, for example, without the need for protein degradation, a process which would have inactivated the other domains as well.

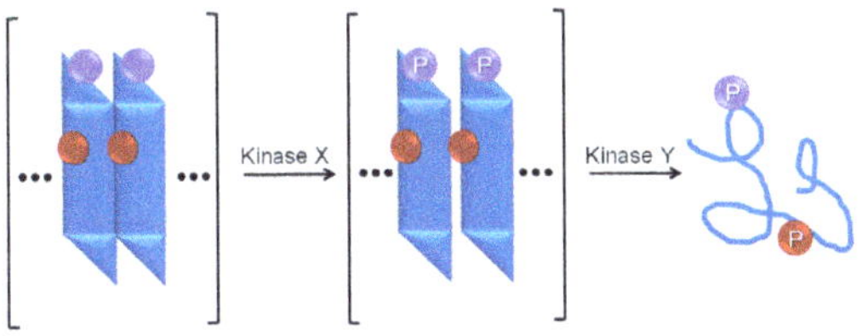

Figure 5. A sequential phosphorylation mechanism mediates a structural switch from order to disorder in Npm-N. Ordered, pentameric Npm-N is phosphorylated at surface exposed sites, which results in destabilization and increased dynamics, transiently exposing buried segments of the polypeptide (Step 1). This enables phosphorylation at the buried sites which introduces severe strain within the folded structure, leading to global unfolding (Step 2).

What is the biological relevance of our findings for Npm-N? Npm is a highly abundant, multi-domain protein and the N-terminal oligomerization plays important roles in modulating its biological function, as evidenced by alteration of normal function through mutagenesis within the N domain [2,42] or treatment with agents which disrupt oligomerization [19,20]. Npm function is modulated in a cell cycle-dependent manner. For example, several different cyclin-dependent kinases phosphorylate Npm, with some of these sites identified within the N domain, and modulate its nucleolar localization and interactions with Crm1-Ran at centrosomes [5,43]. This latter interaction controls centrosome duplication, allowing only one such event per division. Based upon the experimentally characterized polymorphic behavior of Npm-N and our growing understanding of the thermodynamic underpinnings for this phenomenon, we hypothesize that Npm, and possibly oligomeric proteins, have evolved to exhibit two dramatically different structural states—order and disorder— to diversify their biological functionality. Here we propose a phosphorylation mechanism which causes a switch from order to disorder. Phosphatases are likely to reverse the switching action of cell cycle kinases, thus resetting the equilibrium in favor of the ordered state. Folded, ordered Npm-N may participate in inter-molecular interactions (e.g., with Crm1/Ran, histone chaperones) that are disrupted when switched to the monomeric, disordered state. Furthermore, phosphorylation may not only cause the order-disorder switch, but also create phosphorylation-dependent recognition motifs that become exposed through the phenomenon of order-disorder switching. By including a disordered state in its structural repertoire, Npm has evolved to possess complex and diverse functionality. However, the biological implications of its disordered features, which have been appreciated for many years, have remained cryptic.

The approach described herein uses existing energy predictors to analyze the effect of phosphorylation at experimentally validated sites. The kinases responsible for postranslational modifications within the structured part of the Npm-N are unknown, thereby making the experimental measurements quite problematic. Even in the absence of this information, our method allows us to assess the potential repercussions of phosphorylation events on the protein's stability and gain insight into the phosphorylation-dependent regulatory mechanism. We propose that by combining structural information with motif prediction algorithms and energy calculations, the cryptic polymorphic behavior of other biologically important proteins may be revealed. This would add a new level of complexity in protein regulatory mechanisms and further expand our knowledge of the roles of disordered protein/domains in biology.

6. Materials and Methods

FoldX simulations. All thermodynamic calculations were performed using the pentameric structure of human Npm-N determined using X-ray crystallography [2]. The total energy of the structure was first optimized using the RepairPDB command. All simulations, unless otherwise noted, were performed at 298 K, 0.15 M ionic strength and pH 7.0. Each calculation was performed in triplicate, with identical outputs.

References:

1. Herrera JE., Jones AE., and Olson MO., *Biochemistry* **35**, 2668 (1996).
2. Kriwacki RK et al. Mitrea DM, (manuscript in preparation).
3. Nielsen JE, Guerois R, Serrano L., *J. Mol. Biol.* **320** (2), 369 (2002).
4. Rousseau F., Schymkowitz JW., Martins IC., Ferkinghoff-Borg J., Stricher F., Serrano L., *PNAS* **102**, 10147 (2005).
5. Budhu A., Wang W., Forgues M. and Wang XW, *Nature Cell Biology* **7** (8), 823 (2005).
6. Mecucci C., Grisendi S, Falini B., Pandolfi PP. , *Nature Reviews* **6**, 493 (2006).
7. Lindstrom M., *Biochemistry Research International* **2011** (Article ID 195209) (2010).
8. Alcalay M., Colombo E., Pelicci PG., *Oncogene* **30** (23), 2595 (2011).
9. den Besten W., Kuo ML., Bertwistle D., Roussel MF., Sherr CJ., *Genes Dev.* **18** (15), 1862 (2004).
10. Kuo ML., den Besten W., Williams RT., Sherr CJ., *Cell Cycle* **4** (11), 1593 (2005).
11. Shan J., Chen D., Zhu WG., Qin J., Gu W., *Nature* **464** (7288), 624 (2010).
12. Szebeni A., Hingorani K., Olson MO., *J. Biol. Chem.* **275** (32), 24451 (2000).
13. Townsley FM., Grummitt CG., Johnson CM., Warren AJ., Bycroft M., *J. Biol. Chem.* **283** (34), 23326 (2008).
14. Dunbrack RL., Xue B., Williams RW., Dunker AK., Uversky VN., *Biochim. Biophys. Acta* **1804** (4), 996 (2010).
15. Kim HS., Lee HH., Kang JY., Lee BI., Ha JY., Yoon HJ., Lim SO., Jung G., Suh SW., *Proteins* **69** (3), 672 (2007).
16. Akey IV., Dutta S., Dingwall C., Hartman KL., Laue T., Nolte RT., Head JF., Akey CW., *Mol. Cell* **8** (4), 841 (2001).
17. Akey IV., Namboodiri VM., Schmidt-Zachmann MS., Head JF., Akey CW., *Structure* **12** (12), 2149 (2004).
18. Dutta S., Namboodiri VM., Akey IV., Head JF., Akey CW., *Structure* **11** (2), 175 (2003).
19. Shakalya K., Qi W., Stejskal A., Goldman A., Beeck S., Cooke L., Mahadevan D., *Oncogene* **27** (30), 4210 (2008).

20. Gao Z., Jian Y., Sun J., Shen Q., Feng F., Jing Y., Yang C., *Oncogene* **28** (47), 4201 (2009).
21. http://scratch.proteomics.ics.uci.edu/.
22. http://www.predictprotein.org/.
23. Hoffmann I., Krause A., *PLoS One* **5** (3) (2010).
24. Vermeulen M., Olsen JV., Santamaria A., Kumar C., Miller ML., Jensen LJ., Gnad F., Cox J., Jensen TS., Nigg EA., Brunak S., Mann M., *Science Signaling* **3** (104) (2010).
25. Zhou Y., Du W., Pike S., Pang Q., *Carcinogenesis* **31** (2), 302 (2010).
26. Ren J., Xue Y., Gao X., Jin C., Wen L., and Yao X., *Mol. Cell. Proteomics* 7, 1598 (2008).
27. http://kinasephos.mbc.nctu.edu.tw/.
28. http://www.phosida.de/.
29. http://www.cbs.dtu.dk/services/NetPhosK/.
30. Zhou C., Dephoure N., Villen J., Beausoleil S.A., Bakalarski C.E., Elledge S.J., Gygi S.P., *PNAS* **105**, 10762 (2008).
31. Yi W., Cantin G.T., Lu B., Park S.K., Xu T., Lee J.-D., Yates J.R. III, *J. Proteome Res.* 7, 1346 (2008).
32. Gnad F., Oppermann F.S., Olsen J.V., Hornberger R., Greff Z., Keri G., Mann M., Daub H., *Mol. Cell. Proteomics* **8**, 1751 (2009).
33. Vihinen M., Khan S., *Human Mutation* **31** (6), 675 (2010).
34. Serrano L., Nadra AD., Alibes A., *Methods in Enzymology* **498**, 3 (2011).
35. Iino R., Okuno D., Noji H., *J. Biochem.* **49** (6), 655 (2011).
36. Grabarek Z., *Biochim. Biophys. Acta* **1813** (5), 913 (2011).
37. Muñoz IG., Taneva SG., Franco G., Falces J., Arregi I., Muga A., Montoya G., Urbaneja MA., Bañuelos S., *Biochemistry* **47** (52), 13897 (2008).
38. Hollingworth D., Díaz-Moreno I., Frenkiel TA., Kelly G., Martin S., Howell S., García-Mayoral M., Gherzi R., Briata P., Ramos A., *Nat Struct Mol Biol* **16** (3), 238 (2009).
39. Wenta N., Nardozzi J., Yasuhara N., Vinkemeier U., Cingolani G., *JMB* **402** (1), 83 (2010).
40. Ding F., Proctor EA., Dokholyan NV., *JMB* **408** (3), 555 (2011).
41. Traaseth NJ., Gustavsson M., Karim CB., Lockamy EL., Thomas DD., Veglia G., *JMB* **408** (4), 755 (2011).
42. Lindstrom MS., Enomoto T., Jin A., Ke H., Zhang Y. , *J. Biol. Chem.* **281** (27), 18463 (2006).
43. Brancaccio M., Hanashiro K., Fukasawa K., *Oncogene* **30** (19), 2188 (2011).

FUNCTIONAL ANNOTATION OF INTRINSICALLY DISORDERED DOMAINS BY THEIR AMINO ACID CONTENT USING IDD NAVIGATOR

ASHWINI PATIL

Human Genome Center, Institute of Medical Science, University of Tokyo, 4-6-1 Shirokanedai Minato-ku, Tokyo 108-8639, Japan
Email: ashwini@hgc.jp

SHUNSUKE TERAGUCHI*

Host Defense Lab, WPI Immunology Frontier Research Center (IFReC), Osaka University,3-1 Yamadaoka, Suita, Osaka 565-0871, Japan
Email: teraguch@ifrec.osaka-u.ac.jp

HUY DINH

Systems Immunology Lab, WPI Immunology Frontier Research Center (IFReC), Osaka University, 3-1 Yamadaoka, Suita, Osaka 565-0871, Japan
Email: dinh@ifrec.osaka-u.ac.jp

KENTA NAKAI

Human Genome Center, Institute of Medical Science, University of Tokyo, 4-6-1 Shirokanedai Minato-ku, Tokyo 108-8639, Japan
Email: knakai@ims.u-tokyo.ac.jp

DARON M STANDLEY

Systems Immunology Lab, WPI Immunology Frontier Research Center (IFReC), Osaka University, 3-1 Yamadaoka, Suita, Osaka 565-0871, Japan
Email: standley@ifrec.osaka-u.ac.jp

Function prediction of intrinsically disordered domains (IDDs) using sequence similarity methods is limited by their high mutability and prevalence of low complexity regions. We describe a novel method for identifying similar IDDs by a similarity metric based on amino acid composition and identify significantly overrepresented Gene Ontology (GO) and Pfam domain annotations within highly similar IDDs. Applications and extensions of the proposed method are discussed, in particular with respect to protein functional annotation. We test the predicted annotations in a large-scale survey of IDDs in mouse and find that the proposed method provides significantly greater protein coverage in terms of function prediction than traditional sequence alignment methods like BLAST. As a proof of concept we examined several disorder-containing

* The authors wish it to be known that, in their opinion, the first two authors contributed equally to this work.

proteins: GRA15 and ROP16, both encoded in the parasitic protozoa *T. gondii*; Cyclon, a mostly uncharacterized protein involved in the regulation of immune cell death; STIM1, a protein essential for regulating calcium levels in the endoplasmic reticulum. We show that the overrepresented GO terms are consistent with recently-reported biological functions. We implemented the method in the web server IDD Navigator. IDD Navigator is available at http://sysimm.ifrec.osaka-u.ac.jp/disorder/beta.php.

1. Introduction

Intrinsically disordered domains (IDDs) make up a significant portion of many eukaryotic proteins but present a unique challenge to structural bioinformatics-based functional annotation methods, which generally operate on the structure-function paradigm. For example, several groups, including ours, have used 3D modeling in combination with structural alignment to infer biological or biochemical functions of query sequences[1-3]. While it is in principal possible to follow this approach even in the absence of a well-defined structure by simply omitting the intermediate structural modeling step, and substituting structural alignment with sequence alignment, in practice this is problematic; the high rates of evolution coupled with higher than expected correspondence between low complexity regions (LCRs) and IDDs[4], along with the high false positive rate for LCRs in conventional sequence similarity scores make it difficult to quantify the significance of the resulting hits.

We recently investigated two alternatives to conventional sequence alignment in order to extract functional information from IDD sequences[5]. First, we examined short sequence motifs that were over-represented in a set of predicted IDDs. The over-represented motifs corresponded to amino acid repeats in general, but it was not possible to associate them with known functions using Pfam domains or Gene Ontology (GO) annotations. Second, we investigated the similarity between the overall amino acid composition of IDDs using a simple histogram distance method. Surprisingly, some structured Pfam domains associated with proteins that had been clustered only by the amino acid composition of their IDD regions were highly overrepresented compared with a random grouping of IDDs. This presents the possibility of IDDs with similar amino acid composition sharing a related function. Motivated by these results, we constructed a web server that allows a database of predicted IDDs to be searched by several similarity scores that are functions of the amino acid composition for a given IDD. Highly similar IDDs are returned along with their Pfam and GO annotations. The significance of the occurrence of each annotation can then be quantified by comparison with a sample of background IDDs. While the proposed approach is not expected to be as sensitive or as specific as is sequence alignment of structured domains, it can nevertheless have practical utility, especially when used in conjunction with structure-based functional annotations or in cases where no alternatives for function prediction exist. To demonstrate its utility, we carried out a large-scale survey of the predicted GO annotations as well as small-scale analysis of 4 IDD-containing proteins.

2. Methodology

2.1. *Preparation of IDD dataset*

64,322 amino acid sequences of mouse proteins were downloaded from UniProtKB[6]. A representative set of 18,126 non-redundant protein sequences was prepared by clustering at 40% sequence identity using the cd-hit program[7]. For each sequence, IDDs were predicted using the Disopred2 program[8] and 22,057 predicted IDDs of length greater than 30 were retained for analysis. In IDD Navigator, these IDD sequences and the annotations of the corresponding proteins were used as a database for finding similar IDDs and predicting functions for each query IDD.

2.2. *Similarity scores*

In this work, we predicted functions of each query IDD based on the most similar IDDs in the above IDD dataset. We defined similarities for a pair of IDDs through similarity scores. In our previous work, we introduced a similarity score between IDDs based on the frequency of amino acid residues[5]. In the current study, we have defined two similarity scores, which are described below. For each query, IDD Navigator calculates similarity scores against the above 22,057 stored IDDs and by default shows the top 100 most similar IDDs.

2.2.1. *Similarity score based on Euclidean distance*

The frequency $f_i(a)$ of an amino acid residue a in the sequence of IDD i is defined as follows:

$$f_i(a) = \frac{N(a)}{len_i} \tag{1}$$

Here, $N(a)$ is the number of the amino acid residues of type a in the sequence and len_i is the length of the sequence. The Euclidean distance based similarity score between sequences i and j is defined by

$$sim_{i,j} = -100\sqrt{\sum_{a=1}^{20}\left(f_i(a) - f_j(a)\right)^2} \tag{2}$$

The distance was multiplied by a negative number in order to convert it to a similarity score. This similarity score performs comparably with the Gaussian similarity used in the previous work (data not shown).

2.2.2. *BLAST score*

Additionally, we used the protein BLAST (blastp) score for comparison purposes. Blastp was run on each IDD against the database of all 22,057 IDDs with an e-value cutoff of 0.01. The BLAST score was used in the same manner as the Euclidian distance-based score.

2.3. *Pfam domain and Gene Ontology term prediction*

By assuming that there is a correlation between the similarity scores defined above and similarity in functions of the corresponding proteins, IDD Navigator provides a prediction of the functions of a query IDD. We counted the numbers of Pfam domains and GO terms of unique proteins associated with a list of the most similar IDDs (top 100) evaluated by one of the similarity scores. The significantly overrepresented Pfam domains and GO terms constitute the primary output of IDD Navigator. The statistical significance of the predicted functions was estimated by using hypergeometric distribution functions. Here, we modeled the probability of having a particular function by randomly selecting the same number of proteins from 18,126 representatives, counting the fraction annotated by the function in question, and then calculating the corresponding p-values. Thus, the annotations of all proteins with IDDs were used as a reference dataset.

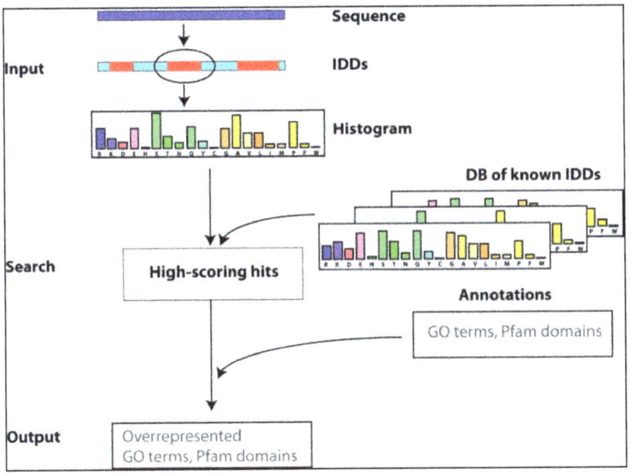

Figure 1. Flowchart of the of the web server. An IDD can be input or IDD prediction can be performed on a full-length sequence. A histogram is then computed for the input IDD, and it is compared with a database of known IDDs. High-scoring hits are then extracted and annotations from GO and Pfam are collected. The overrepresented GO terms and Pfam domains are then identified and output.

2.4. *Evaluation of function prediction*

From the initial dataset of 18,126 mouse proteins, 5,164 had predicted IDDs and Gene Ontology term annotations in UniProt and were used in evaluating the prediction performance of IDD Navigator. For each IDD within the 5,164 proteins, GO terms were predicted as those significantly enriched ($p < 0.01$) among the top 100 most similar IDDs based on a particular (Euclidean or BLAST) score. We tested the similarity of GO terms predicted for each IDD in a protein with the actual GO terms assigned to it. For proteins with multiple IDDs, unique predicted GO terms from all IDDs were combined and compared to the actual GO annotations of the protein. In order to calculate the significance of our results, we compared the predicted GO terms with random terms. We prepared a list of unique GO terms from the actual GO annotations of the 5,614 mouse proteins. We then randomly picked GO terms for each IDD that were equal in number to the predicted GO terms for that IDD. The semantic similarity was calculated between the actual GO terms and the list of randomly selected GO terms.

Clusters of IDDs with similar amino acid content were made based on the Euclidean distance similarity score using hierarchical clustering with Ward's method. The number of clusters was empirically set at 10. Average semantic similarity between predicted and real GO terms was calculated for each cluster and compared to those of random GO terms. Amino acid propensity for amino acid a in the ith cluster was calculated as follows:

$$P_{ai} = \frac{N_{ai}}{N_a} \tag{3}$$

where N_{ai} is the count of the amino acid a in cluster i and N_a is the total count of amino acid a in all clusters. The z-score for the propensity of each amino acid was calculated over its average propensity in all clusters and plotted as a heatmap in figure 4.

Semantic similarity between lists of GO terms was calculated using the R package GOSemSim[9]. GO terms are arranged in the form of a directed acyclic graph (DAG), and GOSemSim calculates the similarity between two lists of GO terms using their location in the DAG and their relationship with the ancestor terms. GOSemSim assigns a value between 0 and 1 with higher values indicating greater similarity between groups of GO terms. The predictions of Cellular Component (CC), Molecular Function (MF) and Biological Process (BP) terms were evaluated independently.

2.5. *Web server*

IDD Navigator can be accessed at http://sysimm.ifrec.osaka-u.ac.jp/disorder/beta.php. A flowchart of steps performed by the web server is shown in Figure 1. The histograms used to score the similarity between IDDs are represented as colored bar graphs. Several methods have been provided to evaluate the similarity score between IDDs based on amino acid content.

3. Results and Discussion

3.1 IDD Navigator Function prediction

In order to determine if the significantly enriched GO terms given by IDD Navigator for a query IDD can be used as function predictions, we first evaluated the accuracy of the predictions with p-value < 0.01. We compared the semantic similarity between the real GO terms and those predicted by IDD Navigator for a set of 5,164 mouse proteins using the Euclidean distance score. Since some proteins have several IDDs, we also tested the performance of the combined list of GO terms predicted for all IDDs within a protein. Finally we compared these results with the similarity of an equal number of random GO terms assigned to each protein.

Figure 2 shows the results of the evaluation. GO terms predicted by the Euclidian distance score show greater semantic similarity to the real GO terms of proteins, on average, than randomly assigned GO terms ($p \ll 0.001$, t-test). This result is consistent across all categories of GO terms. Combined predictions from all IDDs in a protein performed best, though those predicted using single IDDs showed an almost equivalent performance. Among the categories of GO terms, CC terms were predicted with the greatest accuracy, followed by MF and BP terms, respectively. The lower performance of BP terms was possibly due to the large number of GO terms present in this category. Based on these results we concluded that the GO terms given by IDD Navigator could be used as valid function predictions for unannotated proteins. We further, checked the relationship between the length of the IDD and the accuracy of prediction of the GO terms but were unable to find a significant correlation (GO BP: 0.10; GO MF: 0.08; GO CC: 0.12).

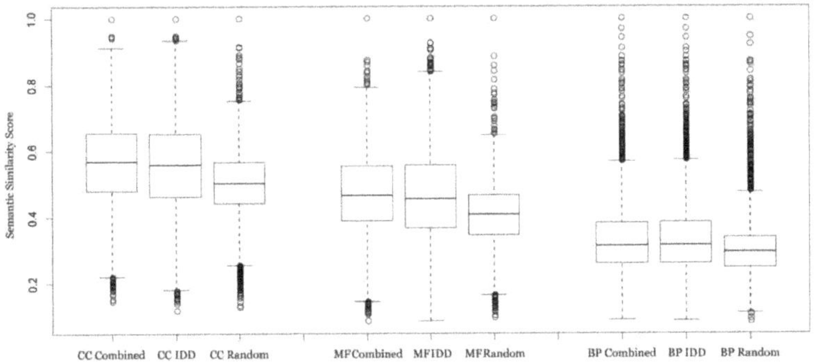

Figure 2. Average semantic similarity between Euclidian distance-predicted and actual GO terms in Cellular Component (CC), Molecular Function (MF) and Biological Process (BP) categories. Combined indicates a combination of all unique GO terms predicted for all IDDs within a protein. IDD denotes GO terms predicted for a single IDD in a protein. Random shows the performance of the same number of random GO terms assigned to a protein.

3.2 Comparing different methods in IDD Navigator

We compared the function prediction performance of the Euclidean distance method and the BLAST score in IDD Navigator. The Euclidean distance method represents the similarity of IDDs by amino acid content, irrespective of the sequence. On the other hand, the BLAST score identifies similar IDDs by direct sequence alignment to the query IDD. The BLAST score option in IDD Navigator is equivalent to performing a BLAST search using an IDD against all the IDDs in the current dataset with low complexity filtering turned off and predicting a function based on the top 100 hits obtained.

In Figure 3, the Euclidian distance score is compared with the BLAST score for each GO term. For a given GO category (CC, MF, BP), Figure 3 shows the combined score (combining all the IDDs from one protein) for Euclidian distance and BLAST, the IDD scores (where IDDs for a given protein are treated separately), and a randomized score. The randomized score corresponds to the semantic similarity between the actual GO terms of the proteins and randomly chosen GO terms that are equal in number to those predicted by the IDD Navigator using the Euclidean and BLAST scoring schemes. The two methods have different randomized scores because of the differences in the number of GO terms predicted by each. IDD Navigator using the BLAST score predicts considerably fewer GO terms as significantly over-represented than using the Euclidean score. Due to the arrangement of the GO terms in the form of a DAG and the method used by GOSemSim to calculate the semantic similarity, the similarity score is dependent on the number of terms, often increasing as the number of terms increases. In order to address this drawback of the semantic similarity scoring, we calculated separate randomized similarity scores for predictions made using the two methods.

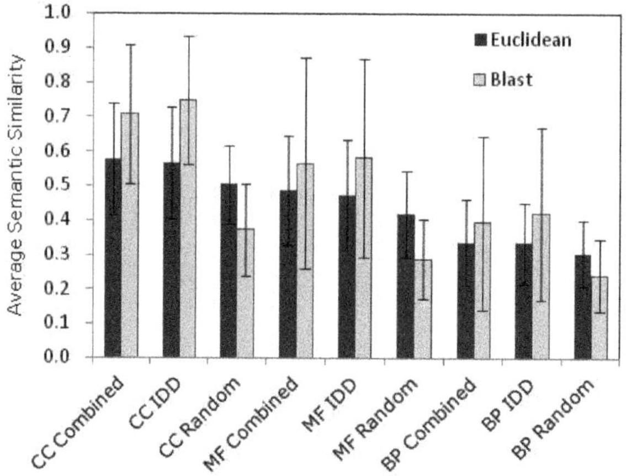

Figure 3. Comparison between Euclidian distance and BLAST scores. Average semantic similarities between predicted and real GO terms for mouse proteins are shown for the two scoring methods along with randomized values. Euclidean: Euclidean distance score for amino acid composition similarity, BLAST: BLAST score for IDD sequence similarity. Term definitions same as in Figure 2.

The BLAST semantic similarity scores are closer to the actual terms than those predicted by the Euclidean distance method. However, there is a large difference in the coverage of proteins for the two types of methods. Significantly enriched GO terms are obtained for 4,383 proteins using the Euclidean method, while over-represented GO terms were identified for IDDs in only 27 proteins using the BLAST score at a significance threshold of 0.01. The very small number of BLAST hits is not surprising. Proteins within the mouse genome that are not obvious homologs (i.e., do not form clusters in the cd-hit step) do not, generally, have sequentially similar IDDs. The high coverage of proteins for which GO terms can be predicted using the Euclidian distance score, on the other hand, highlights the utility of IDD Navigator in novel function prediction.

Several methods have been proposed to assign probable GO terms to proteins without annotations[10-12]. However, these methods rely on BLAST or PSI-BLAST either partially or completely. Predicting the function of proteins with large disordered domains in the absence of conserved or annotated domains and annotated homologs is currently an open problem. Figure 2 indicates that IDD Navigator can provide function associations in at least a subset of such cases. It can also potentially be used to assign new functions to proteins based on the IDDs that they contain. Although functional annotations have been associated with specific IDDs[13, 14] and function prediction using intrinsic disorder in proteins has been studied before[15-17], IDD Navigator uses a novel strategy based on amino acid content to assign probable GO terms to unannotated proteins, and is thus complementary to existing approaches.

3.3 Function prediction for IDD clusters

Our method can also be used to cluster the IDDs into groups that may potentially be associated with specific functions similar to a previous study identifying "flavors of disorder" [18]. To see if this is the case, we clustered the 22,057 IDDs into 10 clusters based on their amino acid similarity. Figure 4 shows the amino acid enrichment for each cluster. Each cluster has a distinct pattern of over-represented amino acids. For each cluster, we calculated the average semantic for each IDD similarity between the predicted GO terms and its observed GO terms. Table 1 shows the numerical results.

The GO term similarity between those predicted by IDD Navigator compared to the observed terms is better than random in all the clusters for all 3 types of GO terms. The GO similarity score in Cluster 4 are particularly high. The IDDs in this cluster are enriched in charged residues, especially Lys, and are primarily parts of ribosomal proteins based on GO term enrichment analysis [19]. However, these results also indicate that the performance of function prediction varies with the amino acid content of the IDD. A good performance for certain clusters may be attributed to the fact that the clusters contain fewer IDDs or proteins containing the IDDs in these clusters are better annotated. On the other hand, a poor similarity score between the observed and predicted GO terms may be the result of the IDDs participating in multiple functions despite similar amino acid content. The performance of the function prediction by IDD Navigator may also change with the number of clusters used to partition the dataset. We did not, for example, find 3 major clusters using our methods, so a direct comparison with the 3 flavors of disorder proposed by Vucetic et al. was not straightforward.

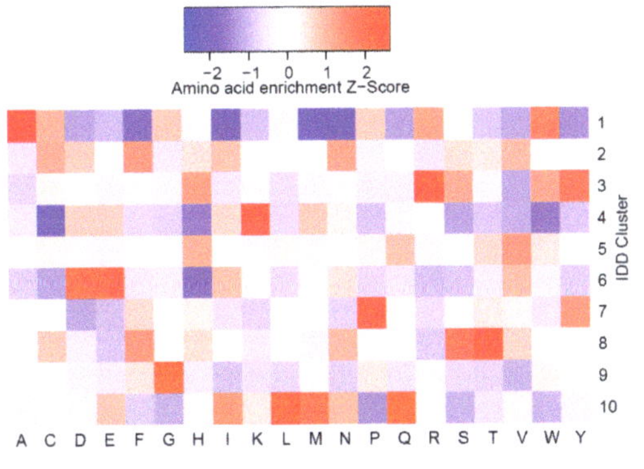

Figure 4. Amino acid enrichment in 10 clusters of IDDs. Each square represents the amino acid enrichment in the form of a z-score.

Cluster	*Average Semantic Similarity*		
	BP	**MF**	**CC**
1	0.339	0.463	0.562
2	0.329	0.465	0.562
3	0.330	0.448	0.528
4	0.354	0.512	0.627
5	0.325	0.468	0.548
6	0.354	0.456	0.588
7	0.346	0.511	0.586
8	0.337	0.473	0.566
9	0.347	0.480	0.548
10	0.345	0.490	0.603
Average	0.335	0.474	0.565
Random	0.305	0.419	0.504

Table 1. Average semantic similarity scores between predicted and real GO terms for IDDs in mouse proteins assigned to 10 clusters based on amino acid content similarity by Euclidean distance score. Average and random values shown in gray.

3.4 Case Studies

The ultimate test of IDD Navigator is to analyze a functionally uncharacterized sequence *de novo* and validate the predicted function experimentally. Since such a test is beyond the scope of the current work, we have selected 4 IDD-containing proteins whose biological function has only recently been reported. Each protein was submitted using default options. In the case of GRA15, some subsequent adjustment of the Disopred false-positive threshold was also used.

3.4.1 GRA15 from T. gondii

GRA15 is a ~500 amino acid polymorphic protein localized in the dense granules of the parasitic protozoa *T. gondii*, and secreted into host cells upon infection. The effect of GRA15 on the host immune system depends on the particular *T. gondii* strain (types I-III). Recently, the effects of type II GRA15 (GRA15-II) on host immune signaling pathways have been described[20, 21]. No hits to GRA15 were found using conventional sequence alignment searches[21]. The sequence of GRA15-II lacking a 51-residue N-terminal signaling peptide was submitted to IDD Navigator using the Disopred option. A single 475-residue IDD was predicted for residues 24-499. The top hits were associated with transcription factors PAX2, PAX7, and A2A410. The next most significant hit was to an AXH-containing protein, which interacts with CIC, a transcriptional repressor. There were also a number of hits involved in signal transduction. The GO molecular function terms "protein binding" (p-value 2.4e-4) and "kinase activity" (p-value 4.9e-4) scored highly, while "cell fate determination" scored highest in the biological process category (p-value 0.8e-3). Taken together, the GRA15-II results are consistent with a role in signal transduction, transcription, or both. Although the biochemical function of GRA15-II has not yet been elucidated, it was recently revealed that its biological role is to activate the transcription factor NF-κB in infected hosts[16]. GRA15-II-mediated NF-κB activation in macrophages inhibited apoptosis, and increased cell migration. Moreover, it was argued that GRA15-II is likely to act in a complex with the kinase IKK and signaling protein TRAF6[21], which are important regulators of the NF-κB signaling pathway. GRA15-II-mediated NF-κB activation in macrophages inhibited apoptosis, and increased cell migration[20, 21]. These known biological functions of GRA15-II are overall consistent with the IDD Navigator results. One potential application of the IDD Navigator query would be to suggest putative protein-protein interaction sites. However, in order to do this, we had to first lower the Disopred2 false-positive threshold from the default value of 5% to 2%, in order to obtain several smaller, but higher-confidence IDDs. This resulted in three distinct IDDs (residues 34-206, 209-389, 402-499). Of these, the last one was of particular interest, as the top GO annotation was "apoptosis" which includes both positive and negative regulators of cell death. The specific hits were comprised of 6 proteins, including "NF-κB-interacting protein 1". Interestingly, the region of greatest divergence between type I/III and II GRA15 consists of an 84–amino acid indel near the C-terminus[21]. This finding is consistent with the observation that type I/II *T. gondii* has been found to *suppress* NF-κB activity, suggesting that the C-terminal IDD may be critical for differentiating effects on host immune cell responses. This result shows that IDD Navigator results are sensitive to the domain boundaries, which is also true for structure-based function prediction. Fortunately, the false-positive threshold in the Disopred program is a convenient way to vary the boundaries.

3.4.2 Cyclon from M. musculus

Cyclon is a coiled coil-containing protein that is known to positively regulate expression of Fas, an immune cell surface protein that mediates activation-induced cell death [22, 23]. However, a direct role in Fas transcription has not been shown, and Cyclone lacks any known DNA binding domains. When we submitted the Cylon sequence to IDD Navigator, we found that GO terms relating to epigenetic regulation dominated the list. For example, in the molecular function Category, the GO top terms and associate p-values included "histone methyltransferase activity (H3-K4 specific)" (2.4e-4); "transcription corepressor activity" (1.4e-3); "chromatin binding" (2.2e-3); "histone-lysine N-methyltransferase activity" (2.8e-3). The top biological process terms were: "chromatin modification" (1.2e-4); "cellular component organization" (1.6e-3); "peptidyl-lysine methylation" (1.6e-3). The top cellular component terms were "histone deacetylase complex" (0.7e-5) and "histone methyltransferase complex" (0.8e-4). A BLAST search against the nr database resulted in a number of hits to uncharacterized proteins in vertebrates and insects. The one exception was the chromatin assembly factor-I p150 from Culex quinquefasciatus (southern house mosquito). Although this annotation could not be confirmed by further BLAST or literature searches, it agreed qualitatively with the IDD Navigator result. The coiled-coil domains are most likely involved in protein-protein interactions, so Cyclon may act as an adaptor that recruits proteins involved in protein remodeling or histone modification. A potentially useful direction would be to systematically screen all such proteins for cyclon-like IDD regions, as this subset may contain epigenetic regulators of Fas gene expression.

3.4.3 STIM1 from M. musculus

STIM1 is a calcium-sensing protein that spans the ER membrane. The N-terminal luminal portion contains the calcium-binding domain and the cytoplasmic C-terminal portion is predicted to contain coiled-coil domains and two IDDs. The first IDD, but not the second, was found to have a significant (p-value 4e-11) over-representation of the GO cellular component term "Cytoskeleton". This result is potentially interesting, as a recent report showed that microtubules affect STIM1-mediated calcium entry[24] and in another report, STIM1 was shown to affect the organization of microtubules [25]. When the second IDD was submitted to IDD Navigator, the cellular component term "Cytoplasm" was over-represented with a p-value of 4e-7, which is consistent with the known localization of this domain. Both findings are relevant to one of the key functions of STIM1, the formation of ER projections known as 'puncta' that grow toward the plasma membrane when calcium stores are depleted. It is already known that calcium depletion promoted STIM1-STIM1 interactions on the luminal side of the ER membrane [26]. Although speculative at this point, the coiled-coils and IDDs may interact directly with components of the cytoskeleton upon calcium depletion in order to induce ER membrane remodeling.

3.4.4 ROP16 from T. gondii

Like GRA15, Rhoptry protein 16 (ROP16) is encoded in *T. gondii*, and secreted into host cells upon infection as part of the parasite's counter-defense against host immunity.

The N-terminal 357 residues of ROP16 are predicted to be disordered. When we submitted residues 1-357 to IDD Navigator, we found no overrepresented GO terms for biological function. However, the cellular component term "cell junction" was over-represented with a p-value of 2.6e-8. This result is potentially relevant since ROP proteins are excreted through cell-cell junctions formed between host and parasite. The N-terminal portion of ROP16 has not been characterized, and the IDD Navigator results suggest that the IDD region may be involved in the initial secretion into the host cell.

4. Conclusions

In this report we have described an extension of our earlier work on IDD clustering toward the goal of functionally annotating IDDs. The large-scale survey results indicate clearly that, on average, GO terms that are overrepresented in IDD Navigator hits compared with background values are closer to true GO terms than a random selection of IDDs. This, in turn, implies that the underlying scoring function can pick up some information that is encoded in the IDD sequences without using sequence alignment. Distinguishing useful information from background noise, however, requires careful interpretation, as the 4 examples show. Submitting the whole GRA15-II sequence and running with default parameters produced results that are, in general, consistent with the known biological function of GRA15-II: the protein appears to function as an adaptor and may effect transcription or signal transduction pathways; however, the IDD Navigator hits did not give a clear clue as to which pathways are effected, or which parts of the protein are involved in specific protein-protein interactions. When the Disopred2 false-positive threshold was lowered from 5% to 2%, however, individual domains appeared with sequence-specific associated GO terms and Pfam domains. The appearance of apoptosis as the top GO term, with a specific hit to an NF-κB-interacting protein was encouraging, given recent reports that GRA15 suppresses apoptosis by activating NF-κB. However, it must be acknowledged that, in the absence of some knowledge of basic biological function, it would be difficult to isolate this hit from the others. The cylon result was somewhat more straightforward to interpret due to the dominance of hits related to epigenetic control of gene expression. Interaction with histone modifying proteins would explain the protein's role in regulation of Fas. However, this hypothesis requires further experiment before it can be validated. Similarly, the hits to STIM1 and ROP16 help to generate hypotheses. In the case of STIM1, the first IDD might interact with microtubules, which would be consistent with the coiled-coil domains predicted nearby. In the case of ROP16, the IDD might be is necessary for migrating through the cell-cell junction. Fortunately, each of these hypotheses can be experimentally tested, and we are currently collaborating with experimental groups to validate our predictions.

Future versions of IDD Navigator will include all sequences from Uniref100, allowing better function predictions across multiple species. In order to prepare and maintain such a large database of IDDs and also to process query sequences rapidly, the disorder prediction calculation time must be reduced. For this reason, in the future, we plan to incorporate faster disorder prediction methods[27]. We also aim to improve the scoring functions, which are highly simplistic at this stage.

5. Acknowledgements

We would like to thank M. Yamamoto for helpful discussions concerning GRA15, and H. Fujii for advice regarding Cyclon function. Computation time was partly provided by the Super Computer System, Human Genome Center, Institute of Medical Science, University of Tokyo. This research was partially supported by the Japan Society for the Promotion of Science (JSPS) through its "Funding Program for World-Leading Innovative R&D on Science and Technology (FIRST Program).

References

1 J. Dundas, L. Adamian, and J. Liang, J Mol Biol **406**, 713 (2010).
2 A. Roy, A. Kucukural, and Y. Zhang, Nat Protoc **5**, 725 (2010).
3 D. M. Standley, R. Yamashita, A. R. Kinjo, et al., Bioinformatics **26**, 1258 (2010).
4 M. A. DePristo, M. M. Zilversmit, and D. L. Hartl, Gene **378**, 19 (2006).
5 S. Teraguchi, A. Patil, and D. M. Standley, BMC Bioinformatics **11 Suppl** 7, S7 (2010).
6 M. Magrane and U. Consortium, Database (Oxford) **2011**, bar009 (2011).
7 W. Li and A. Godzik, Bioinformatics **22**, 1658 (2006).
8 J. J. Ward, J. S. Sodhi, L. J. McGuffin, et al., J Mol Biol **337**, 635 (2004).
9 G. Yu, F. Li, Y. Qin, et al., Bioinformatics **26**, 976 (2010).
10 M. Chitale, T. Hawkins, C. Park, et al., Bioinformatics **25**, 1739 (2009).
11 T. Hawkins, S. Luban, and D. Kihara, Protein Sci **15**, 1550 (2006).
12 W. T. Clark and P. Radivojac, Proteins **79**, 2086 (2011).
13 H. Xie, S. Vucetic, L. M. Iakoucheva, et al., J Proteome Res **6**, 1882 (2007).
14 A. K. Dunker, C. J. Brown, J. D. Lawson, et al., Biochemistry **41**, 6573 (2002).
15 A. Lobley, M. B. Swindells, C. A. Orengo, et al., PLoS Comput Biol **3**, e162 (2007).
16 B. Meszaros, I. Simon, and Z. Dosztanyi, PLoS Comput Biol **5**, e1000376 (2009).
17 A. Mohan, C. J. Oldfield, P. Radivojac, et al., J Mol Biol **362**, 1043 (2006).
18 S. Vucetic, C. J. Brown, A. K. Dunker, et al., Proteins **52**, 573 (2003).
19 W. Huang da, B. T. Sherman, and R. A. Lempicki, Nucleic Acids Res **37**, 1 (2009).
20 K. D. Jensen, Y. Wang, E. D. Wojno, et al., Cell Host Microbe **9**, 472 (2011).
21 E. E. Rosowski, D. Lu, L. Julien, et al., J Exp Med **208**, 195 (2011).
22 A. Hoshino and H. Fujii, FEBS Lett **581**, 975 (2007).
23 S. Saint Fleur, A. Hoshino, K. Kondo, et al., Blood **114**, 1355 (2009).
24 C. Galan, N. Dionisio, T. Smani, et al., Biochem Pharmacol **82**, 400 (2011).
25 Z. Hajkova, V. Bugajev, E. Draberova, et al., J Immunol **186**, 913 (2010).
26 P. B. Stathopulos, L. Zheng, G. Y. Li, et al., Cell **135**, 110 (2008).
27 B. Xue, R. L. Dunbrack, R. W. Williams, et al., Biochim Biophys Acta **1804**, 996 (2010).

ON THE COMPLEMENTARITY OF THE CONSENSUS-BASED DISORDER PREDICTION

ZHENLING PENG AND LUKASZ KURGAN

Electrical and Computer Engineering Department, University of Alberta, Edmonton, AB, Canada
Emails: zhenling@ualberta.ca, lkurgan@ece.ualberta.ca

Intrinsic disorder in proteins plays important roles in transcriptional regulation, translation, and cellular signal transduction. The experimental annotation of the disorder lags behind the rapidly accumulating number of known protein chains, which motivates the development of computational predictors of disorder. Some of these methods address predictions of certain types/flavors of the disorder and recent years show that consensus-based predictors provide a viable way to improve predictive performance. However, the selection of the base predictors in a given consensus is usually performed in an ad-hock manner, based on their availability and with a premise that more is better. We perform first-of-its-kind investigation that analyzes complementarity among a dozen recent predictors to identify characteristics of (future) predictors that would lead to further consensus-based improvements in the predictive quality. The complementarity of a given set of three base predictors is expressed by the differences in their predictions when compared with each other and with their majority vote consensus. We propose a regression-based model that quantifies/predicts quality of the majority-vote consensus of a given triplet of predictors based on their individual predictive performance and their complementarity measured at the residue and the disorder segment levels. Our model shows that improved performance is associated with higher (lower) similarity between the three base predictors at the residue (segment) level and to their consensus prediction at the segment (residue) level. We also show that better consensuses utilize higher quality base methods. We use our model to predict the best-performing consensus on an independent test dataset and our empirical evaluation shows that this consensus outperforms individual methods and other consensus-based predictors based on the area under the ROC curve measure. Our study provides insights that could lead to the development of a new generation of the consensus-based disorder predictors.

1. Introduction

The intrinsically disordered proteins (IDPs) are characterized by the lack of stable tertiary structure when their isolated chains are under physiological conditions *in vitro*.[1] IDPs include random coil-like regions, partially folded or molten/pre-molten globule-like domains with poorly packed side chains, and dynamic structural ensembles.[2,3] They implement important functional roles in transcriptional regulation, translation, and cellular signal transduction,[4] and they are relatively common.[5] The prevalence of disorder was implicated in various human diseases and they were suggested as important targets for drug discovery.[6,7] The above motivates research towards improved understanding of the principles and mechanisms of IDPs. Some studies show that IDPs are characterized by relatively unique sequence signatures. For example, they often have a low content of bulky hydrophobic amino acids and a high proportion of polar and charged residues, a low content of predicted secondary structure, low complexity, and unique evolutionary and solvent accessibility profiles.[8-13] This implies that the disorder is predictable from the sequence. Therefore, a number of computational predictors were developed over the past decade. These efforts intensified after the disorder prediction was introduced into the biannual CASP experiments in 2002.[14-16] The disorder predictors are categorized into four types:[17]

1. *propensity-based* methods based on relative propensity of amino acids to form disorder/ordered regions: GlobPlot,[18] FoldIndex,[19] IUPred,[20] and Ucon;[21]
2. *machine learning-based* predictors: DISOPRED2,[22] DISpro,[23] RONN,[24] ProfBval,[25,26] PONDR predictors,[9,14,27-30] PreDisorder,[23,32] NORSnet,[21] DisEMBL,[18] and Spritz;[31]
3. *consensus-based* methods that combine predictions from multiple base predictors: metaPrDOS,[33] GS-MetaServer,[34] MD,[35] PONDR-FIT,[36] and MFDp;[17]
4. *structural models-based* approaches that make use of predicted tertiary structure models: PrDOS [37] and DISOCLUST.[38]

The results from a recent comparative review [39] and the CASP8 competition,[16] demonstrate that the consensus-based methods, such as the GS-MetaServer,[34] MD,[35] and MFDp,[17] generally outperform other methods. This is perhaps due to the fact that certain disorder predictors target specific flavors/types of the disorder and thus they are suitable to form a well-performing consensus.[35] One of the desired characteristics of the consensus-based methods is that the base predictors that are being combined should be complementary to each other.[17,35,40] However, the existing methods select the base predictors in an ad hock manner: based on their availability, differences in the architecture of the base methods and/or their targeted type of the disorder, and utilizing a premise that inclusion of a larger number of base predictors is beneficial.[17,34-36] We perform a first-of-its-kind empirical investigation into the complementarity among disorder predictors. For a given triplet of methods, we measure complementarity based on the differences in their predictions, by comparing them with each other and to their majority vote consensus (MVC). We build a regression-based model that predicts the quality of the MVC using the information concerning complementarity measured at the residue- and segment-level and the predictive performance of the base methods. We empirically demonstrate that a proper selection of the base methods leads to the improvements in the disorder predictions, when compared with the current single and consensus-based disorder predictors.

2. Methods

2.1. *Considered disorder predictors*

We consider 17 disorder predictors, which are accessible to the end user via web server or a standalone program, including 16 that were discussed in a recent review [39] and DRIPPRED. Since our study requires that a given predictor outputs real-value disorder propensity, we could not include the predictors that only produce the binary outputs, namely GlobPlot,[18] FoldIndex,[19] DisEMBL,[18] and Spritz.[31] Furthermore, the IUPred method is used in both of its modes, one for the prediction of short disordered segments, named IUPredS, and the other for long segments, named IUPredL. Consequently, the 14 disorder predictors are considered; see Table 1.

Table 1. Summary of disorder predictors included in this study. Methods are sorted by the year of publication. Acronyms in column "Type": propensity-based (PB), machine learning-based (MLB), consensus-based (CB), and structural models-based (SB); in column "Algorithm": support vector machine (SVM), neural network (NN), scoring function (SF) and self-organizing map (SOM).

Prediction method				URL
Name	Published	Type	Algorithm	
MFDp	2010	CB	SVM	http://biomine-ws.ece.ualberta.ca/MFDp.html
PONDR-FIT	2010	CB	NN	http://www.disprot.org/pondr-fit.php
MD	2009	CB	NN	https://rostlab.org/owiki/index.php/Metadisorder
DISOCLUST	2008	SB	SF	http://www.reading.ac.uk/bioinf/DISOclust/
NORSnet	2007	MLB	NN	https://rostlab.org/owiki/index.php/Norsnet
Ucon	2007	PB	SF	https://rostlab.org/owiki/index.php/UCON
ProfBval	2006	MLB	NN	https://rostlab.org/owiki/index.php/Profbval
VSL2B	2006	MLB	SVM	http://www.dabi.temple.edu/disprot/Predictors.html
DISpro	2005	MLB	NN	http://scratch.proteomics.ics.uci.edu/
IUPred	2005	PB	SF	http://iupred.enzim.hu/
RONN	2005	MLB	NN	http://www.strubi.ox.ac.uk/RONN
DISOPRED2	2004	MLB	SVM	http://bioinf.cs.ucl.ac.uk/disopred/
DRIPPRED	2004	PB	SOM	http://www.sbc.su.se/~maccallr/disorder/

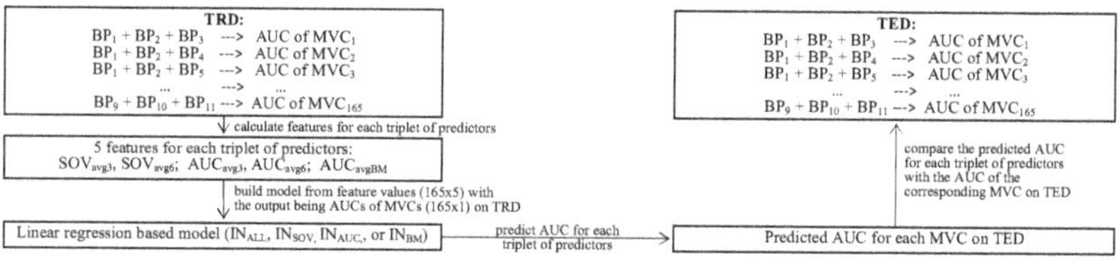

Figure 1. Overview of the prediction model. BP_i and *MVC* denote the ith base predictor and the majority vote consensus, respectively. Sections 2.4 and 2.5 provide detailed explanations.

2.2. *Datasets*

We utilize a recently-built MxD dataset that was developed by Mizianty *et al.*,[17] and we further improve the disorder annotation using the SL dataset-based procedure.[41] The protein chains were originally collected from the Protein Data Bank (PDB)[42] and the release 4.9 of the DisProt database[43]. The resulting MxD_SL dataset includes 494 out of 514 proteins from the MxD; the remaining 20 chains could not be predicted by the MD and DISpro methods. The SL dataset-based procedure combines the disorder annotations from the DisProt with the disorder/order annotations based on the corresponding structural domains in PDB. Consequently, the DisProt derived chains include the DisProt's annotation of disorder, while the remaining residues that were not annotated in this database are annotated using the PDB domains. This means that some of these residues are left without any annotation (if they cannot be found in PDB); we did not use them to perform design and evaluations. However, we use all annotated residues, even if they are in the partially annotated chains. The MxD_SL dataset includes 196,434 residues, where 51,733 are without annotation. Similarly as in CASP8, we discard the native disordered segments with 3 or fewer consecutive disordered residues.[16] The MxD_SL dataset is randomly partitioned into equal-size training dataset (TRD) and test dataset (TED), which have 23,406 and 23,074 disordered residues, respectively. Importantly, chains in MxD_SL are characterized by pair-wise sequence identity below 25%,[17] which means that TRD and TED are independent at the 25% similarity level. The datasets can be downloaded from http://biomine.ece.ualberta.ca/MVCdisorder/MVCdisorder.htm.

2.3. *Overview of the prediction model*

We investigate complementarity between disorder predictors based on a majority vote consensus (MVC). We combine three base predictors to model the relation between the complementarity in their predictions, which is measured using an approach described in section 2.4, and the predictive quality of the resulting MVC using linear regression on the TRD. Next, we apply regression to predict the best combination of methods on the TED, and we compare this consensus with the existing solutions (see Figure 1). We do not use cross-validation or another sampling since we do not perform parameterizations that could lead to overfitting. Furthermore, we have sufficient amount of data in the training dataset, and training and test datasets are independent.

2.4. *Inputs for the prediction model*

We assess the complementarity of the base predictors at the residue and disordered segment levels, i.e. we study the complementarity between the individual predicted propensities and between the binary predictions that form segments of predicted consecutive disordered residues:

1. The area under the ROC curve (AUC)[26] is usually used to examine the predicted residue-level propensities. For a pair of predictions, we quantify their residue-level complementarity by computing two AUC values between them, i.e., the first AUC when one prediction is assumed to be the true outcome and the other to be the prediction, and another AUC when the second prediction is assumed to be the true outcome. Consequently, higher AUC values indicate lower complementarity. Given three base methods, we compute the average of the six corresponding AUC values, and we call this input AUC_{avg6}. We also quantify complementarity between the three base predictions and their MVC. We calculate the average of the three AUC values when using the consensus as the true outcome, and we refer to this input as AUC_{avg3}.
2. The segment overlap (SOV)[45] measures the amount of overlap between the disordered segments.[39] Given three base methods, we compute average of the six SOV values (SOV_{avg6}), i.e., two SOV values for each pair of methods when one prediction is assumed to be the true outcome and the other to be the prediction, and when second prediction is assumed to be the true outcome. Similarly as for the AUC, we also compute the average SOV between the base predictions and their MVC (SOV_{avg3}). Higher SOV values correspond to predictions with more overlapping disordered segments, i.e., weaker complementarity at the segment-level.

The performance of the individual base methods is also likely to have impact on the predictive quality of their consensus. We quantify their performance based on the average of three AUC values (AUC_{avgBM}), where predictions from each method are compared against the native disorder.

We group these inputs into four sets: IN_{ALL} (with all 5 inputs), IN_{SOV} (SOV_{avg6} and SOV_{avg3}), IN_{AUC} (AUC_{avg6} and AUC_{avg3}), and IN_{BM} (AUC_{avgBM}), to investigate whether combining the 3 input types (segment-, residue-, and base method-based) provides a better model when compared with the use of the individual input types. The calculation of the AUC_{avgBM} involves the native annotations and we use AUC_{avgBM} values calculated on TRD to perform predictions on the TED.

2.5. *Majority vote and linear regression*

We normalize the propensities generated by the 14 considered methods so that they use the same threshold = 0.5 to binarize their predictions. Specifically, for a given method that uses threshold p, we linearly map the probability within the range $[0, p)$ and $[p, 1]$ to $[0, 0.5)$ and $[0.5, 1]$, respectively. We next build the MVC for every combination of three methods. Since MD,[35] MFDp,[17] and PONDR-FIT[36] are already consensuses, they are not utilized as the base methods. Consequently, using the remaining 11 methods we generate $(11\times10\times9)/6 = 165$ combinations. The predictions from the base methods are combined using majority vote; as shown in Table 2.
Following Figure 1, we build a linear regression-based model on TRD to express the relation between the AUC of the MVC and the complementarity and quality of the three base predictors, which is quantified using IN_{ALL} feature set. The model outputs real values that predict the AUC of a given MVC. The regression coefficients demonstrate the relation between each of the inputs and the predictive quality (expressed with AUC), and can be used to investigate how the residue- and segment-level complementarity among the base methods affects the performance of the consensus.

Table 2. Implementation of MVC with 3 base predictors BP_i where i=1,2,3. The outputs include the propensity calculated from normalized propensities $p(BP_i)$ generated by BP_i, and binary prediction, where D and O are disordered and ordered residues, respectively.

BP_1	binary	O	D	O	O	D	D	O	D
BP_2	binary	O	O	D	O	D	O	D	D
BP_3	binary	O	O	O	D	O	D	D	D
MVC	binary	O				D			
	probability	$\min_{i=1,2,3}\{p(BP_i)\}$				$\max_{i=1,2,3}\{p(BP_i)\}$			

2.6. *Evaluation criteria*

We compare AUC values predicted by the linear regression with AUCs of the MVC using Pearson Correlation Coefficient (*PCC*), Mean Absolute Error (*MAE*), and Mean Squared Error (*MSE*).

We use the regression to predict the top performing MVC on the TED. The quality of disorder predictions using this MVC, two runner-up predicted MVCs, and existing disorder predictors is evaluated using Matthews Correlation Coefficient (MCC), S_w, AUC, and SOV. The first 3 criteria, which were used in CASP,[15,16] assess the per-residue predictions; SOV evaluates prediction of disordered segments.[39] The binary per-residue predictions are assessed with MCC and S_w:

$$MCC = \frac{TP\cdot TN - FP\cdot FN}{\sqrt{(TP+FP)\cdot(TP+FN)\cdot(TN+FP)\cdot(TN+FN)}}$$

$$S_w = \frac{w_{disorder} \cdot TP - w_{order} \cdot FP + w_{order} \cdot TN - w_{disorder} \cdot FN}{w_{order} \cdot N_{disorder} + w_{disorder} \cdot N_{order}}$$

where *TP* is *true positive* (number of correctly predicted disordered residues), *FP* denotes *false positive* (number of ordered residues predicted as disordered), *TN* denotes *true negative* (number of correctly predicted ordered residues), *FN* stands for *false negative* (number of disordered residues predicted as ordered), $W_{disorder}$ (N_{order}) and W_{order} ($N_{disorder}$) are the percentages (numbers) of the ordered and the disordered residues, respectively. The S_w and MCC values range between -1 and 1; they are equal to zero when all residues are predicted as ordered or disordered and their higher values indicate better predictions.

The quality of the predicted per-residue propensities is assessed with the ROC curve. For each value of propensities *p* achieved by a given method (between 0 and 1), the residues with propensities $\geq p$ are set as disordered, and all other residues are set as ordered. Next, the *TP-rate* = *TP* / (*TP* + *FN*) and the *FP-rate* = *FP* / (*FP* + *TN*) are calculated and we use the area under the resulting ROC curve (*AUC*) to quantify the predictive quality.

The segment-level evaluation is based on the SOV,[39,45] which quantifies the overlap between the segments formed by the binary per-residue predictions and the native disorder segments. We compute the SOV values only for the chains with the complete native disorder annotation.

Table 3. The coefficients of the linear regression models that predict AUC of the majority-vote consensuses which are based on the 5 input features (IN_{ALL}), the 2 SOV-based inputs (IN_{SOV}), the 2 AUC-based inputs (IN_{AUC}), and the average AUC of the base methods (IN_{BM})

Model	SOV_{avg3}	SOV_{avg6}	AUC_{avg3}	AUC_{avg6}	AUC_{avgBM}	Constant
IN_{ALL}	0.013	-0.023	-0.089	0.024	0.109	0.823
IN_{SOV}	0.055	-0.037				0.835
IN_{AUC}			-0.051	0.070		0.837
IN_{BM}					0.039	0.820

3. Results

3.1. *Modelling the complementarity in disorder prediction*

We generate linear regression-based model on the training dataset (TRD), using the inputs defined in section 2.4 and following Figure 1. We compare the models based on all 5 inputs (IN_{ALL}) with the models based on the IN_{SOV}, IN_{AUC}, and IN_{BM} features; see Table 3. We observe that signs of the coefficients are consistent between different input sets, i.e., the SOV_{avg3}, AUC_{avg6}, and AUC_{avgBM} are positively correlated with the AUC of the MVCs while SOV_{avg6} and AUC_{avg3} are negatively correlated. The model reveals that the improved performance of a MVC is associated with: (1) higher quality of the base methods (positive coefficient for AUC_{avgBM}); (2) lower complementarity/higher similarity between the base predictors (positive coefficients for AUC_{avg6}) and their higher complementarity/lower similarity to the consensus prediction (negative coefficient for AUC_{avg3}) at the residue level; and (3) their lower complementarity/higher similarity to the consensus prediction (positive coefficient for SOV_{avg3}) and higher complementarity/lower similarity between them (negative coefficient for SOV_{avg6}) at the segment level.

Table 4. Comparison of the predictive quality of the regression models that utilize 5 input features (IN_{ALL}), the 2 SOV-based inputs (IN_{SOV}), the 2 AUC-based inputs (IN_{AUC}), and the average AUC of the base methods (IN_{BM}), and the random model on the training and the independent test sets.

Model	Evaluation on training dataset			Evaluation on test dataset		
	MAE	MSE	PCC	MAE	MSE	PCC
IN_{ALL}	0.007	0.105	0.813	0.007	0.112	0.785
IN_{SOV}	0.011	0.167	0.381	0.009	0.144	0.444
IN_{AUC}	0.010	0.158	0.476	0.008	0.133	0.571
IN_{BM}	0.009	0.145	0.597	0.008	0.125	0.640
random	0.019	0.290	0.008	0.018	0.282	0.009

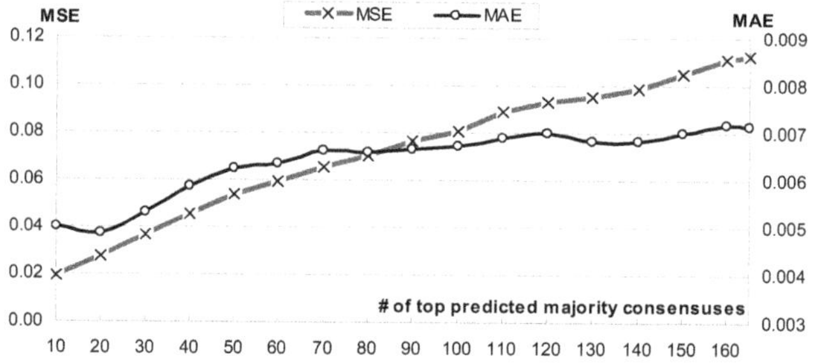

Figure 2. Relation between the number of evaluated consensuses that are sorted by their predicted AUC on TED (*x*-axis), and MSE and MAE (*y*-axis) which are based on the predictions from the regression that utilizes 5 inputs (IN_{ALL}) generated using TRD and tested on TED.

Next, we evaluate each regressions model (based on the IN_{ALL}, IN_{SOV}, IN_{AUC}, and IN_{BM} features) and compare them with a model that generates AUC values in the actual range of these values in the TRD at random. The random prediction is repeated 100 times and we report the average results; we use that as a baseline to examine the performance of our regression models. The predictive quality is measured by MAE, MSE, and PCC, on both TRD and the independent test dataset (TED); see Table 4. The four regression models outperform the random predictor by a wide margin. The model that utilizes the five inputs (IN_{ALL}) improves over the other regressions that use a subset of inputs, which demonstrates that all inputs contribute to the predictions. The IN_{ALL}-based model obtains the lowest MAE (0.007 on TRD and TED), lowest MSE (0.105 on TRD and 0.112 on TED) and the highest PCC (0.81 and 0.78 on TRD and TED, respectively). Moreover, the quality of the regression predictions is better for the consensuses with higher predicted AUCs and it progressively gets worse for the lower predicted AUCs; see Figure 2. The MAE and MSE are at about 0.005 and 0.02, respectively for the top 10 predicted consensuses, which indicates that our predictions are quite accurate when predicting the top-performing consensuses. The model successfully predicts the top-performing consensus, which consists of DISOCLUT, DISpro, and VSL2B, on the TED; the predicted and actual AUCs for this consensus are 0.871 and 0.872, respectively. To compare, the worst performing consensus with DRIPPRED,

NORSnet, and ProfBval, which obtains the AUC = 0.809, is predicted to obtain AUC = 0.819. The relatively high predictive quality of our regression model suggests that the observations made based on this model should be accurate.

For the top predicted MVC, the SOV_{avg3} and SOV_{avg6} values are 3.7% and 10.7% smaller than the average values of SOV_{avg3} (0.521) and SOV_{avg6} (0.546), respectively, which are calculated over all 165 MVCs; The AUC_{avg3}, AUC_{avg6}, and AUC_{avgBM} values are 8.4%, 18.2%, and 28.3% higher than the corresponding averages (0.582, 0.510, and 0.59), respectively. The feature values were normalized to make these calculations, i.e., the original feature x was normalized as $(x-a)/(b-a)$, where a and b are the minimum and maximum value of x on TRD, respectively; if (for TED) $x<a$, we set $x=a$; if $x>b$, we set $x=b$. The above differences indicate that the corresponding three base predictors have high predictive quality (high AUC_{avgBM}), above average complementarity at the segment level (low SOV_{avg}), and below average complementarity at the residue level (high AUC_{avg}). This shows that complementarity between base predictors affects performance of their majority vote consensus. The 2nd and 3rd top predicted MVCs have all feature values above the corresponding averages, with their AUC_{avgBM} values higher by wide, 39% and 41%, margins, respectively. This indicates that the high quality of these MVCs stems mostly from the good predictive performance of their base methods.

We also investigate whether the quality of a given MVC depends on the types of its base predictors (see Table 1). The top predicted MVC includes structural models-based and two different machine learning-based predictors, one that utilizes neural networks (NNs) and another that uses Support Vector Machines (SVMs). The 2nd top predicted MVC includes three different machine learning-based predictors, two of which use SVMs and one based on NN. The 3rd top predicted consensus uses one propensity-based method and two machine learning-based predictors. Our analysis suggests that use of diverse types of base predictors may have an effect on the complementarity and may lead to an improved consensus.

3.2. *Evaluation of consensus-based and individual disorder predictors*

We evaluate disorder predictions, which were generated by the 14 considered predictors and using the top 3 MVCs predicted by the linear regression (MVCs with the highest predicted AUCs), on the independent test dataset (TED), see Table 5. MD[35] and MFDp[17] are based on a consensus approach; however, both of these methods utilize complex, non-linear predictors to combine the outputs of base classifiers with additional inputs, including multiple alignment, predicted flexibility, secondary structure, solvent accessibility, etc. Since such complex consensuses were shown to outperform the simple MVC,[35] Table 5 includes results of the MVCs which use the same base methods as in MD and MFDp to allow for a fair comparison to the MVCs that were predicted with the regression. The MD-based MVC includes NORSnet, ProfBval, Ucon, and DISOPRED2, while the MFDp-based MVC incorporates DISOPRED2, IUPredL, IUPredS, and DISOCLUST. In both cases the consensus has 4 methods, and thus we resolve the ties by using the prediction from the DISOPRED2 that has the highest AUC on the TRD. We also rebuilt the MFDp method by substituting the original base predictors with the methods included in the top MVC predicted by the regression, i.e., DISOCLUST, DISpro, and VSL2B. The other consensus method is PONDR-

FIT,[36] which uses a relatively simple approach to combine the predictions from six base methods and does not use additional inputs. Thus, we decided to directly compare with this method, without building the MVC. We note that our 3rd best predicted MVC corresponds to the combination of three methods (DISOPRED2, IUPredL, and VSL2B) that obtain the highest AUC and MCC values on TRD.

Table 5. Predictive quality, measured with MCC, S_w, and AUC on TED for the 14 considered disorder predictors and the top 3 majority-vote consensuses (MVCs) predicted by the linear regression. The consensus-based predictors include MFDp, MD, PONDR-FIT, and the MVCs based on our top 3 predictions and using the base methods from MFDp and MD. The consensus-based and the single predictors are sorted by their AUC in the descending order, respectively. The SOV is calculated based on 176 proteins in TED that are fully annotated.

Type	Predictor	MCC	S_w	SOV	AUC
Consensus predictors	MFDp rebuild using the top predicted base methods	0.637	0.636	0.620	0.8895
	MFDp	0.633	0.625	0.638	0.8875
	MD	0.625	0.596	0.478	0.8784
	top predicted MVC (DISOCLUST, DISpro, VSL2B)	0.613	0.602	0.609	0.8716
	3rd predicted MVC (DISOPRED2, IUPredL, VSL2B)	0.634	0.604	0.558	0.8633
	2nd predicted MVC (DISOPRED2, DISpro, VSL2B)	0.629	0.568	0.524	0.8607
	MVC using 4 base methods from MFDp	0.627	0.600	0.544	0.8606
	MVC using 4 base methods from MD	0.618	0.582	0.462	0.8569
	PONDR-FIT	0.592	0.556	0.590	0.8549
Single predictors	DISOPRED2	0.608	0.570	0.514	0.8574
	VSL2B	0.557	0.576	0.644	0.8568
	DISOCLUST	0.504	0.533	0.608	0.8497
	DISpro	0.458	0.304	0.361	0.8391
	IUPredL	0.575	0.514	0.341	0.8390
	IUPredS	0.544	0.472	0.511	0.8314
	RONN	0.524	0.517	0.517	0.8190
	NORSnet	0.569	0.496	0.184	0.8146
	DRIPPRED	0.480	0.470	0.500	0.7874
	Ucon	0.423	0.405	0.297	0.7844
	ProfBval	0.273	0.279	0.527	0.7353

Table 5 shows that MFDp and MD have the highest AUC, but this is since they use the complex consensuses. The MVCs built using their base predictors have AUC equal 0.86 and 0.85, respectively, which is lower than the AUCs of the top three regression predicted MCVs. Among these three predictions, the highest AUC = 0.87 is obtained by our top prediction. Moreover, the rebuilt MFDp that uses the base methods from the top prediction slightly improves over the original MFDp in spite of the fact that it uses fewer base methods (3 vs. 4). Importantly, the top three predicted MCVs improve over all other methods, including the simple consensus-based PONDR-FIT and the 11 modern disorder predictors. The AUC is higher by 1.5% when comparing the top predicted MVC and the best result from among these 12 methods. Similarly, the MCC and S_w values are higher and they equal 0.61 and 0.6 for the top prediction generated by the regression.

However, the top 3 predicted ensembles are outperformed by VSL2B in the disordered segment prediction measured with SOV; this is consistent with prior work that shows that segment predictions from VSL2B outperform other methods, including consensus approaches.[39] At the same time, our top prediction improves over VSL2B by 5.5% in MCC, 2% in S_w, and 1.5% in AUC. Moreover, the segment predictions from our top predicted consensus are competitive or better than the predictions from the remaining 11 methods, MD, and the MVCs based on the base predictors from the MD and from the MFDp; this is likely since our consensus includes VSL2B.

The top three predicted MVCs are better than their base methods by a relatively wide margin. Table 5 shows that the top predicted ensemble improves the MCC, S_w, and AUC by at least 5.6%, 2.7%, and 1.5%, respectively, when compared to the best base method. The 3rd best predicted ensemble is better by 2.7% in MCC, 2.8% in S_w, and 0.6% in AUC.

Overall, our analysis shows that the top MVC predicted using linear regression outperforms or is at least competitive when compared with the existing single and consensus-based predictors.

```
Native    ------------------*--------------*-----------------------------*--DDDDDDDDDDDDD-------*--------DDDDDDDDD
Top1      DDDDD-------------*---DDDDD------*-----------------------------*---DDDDDDDDDDDDDDDD---*-DDDDDDDDDDDDDDDD
VSL2B     DDDDD-------DDDDD-*---DDDDD------*-----------------------------*---DDDDDDDDDDDDDDDDDD-*----------DDDDDDD
DISOCL    DDDDDDDDDDD-------*-DDDDDDDDDDDD-*-D----DDDDDDDDDDDDDDDDDDDDDD-*-DDDDDDDDDDDDDDDDDD---*-DDDDDDDDDDDDDDDD
DISpro    DD----------------*--------------*-----------------------------*----------------------*-DDDDDDDDDDDDDDDD
MVCMD     ------------------*--------------*-----------------------------*---------DDDDDD-------*-----------------
MVCMFDp   DDDD---D----------*--------------*-----------------------------*---------DDDDDD-------*---------DDDDDDDD
PONDRF    DDDDDDDDDDDDDDDD--*--------------*-----------------------------*----DDDDD-------------*---------DDDDDDDD
```

Figure 3. Comparison of the predictions for the Glutamyl-Q tRNA(Asp) synthetase protein (PDB id: 1NZJA) based on the top MVC predicted by the regression (Top1 line), its base methods DISOCLUST (DISOCL line), DISpro, and VSL2B, the MVC using the base methods from MD (MVC_{MD} line) and from MFDp (MVC_{MFDp} line), and PONDR-FIT (PONDRF line). 'D' and '–' denote the disordered and ordered residues, respectively, and '*' stands for a segment of ordered residues. The first line shows the native annotation.

3.3. *Case study*

We use the Glutamyl-Q tRNA(Asp) synthetase protein (PDB id: 1NZJA) from the TED that has two relatively short disordered segments, including one at the C-terminus and one inner-chain segment, as a case study to compare disorder predictions. We compare predictions from the top MVC predicted by the regression, its three base methods: DISOCLUST, DISpro, and VSL2B, and based on the three most relevant other consensuses: the MVCs using the base methods from MD and from MFDp, and PONDR-FIT; see Figure 3. Majority of the predictors, except for the DISpro and the MVC using the base methods from MD, find both native disordered segments; however, most of the methods find additional segments towards and at the N-terminus. Our top MVC reduces the over-prediction of the disorder when compared with two of its base methods, VSL2B and DISOCLUST, and finds one native segments that was missed by the third base method DISpro. The MVC using the base methods from MFDp under-predicts the disorder, especially the inner segment, and still produces disorder predictions at the N-terminus. PONDR-FIT also under-predicts the inner-chain segment and generates a relatively long disordered segment at the N-terminus. Although the predictions in this case study should not be assumed typical, we conclude that they demonstrate that the proper selection of the base methods can result in improvements.

4. Conclusion

Our study provides insights that could help in development of a new generation of consensus-based disorder predictors. We use linear regression to model relation between certain aspects that quantify complementarity between the base methods and the predictive quality of the resulting majority-vote based consensus. Our modeling reveals that the complementarity has to be evaluated at the residue and the disordered segment levels, i.e., models that ignore one of these aspects are shown to provide inferior results. The model shows that improved predictive performance is associated with inclusion of accurate base predictors that are similar to each other at the residue level and which complement each other at the segment level, and which complement the consensus at the residue level. We also observe that use of different types of base predictors to implement the consensus seems to be beneficial. We empirically demonstrate the top majority-based consensuses predicted by our model on an independent test dataset outperform existing predictors, including consensus-based methods, which suggests that our observations have practical value.

Acknowledgments

This work was supported by the Alberta Innovates Scholarship in Omics to ZP and the NSERC Discovery grant to LK. We thank Marcin Mizianty for help with the datasets and MFDp method.

References

1. V. N. Uversky, A. K. Dunker, *Biochim Biophys Acta* **1804**(6), 1231(2010).
2. A. K. Dunker, J. D. Lawson, C. J. Brown, et al., *J Mol Graph* **19**(1), 26 (2001).
3. V. N. Uversky, *Prot Sci* **11**, 739 (2002).
4. A. K. Dunker, C. J. Oldfield, J. W. Meng, P. Romero, J. Y. Yang, J. W. Chen, V. Vacic, Z. Obradovic and V. N. Uversky, *BMC Genomics* **9**(S2), S1 (2008).
5. A. K. Dunker, Z. Obradovic, P. Romero, E. C. Garner and C. J. Brown, *Genome Inform Ser Workshop Genome Inform* **11**, 161 (2000).
6. V. N. Uversky, C. J. Oldfield, U. Midic, H. Xie, B. Xue, S. Vucetic, L. M. Iakoucheva, Z. Obradovic and A. K. Dunker, *BMC Genomics* **10**(S1), S7 (2009).
7. Y. Cheng, T. LeGall, C. J. Oldfield, J. P. Mueller, Y. Y. Van, P. Romero, M. S. Cortese, V. N. Uversky and A. K. Dunker, *Trends Biotechnol* 24(10):435 (2006).
8. V. N. Uversky, J. R. Gillespie and A. L. Fink, *Proteins* **41**, 415 (2000).
9. P. Romero, Z. Obradovic, X. Li, E. C. Garner, C. J. Brown and A. K. Dunker, *Proteins* **42**, 38 (2001).
10. J. Liu, H. Tan and B. Rost, *J Mol Biol* **322**, 53 (2002).
11. H. J. Dyson and P. E. Wright, *Nat. Rev. Mol. Cell Bio.* **6**, 197 (2005).
12. Z. Dosztányi, B. Mészáros and I. Simon, *Brief. Bioinform* **11**(2), 225 (2010).
13. M. J. Mizianty, T. Zhang, B. Xue, Y. Zhou, A. Keith Dunker, V.N. Uversky and L. Kurgan, *BMC Bioinformatics* **12**, 245 (2011).
14. S. Vucetic, C. J. Brown, A. K. Dunker and Z. Obradovic, *Proteins* **52**, 573 (2003).
15. L. Bordoli, F. Kiefer and T. Schwede, *Proteins* **69**(S8), 129 (2007).

16. O. Noivirt-Brik, J. Prilusky and J. L. Sussman, *Proteins* 77(S9), 210 (2009).
17. M. J. Mizianty, W. Stach, K. Chen, K. D. Kedarisetti, F. M. Disfani and L. Kurgan, *Bioinformatics* **26**(18), i489 (2010).
18. R. Linding, L. J. Jensen, F. Diella, P. Bork, T. J. Gibson and R. B. Russell, *Structure* **11**, 1453 (2003).
19. J. Prilusky, C. E. Felder, T. Zeev-Ben-Mordehai, E. H. Rydberg, O. Man, J. S. Beckmann, I. Silman and J. L. Sussman, *Bioinformatics* **21**, 3435 (2005).
20. Z. Dosztányi, V. Csizmok, P. Tompa and I. Simon, *Bioinformatics* **21**, 3433 (2005).
21. A. Schlessinger, J. Liu and B. Rost, *PLoS Comput Biol* **3**, e140 (2007).
22. J. J. Ward, L. J. McGuffin and Bryson K, *Bioinformatics* 20, 2138 (2004).
23. J. Cheng, M. Sweredoski and P. Baldi, *Data Min Knowl Disc* **11**(3), 213 (2005).
24. Z. R. Yang, R. Thomson, P. McNeil and R. M. Esnouf, *Bioinformatics* 21, 3369 (2005).
25. A. Schlessinger and B. Rost, *Proteins* **61**(1), 115 (2005).
26. A. Schlessinger, G. Yachdav and B. Rost, *Bioinformatics* 22, 891 (2006).
27. Z. Obradovic, K. Peng, S. Vucetic, P. Radivojac, C. J. Brown and A. K. Dunker, *Proteins* 53(S6), 566 (2003).
28. K. Peng, S. Vucetic, P. Radivojac, C. J. Brown, A. K. Dunker and Z. Obradovic, *J Bioinform Comput Biol* 3, 35 (2005).
29. Z. Obradovic, K. Peng, S. Vucetic, P. Radivojac and A. K. Dunker, *Proteins* 61(S7), 176 (2005).
30. K. Peng, P. Radivojac, S. Vucetic, A. K. Dunker and Z. Obradovic, *BMC Bioinformatics* **7**, 208 (2006).
31. A. Vullo, O. Bortolami, G. Pollastri and S. C. Tosatto, *Nucleic Acids Res* **34,** W164 (2006).
32. X. Deng, J. Eickholt and J. Cheng, *BMC Bioinformatics* 10, 436 (2009).
33. T. Ishida and K. Kinoshita, *Bioinformatics* 24, 1344 (2008).
34. M.A. Kurowski and J.M. Bujnicki, *Nucleic Acids Res* 31, 3305-3307 (2003).
35. A. Schlessinger, M. Punta, G. Yachdav, L. Kajan and B. Rost, *PLoS One* **4**, e4433 (2009).
36. B. Xue, R. L. Dunbrack, R. W. Williams, A. K. Dunker and V. N. Uversky, *Biochim Biophys Acta* **1804**, 996 (2010).
37. T. Ishida and K. Kinoshita, Nucleic Acids T. Ishida and K. Kinoshita, *Nucleic Acids Res* 35,W460 (2007).
38. L. J. McGuffin, *Bioinformatics* 24, 1798 (2008).
39. Z. L. Peng and L. Kurgan, *Curr. Prot. Pept. Sci.*, accepted on October 2010.
40. C. J. Oldfield, *Biochemistry* 44, 1989 (2005).
41. F. L. Sirota, H. S. Ooi, T. Gattermayer, G. Schneider, F. Eisenhaber and S. Maurer-Stroh, *BMC Genomics* 11(S1), S15 (2010).
42. H.M. Berman, J. Westbrook, Z. Feng, G. Gilliland, T.N. Bhat, H. Weissig, I.N. Shindyalov, and P.E. Bourne, *Nucleic Acids Res.* **28**, 235-242 (2000).
43. M. Sickmeier, J.A. Hamilton, T. LeGall, V. Vacic, M.S. Cortese, A. Tantos, B. Szabo, P. Tompa, J. Chen, V.N. Uversky, et al., *Nucleic Acids Res.* **35**, D786-793 (2007).
44. T. Fawcett, *Pattern Recogn Lett* 27, 861 (2006).
45. A. Zemla, C. Venclovas, K. Fidelis and B. Rost, *Proteins* **34**(2), 220 (1999).

MODULATING PROTEIN–DNA INTERACTIONS BY POST-TRANSLATIONAL MODIFICATIONS AT DISORDERED REGIONS

DANA VUZMAN
Department of Structural Biology, Weizmann Institute of Science, Rehovot, 76100, Israel
Email: dana.golbin@weizmann.ac.il

YONIT HOFFMAN
Department of Structural Biology, Weizmann Institute of Science, Rehovot, 76100, Israel
Email: yonit.hoffman@weizmann.ac.il

YAAKOV LEVY*
Department of Structural Biology, Weizmann Institute of Science, Rehovot, 76100, Israel
Email: Koby.Levy@weizmann.ac.il

Intrinsically disordered regions, particularly disordered tails, are very common in DNA-binding proteins (DBPs). The ability of disordered tails to modulate specific and nonspecific interactions with DNA is tightly linked to their being rich in positively charged residues that are often non-randomly distributed along the tail. Perturbing the composition and distribution of charged residues in the disordered regions by post-translational modifications, such as phosphorylation and acetylation, may impair the ability of the tail to interact nonspecifically with DNA by reducing its DNA affinity. In this study, we analyzed datasets of 3398 and 8943 human proteins that undergo acetylation or phosphorylation, respectively. Both modifications are common on the disordered tails of DBPs (3.1 ± 0.2 (0.07 ± 0.007) and 2.0 ± 0.2 (0.02 ± 0.003) acetylation and phosphorylation sites per tail (per tail residue), respectively). Phosphorylation sites are abundant in disordered regions and particularly in flexible tails for both DBPs and non-DBPs. While acetylation sites are also frequently occurred in the disordered tails of DBPs, in non-DBPs they are often found in ordered regions. This difference may indicate that acetylation has different function in DBPs and non-DBPs. Post-translational modifications, which often take place at disordered sites of DBPs, can modulate the interactions of proteins with DNA by changing the local and global properties of the tails. The effect of the modulation can be tuned by adjusting the number of modifications and the cross-talks between them.

1. Introduction

Post-translational modifications (PTMs) are widely used to modulate protein function in the cell. PTMs can therefore be viewed as covalent modifications that increase the structural and biophysical diversity of proteins and thus enrich the information stored in the genomes. There are dozens of different PTMs that are incorporated by the cell. To achieve the required effect, a protein may undergo a single PTM or several PTMs that may engage in cross-talk. In many cases, a single position on the protein can be altered by different modifications so that switching between several effects (or functions) can be regulated by the identity of the PTM at that position. PTMs are often classified according to the mechanisms involved: the addition of functional groups (e.g., phosphorylation and glycosylation); attachment of other polypeptides (e.g., ubiquitination and SUMOylation); changing of the chemical nature of amino acids (e.g., acetylation, deamidation and oxidation); and cleavage of the backbone by proteolysis [1, 2]. PTMs can be categorized according to the conformational preference of the modification sites; namely, if the PTM occurs on a structured or disordered region. PTMs at structured domains are crucial, for example, for modifying enzymatic activities or stabilizing protein structure. PTMs at disordered regions are advantageous because of the high-exposure of these sites, which enables them to be accessed easily by the modifying enzymes through high-specificity and low-affinity interactions that also often involve a disorder-to-order transition. Indeed, the abundance of disordered sites as targets for PTMs is evident from many experimental studies [3] and various predictive approaches indicate that sequences surrounding PTMs have features similar to those of disordered regions [4].

In this study, we focus on two widespread PTMs, acetylation and phosphorylation, that play important roles in modulating various regulatory processes. We focus on the analysis of these two PTMs in DNA-binding proteins (DBPs), which are a unique but highly central group of proteins that mediate numerous genetic processes, such as transcription, repression, and replication via interaction with DNA. Very often, DBPs include a patch of positively charged residues that serve as the binding site for interacting with the negatively charged DNA. The location of the positively charged residues (Lys and Arg) in the 3D structure supports the participation in hydrogen bonding with the bases of the nucleotides that constitute the cognate site. The positive patches on DBPs not only contribute to specific recognition of the protein residues and the DNA nucleotides, but also play a pivotal role in allowing DBPs to interact non-specifically with DNA. While the affinity of DBPs for non-specific DNA is relatively low and can be strongly affected by salt concentrations that mask the electrostatic complementarity, non-specific interactions are nevertheless central to the achievement of fast recognition.

Electrostatic interactions between DBPs and DNA mediate the search mechanism for the cognate target site, which is surrounded by many alternative sites. Non-specific electrostatic interactions govern the one-dimensional sliding of DBPs along the linear contour of the DNA [5]. The time a DBP spends in the different search modes (sliding, hopping, inter-segment transfer, and 3D diffusion) is dictated by the electrostatic features of the protein as well as by the salt concentration. It was shown that DBPs may not always slide using the patch that is used for specific interactions [6]. The non-specific and specific interactions of DBPs with DNA can be supported by disordered regions, often at either or both the C- and N-termini. The disordered tail on DBPs is positively charged and can be viewed as another sub-domain that interacts with DNA. The high specificity and high selectivity of the recognition motif at the structured domain involved in interacting with DNA can be enhanced by the disordered sub-domain. The importance of the N-terminal disordered tail for protein folding and for specific or non-specific interactions with DNA was studied recently for several homeodomain transcription factors, both experimentally [7, 8] and computationally [9, 10]. These disordered tails, which are positively charged and highly disordered in solution, can cause thermodynamic destabilization or stabilization of the protein in the absence or presence of DNA, respectively [8, 11, 12]. The tails of the homeodomain may fold upon binding to the DNA in the minor groove or remain partly flexible or disordered [13]. A number of studies have demonstrated that N-tails can contribute to selective sequence binding specificity [8, 13-16]. A recent computational study of three homeodomain proteins with different tail lengths and net charges demonstrated the role of the disordered tail in facilitating DNA search [9], in agreement with kinetic NMR studies [17]. The presence of an N-tail increases the affinity of the protein to the DNA, and therefore enhances its sliding propensity at the expense of hopping and 3D diffusion. However, a higher propensity for sliding has its price: the linear diffusion coefficient of the protein moving along the DNA is lower, which results in a slower search [9]. The disordered tail can also assist intersegment transfer[18, 19] via the monkey-bar mechanism without accumulation of free protein during the jump from one segment of DNA to another [9, 10].

Clearly, PTMs, such as acetylation, which masks the positive charge of Lys, or phosphorylation, which introduces negative charges to Ser, Thr, or Tyr, can reversibly change the electrostatic potential of DBPs and therefore their nonspecific and specific interactions with DNA. Indeed, many DBPs, especially transcription factors, undergo acetylation[20] and phosphorylation[21]. Two well known examples of the important role played by a PTM on a disordered tail in modulating interactions with DNA are the modifications on p53 transcription factor C-tail and on the histone tails. The p53 protein is regulated mostly through covalent modifications[22] that occur on both the C- and N-tails[23, 24]. Acetylation events within the C-tail, which is positively charged and strongly interacts with the DNA, significantly enhance site-specific DNA-binding activity and the degree of acetylation is well correlated with p53-mediated activation *in vivo*[25]. Phosphorylation of the C-tail positively regulates DNA binding and tetramerization of p53[26]. The histone tails are subject to a variety of PTMs that increase the accessibility of nucleosomal DNA by weakening histone–DNA interactions[27-30] and consequently influence various cellular processes that depend on the state of the chromatin [28].

To quantify the abundance of acetylation and phosphorylation in DBPs and in particular at ordered or disordered sites, we analyzed human DBPs with experimentally characterized PTM sites. We then refined our analysis by investigating the occurrence of acetylation and phosphorylation on disordered tails and illustrated the effects of gradual modifications on DNA search dynamics by coarse-grained molecular dynamics simulations of the Ets-1 transcription factor..

2. Methods

2.1. Statistical analysis of acetylation and phosphorylation in DNA-binding proteins

Datasets of acetylated and phosphorylated human proteins were constructed using PhosphoSite (http://www.phosphosite.org). The Uniprot site (http://www.uniprot.org)

was used to download the sequences of annotated human proteins and to obtain a subset of DNA-binding proteins (DBPs). Next, the sequences of the proteins from the PhosphoSite datasets were retrieved from the Uniprot sequences. The charges on the protein were calculated by assigning a single positive charge (+1) to each Lys or Arg and a single negative charge (-1) to each Glu or Asp. Acetylated Lys was considered a neutral residue and phosphorylated Ser, Thr, or Tyr were analyzed as negative residues (with a charge unit of -2). Determination of disordered regions was performed using IUPred in the long-range [31]. A protein was considered to have a disordered tail if an unstructured segment of at least five consecutive amino acids was predicted at either its N-terminus or its C-terminus (if the protein was predicted to have unstructured segments at both ends, two separate tails were counted). The ordered region of a protein was defined by excluding the disordered segments from the whole protein sequence. The effects of acetylation and phosphorylation on the composition and distribution of charges on the disordered segments or on the ordered parts of DBPs and non-DBPs can be characterized by various measures that aim to capture the complexity introduced to the protein sequence by the PTMs. In the current study, we focused on two measures: the number of PTMs, calculated as the sum of all PTMs in the tail or globular part and the density of PTMs, calculated as the number of the PTMs normalized by tail length. We emphasize that the influence of the number of each type of PTMs, their location on the protein sequence, and, in particular, the pattern of the charged residues can be further characterized by other parameters.

2.2. Simulation model for searching the DNA of modified proteins

We studied the dynamic nature of DNA search by phosphorylated variants of the Ets-1 protein using a reduced model [5, 9] that allows sampling of long timescale processes, such as sliding, hopping, 3D diffusion, and intersegment transfer. We modeled the DNA as having three beads per nucleotide, representing phosphate, sugar, and base. Each bead was located at the geometric center of the group it represents and a negative point charge was assigned to beads representing the DNA phosphate groups. In the simulations, a 100 bp B-DNA molecule was used to study protein diffusion and the DNA remained in-place and rigid throughout the simulations.

The protein was represented by a single bead for each residue located at the Cα of that residue. Beads representing charged amino acids (Lys, Arg, Asp, and Glu) were charged in the model. Non-specific protein–DNA interactions were modeled by electrostatic interactions between all charged residues of the protein and the phosphate bead of the DNA using the Debye–Huckel potential, which accounts for the ionic strength of a solute immersed in aqueous solution[5]. The dynamics of Ets-1 was studied at salt concentrations in the range of 0.01–0.12 M using a dielectric constant of 80 and a temperature at which the protein is completely folded. More details of the simulations can be found in refs. [5, 9, 10, 32]. Using this model, we studied the interaction of several phosphorylated variants of Ets-1 in which a phosphorylated residue was modeled as a point charge of -2. The non-specific interactions between the protein and DNA were quantified by classifying each snapshot as performing 1D diffusion (either sliding or hopping) or 3D diffusion.

3. Results

3.1. The occurrence of acetylation and phosphorylation at disordered regions

Intrinsically disordered regions are often found as primary sites for PTMs[33-35]. Some PTMs are more frequently found at disordered sites than others[36] (e.g., acetylation and phosphorylation are abundant in the disordered regions of DBPs[37, 38]). In this study, we focus on quantifying the existence of acetylation and phosphorylation modifications in DBPs as these PTMs involve changes to the electrostatic charges of the modified sites in a way that may affect interactions with the negatively charged DNA molecules. We found that among 3398 human proteins that undergo acetylation, 65% of the acetylated Lys are located in intrinsically disordered regions. Among the 8943 human proteins that undergo phosphorylation, 79% of the modified Ser, Thr, or Tyr are located in intrinsically disordered regions. The occurrence of acetylation sites in disordered regions is much smaller in non-DBPs (34%), while phosphorylation sites are still widespread in disordered sites (69%) (Fig. 1).

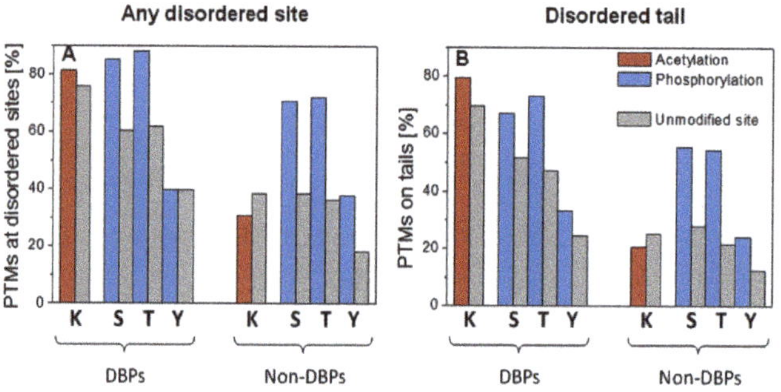

Fig. 1. The occurrence of acetylation and phosphorylation modification on proteins. The occurrence of acetylation (on Lys) and phosphorylation (on Ser, Thr, or Tyr) (shown in red and blue, respectively) post-translational modifications (PTMs) at the disordered sites of proteins (A) and disordered tails (B) is analyzed for two datasets of human acetylated and phosphorylated proteins (including 671 and 1673 proteins, respectively). Each dataset of modified proteins was divided into groups of DNA-binding proteins (DBPs; 81 and 132 acetylated and phosphorylated, respectively) and non DNA-binding proteins (non-DBPs; 590 and 1541 acetylated and phosphorylated, respectively). The values for the occurrence of the corresponding unmodified residues at disordered or tail regions are shown in gray.

To simplify the analysis, we focused on smaller proteins that are composed of <300 amino acids. This dataset includes 671 and 1673 human proteins that undergo acetylation and phosphorylation, respectively. Among the acetylated proteins, 8% are DBPs, which account for 30% of the acetylated Lys in the dataset (517 modified Lys). Among the phosphorylated proteins, 8% are DBPs, which account for 13% of the modified phosphor-sites (298 Ser, 127 Thr, and 196 Tyr) in the dataset. The involvement of acetylation and phosphorylation in proteins that are not DBPs provides further illustration of the diversity of their cellular function in addition to their modulatory interactions with DNA.

Importantly, the majority of the modification sites in the DBPs (about 81% and 77% of the acetylation and phosphorylation sites, respectively) are in disordered regions. In non-DBPs, the opposite trend is observed for acetylation, as only 30% of the acetylation sites in these proteins are in disordered sites. This is in agreement with analysis of the mouse acetylome, which showed the high preference of acetylation sites to locate in structured regions [39]. This trend was also reported recently in an analysis of acetylation using a dataset of ubiquitinated proteins [36]. For phosphorylation, whose sites generally show a high preference for disordered regions [3], there is a smaller difference between DBPs and non-DBPs. However, a larger fraction of phosphorylation sites is found in disordered sites on DBPs than non-DBPs (63% on DBPs vs 46% on non-DBPs) (Fig. 1).

3.2. The occurrence of acetylation and phosphorylation at disordered tails

In addition to analyzing the occurrence of PTMs at disordered regions, it is of interest to focus specifically on disordered tails (at either the C- or N-termini) because they are quite common in proteins, especially DBPs (Fig. 2). We have previously shown that ~70% of DBPs and only ~50% of non-DBPs in humans posses disordered tails [10, 32]. The tails of human DBPs are longer than those of other human proteins (~70% of the tails of DBPs are longer than 5 residues while only 40% of all human proteins have tails longer than 5 residues, Fig. 2B). The disordered tails of DBPs have a larger net positive charge than the tails of all human proteins and their charges tend to be much more clustered (Fig. 2C). We note that the composition of positively charged residues and their degree of clustering in the structured parts of DBPs and all human proteins set are very similar[37]. We have previously reported that the lengths and the net charges of the disordered tail are correlated with the efficiency of the DNA search (e.g., the ability of the tail to promote sliding or jumping between different stretches of DNA).

The analysis of PTM sites on disordered tails (Figs. 1A and 1B) shows that the tails of DBPs are richer in acetylation and phosphorylation sites than the corresponding tails of non-DBPs. The occurrence of acetylation and phosphorylation sites on DBP tails is similar to that in any disordered protein region, indicating that most of the modification sites are located at the tails (Fig. 1). In the non-DBPs, most acetylation and phosphorylation sites are located at structured sites, and when they are located at disordered regions these may be tails or other disordered regions (e.g., internal loops).

Since the interactions of the disordered tails with DNA are governed by the number of positive and negative charges on the tail and their organization, one must also characterize how many acetylation and phosphorylation sites there are in the tails of DBPs and non-DBPs. The average number of acetylation sites is 3.1 ± 0.2 on the tails of DBPs and 0.6 ± 0.1 on the tails of non-DBPs. On the globular parts of DBPs and non-DBPs, the average numbers of acetylation sites are 1.8 ± 0.2 and 1.6 ± 0.2, respectively (Fig. 3). The trend remains when the number of modification sites is normalized by the length of the domain (or by the disordered tail of the globular domain). Thus, the acetylation density is much higher in the tails of DBPs than in the tails of non-DBPs and the density is lower and very similar in the globular domains of DBPs and non-DBPs (Fig. 3a and 3c (inserts)).

Similarly to acetylation, phosphorylation is ubiquitous on disordered tails and there are DBPs with even 10 phosphorylation sites in the tail. However, the occurrence of phosphorylation in the disordered tails of DBPs is very similar to that in non-DBPs (the average numbers of phosphorylation sites in the tails of DBPs and non-DBPs are 2.0 ± 0.2 and 1.7 ± 0.1, respectively). The occurrence of phosphorylation sites in the globular part of DBPs and non-DBPs is similar and resembles that found in the tail sub-domains.

Clearly, the number of acetylation and phosphorylation sites in the disordered tail provides only partial information about the potency of the PTMs to modulate function (e.g., to affect the interaction with DNA). It is important to quantify how the modifications communicate with the charged residues of the disordered region. The location of acetylation disrupts charge clustering patterns in a way that not only reduces the net charge but also reduces the local charge density in some regions of the tail more than in other regions.

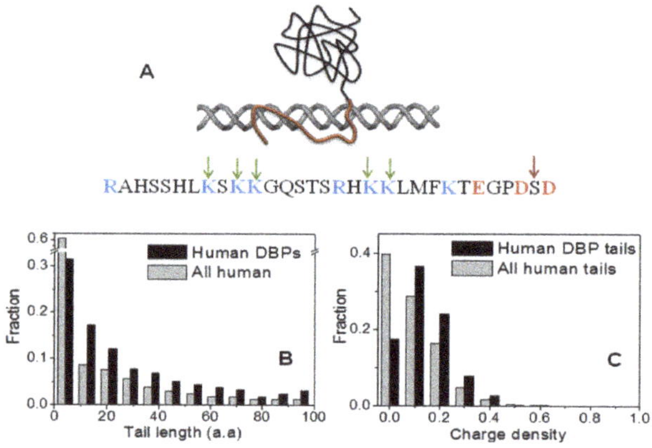

Fig. 2. Analysis of the properties of the disordered tails of DNA-binding proteins in the human genome. A). A schematic illustration of the disordered tail (in orange) of a DNA-binding protein. The tail can interact either non-specifically or specifically with DNA and affect various biophysical properties of the interaction. The sequence of the C-tail of p53 is shown as an example with the positively and negatively charged residues shown in blue and red, respectively. Acetylation sites (green arrows) and phosphorylation sites (red arrow) are shown as well. B). Histograms of tail lengths in human DNA-binding proteins (DBPs; a total of 864 proteins) and in all human proteins (which total 20,334 proteins). A tail is defined as a disordered region predicted by IUPred. C). Distribution of the positive charge density in the tails of DBPs.

To illustrate the importance of the location of the PTMs, we focus on the disordered C-terminus of p53, which is positively charged and thus strongly interacts with DNA and can affect sliding features very significantly. Recent single-molecule experiments [40] and a coarse-grained simulation study [41] showed that the C-tails of p53 increase its sliding speed along DNA. The C-tail of p53 protein undergoes several covalent modifications [22, 42] (Fig. 2) [23, 24]. Dynamic p53 acetylation and deacetylation events were observed in response to DNA damage [25]. The acetylation events within the C-tail significantly enhance site-specific DNA-binding activity [25, 43]. Moreover, C-tail acetylation levels are well correlated with p53-mediated activation *in vivo* [25]. Phosphorylation of the C-tail positively regulates DNA binding and tetramerization of p53 [26]. The C-tail of p53 can undergo acetylation at 5 different sites and this can affect the number of clusters of consecutive charges and the local effective charge along the tail. The single phosphorylation site on the C-tail is isolated from the acetylation sites and apparently does not interfere with them and its effect is to increase the negative charges at the end of the C-tail.

We propose that acetylation may directly modulate DNA-binding while phosphorylation may indirectly influence attraction to DNA for example, via intra- or inter-molecular protein-protein interactions. The different biological functions of acetylation and phosphorylation may explain the different occurrence of these PTMs in DBPs and non-DBPs. While phosphorylation, which serves in protein function and interactions, is abundant in the tails of all proteins, acetylation, which modulates DNA-binding, is highly pronounced in the tails of DBPs.

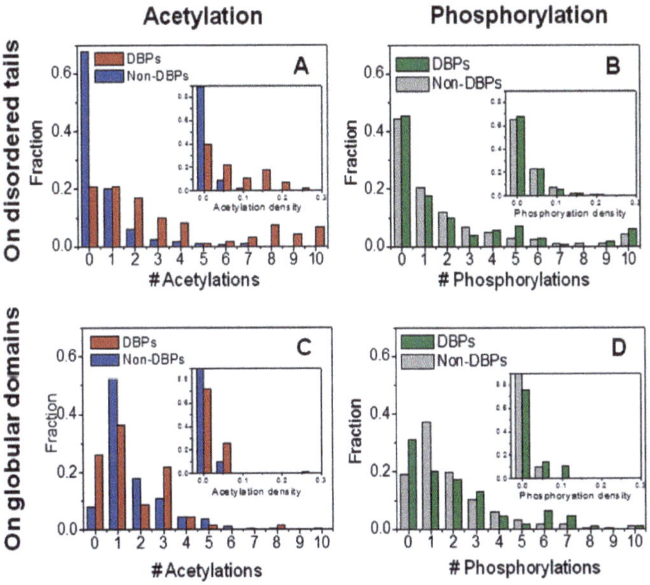

Fig. 3. Bioinformatic analysis of acetylation and phosphorylation on DNA-binding proteins (DBPs). A and B) The number of acetylation (A) and phosphorylation (B) sites on the disordered tails of DBPs (red and green, respectively) and non-DBPs (blue and gray, respectively). The insets indicate the density of acetylation and phosphorylation sites (i.e., the number of sites normalized by the length of the tail sub-domain). C and D). Similar to A and D but for the globular domains of DBPs and non-DBPs, respectively.

3.3. The interplay between post-translational modifications of disordered tails

In many cases, several modifications are needed to achieve the required effect. For example, the 31- and 40-residue long C-tails of p53 and H3 histone tail involve 6 modifications (5 acetylations and a single phosphorylation) and 10 modifications (5 acetylations, 6 methylations, and 3 phosphorylations)[44], respectively. Various types of cross-talks may come into play between different PTMs. Some modifications require the existence of other modifications. In some cases, two PTMs may compete for the same modification site so that the timing of the effect of one PTM is determined by the removal of the other PTM. In other cases, the effects of multiple PTMs accumulate when the level of modification is incrementally changed. The latter scenario suggests that each PTM may not be sufficient to achieve the needed outcome alone but that several PTMs of the same type gradually achieve the required outcome.

A gradual change by successive PTMs was reported for the Ets-1 transcription factor, whereby gradual phosphorylation of its disordered tail results in gradual attenuation of its binding affinity to DNA. This ability to gradually tune the binding affinity was viewed as an "incremental rheostat" rather than as a switch on/off by the modifications[45]. It was shown for Ets-1 that phosphorylation of its disorder tail interferes with formation of intramolecular interactions, which results in a lower binding affinity to DNA[38, 45]. The five phosphorylation events on the disordered tail of Ets-1 were reported to attenuate affinity to DNA not by directly affecting its interactions with DNA but by modulating its internal flexibility. Nevertheless, it is likely that the phosphorylated tail interacts directly with non-specific DNA. Here, we followed this scenario and studied the non-specific interactions of the Ets-1 transcription factor with a disordered tail in which different numbers of phosphorylation sites were phosphorylated.

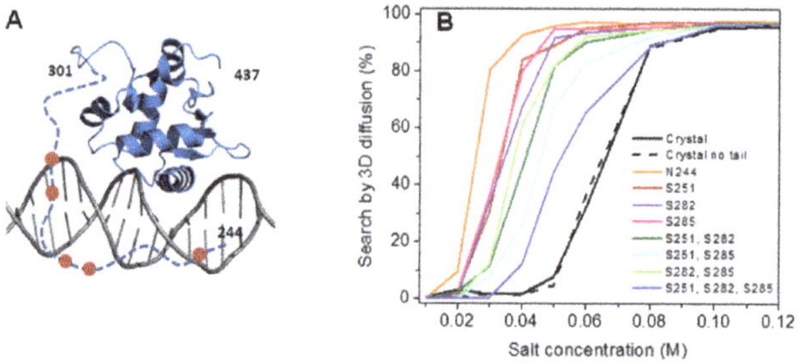

Fig. 4. The effect of incremental phosphorylation on non-specific protein–DNA interactions. A). The Ets-1 transcription factor includes a long disordered region with five phosphorylation sites that were found to attenuate specific binding to DNA indirectly by affecting internal protein dynamics so as to weaken specific affinity. The disordered tail can also interact directly with DNA and the strength of the interaction might be dependent on the degree of phosphorylation. B). The populations of 3D diffusion for 10 variants of Ets-1 as a function of salt concentration indicate a clear linkage between the number of phosphorylations and the salt concentration needed to detach the protein from the DNA to the bulk. As phosphorylation sites are mutated, a higher salt concentration is needed to populate the 3D search (at the expense of linear diffusion).

Phosphorylation of the disordered tail of the Ets-1 transcription factor results in weaker non-specific binding to DNA. This weaker binding is reflected in the lower salt concentration needed to detach the protein from the DNA and to switch from DNA search via linear diffusion to 3D diffusion (Fig. 4). The attraction of the folded part of Ets-1 to the DNA is gradually attenuated as the level of phosphorylation increases (namely, gradually mutating the phosphorylation sites results in search characteristics that are more similar to that of Ets-1 with a truncated disordered tail). Interestingly, similarly to the gradual and indirect effect of phosphorylation on specific binding that arises from modulating the internal protein dynamics, our results suggest that phosphorylating the disordered tail may directly perturb nonspecific binding to DNA.

4. Conclusions

PTMs, which are widely used by the cell to increase the structural and biochemical diversity of proteins, often take place at intrinsically disordered sites. These disordered regions, particularly at the protein termini (namely, tails), are more abundant in DNA-binding proteins than in other proteins and serve as potential sites for PTMs. In this study, we analyzed the occurrence of acetylation and phosphorylation in human DBPs in comparison to non-DBPs at disordered versus ordered sites. There is a strong preference for acetylation to occur at either disordered sites or in the tail regions of DBPs while in non-DBPs an opposite trend is seen in which acetylation sites are mostly located in structured regions. With respect to phosphorylation, there is general tendency to find phosphorylated Ser and Thr in disordered regions, irrespective of whether the protein binds DNA or not. However, the occurrence of phosphorylation sites in disordered sites is still higher for DBPs than for non-DBPs. While both PTMs have similar effect of reducing the net charge, phophorylation is known to be important in protein activation and protein-protein interaction and acetylation is thought to modulate DNA binding such as serving as primary device of basic chromatin modification leading to transcription activation. The difference between the two PTMs may explain the widespread occurrence of phosphorylation in the disordered regions of all human proteins, and the high occurrence of acetylation sites mainly in tails of DBPs, where the acetylation primarily affect the protein-DNA binding.

Both acetylation and phosphorylation decrease the number of positive charges (by masking the positive charge of Lys in acetylation and introducing a negative charge on either Ser, Thr, or Tyr in phosphorylation). Clearly, tuning the positive potential of the protein will affect association with DNA. This tuning can be achieved in principle on either the ordered or disordered subdomains of DBPs. The observation that most acetylation and phosphorylation sites in DBPs are located in the tails provides additional evidence for the role of the tail as an affinity tuner for interactions with DNA that can be modulated by PTMs[37].

PTMs may exert a complex effect when they occur at multiple sites. In some cases, the effects of PTMs accumulate additively to regulate an outcome in a graded manner (e.g., the regulation of DNA-binding affinity in a progressive manner by multiple phosphorylations of the Ets-1 transcription protein[45]). One can easily envision cases in which the effect of multiple PTMs at disordered region will not be simply additive but will depend on the local environment of each PTM.

Acetylation and phosphorylation affect the composition of the charges of the disordered tail. The global and local changes in the net charge of the disordered tail may directly affect not only its attraction to nonspecific DNA, but also the internal properties of the tail itself (such as its conformational preferences). Recent findings call for a clear linkage between the degree of structure of intrinsically disordered proteins and the charge content [10, 46-48]. The effect of the PTMs on the conformational ensemble of the disordered region may depend both on the net charges and on how they communicate with the original charges in the sequence and thus the local effective charges. Our future goals are to characterize the effect of the PTMs on the charge pattern of the disordered tails and to analyze the evolutionary conservation of the PTMs sites on tails.

In summary, the disordered regions and particularly disordered tails of DBPs have important features that allow them to modulate interactions with DNA. Some of these features can be manipulated in an incremental manner by PTMs that gradually shift the composition and distribution of charges and consequently fine tune the strength of protein–DNA interactions but in others the cross-talks between the different PTMs sites might be more complex.

5. Acknowledgments

This work was supported by the Kimmelman Center for Macromolecular Assemblies and the Minerva Foundation with funding from the Federal German Ministry for Education and Research. Y.L. is the incumbent of the Lillian and George Lyttle Career Development Chair. We thank Tzachi Hagai for the inspiration of studying acetylation pattern in DBPs.

1 H. Xie, S. Vucetic, L. M. Iakoucheva, C. J. Oldfield, A. K. Dunker, Z. Obradovic and V. N. Uversky, *Journal of proteome research*, 2007, **6**, 1917-1932.
2 V. N. Uversky, C. J. Oldfield, U. Midic, H. Xie, B. Xue, S. Vucetic, L. M. Iakoucheva, Z. Obradovic and A. K. Dunker, *BMC genomics*, 2009, **10 Suppl 1**, S7.
3 C. R. Landry, E. D. Levy and S. W. Michnick, *Trends Genet*, 2009, **25**, 193-197.
4 L. M. Iakoucheva, P. Radivojac, C. J. Brown, T. R. O'Connor, J. G. Sikes, Z. Obradovic and A. K. Dunker, *Nucleic acids research*, 2004, **32**, 1037-1049.
5 O. Givaty and Y. Levy, *Journal of the molecular biology*, 2009, **385**, 1087-1097.
6 A. Marcovitz and Y. Levy, *Proc Natl Acad Sci USA, In press*
7 A. I. Dragan, Z. L. Li, E. N. Makeyeva, E. I. Milgotina, Y. Y. Liu, C. Crane-Robinson and P. L. Privalov, *Biochemistry*, 2006, **45**, 141-151.
8 C. Crane-Robinson, A. I. Dragan and P. L. Privalov, *Trends in Bioch Sci*, 2006, **31**, 547-552.
9 D. Vuzman, A. Azia and Y. Levy, *J Mol Biol*, 2010, **396**, 674-684.
10 D. Vuzman and Y. Levy, *Proc Natl Acad Sci USA*, 2010, **107**, 21004-21009.
11 A. Tóth-Petróczy, M. Fuxreiter and Y. Levy, *J Amer Chem Soc*, 2009, **131**, 15084-15085
12 P. L. Privalov, A. I. Dragan, C. Crane-Robinson, K. J. Breslauer, D. P. Remeta and C. A. S. A. Minetti, *J Mol Biol*, 2007, **365**, 1-9.

13 J. M. Gruschus, D. H. Tsao, L. H. Wang, M. Nirenberg and J. A. Ferretti, *Biochemistry*, 1997, **36**, 5372-5380.
14 R. Joshi, J. M. Passner, R. Rohs, R. Jain, A. Sosinsky, M. A. Crickmore, V. Jacob, A. K. Aggarwal, B. Honig and R. S. Mann, *Cell*, 2007, **131**, 530-543.
15 P. L. Privalov, A. I. Dragan, C. Crane-Robinson, K. J. Breslauer, D. P. Remeta and C. A. Minetti, *J Mol Biol*, 2007, **365**, 1-9.
16 A. I. Dragan, Z. Li, E. N. Makeyeva, E. I. Milgotina, Y. Liu, C. Crane-Robinson and P. L. Privalov, *Biochemistry*, 2006, **45**, 141-151.
17 J. Iwahara and G. M. Clore, *J Am Chem Soc*, 2006, **128**, 404-405.
18 T. Hu and B. I. Shklovskii, *Physical Review E*, 2007, **76**, 051909.
19 M. Doucleff and G. M. Clore, *Proc Natl Acad Sci USA*, 2008, **105**, 13871-13876.
20 A. J. Bannister and E. A. Miska, *Cell Mol Life Sci*, 2000, **57**, 1184-1192.
21 A. J. Whitmarsh and R. J. Davis, *Cell Mol Life Sci*, 2000, **57**, 1172-1183.
22 A. M. Bode and Z. G. Dong, *Nature Reviews Cancer*, 2004, **4**, 793-805.
23 E. Appella and C. W. Anderson, *European Journal of Biochemistry*, 2001, **268**, 2764-2772.
24 X. Yang, *Cell Death and Differentiation*, 2003, **10**, 400-403.
25 J. Y. Luo, M. Y. Li, Y. Tang, M. Laszkowska, R. G. Roeder and W. Gu, *Proc Natl Acad Sci USA*, 2004, **101**, 2259-2264.
26 K. Sakaguchi, H. Sakamoto, D. Xie, J. W. Erickson, M. S. Lewis, C. W. Anderson and E. Appella, *J Prot Chem* 1997, **16**, 553-556.
27 A. Eberharter and P. B. Becker, *Embo Reports*, 2002, **3**, 224-229.
28 S. Y. Roth, J. M. Denu and C. D. Allis, *Annual Review of Biochemistry*, 2001, **70**, 81-120.
29 Z. Y. Yang, C. Y. Zheng and J. J. Hayes, *Journal of Biological Chemistry*, 2007, **282**, 7930-7938.
30 B. M. Turner, *Bioessays*, 2000, **22**, 836-845.
31 Z. Dosztanyi, V. Csizmok, P. Tompa and I. Simon, *Bioinformatics*, 2005, **21**, 3433-3434.
32 D. Vuzman, M. Polonsky and Y. Levy, *Biophys J*, 2010, **99**, 1202-1211.
33 P. Tompa, *Structure and function of intrinsically disordered proteins*, Boca Raton, 2010.
34 A. K. Dunker, I. Silman, V. N. Uversky and J. L. Sussman, *Curr Opin Struct Biol*, 2008, **18**, 756-764.
35 V. N. Uversky, C. J. Oldfield and A. K. Dunker, *Annual Rev of Biophysics*, 2008, **37**, 215-246.
36 T. Hagai, A. Azia, A. Tóth-Petróczy and Y. Levy, *J Mol Biol*, 2011, 412, 319-324, (2011)
37 D. Vuzman and Y. Levy, *Submitted*, Mol BioSystems, In press.
38 M. Fuxreiter, I. Simon and S. Bondos, *Trends in Bioch Sci*, 36, 415-423, (2011)
39 C. Choudhary, C. Kumar, F. Gnad, M. L. Nielsen, M. Rehman, T. C. Walther, J. V. Olsen and M. Mann, *Science*, 2009, **325**, 834-840.
40 A. R. Fersht, A. Tafvizi, F. Huang, L. A. Mirny and A. M. van Oijen, *Proc Natl Acad Sci USA*, 2011, **108**, 563-568.
41 N. Khazanov and Y. Levy, *J Mol Biol*, 2011, **408**, 335-355.
42 L. Jayaraman and C. Prives, *Cellular and Molecular Life Sciences*, 1999, **55**, 76-87.
43 W. Gu and R. G. Roeder, *Cell*, 1997, **90**, 595-606.
44 S. J. Nowak and V. G. Corces, *Trends Genet*, 2004, **20**, 214-220.
45 M. A. Pufall, G. M. Lee, M. L. Nelson, H. S. Kang, A. Velyvis, L. E. Kay, L. P. McIntosh and B. J. Graves, *Science*, 2005, **309**, 142-145.
46 A. H. Mao, S. L. Crick, A. Vitalis, C. L. Chicoine and R. V. Pappu, *Proc Natl Acad Sci USA*, 2010, **107**, 8183-8188.
47 M. Borg, T. Mittag, T. Pawson, M. Tyers, J. D. Forman-Kay and H. S. Chan, *Proc Natl Acad Sci USA*, 2007, **104**, 9650-9655.
48 D. Potoyan and G. Papoian, *J Amer Chem Soc*, 2011, **133**, 7405-7415.

MICROBIOME STUDIES: ANALYTICAL TOOLS AND TECHNIQUES*

JAMES A. FOSTER

Department of Biological Sciences, University of Idaho, Moscow, ID 83844-3051
USA Initiative for Bioinformatics and Evolutionary STudies (IBEST)
BEACON Center for the Study of Evolution in Action
Email: foster@uidaho.edu

JASON H. MOORE

Institute for Quantitative Biomedical Sciences, Departments of Genetics and Community and Family Medicine, Dartmouth Medical School Lebanon, NH 03756 USA
Email: jason.h.moore@dartmouth.edu

JACK A. GILBERT

Argonne National Laboratory, 9700 South Cass Avenue, Argonne, IL 60439, USA
Department of Ecology and Evolution, University of Chicago, 5640 South Ellis Avenue, Chicago, IL 60637, USA
Email: gilbertjack@anl.gov

JOHN BUNGE

Department of Statistical Science, Cornell University, Ithaca, NY 14853 USA
Email: jab18@cornell.edu

Bacteria and Archaea are major factors in human health and environmental well-being, in ways that we are only now discovering. With new sequencing technologies it is now possible to take deep samples of the microbial species present in a given environment, resulting in very large sequence datasets that require ecologically aware analysis. However, the bioinformatics and statistics this task requires are not fully developed, constituting an active and important research area. This session, the second annual Microbiome Studies session at PSB, presents recent work on the computational and statistical analysis of microbiome data.

---

* This work was funded in part by: NIH COBRE grant P20RR16448, NIH INBRE grant P20RR016454, and NSF STC "BEACON Center for the Study of Evolution in Action" NSF STC DBI-0939454 to JAF; National Science Foundation grant DEB-08-16638 to JB; U.S. Dept. of Energy Contract DE-AC02-06CH11357 to JAG; and NIH grants LM009012, LM010098 and AI59694 to JHM. We also thank the organizers of the Pacific Symposium on Biocomputing.

Introduction to Microbiome Studies

This year's special session on Microbiome Studies at the annual Pacific Symposium on Biocomputing emphasizes practical solutions to the very large data interpretation problems arising from next generation sequencing and microbiome studies.

This is the second year for this special session. Last year's session was well attended, with especially enthusiastic tutorial and discussion sessions. In response to participant feedback, we narrowed the scope of the session to computational and statistical advances in techniques for processing next generation microbial survey data. We also expanded the organizing committee to include more expertise on environmental microbiomes and on statistical approaches. We received several high quality submissions again this year, and anticipate another informative and useful session.

As is only now becoming appreciated, Bacteria and Archaea dominate the Earth's ecology. They are a major factor in global processes such as the carbon cycle and climate. On a more personal level, the human microbiome, including non-pathogenic commensals, affects our health and behavior in surprising ways. Unfortunately, most current tools for analyzing microbial ecosystems were developed first for macro-biology. And most microbial studies have targeted isolates and potential pathogens. This has significantly biased the choice of model systems such as *E. coli*, microbial genome sequencing projects, diversity statistics, and bioinformatics. It is no surprise that many common analytical techniques appear to be ill suited to studying the microbial world.

In the meantime, sequencing technologies continue to improve, with new protocols and methodologies as well as incremental improvements in existing ones. It is no longer wild speculation to talk about sequencing "everything" in a given environment, or to raise the possibility of analyses that reject most of the data. There are now ambitious projects in place to characterize the entire human microbiome and all of the microbial life on earth.

The number of relevant publications continues to increase, along with potential funding. Community resources, both databases and software, are becoming more sophisticated and reliable.

Papers in this session

The papers in this session represent a prospective or leading-indicators view of the direction of the field: away from pure description and toward understanding the functional structure of communities, both internally (i.e., within a given dataset), and externally, with respect to metadata or covariates (the latter being measured directly or derived from database searches).

The paper "MetaDomain: A protein domain classification tool for short sequences" by Zhang and Sun addresses protein homology searching, which supports functional profiling in metagenomic annotation. Since the sensitivity of this process declines with decreasing read length, the authors present a tool specifically designed for short reads (as produced by next-generation sequencing technologies), which compensates for some of the shortcomings and achieves significantly improved sensitivity, even relative to the state-of-the-art profile HMM alignment tool.

The paper "Artificial functional difference between microbial communities caused by length difference of sequencing reads" by Zhang, Doak and Ye also analyzes homology-based approaches to annotation of microbial community data. The authors note that apparent functional differences may be the result of differences in read lengths rather than of differences in, for example, sequencing techniques. They show that certain functional categories are under-annotated and that the accuracy of annotation of short reads varies by function. They present an improved method to address these issues.

The paper "SEPP: SATé-Enabled Phylogenetic Placement" by Mirarab, Nguyen, and Warnow deals with the insertion of short molecular "query" sequences into an existing phylogenetic tree (phylogenetic placement), and subsequent downstream analysis. Using real and simulated data they show that existing methods work on well-behaved inputs (accurate alignments and trees for full-length sequences along with relatively small and not overly diverse sets of sequences), but for more general conditions accuracy declines. They therefore present a "boosting" technique which places metagenomic sequences more accurately when there is a large evolutionary diameter, and also shortens computing time for simpler cases.

The paper "Comparisons of distance methods for combining covariates and abundances in microbiome studies" by Fukuyama, McMurdie, Dethlefsen, Relman and Holmes compares different nonparametric methods for combining abundance and tree data with other metadata or covariates (e.g., clinical information). They look at the use of principal coordinates analysis on UNIFRAC output and consider the power of signal extraction in various settings, based on both real and simulated data. Since this form of analysis is becoming widespread, this is a timely study.

The paper "Estimating population diversity with unreliable low frequency counts" by Bunge, Böhning, Allen and Foster discusses the "singleton problem": when are low-frequency "species" counts derived from high-throughput sequencing real rather than artifactual? The authors present various *ex post facto* statistical adjustments for potential errors in the singletons or low frequency sample counts, while noting that correction at the data source is generally preferable to downstream statistical adjustment.

ESTIMATING POPULATION DIVERSITY WITH UNRELIABLE LOW FREQUENCY COUNTS

JOHN BUNGE*

Department of Statistical Science, Cornell University,
Ithaca, NY 14853, USA
**E-mail: jab18@cornell.edu*
www.northeastern.edu/catchall

DANKMAR BÖHNING

Department of Mathematics and Statistics, University of Reading,
Reading RG6 6BX, UK
E-mail: d.a.w.bohning@reading.ac.uk

HEATHER ALLEN

Food Safety and Enteric Pathogens Research Unit, National Animal Disease Center,
Agricultural Research Service, Ames, Iowa, 50010, USA
E-mail: heather.allen@ars.usda.gov

JAMES A. FOSTER

Department of Biological Sciences, University of Idaho,
Moscow, ID 83844, USA
E-mail: foster@uidaho.edu

We consider the classical population diversity estimation scenario based on frequency count data (the number of classes or taxa represented once, twice, etc. in the sample), but with the proviso that the lowest frequency counts, especially the singletons, may not be reliably observed. This arises especially in data derived from modern high-throughput DNA sequencing, where errors may cause sequences to be incorrectly assigned to new taxa instead of being matched to existing, observed taxa. We look at a spectrum of methods for addressing this issue, focusing in particular on fitting a parametric mixture model and deleting the highest-diversity component; we also consider regarding the data as left-censored and effectively pooling two or more low frequency counts. We find that these purely statistical "downstream" corrections will depend strongly on their underlying assumptions, but that such methods can be useful nonetheless.

Keywords: microbial diversity, mixture model, species problem, capture-recapture, left-censored data.

1. Introduction

Classical population diversity estimation is based on the assumption that the observed data counts — so many taxa observed once, twice, etc., in the sample — are correctly recorded. One then uses these "frequency count" data $(i, f_i), i = 1, 2, 3, \ldots$, where f_i is the number of taxa occurring i times in the sample, to estimate the total number of taxa in the population. There is a substantial literature on this problem, which can be framed either in terms of taxa or species, or in terms of the number of individuals in the population. In the latter case we have the capture-recapture situation, where one wishes to estimate the population size. Here we will take diversity to be the total number of taxa or classes, denoted by C, but our discussion is equally applicable to the capture-recapture problem.

It may be argued that "bounds on this quantity [C] may be highly uncertain because a small fraction of the environment could be composed of a vast number of different species."[1] We will proceed nonetheless under the assumption that such estimation is in principle statistically feasible;[2] in this case, however, the quality of the observed data counts is crucial to the accuracy and precision of the diversity estimate. Ordinarily one assumes that the sample consists of an (approximately) random selection of individuals from the population; the sampled individuals are then sorted or divided into classes, and the sizes of the sample classes are recorded. The numbers of sample classes of various sizes constitute the frequency count data. In some situations, though, the initial identification of the sampled individuals may be at fault, or the classification into sample classes may be questionable. This is the case, for example, when dealing with high-throughput DNA sequencing data, which are prone to errors of various types. These errors may arise at various stages, in particular the identification of the sequences and the clustering algorithms used to combine sequences into clusters or taxa may be questionable. The end result is that the number of low frequency counts — especially the singletons (f_1) — may be "artificially" inflated compared to what would be obtained by a data-collection process with a lower error rate.

The question then arises: Is it possible to *statistically* discount or down-weight the low frequency counts in the data (which are quite influential in estimating total diversity)? Could such a procedure be based solely on the available data, without knowledge of the mechanisms that produced the errors? Put this way the answer would seem to be No; and certainly the best procedure would be to correct the errors at the source before they enter the data stream. However, there are some *ex post facto* statistical analyses that are worth considering, if not as a complete fix then at least as a benchmark, to provide diversity estimates under different sets of assumptions. Such adjustments could also provide lower bounds for the total diversity that are comparatively insensitive to the errors in question. In this paper we describe the spectrum of available statistical approaches to this issue, examining two in detail: fitting a model with several diversity components and deleting the component corresponding to highest diversity; and declaring the low frequency counts uncertain in measurement and "left-censoring" the data. We examine these methods using a high-diversity dataset derived from a phage metagenome, and we compare numerical results based on this dataset. We conclude that some of these procedures may be useful, but that the investigator must take careful account of their underlying assumptions and the implications thereof.

2. Error correction at the source

The ideal way to deal with uncertain low frequency counts is to render them certain; i.e., to correct any errors in the counts before they are analyzed. This would naturally take place at the initial data collection or generation stage. In the case of high-throughput sequencing, there have been remarkable recent developments in this regard such as the programs PyroNoise and AmpliconNoise, which aim at "separately removing 454 sequencing errors and PCR single base errors."[3] It is arguably true that corresponding advances, or at least standardization, are also needed at the computational processing stage (alignment and clustering of sequences). These matters are non-statistical and hence we do not pursue them further here, but they are the subject of current research in bioinformatics.

3. Lower bounds

One conceptually simple approach is to require only lower bounds for the total diversity. Indeed since the beginning of statistical population diversity estimation, long before the appearance of high-throughput sequencing, researchers have been interested in comparing formal estimates of diversity (based on frequency count data regarded as error-free) with lower bounds or benchmarks for the total diversity. In fact, it has been argued mathematically that under the least restrictive nonparametric assumptions, only lower bounds (as opposed to estimates or upper bounds) are possible. Here we assume that we are willing to make the minimal structural modeling assumptions required to estimate C (and we do not enter further into this particular mathematical issue). However, it is possible to define a maximally biased model, namely, the equal-class-sizes model, i.e., the assumption that the population is *equally* subdivided into taxa or classes C. It is known that estimates of C made under this assumption are maximally biased downward, in a certain sense.[4] At least two such estimates are available: parametric, which is the maximum likelihood estimator of C under the unmixed Poisson model (each class contributes a Poisson number of representatives to the sample, and the mean number of representatives is equal across classes); and the nonparametric Good-Turing estimate, which is derived differently but still falls under the equal-sizes assumption. These estimates are typically reported along with the standard errors that would be appropriate under the equal-sizes assumption. It may be of some value to consider these downwardly-biased estimates when the low frequency counts are questionable, even though they are based on different modeling assumptions (correct data, equal class sizes). We look at numerical results below for our example dataset.

4. Deleting the high-diversity component of a mixture model

A promising but potentially rather extreme approach to the problem of uncertain low frequency counts consists of statistically reducing the lowest-abundance counts. This is based on a finite mixture model for estimating C, i.e., a model consisting of a convex combination (weighted average) of several components, which correspond to several different levels of diversity in the population.[2] A three-component model, for example, is more flexible than a simple one-component model, because it essentially allows for three categories or levels of diversity. Each class in the population contributes representatives to the sample according to one of three (mean) rates: the lowest rate corresponds to the rarest or smallest classes and the highest rate to the largest or most abundant classes. (Some mathematical details for the particular finite mixture model used here are given in the Appendix.) Such a model can be fitted directly to frequency count data by maximum likelihood, and each of the three components (in this example) is then identified by its mean rate and by the proportion of the population to which it corresponds. If the low frequency counts are supposed to be uncertain, one can then delete (mathematically subtract) the highest diversity or lowest abundance component from the fitted model, and base the projection of total diversity on the remaining, lower diversity components. This is mathematically straightforward, but as we will see in the numerical example below, the resulting reduction in the estimate of total diversity may be as much as two orders of magnitude, leading to conceptual questions about what exactly is the target of estimation under the reduced (highest diversity component deleted) model.

5. Left-censored data

The final approach we considered is left-censoring the frequency count data, as follows. Suppose we assume that the number of classes observed in the sample is correct, but that there is uncertainty regarding the exact values of the counts for certain classes, especially the low frequency counts. That is, for each observed class in the sample we declare the possibility of "measurement error," so that a singleton could perhaps have been a doubleton, or a doubleton a singleton. In other words, we separate the observed frequency counts at some level L into a pooled or total low frequency count, say $f_L^* = f_1 + f_2 + \ldots + f_L$, vs. the higher frequency counts $f_{L+1}, f_{L+2}, f_{L+3}, \ldots$. In general statistical terms this is called left-censored data (at L). Under this scenario the observed number of classes in the sample, say c, is preserved exactly, but we suppose that (for sample class sizes L and below) the number of individuals contributing to the observation of a given class is uncertain. This has the effect of essentially reducing the apparent diversity of the sample, because although there are still c classes, the number of counts of size L and below are combined. In particular, the number of singletons is no longer "known" and contributes only to the total f_L^*. This seems not unreasonable, but on the other hand it has the effect of rendering the total number of "individuals," namely $1f_1 + 2f_2 + 3f_3 + \ldots$, uncertain, which may or may not be logical in a given application. Maximum likelihood estimation of C based on left-censored data is relatively straightforward in this problem, although it does not admit a closed-form formula but requires numerical optimization (see the Appendix for a brief mathematical outline under the single exponential model). Robb and Böhning[5] dealt in detail with a closely related version of this problem for the capture-recapture setting. The reduction in the estimate of total diversity is less dramatic relative to the unadjusted estimate than under the previous approach (deleting the high-diversity component). However, the structural assumptions underlying this procedure may or may not be reasonable from the investigator's point of view.

6. Possible Bayesian extensions

It is possible to lower the estimate of total diversity by establishing a prior distribution on the population diversity C. This is (part of) the Bayesian approach, and there are two basic viewpoints. Some investigators prefer a method that puts minimal constraints on C, and in fact mathematically minimizes the prior information assigned to C. Such a method is called noninformative or objective Bayesian, and has been studied in this context by Barger and Bunge,[6] Farcomeni and Tardella,[7] and others. We do not pursue this further here because in our present setting we are interested in specifying prior information (also, no relevant computer software is readily available). The alternative, then, is an informative or even subjective Bayesian approach, in which the investigator specifies more or less strong prior assumptions regarding C. Perhaps the simplest version of this is to set an upper bound $C_{\max}$ for C, which can sometimes be done on the basis of biological or chemical considerations; for example, the taxonomic diversity of a microbial population cannot exceed the number of cells in it. The prior distribution of C is typically taken to be approximately flat, meaning that all values up to $C_{\max}$ are equally likely *a priori*. This has been explored by Manrique-Vallier and Fienberg[8] and others. Alternatively one may opt for an informative or even subjective prior distribution on C with no upper bound. This may decrease rapidly, for example, meaning that the larger the value of C, the less likely it is thought to be *a priori*. In this case the influence of the prior on the final diversity estimate may be considerable, and corresponding sensitivity analysis is called for. Again such methods have been dealt with to some extent in theory,[6] but no software is readily available. A further extension would be to apply an informative prior to the components of a mixture model, partially but not entirely downweighting the high-diversity component. This remains to be explored. In summary, the Bayesian approach may be promising in terms of controlling the estimate of C, but it is an open area for research in this regard and is beyond our scope here.

7. Example, numerical results and remarks

Our sample dataset is based on a phage metagenome.[9] More specifically, the frequency counts were derived from a contig spectrum from a swine fecal metagenome; the contig spectrum was generated using Circonspect via the CAMERA pipeline.[10] For the purpose of discussion here we can assume that we are interested in estimating the taxonomic diversity of this metagenome. The frequency counts are given in Table 1. The total number of observed taxa

Table 1. Phage metagenome frequency count data

i	f_i	i	f_i	i	f_i	i	f_i
1	4736	12	8	23	2	34	1
2	521	13	7	24	3	35	1
3	152	14	6	25	3	36	1
4	69	15	5	26	1	37	1
5	46	16	4	27	2	38	1
6	27	17	4	28	1	39	1
7	21	18	3	29	2	40	1
8	18	19	3	30	2	41	1
9	16	20	3	31	1	43	1
10	10	21	3	32	1	45	1
11	9	22	2	33	1	52	1

is $c = 5703$. It is clear even without graphing the data that the sample diversity is high: for instance, the number of singletons is almost an order of magnitude higher than the number of doubletons. There is some basis to believe that the experimental and bioinformatic procedures that generated these data are prone to erroneous inflation of the low frequency counts,[3] so this dataset is a good test-bed on which to compare the damping approaches described above. The output from the program CatchAll v.3.0[11] that is relevant for our purposes is shown in Table 2. (Here for simplicity we analyze the full dataset, i.e., all frequency counts up to and including the maximum f_{52}, rather than checking for right-hand outlier cutoff values τ as discussed elsewhere.[11]) Note first that the best estimate selected by CatchAll, under the

Table 2. Phage data analysis. Est Div = estimated total diversity; SE = standard error; LCB/UCB = lower/upper 95% confidence bounds

Method	Est Div	SE	LCB	UCB
Poisson	8730	103	8535	8938
GoodTuring	11690	346	11050	12407
ThreeMixedExp	67792	8656	53009	87195
Discounted: TwoMixedExp	1727	221	1410	2305

assumption that the data are error-free, is the (finite) mixture of three exponentially-mixed Poisson components, or equivalently a mixture of three geometric distributions. This yields an estimated total diversity of 67792 (SE 8656). This value is believed on scientific grounds to be too high, resulting from an unknown number of possibly artifactual or erroneous singleton (and perhaps doubleton and tripleton) counts. Considering lower bounds as discussed in Section 3 above, Table 2 displays the results for the Poisson maximum likelihood estimate (MLE) and the Good-Turing nonparametric estimate. Both of these are based on the structural assumption of equal class sizes in the population, but the Poisson estimate is the formal MLE under that model and the Good-Turing estimate is a simple nonparametric approximation thereof (which may possess certain robustness properties under mild departures from the equal-class-sizes model). Both of these estimators assume that (all of) the frequency count data are correct, but are maximally biased downward if the true population does not have equal class sizes — which it certainly does not, irrespective of the correctness of the data. Thus they constitute lower bounds for the total diversity if the data are correct, and the same is true *a fortiori* if the sample diversity is incorrectly inflated. It is interesting to note that the two estimates do not agree exactly; this is due to the different assumptions of the two estimators (parametric/nonparametric), and also to the very high number of singletons in this particular dataset. (The SE's and 95% confidence intervals associated with these estimates are not too meaningful in this case: they are values obtained under the assumption of equal class sizes in the population.)

The best parametric model selected for the phage data by CatchAll's selection routine is the mixture of three components. Figure 1 shows the frequency count data and the fitted model. This model specifies three levels of abundance in the population: high, with mean sampling rate 9.66 and proportion 0.004; medium, with mean sampling rate 1.160 and proportion 0.022; and low, with mean sampling rate 0.076 and proportion 0.975. This means that the highest abundance classes enter the sample at approximately $9.66/0.076 \approx 127$ times the rate of the lowest abundance classes on average. The high-abundance classes account for 0.4% of the population, the medium-abundance classes 2.2%, and the low-abundance classes 97.5% (allowing for rounding error). Figure 2 shows these three components with a logarithmic vertical axis for frequencies 0–10 (f_0 is the estimated zero count, i.e., the estimated number of unobserved taxa). If we delete the low-abundance (high diversity) component of the fitted

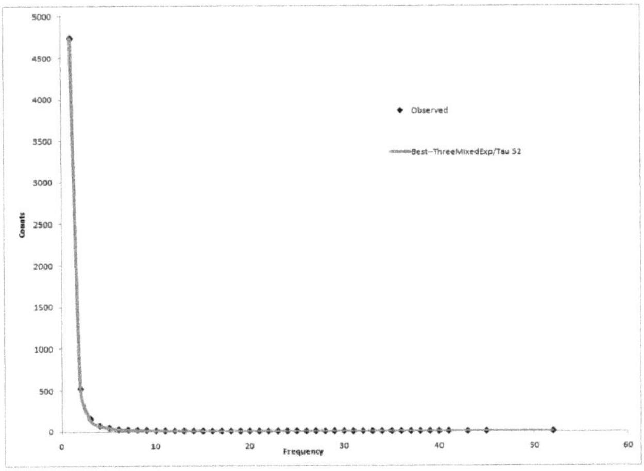

Fig. 1. Observed phage frequency count data vs. best fitted model.

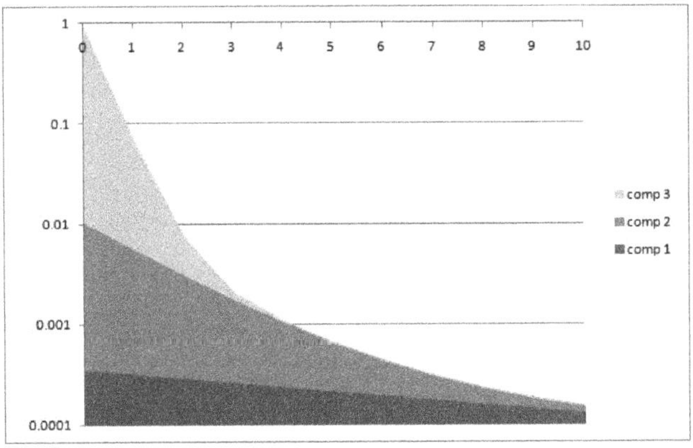

Fig. 2. Three components of fitted mixture model, logarithmic vertical axis, frequencies 0–10.

model (component 3 in Figure 2) and project the total population diversity based on the two remaining components, we obtain the results shown in the last row of Table 2. The estimated total diversity is now 1727, corresponding to the (sum of) the two higher-abundance components of the model, a reduction of 97.5% from the original unadjusted estimate, or more than an order of magnitude. This reduction may be too severe, since it deletes the entire low-abundance component, but there is no way, based on the data alone, to distinguish between real and erroneous low frequency counts.

Finally we consider left-censoring. No general software has yet been developed for this so for simplicity in exploring the procedure we use a single-component exponential (geometric) parametric model, as the computational complexity for the higher-order mixture models is considerable. This is a low but not minimum diversity model for the frequency count data, so it produces estimates that are above the Poisson and Good-Turing values; however like those models it tends to fit real data poorly (because it allows for insufficient diversity), and produces estimates below those of the best selected model (the three-component mixture in this case). It is however sufficient for our demonstration here. Table 3 shows the estimated total diversity at several different censoring points. As noted above $f_L^* = f_1 + f_2 + \ldots + f_L$, so for $L = 1$ there

Table 3. Estimated total diversity at different left censoring points under single geometric model. L = censoring point, Est Div = estimated total diversity.

L	1	2	3	4	5	10
Est Div	14880	12693	11267	10339	9712	8202

is no censoring; for $L = 2$ the singletons and doubletons are pooled, for $L = 3$ the singletons, doubletons and tripletons are pooled, and so on. The unadjusted estimate under the single geometric model (14880) is between those derived from the equal-abundance (8730) and the selected three-component mixture models (67792), as expected. Also as expected the estimate declines as the censoring point L increases, i.e., as low frequency information is removed from the data. Figure 3 displays the fitted values for the first 10 frequency counts $(f_1, \ldots, f_{10})$ for the best model (mixture of three exponentials/geometrics), the single geometric with no censoring, the single geometric with censoring point $L = 2$, and the equal-abundance unmixed Poisson models.

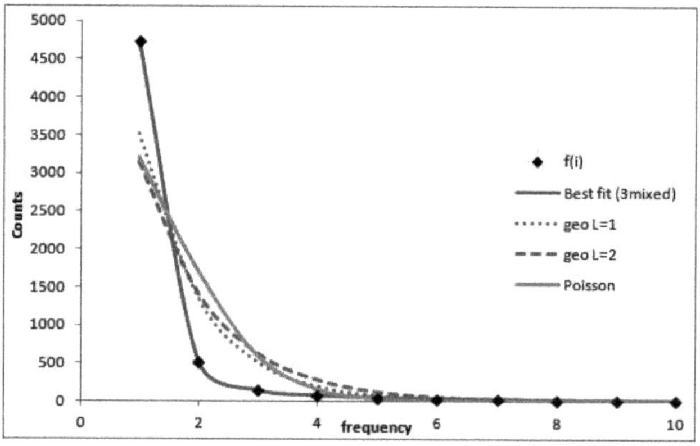

Fig. 3. Best fitted vs. discounted models, frequencies 1–10.

It is tempting to look for convergence of the estimates as L increases, and this is logical in a sense since if L is set to its maximum (the maximum frequency in the data, here 52) then all of the data are pooled into a single class. However it is not clear how we would interpret such convergence statistically. Thus this method presents us with the need to choose a cutoff L more or less arbitrarily. Furthermore it is not obvious that this method is reasonable in the microbial diversity application, even though it does damp the estimate of total diversity (as a function of L), because it essentially declares that every observed taxon does exist, and that in fact the low sample frequency classes might have been observed with higher frequency had the data been correctly recorded. On the other hand the censoring approach might be reasonable in the capture-recapture setting, where one might suppose that apparent singletons (say) should have been recorded more often but were inadvertently missed. In summary, this method is intriguing but presents conceptual and statistical difficulties at present, especially for microbial diversity estimation.

In summary, then, it is not surprising that any proposed *ex post facto* statistical approach to compensate for incorrect or uncertain low frequency counts will depend strongly on its underlying structural assumptions. Another approach, which has not yet been attempted, would be to construct a probabilistic model for the process by which errors in the low frequency counts are generated, and to incorporate this in diversity estimation. However, in models of this type any error-generation process parameter, say $p := P(\text{incorrect generation of a singleton})$, will typically be statistically confounded with the parameters of the model used to estimate diversity (the mean sampling rates, etc.), leading again to the problem of being unable to separate true from spurious counts. Still, it is not impossible that progress could be made, either in the Bayesian direction mentioned in Section 3, or via error-process modeling, or by some combination of these, and this is a topic for further research. The gold standard remains, however, to obtain correct data at the outset, or to correct it at the source.

References

1. M. Lladser, R. Gouet and J. Reeder, *PLoS ONE* **6**, (2011). doi:10.1371/journal.pone.0021105
2. J. Bunge and K. Barger, *Biometrical Journal* **50**, 971 (2008).
3. C. Quince, L. Anders and R. J. Davenport, *BMC Bioinformatics* **12**, 38 (2011).
4. D. Böhning and D. Schön, *J. Roy. Statist. Soc. Ser. C* **54**, 721 (2005)
5. M. Robb and D. Böhning, *Biometrical Journal* **53**, 75 (2011).
6. K. Barger and J. Bunge, *J. Bayesian Analysis* **5**, 765 (2010).
7. A. Farcomeni and L. Tardella, *Test* **19**, 187 (2010).
8. D. Manrique-Vallier and S. Fienberg, *Biometrical Journal* **50**, 1051 (2008).
9. H. Allen, T. Looft, D. Bayles, S. Humphrey, U. Levine, D. Alt, and T. Stanton (2011). Submitted.
10. S. Sun *et al.*, *Nucleic Acids Research* **39**, 546 (2011).
11. J. Bunge, *Pac. Symp. Biocomput.*, 121 (2011).

Acknowledgments

We thank Rob Knight for his ideas in regard to discounting uncertain low frequency counts. This research was funded in part by National Science Foundation grant DEB-08-16638 to JB. This research was conducted using the resources of the Cornell Center for Advanced

Computing, which receives funding from Cornell University, the National Science Foundation, and other leading public agencies, foundations, and corporations. Mention of trade names or commercial products in this publication is solely for the purpose of providing specific information and does not imply recommendation or endorsement by the U.S Department of Agriculture. USDA is an equal opportunity provider and employer.

Appendix A. Some mathematical details

Deleting the high-frequency component. The three-component mixture model used in the numerical example is

$$P(X=j;\theta)=\theta_4\left(\frac{1}{1+\theta_1}\left(\frac{\theta_1}{1+\theta_1}\right)^j\right)+\theta_5\left(\frac{1}{1+\theta_2}\left(\frac{\theta_2}{1+\theta_2}\right)^j\right)+(1-\theta_4-\theta_5)\left(\frac{1}{1+\theta_3}\left(\frac{\theta_3}{1+\theta_3}\right)^j\right),$$

$j=0,1,2,\ldots,,\theta_1,\theta_2,\theta_3>0,\ 0<\theta_4,\theta_5<1$, where X is the number of representatives contributed to the sample by an arbitrary species. Assuming without loss of generality that the highest-diversity component is the first listed, then the discounted estimate of diversity is $(1-\hat{\theta}_4)\hat{C}$, with associated standard error equal to $(1-\hat{\theta}_4)$SE, where $\hat{\theta}_4$ is the MLE of θ_4, $\hat{C}$ is the unadjusted (conditional) MLE of C, and SE is the unadjusted SE of $\hat{C}$.[11]

Left-censoring. The single geometric model discussed in the text is $P(X=j;p)=(1-p)p^j$, with j and X as above, $0<p<1$. The zero-truncated version of this is $P(Y=j;p)=(1-p)p^{j-1}$, $j=1,2,\ldots$. The likelihood of a dataset $f_1,f_2,\ldots$ censored at L is then

$$\ell(p):=\left((1-p)p^0+(1-p)p^1+\ldots+(1-p)p^{L-1}\right)^{f_1+f_2+\ldots+f_L}\prod_{j>L}\left((1-p)p^{j-1}\right)^{f_j},$$

and the estimates given in the example are derived by maximizing ℓ with respect to p for the given dataset.

COMPARISONS OF DISTANCE METHODS FOR COMBINING COVARIATES AND ABUNDANCES IN MICROBIOME STUDIES

JULIA FUKUYAMA, PAUL J. MCMURDIE
Statistics Department, Stanford University,
Stanford, CA 94305, USA
**E-mail:{jfukuyama,mcmurdie}@stanford.edu*

LES DETHLEFSEN, DAVID A. RELMAN
Departments of Microbiology & Immunology, and Medicine
Stanford University, VA Palo Alto Health Care System 154T
3801 Miranda Avenue, Palo Alto, CA 94304

SUSAN HOLMES*
Statistics Department, Stanford University,
Stanford, CA 94305, USA
**E-mail:susan@stat.stanford.edu*
www-stat.stanford.edu/~susan/

This article compares different methods for combining abundance data, phylogenetic trees and clinical covariates in a nonparametric setting. In particular we study the output from the principal coordinates analysis on UNIFRAC and WEIGHTED UNIFRAC distances and the output from a double principal coordinate analyses DPCOA using distances computed on the phylogenetic tree.

We also present power comparisons for some of the standard tests of phylogenetic signal between different types of samples. These methods are compared both on simulated and real data sets. Our study shows that DPCoA is less robust to outliers, and more robust to small noisy fluctuations around zero.

Keywords: Distances. DPCOA. UNIFRAC. Phylogenetic Signal. Nonparametric testing.

1. Introduction

The activities of complex microbial communities are increasingly recognized as vital aspects of the biosphere, often with direct relevance for humans. Since the transition from cultivation- and microscopy-based techniques of traditional microbiology to the nucleotide sequence-based, cultivation-independent techniques predominant today, and especially with the advent of high throughput next generation sequencing technologies, microbial ecologists have had to contend with high dimensional datasets representing the abundance of hundreds or thousands of marker gene sequence variants across tens or hundreds of samples.

This challenge has resulted in the development of preprocessing pipelines such as `qiime`[1] and `mothur`[2] that deliver output in the form of abundance tables of various *phylotypes* or OTUs (An Operational Taxonomic Unit is defined only by sequence data, lacking the physiological characterization required to establish and name a traditional taxon, but serving as a proxy for a microbial species, genus or other taxonomic entity). We will not go into the difficulties of correctly assigning the sequence reads to OTUs. This is most often done by using a x% similarity rule of thumb, e.g., x in the range 97-99 for the 16S rRNA gene sequences. The assignments often include a complex denoising procedure dependent on the actual sequencing technology (454 Life Sciences, Illumina). The goals of these studies include the comparison of bacterial communities from samples subjected to an experimental intervention, or chosen to represent a natural contrast of interest. Thus, values for a number of clinical or environmental covariates are generally associated with each sample. The analysis of these studies needs to be multivariate to capture complex high dimensional interactions. All current methods rely on computations of relevant distances between communities and their representation using standard multidimensional scaling (MDS/PCoA).

1.1. Challenges in including side information for contingency tables

The output from the standard pipelines mentioned above are contingency tables with abundances in the cells, *species*, *phylotypes* or OTUs are the rows of the tables and columns of the table represent the sampling locations, often different patients at different times. However, complementary side information is available, both about the relationships between the OTUs and about clinical/environmental covariates measured on the sampling locations. From a practical perspective this is handled by the `phyloseq`[3] package. The advantage of a specific structured data approach is that we can use many of the packages already developed for ecological and multivariate data analysis; the `R`[4] packages `ape`,[5] `picante`,[6] `ade4`,[7] `vegan`,[8] and `phyloseq`[3] were used in the current paper. Here, we will show a comparative study of some of the multivariate visualization and testing procedures available in these packages, concentrating on ways to incorporate side information effectively.

The most common statistical approach to date is to use either unweighted or weighted UniFrac[9] distances between communities. Here we compare these to DPCoA[10] (defined in section 2.2) a two step process that combines phylogenetic and abundance data for PCR sequenced phylotypes in a geometric framework.

We also provide complementary visualizations of multivariate biases and a review of available nonparametric testing procedures in the presence of important covariates.

2. Description of Techniques

2.1. UniFrac

UniFrac[11] is a distance between microbial communities for which phylogenetic information about the OTUs is available. The UniFrac distance between community A and community B is defined as the fraction of branches of the phylogenetic tree that lead to members of community A or community B but not both. This definition only considers whether an OTU is present or absent in a community and not how abundant it is.

Weighted UniFrac[12] incorporates abundances and is defined as $wUF(A, B) = \sum_i b_i|A_i/A_T - B_i/B_T|$ where the sum is over the branches of the phylogenetic tree, b_i is the length of the ith branch, A_T is the overall abundance of OTUs in community A, and A_i is the number of OTUs in community A that correspond to descendants of branch i. It is also possible to normalize weighted UniFrac by the average distance of members of the two communities to the root. This normalization can help correct for unequal sampling effort or different evolutionary rates between taxa, but for the purposes of this paper, we will take weighted UniFrac to be the raw (unnormalized) weighted UniFrac distance given above.

2.2. Double Principal Coordinates Analysis

DPCoA[13] is based on work by Rao[14] that aimed to integrate diversity and dissimilarity measures. Rao's description of diversity and distance starts with a distance between individuals and builds up to a measure of the diversity of a distribution and a dissimilarity between distributions. Briefly, he defines the diversity within a population to be the average distance between members of that population. Similarly, the diversity between two populations is the average distance between members of the two populations. He then notes that we expect the average distance between members of two different communities to be larger than the average distance between members within the individual communities. Therefore a natural distance between community i and community j is

$$RD_{ij} = H_{ij} - (H_i + H_j)/2 \tag{1}$$

where H_{ij} is the average distance between members of community i and community j, and H_i is the average distance between members of community i. These definitions of diversity within a community and distance between communities are fairly natural, and they are useful because they allow for a decomposition of diversity of a group of communities similar to the decomposition of variance in ANOVA. If we have k communities with frequencies $\lambda_1, \ldots, \lambda_k$, the diversity of all k communities taken together can be written as

$$H_0 = \sum_{i=1}^{k} \lambda_i H_i + \sum_{i=1}^{k} \sum_{j=1}^{k} \lambda_i \lambda_j D_{ij} = H(w) + D(b) \tag{2}$$

where the first term is the weighted sum of the internal diversities of the communities and the second term is the weighted sum of the pairwise distances between the communities.

DPCoA is based on the distance in formula (1). The idea is to define a typology for which the inertia of the cloud of points representing the OTUs and communities decomposes the same way the quadratic entropy does. To do this, the OTU points are first positioned in a high-dimensional space in such a way that the distances between them are the same as the patristic distances defined by the tree. Then, if the community points are put at the barycenter of their OTU profiles, the distance between the community points will be the square root of Rao's distance, formula (1).

Distances between individuals in Rao's definition of diversity and distance do not have to come from a tree, so in some sense, DPCoA is more general than UniFrac and weighted UniFrac. Here we will compare the two step DPCoA approach using the patristic distance between OTUs to the standard approach of computing the weighted/unweighted UniFrac and then using this in a MDS/PCoA plot.

2.3. A single framework

Although UniFrac, weighted UniFrac, and the community distances in DPCoA come from very different theoretical perspectives, we find it useful to present them in a single framework as weighted sums over the branches of a tree. Recall that the weighted UniFrac distance is defined as

$$wUF(A, B) = \sum_i b_i |A_i/A_T - B_i/B_T| \tag{3}$$

Method	Original description	New formula	Properties
DPCoA	square root of Rao's distance based on the square root of the patristic distances	$[\sum_i b_i (A_i/A_T - B_i/B_T)^2]^{1/2}$	Most sensitive to outliers, least sensitive to noise, upweights deep differences, gives OTU locations
wUniFrac	$\sum_i b_i \lvert A_i/A_T - B_i/B_T \rvert$	$\sum_i b_i \lvert A_i/A_T - B_i/B_T \rvert$	Less sensitive to outliers/more sensitive to noise than DPCoA
UniFrac	fraction of branches leading to exactly one group	$\sum_i b_i \mathbf{1}\{\frac{A_i/A_T - B_i/B_T}{A_i/A_T + B_i/B_T} \geq 1\}$	Sensitive to noise, upweights shallow differences on the tree

Table 1: Summary of the methods under consideration. "Outliers" refers to highly abundant OTUs, and noise refers to noise in detecting low-abundance OTUs (see the text for more detail).

Evans and Matsen[15] showed that weighted UniFrac was the first Wasserstein distance on a tree and that the second Wasserstein distance on a tree was the quantity in equation (1), although they did not link that observation to Rao or to DPCoA. However, it follows from their work that we can write the distances between communities in DPCoA as

$$[DPCoA(A, B)]^2 = \sum_i b_i (A_i/A_T - B_i/B_T)^2 \tag{4}$$

with one caveat: the branch lengths of the tree that we sum over are slightly different in DPCoA compared to weighted UniFrac. This comes from the fact that in DPCoA, the inertia is supposed to decompose the same way that quadratic entropy does.

Finally, it is informative to rewrite the formula for unweighted UniFrac so that it is as similar as possible to our formulae for weighted UniFrac and DPCoA. Weighted UniFrac and DPCoA have been written in terms of b_i, the branch lengths, and A_i/A_T and B_i/B_T. Unweighted UniFrac can be written, in terms of those same variables, as

$$UF(A, B) = \begin{cases} \sum_i b_i \mathbf{1}\{\lvert \frac{A_i/A_T - B_i/B_T}{A_i/A_T + B_i/B_T} \rvert \geq 1\} & A_i/A_T + B_i/B_T > 0 \\ 0 & A_i/A_T + B_i/B_T = 0 \end{cases} \tag{5}$$

where $\mathbf{1}$ is the indicator function (i.e. evaluates to 1 if its argument is true and 0 otherwise). This seems like an unnecessarily complicated way of writing unweighted UniFrac, but it is useful because it puts unweighted UniFrac in a form that is more comparable to weighted UniFrac and DPCoA.

Comparing these three formulae (see table 1) makes clear the differences between the three ordination methods. Weighted UniFrac and DPCoA are quite similar: DPCoA is slightly less robust to outliers (in our case, an OTU is an outlier if it is much more abundant than the other OTUs) than weighted UniFrac, but both suppress small "noisy" fluctuations around zero. This noise can be thought of as the noise inherent in measuring OTUs that are present in abundances near the detection limit. For example, if an OTU is present in all samples but has a very low abundance, it might be only be detected in half the samples. Unweighted UniFrac, on the other hand, is quite different. It puts much more weight on shallow branches than either DPCoA or weighted UniFrac. This makes it more sensitive to the kind of noise discussed before, but also allows it to pick up shallower differences that the other two methods suppress. We will see these properties illustrated in simulated and real data sets.

2.4. Runtimes

DPCoA's runtime is quadratic in the number of OTUs but is not that dependent on the number of samples. UniFrac's runtime (as implemented in `picante`), in contrast, is linear in the number of OTUs but super-linear in the number of samples. See figure S4 in the supplementary section. The fact that DPCoA is $O(n^2)$ in the number of OTUs could potentially be a problem, but we have performed DPCoA on abundance matrices with as many as 2,500 OTUs in forty minutes in a 32 CPU linux cluster.

3. Simulations

As a first pass at comparing DPCoA with PCoA using weighted UniFrac, we looked at the results of the two methods on simulated data. For the simulation, we imagine that we have four subjects (A through D), each of whom is sampled in eight locations (1 to 8). For each combination of subject and location, we have abundances for 300 species. The relationship between the species is described by a random coalescent tree. The model is that one clade varies along the location gradient (it is overrepresented in location 1 and underrepresented in location 8) and another clade varies between two groups of patients (it is overrepresented in patients A and B relative to patients C and D). This is a very simple data set, but it allows us to look at both continuous and categorical covariates.

Figure 1 shows the ordination of the simulated data by both PCoA with weighted UniFrac (a) and by DPCoA ((b) shows the community points and (c) shows the species points). MDS/PCoA with weighted UniFrac and DPCoA both show the subject effect (in both cases subjects A and B are to the left of subjects C and D) and the location effect (we see that the locations are arranged roughly in order along the second axis in both methods). DPCoA gives us some additional information, however: the locations of the species points. Since the species points and the community points are built in the same space, the plots of the communities and the species could be positioned on the same figure (we have separated them for readability purposes). In figure 1(c), blue points indicate species that we modeled as being overrepresented in subjects A and B relative to C and D, red points indicate species that we modeled as being over- or under-represented along a gradient according to location, and green indicates all other species. It looks like there are only four points in figure 1(c), but there are actually 300, many of which are located in almost exactly the same place because the structure of the simulated data is particularly simple.

Notice that the vector pointing to the center of the red (location effect) points is nearly orthogonal to the vector pointing to the center of the blue (species effect) points. This is in line with the fact that the species effect and the location effect are independent of each other.

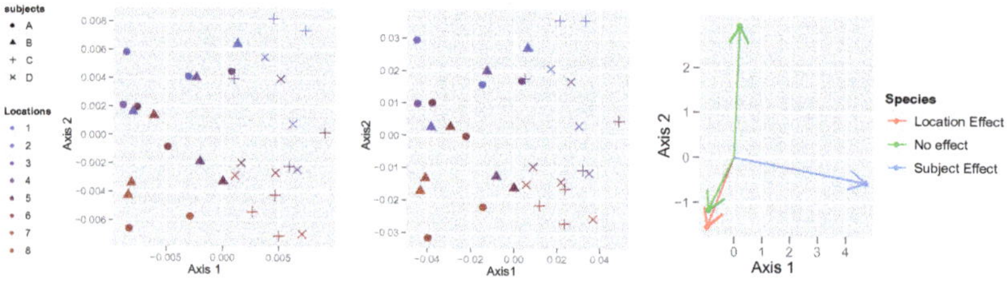

(a) UniFrac + PCoA (b) DPCoA communities (c) DPCoA species

Fig. 1: (a) shows ordination of community points by PCoA with the UniFrac metric; (b) by DPCoA. (c) shows ordination of species points by DPCoA: blue indicates species whose levels change in subjects A/B versus C/D, red indicates species whose levels change along a gradient related to location, and green indicates all other species.

Since we can see from the community plot that the first axis is primarily a subject effect and the second axis is primarily a location effect, we could have guessed that species points that lie close to the first axis would represent species related to the subject effect and species points that lie close to the second axis would represent species related to the location effect. Since we have simulated data, we know this is true.

3.1. Robustness to noise

One form of noise present in OTU abundance tables can be thought of as noise around zero: OTUs that are actually present might not be detected if they are present in very low abundances. To look at the robustness of the two methods to this sort of noise, we looked at another simulated data set. We simulated two groups of locations and 100 species related to each other by a tree, each of which is either present or absent in each location. Splitting the species approximately in half at the root, one half of the species are primarily present in group "a" and the other half are primarily present in group "b". We have three sets of simulations, each one with a different amount of noise (see figure S1).

We then used DPCoA and PCoA/MDS with UniFrac to analyze these data sets. The ordination method should correctly form two clusters and recognize that this is essentially a one-dimensional problem (each community is in one of two groups). The results can be seen in figure 2. Both of the techniques separate the two groups along the first axis for all noise levels, but DPCoA keeps most of the variance along the first axis, while in PCoA/MDS with UniFrac, increasing the noise causes the data to spread out along the higher axes. As predicted by our results from section 2.3, DPCoA is more robust to noise than UniFrac. This simulation demonstrates how strongly UniFrac upweights shallow differences compared to DPCoA. We ran a second simulation to look at the kind of noise that is more likely to come up in 454 sequencing (that is, noise resulting in fake OTUs), and we saw a similar effect (see figure S2). Considering the comparative formulae in table 1, we should not be surprised that many different kinds of noise are suppressed by DPCoA.

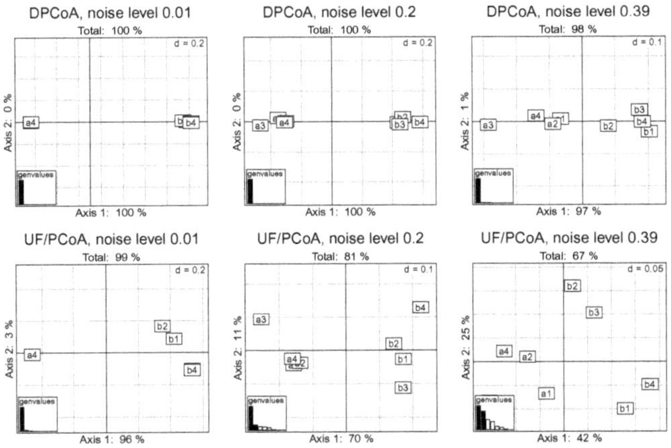

Fig. 2: Comparisons of DPCoA (row 1) and PCoA/MDS with unweighted UniFrac (row 2) for different noise levels. Columns correspond to data sets with noise levels .01, .2, and .39.

4. A Real Dataset

We then looked at the two methods on a real dataset. The data come from stool samples of three subjects, each of whom took two courses of *ciprofloxacin* over the course of ten months. Each patient was sampled about fifty times over those ten months, and we have abundance data for about 2500 OTUs for each time point and patient. The OTUs are related to each other by a phylogenetic tree, and the time points are categorized as `pre-cipro`, `1st cipro`, 1st week post *cipro*, interim, `2nd cipro`, 2nd week post *cipro*, and `post-cipro`. We looked at the data using unweighted UniFrac with PCoA/MDS, weighted UniFrac with PCoA/MDS, and with DPCoA.

4.1. PCoA with UniFrac

We can see from figure 3 that the unweighted UniFrac distance emphasizes different aspects of the data from weighted UniFrac. Unweighted UniFrac separates the subjects into fairly distinct clusters, while weighted UniFrac shows much less of a subject effect. We know that the primary difference between unweighted and weighted UniFrac is that unweighted UniFrac upweights shallow differences, and we can therefore infer from the plots that the differences between the subjects that we see in the unweighted UniFrac plot are probably due to shallow differences in species composition. To show that the difference between unweighted and weighted UniFrac is not merely due to the fact that weighted UniFrac takes into account abundances, we also show the result of using weighted UniFrac on presence/absence data. Comparing figures 3(c) and (b) shows us that the fact that weighted UniFrac can "see" abundances whereas unweighted UniFrac cannot is not the source of the difference between the two methods.

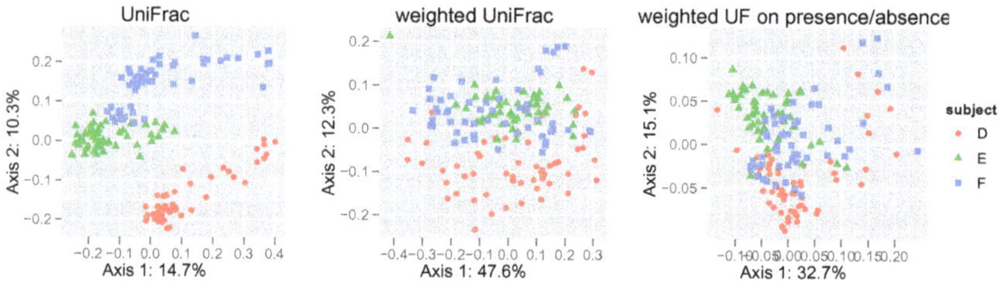

Fig. 3: Comparing the UniFrac variants. From left to right: PCoA/MDS with unweighted UniFrac, with weighted UniFrac, and with weighted UniFrac performed on presence/absence data extracted from the abundance data used in the other two plots.

4.2. DPCoA

The result of DPCoA can be seen in figure 4 (b) and (c). Since, as we saw from equation 4, DPCoA is more sensitive to outliers than either weighted or unweighted UniFrac, we had to remove any outliers, which would otherwise dominate the ordination. The result of DPCoA on the full abundance matrix is given in figure S3 for comparison. The second axis seems to separate the three subjects to a greater extent than weighted UniFrac, but not as much as unweighted UniFrac. The OTU plot can give us some more insight into the ordination given

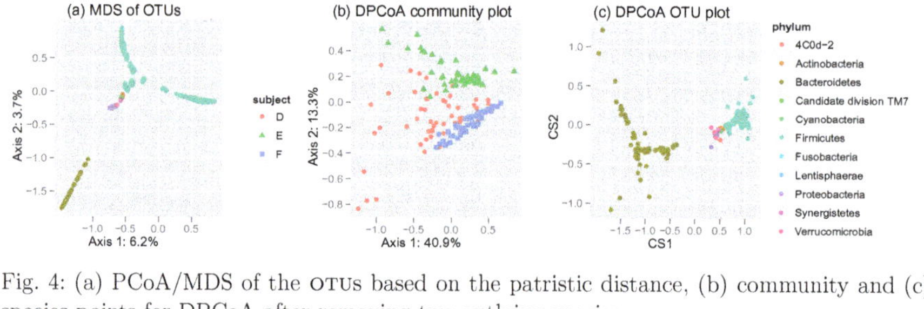

Fig. 4: (a) PCoA/MDS of the OTUs based on the patristic distance, (b) community and (c) species points for DPCoA after removing two outlying species.

by DPCoA, particularly when we compare it to PCoA/MDS of the OTUs. Since the OTU plot in DPCoA shows the OTUs from the perspective that maximizes the inertia of the community points, while a simple PCoA/MDS of the OTUs would show the OTUs from the perspective that maximizes the inertia of the OTUs, groups of OTUs that are more spread out in DPCoA compared to PCoA/MDS are likely to be important to the ordination. When we compare the OTU plot from DPCoA versus the OTU plot from PCoA/MDS, we see that in the DPCoA OTU plot, *Bacteroidetes* are much more spread out in the DPCoA plot, and *Firmicutes* are much closer together. This indicates that the first axis primarily represents the difference betwen *Bacteroidetes* and the rest of the tree, and the second axis primarily represents differences in the specific *Bacteroidetes* OTUs present in the different communities.

4.3. Antibiotic Stress

We next wanted to visualize the effect of the antibiotic. Figure 5 shows the ordinations of the communities due to DPCoA and UniFrac with information about the whether the community was stressed or not stressed (pre cipro, interim, and post cipro were considered "not stressed", while first cipro, first week post cipro, second cipro, and second week post cipro were considered "stressed"). We see that for UniFrac, the first axis seems to separate the stressed communities from the not stressed communities. DPCoA also seems to separate the out the stressed communities along the first axis (in the direction associated with *Bacteroidetes*), although only for subjects D and E. Since UniFrac emphasizes shallow differences on the tree and since

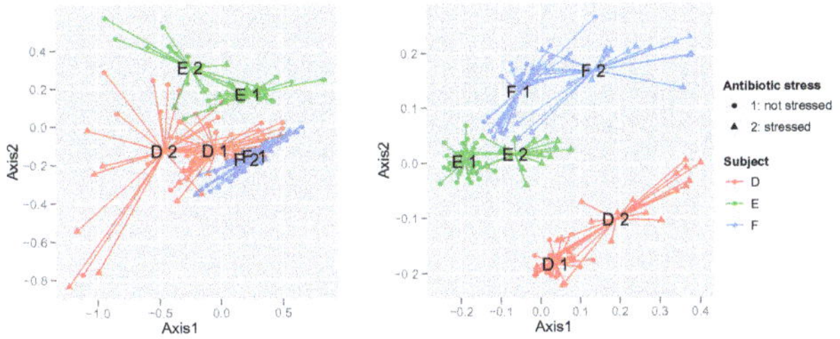

Fig. 5: Community points as represented by DPCoA (left) and PCoA/MDS with unweighted UniFrac (right). The labels represent subject plus antibiotic condition. The ordinations are the same as in figures 4 and 3.

PCoA/MDS with UniFrac seems to separate the subjects from each other better than the other two methods, we can conclude that the differences between subjects are mainly shallow ones. However, DPCoA also separates the subjects and the stressed versus non-stressed communities, and examining the community and OTU ordinations can tell us about the differences in the compositions of these communities.

5. One step further: confirmatory analyses

The previous methods were purely exploratory and allow us to visualize the effect of the covariates in principal coordinate plots. In reality we would like to be able to test the effect of the covariate through a non parametric test. There are a number of methods available for this sort of testing. One example is the UniFrac test, which tests whether two groups have significantly different species compositions by calculating the amount of branch length unique to each environment, permuting the labels, and re-calculating the amount of unique branch length. The p-value is then the fraction of trials for which the permuted unique branch length was longer than the observed unique branch length. The downside of this test is that it only tests differences between individual locations or communities, not groups of communities, and so it is more useful when you are working with small numbers of communities.

`adonis` (in the package `vegan`) provides a nonparametric multivariate analysis of variance using distances. Given a set of distances between observations, it decomposes the weighted sums of distances according to a linear model and calculates p-values associated with those decompositions by permutation of the labels.

Mantel's test[16] and the RV test (both implemented in `ade4`) test correlation between matrices. Mantel's test takes two distance matrices X and Y and uses as a test statistic $\sum_{i<j} X_{ij}Y_{ij}$. The null distribution is generally estimated by a Monte Carlo method: keeping the entries of one matrix fixed, permuting the entries in the rows and columns of the other matrix, and recomputing the test statistic. The RV test is similar, it is based on a multi-table generalization of the correlation coefficient. ($RV(A, B) \propto Trace(t(A)B)$).[17] This nonparametric test estimates the null distribution of the RV coefficient between two matrices by permuting the rows of one matrix and recomputing the RV coefficient for each permutation.

Finally, the Abouheif test tests traits for a phylogenetic signal. It was originally described as a version of a test for serial independence on the leaves of a phylogenetic tree where the null distribution was determined by permutation of the leaves,[18] but was later shown to be a a version of Moran's I with a certain distance on the tree.[19]

5.1. Testing our data

Our ordination of the simulated data suggested two hypotheses to test: first, that subjects A and B are different from subjects C and D, and second, that the species composition of each subject changes along a gradient associated with the location. Since the simulated data looked fairly similar in DPCoA and PCoA with the UniFrac distance, we will do all the testing on the DPCoA results. To test whether the species composition changes along the location gradient, we can use Mantel's test, the RV test, or `adonis`. For Mantel's test and the RV test, we compare the distances between communities given by DPCoA with distances between the covariates (for example, if we were testing a location effect, we could test the DPCoA distances with the distances between the locations). For `adonis`, we will be testing whether location explains a significant amount of the distances between the communities (distances again as given by DPCoA). Figure 6 shows the p-values we get from each test for varying magnitudes of the effect (0 is no location effect, and 5 is a pretty strong location effect), and we can see that the Mantel is the most sensitive here, followed by `adonis`, followed by the RV test.

To test whether species composition is different in the group containing subjects A and B compared to the group containing subjects C and D, we can use adonis, the Abouheif test, or the UniFrac test. For the Abouheif test, we can look for a phylogenetic signal in the difference in the abundance of each species between subjects A/B and C/D. Adonis, as before, will look at the amount of variance explained by the groups. For the UniFrac test, if we just want to test the difference between the two groups of subjects, we need to take the average species composition over all the samples from subjects A and B and compare it to the average species composition over all samples from subjects C and D. If the distance between the two averages is significant, we can say that the difference between the groups is significant. Figure 6(b) shows the p-values for the Abouheif, adonis, and UniFrac tests at different magnitudes of the

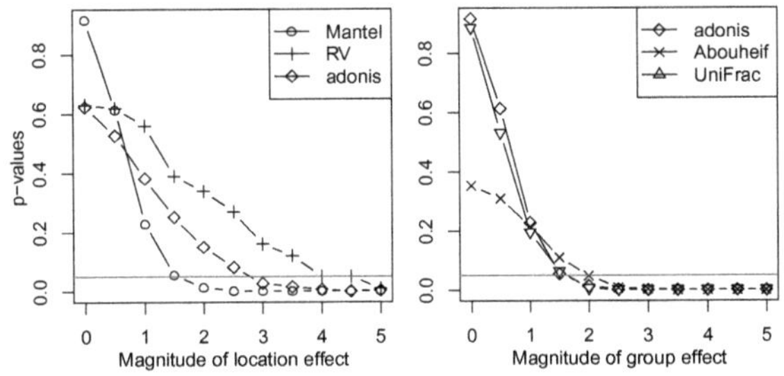

Fig. 6: The left shows the p-values of the Mantel (circles), RV (pluses), and `adonis` (diamonds) tests for a location effect at different magnitudes of the effect. The right shows the p-values of the Abouheif (x's), `adonis` (diamonds), and UniFrac (triangles) tests at different magnitudes of the group effect.

group effect. We see that all three tests have similar performance on these data, and the adonis and UniFrac tests give remarkably similar results.

Our ordination of the real data suggested that we should test whether there is a difference in intestinal bacteria composition between times of antibiotic stress and no stress. Adonis seems to be the most appropriate method in presence of a patient effect, an antibiotic stress effect, and maybe an interaction between the two. Carrying out this analysis gives the result that, for both the distances given by DPCoA and the distances given by UniFrac, the patient effect and the antibiotic stress effect are significant at the .001 level. The interaction term between patient and antibiotic stress is significant at the .05 level for the UniFrac distances and at the .001 level for the DPCoA distances. We can also try the UniFrac test on this data, aggregating by subject or by stress condition as before. We find that the unweighted UniFrac test gives a significant result for both the subject effect and the antibiotic stress effect, but the weighted UniFrac test gives significant results for neither.

6. Summary

This article presents comparisons of some of the current techniques available for analyzing abundance tables in the presence of side information on the OTUs and covariates on the samples. The classical framework for analyzing abundance tables using parametric models or multivariate analysis of variance does not apply in the microbiome studies and recent papers have used UniFrac[20] and DPCoA[21] with success. Both approaches are similar in their use of phylogenetic distances between OTUs and projections using PCoA. However whereas the UniFrac approach allows the comparison of two samples at a time, DPCoA provides a more complex framework enabling different categorical and continuous covariates to be taken into account. In our comparisons of weighted and unweighted UniFrac approaches the unweighted UniFrac seems to distinguish a much clearer separation between subjects for instance, both on simulated and real data. We compared DPCoA and unweighted UniFrac on simulated data to test their performances in the presence of noise. The results show that when there is a simple one dimensional contrast between the data, DPCoA is impervious to a substantial noise component, whereas the first axis in an unweighted UniFrac PCoA plot has a much smaller eigenvalue. The consequence in more complex studies is that the subsequent components of variation can be interleaved with the main phylogenetic effect leading to difficulties in interpreting the results.

Acknowledgments

We thank Elisabeth Purdom, Sam Pimentel, Kris Sankaran, Yana Hoy, Katie Shelef as well as two anonymous referees for useful comments and motivating questions. J.F. was partially supported by a NSF-DMS-VIGRE grant. P. J. M. and S. H. were supported by the NIH Grant NIH-5-R01GM086884 D. A. R. is funded by the Doris Duke Charitable Trust and a National Institutes of Health Pioneer Award (DP1OD000964), and by the Thomas C. and Joan M. Merigan Endowment at Stanford University.

Supplementary information: `http://stat.stanford.edu/~susan/projects/psb2012.pdf`.

References

1. J. Caporaso, J. Kuczynski, J. Stombaugh, K. Bittinger, F. Bushman, E. Costello, N. Fierer, A. Peña, J. Goodrich, J. Gordon and R. Knight, *Nature methods* **7**, 335 (2010).
2. P. D. Schloss, S. L. Westcott, T. Ryabin, J. R. Hall, M. Hartmann, E. B. Hollister, R. A. Lesniewski, B. B. Oakley, D. H. Parks, C. J. Robinson, J. W. Sahl, B. Stres, G. G. Thallinger, D. J. Van Horn and C. F. Weber, *Appl. and environmental microbiology* **75**, 7537 (November 2009).
3. P. J. McMurdie and S. Holmes, Phyloseq: A bioconductor package for handling and analysis of high-throughput phylogenetic sequence data.
4. R. Ihaka and R. Gentleman, *Journal of Computational and Graphical Statistics* **5**, 299 (1996).
5. E. Paradis, J. Claude and K. Strimmer, *Bioinformatics* **20**, 289 (2004).
6. S. Kembel, P. Cowan, M. Helmus, W. Cornwell, H. Morlon, D. Ackerly, S. Blomberg and C. Webb, *Bioinformatics* **26**, 1463 (2010).
7. D. Chessel, A. Dufour and J. Thioulouse, *R News* **4**, 5 (2004).
8. J. Oksanen, F. G. Blanchet, R. Kindt, P. Legendre, R. G. O'Hara, G. L. Simpson, P. Solymos, M. H. H. Stevens and H. Wagner, *vegan:Community Ecology Package*, (2010).
9. M. Hamady, C. Lozupone and R. Knight, *The ISME Journal* (Jan 2009).
10. E. Purdom, *Annals of Applied Statistics* (Jul 2010).
11. C. Lozupone and R. Knight, *Applied and environmental microbiology* **71**, p. 8228 (2005).
12. C. Lozupone, M. Hamady, S. Kelley and R. Knight, *Applied and environmental microbiology* **73**, p. 1576 (2007).
13. S. Pavoine, A. Dufour and D. Chessel, *Journal of theoretical biology* **228**, 523 (2004).
14. C. R. Rao, *Theoretical Population Biology* **21**, 24 (1982).
15. S. N. Evans and F. A. Matsen, *arXiv* **q-bio.PE** (Jan 2010).
16. N. Mantel, *Cancer research* **27**, p. 209 (1967).
17. P. Robert and Y. Escoufier, *Jour of the Roy. Stat. Soc.. Series C* **25**, 257 (1976).
18. E. Abouheif, *Evolutionary Ecology Research* **1**, 895 (1999).
19. S. Pavoine, S. Ollier, D. Pontier and D. Chessel, *Theoretical population biology* **73**, 79 (2008).
20. R. Ley, M. Hamady, C. Lozupone, P. Turnbaugh, R. Ramey, J. Bircher, M. Schlegel, T. Tucker, M. Schrenzel, R. Knight *et al.*, *Science* **320**, p. 1647 (2008).
21. P. B. Eckburg, E. M. Bik, C. N. Bernstein, E. Purdom, L. Dethlefsen, M. Sargent, S. R. Gill, K. E. Nelson and D. A. Relman, *Science* **308**, 1635 (Jun 2005).

PROTEOTYPING OF MICROBIAL COMMUNITIES BY OPTIMIZATION OF TANDEM MASS SPECTROMETRY DATA INTERPRETATION

ALYS HUGO[1], DOUGLAS J. BAXTER[2], WILLIAM R. CANNON[1], ANANTH KALYANARAMAN[4], GAURAV KULKARNI[4], STEPHEN J. CALLISTER[3]

[1]Computational Biology and Bioinformatics Group,
[2]Molecular Sciences Computing Facility,
[3]Biological Separations Group,
Pacific Northwest National Laboratory
Richland, WA 99352

[4]School of Electrical Engineering and Computer Science,
Washington State University,
Pullman, WA, 99164.

We report the development of a novel high performance computing method for the identification of proteins from unknown (environmental) samples. The method uses computational optimization to provide an effective way to control the false discovery rate for environmental samples and complements *de novo* peptide sequencing. Furthermore, the method provides information based on the expressed protein in a microbial community, and thus complements DNA-based identification methods. Testing on blind samples demonstrates that the method provides 79-95% overlap with analogous results from searches involving only the correct genomes. We provide scaling and performance evaluations for the software that demonstrate the ability to carry out large-scale optimizations on 1258 genomes containing 4.2M proteins.

1. Introduction

Peptide and protein identification from global proteomics studies of environmental samples is challenging because the standard approach used to identify peptides and proteins is difficult to effectively apply. The standard approach involves growing clonal cells in the laboratory, separating the proteins from other cellular components, digesting the proteins into smaller peptides (approximately 6-55 amino acids long), partially separating the peptides using liquid chromatography, and then introducing each peptide into a mass spectrometer.

The peptides are then collided with an inert gas that vibrationally excites the peptide. The vibrational excitation leads to fragmentation of the peptide into smaller chemical components, which are often short sequences of amino acid residues. The most widely used method to identify the peptide from its mass spectrum is to compare a predicted fragmentation pattern of the peptide (a model spectrum) to the experimental spectrum. This approach involves a search of all possible peptides that can be generated from the protein sequences. There are a number of available analysis tools that perform these searches, including Sequest [1], X!Tandem [2], Crux [3], Mascot [4], OLAV [5], InsPecT [6], and MSPolygraph [7-9].

Because the model spectra used in these searches are typically developed by training over diverse sets of peptide spectra, the resulting model spectra are often not good representations of the true spectrum, which can result in low sensitivity [9]. Given the high mass accuracy of many spectrometers, this would not be a problem except for the large number of peptides that can be generated from the protein sequences of a single microorganism, resulting in a high chance of a random match. This, of course, reduces the number of peptides that can be confidently identified. For instance, when controlling the false discovery rate at 5%, often only 15-20% of the spectra can be identified with a peptide.

In an environmental sample, the problem is exacerbated because there may be several to hundreds of different microbes present. The vast number and diversity of possible peptides that have to be searched leads to a very low number of spectra that can be confidently identified at a reasonable false discovery rate.

Approaches to identify peptides and proteins from environmental samples have generally taken one of two approaches for searching protein sequences. One approach is to search against all known protein sequences using sequence databases such as the NR database at NCBI or likewise the Swiss-Prot database at the Swiss Institute of Bioinformatics. However, this approach has not been favored because the large number of protein sequences leads to a high chance of a match simply by chance. Alternatively, recent environmental proteomics studies have sought to limit the number of potential candidate peptides by searching against metagenome sequences from the same environment [10-12]. This is a sensible approach, but it too has many challenges. First, it requires having metagenome sequences to begin with. If one is available, protein coding regions must be determined from the draft sequences, or spectra can be matched to all potential open reading frames from the draft sequences. However, the problem with the latter approach is similar to searching against large databases such as NR or Swiss-Prot: the chance of a random match is quite high due to the large number of candidates and the moderate specificity of the model spectra that are used for identification.

Frequently, metaproteomics studies also supplement the analysis with *de novo* peptide identification, which seeks to determine the identity of the peptide responsible for the spectrum only from the peaks in the spectrum. Unfortunately, peptide MS/MS spectra most often do not provide enough information to obtain full-length peptide sequences and are challenging to apply to spectra from peptides with charges greater than +2.

As an alternative, access to high performance computing resources allowed us to developed an optimization method for identifying peptides and proteins from fully sequenced microbial genomes. The method searches all fully sequenced genomes and optimizes proteome-spectra matches by iteratively eliminating microbes that are not likely to be in the sample. The method has been tested using samples containing blind mixtures of spectra from known microbes and samples containing unknown mixtures of microbes. In the case of five blind mixtures of varying complexity, the method has been able to identify the correct microbes reliably. In addition, the spectra identified with each microbe has a high overlap with spectra identified at a 5% FDR when searching only the protein sequences of the correct organisms.

2. Methods

The method evaluates which known genomes are appropriate for analysis of an environmental sample using the statistics based on the number of top hits observed for each genome in the MS/MS data. Specifically, a database search tool, *MSPolygraph* in this case, is used to evaluate matches between each spectrum and candidate peptides from the protein sequences of each of the fully sequenced genomes available. In the work here, we used the 1258 fully sequenced bacterial and fungal genomes available from NCBI on Oct 4, 2011. Each candidate peptide may correspond to one or more genomes, and genomes can have multiple candidate peptides for each spectrum. However, for a particular spectrum only the highest scoring peptide from a genome is recorded. If the genome has no candidates for the spectrum, the genome's score for the spectrum is zero. Spectra with no matches are removed. The genome having the peptide with the highest score for a spectrum is referred to as the top hit for that spectrum. Also, for each genome, the total number of candidate peptides for that genome in each spectrum is recorded. The significance of each genome is then calculated based on a statistical likelihood. A large number of ties occur in this process, in which a peptide that is a top hit to a spectrum belongs to proteins from multiple genomes. The ties are broken in an iterative manner with an expectation-maximization approach. Details on the method follow.

2.1 *Statistical Null Model*

The probability that a particular genome would appear as the top hit for a particular spectrum by chance is estimated as the number of peptide candidates generated from the protein sequences of the genome divided by the total number of candidates from all genomes for that spectrum. That is, the probability of a random top hit for a particular genome to a spectrum is proportional to the relative number of peptide candidates that can be derived from the protein sequences of that genome. From these probabilities, we estimate the number of top hits that would be expected to be observed for each genome. This is the number of spectra multiplied by the null probability. In addition, the information on the null probabilities is later used to generate a population of top hits that a genome would obtain by chance for each spectrum from a simulation.

2.2 *Sample likelihood ratio and significance*

For each genome, we calculate the sample likelihood ratio, which is the likelihood of observing the genome from the data relative to the likelihood of observing the genome by chance. Accordingly, the sample likelihood is calculated as the number of top hits observed from the data $n_{obs,i}$ relative to the number of top hits expected from the null probabilities $n_{null,i}$,

$$\theta_i = \frac{n_{obs,i}}{n_{null,i}}.$$

To estimate the significance of the sample likelihood ratios, the number of top hits expected by chance for each genome is estimated. As mentioned above, this is the number of spectra multiplied by the null probability. Next, using the null probabilities, a simulation is then used to generate a population of likelihood ratio values that would be expected to be observed by chance. The population of likelihood ratios is then fit to a generalized extreme value distribution. From the fit, the probability of observing the sample likelihood ratio by chance (p-value) is estimated.

2.3 Iterative assessment of likelihood

Initially, distribution of likelihood scores is obtained in which no particular genome stands out, in large part because many top-scoring peptides are shared among them. In this case, it is not clear which genomes are good proxies for the organisms contained in the environmental samples. Ideally, ties are broken by assigning each involved genome a fraction of the top hits proportional to the their abundance in the environmental sample. However, the proportion of each species in the sample is unknown, and the number of observed top hits is a conflation of hits due to the microbe actually being present and random hits to the genome of the microbe. Instead, ties are broken by assigning each genome a fraction of the top hits proportional to the sample likelihood ratio for the genome determined in the previous iteration and scaled by the probability that the sample likelihood ratio was not by chance (1- p-value). Specifically, the quantity C_i, as calculated below, is found for each top scoring genome i and normalized over all top scoring genomes for the spectrum, resulting in each genome's share of the top hit,

$$C_i = \theta_i * (1 - p_i).$$

Here θ_i is the sample likelihood ratio for genome i from previous iteration, and p_i is the p-value of the sample likelihood ratio for genome i, calculated using the results of simulation. New θ and p-values are calculated by totaling the fraction of top hits assigned to each genome and comparing the results to simulated results.

This process is repeated for a predetermined number of iterations or until some convergence criterion is met. (In our case, 150 iterations or until top 50 θ values changed by less than 0.0001.) After the last iteration, ties are reassigned if desired. For each spectrum, the tied genomes with the most top hits overall from the last iteration receives full credit for the hit. This gives a simplified explanation in which the spectra are accounted for with as few genomes as possible. In most cases, this results in consolidation of hits from several closely related genomes (often species from the same genus), leaving one representative that best explains the observed hits. This step is most useful when one expects only single representatives of each genus to be present and can be omitted if multiple species from a genus are expected to be present. *P*-values are calculated for each genome based on the recalculated number of hits using results from a second simulation.

2.4 *False Discovery Rate Estimation*

Finally, genomes are determined to be appropriate or not based on a desired false discovery rate. The false discovery rate is estimated from the final set of *p*-values using a nonparametric estimation with the program *qvality* [13]. If a genome has posterior error probability less than 1%, it is considered to be appropriate for use in the analysis. If a genome fails to meet this requirement, it is considered to be inappropriate.

2.5 *Programming Model*

The statistical methods discussed herein were implemented in Matlab, and run on common workstations. The database search tool *MSPolygraph* had to be modified to handle protein sequences from thousands of genomes and to keep a list of only the top scoring peptides from each genome. As the number of genomes is constantly growing, high performance computing is required.

The parallel version of the *MSPolygraph* is essentially a task-scheduling wrapper around the serial version of the code. In the serial code an input parameter file is read, along with the fragmentation model for generating model spectra, the protein sequences of all organisms to be searched, and the spectra file(s) to be analyzed. The code loops over the spectra scoring each spectrum against all peptides generated from the target protein sequences that are consistent with the observed mass-to-charge ratio of the observed, intact peptide, accumulating high scoring matches and printing them out in a sorted list, sorted by likelihood ratio score. The amount of work required to score a spectrum depends on a variety of factors which include its length (number of peaks), how many peptide candidates were generated from the target protein sequences, and how many peaks from each of those peptide candidates match peaks from the experimental spectrum peaks. The number of peaks from each peptide that will match the spectrum cannot be predetermined without doing 2/3 of the work required to actually score the spectrum, which makes an *a priori* determination of run time for a spectrum impractical. Hence, we use a dynamic scheduling scheme facilitated by a server/client process model.

In the parallel version we use the MPI (Message Passing Interface) standard for communication. The input files are placed on a globally visible file system (mounted on all of the compute nodes). Each processor reads in to its own local memory the input files and we then use a dynamically scheduled server-client model to control which process (mpi rank) scores which spectrum. A processor's behavior is controlled by its mpi rank. One processor is the dedicated server process (mpi rank 0) and all others are considered client processors.

After reading the input data, the server process issues a non-blocking receive to each client. It uses a simple counter to determine which task to send to a requesting client. It polls clients for responses indicating that a spectrum has been completed (or during the first pass that a client has initialized and is ready to start scoring), and replies with another index for a spectrum to be scored if there is one or a quit message if all tasks have been distributed. The manager utilizes non-blocking sends so as not to need to wait for the clients to receive their messages. A different buffer for each client is used. As the outgoing messages are only an index as to which spectra to score, and the incoming messages are a fixed length summary line, only a small amount of space is required even for a very large number of clients. Also, as the server hands out a new index to a "ready to start" or "completed spectrum" message from a client, no more than 1 message per client is ever in flight. After all tasks have been handed out, the server processor continues polling for responses till responses for all spectra have been received.

The client process also reads the input files from the global file system, opens its own output file for printing scoring results for spectra assigned to it. It then issues a non-blocking receive for an index of the next spectra to be score and sends a "ready to start" message to the server processor. The client then enters a communication and scoring loop in which the client repeatedly:

- waits for server message indicating which spectrum to process (or a quit message),
- processes the spectrum (or exits the loop),
- writes data and flushes the results to its output file,
- issues a non-blocking receive for the next spectrum index and,
- sends a summary message for the scored spectrum to the server.

Scaling results are presented below.

3. Results

3.1 Optimization of Artificial Microbial Communities

Five laboratory samples were prepared of varying complexity, ranging from 3 to 15 microbes. The samples were mixed in equal proportions based on estimates of total protein concentration. The results for the lowest complexity sample are shown in Table I. In this sample, three species were present, *Anaeromyxobacter dehalogenans, Geobacter uraniumreducens* and *Salmonella typhimurium*. As shown in Table I, these three species were identified as the top three species by the optimization procedure and were the only species identified at a 1% false discovery rate (q-value). For *Salmonella typhimurium*, however, a strain that is 98% identical to the cultured strain, *Salmonella typhimurium LT2*, was identified (*Salmonella enterica serovar Typhimurium*). For each of these species, the number of spectra associated with the species is shown in the second column of Table I. For comparison, the number of spectra found for each of the species using a standard identification procedure is also shown in the third and fourth columns of Table I. In the standard approach, only the proteomes of the known species are searched, and the number of spectra identified with each of the species at 5% and 10% false discovery rates is shown in these latter columns. Also shown in parentheses in columns three and four are the percent of peptide-spectrum matches occurring in the optimization approach that also are found

Species	Number of Identifications				
Target Species Other Species	Optimization	Overlap 5% FDR	Overlap 10% FDR	LR	Posterior Error Prob
Salmonella enterica serovar Typhimurium	3145	3016 96%	3081 98%	289.20	0.03
Anaeromyxobacter dehalogenans 2CP-C	1256	1199 94%	1249 97%	67.12	0.06
Geobacter uraniumreducens Rf4	539	525 95%	533 96%	39.83	0.08
Below 1% q-value cutoff:					
Salmonella typhimurium LT2	6	0 0%	8 38%	0.57	1

Table I. Optimization results for model microbial community containing three known microbes.

in the standard approach. For this dataset, the overlap in peptide-spectrum matches between the optimization method and the standard search is remarkably high, 94-98%. That is, despite searching 1258 genomes containing over 4 million proteins and several orders of magnitude more peptides, the optimization approach on this data set contained nearly the same quality of identifications as a standard database search of only the proteins correct genomes. Of course, as a model of a microbial community found in the environment, three organisms would be a very simple microbial community.

To test the ability of the method to analyze more complex samples, we also investigated four other datasets consisting of 6, 9, 12 and 15 microbes. The results for the highest complexity data set, derived from the mixture of batch cultures from 15 species, are shown in Table II. For this data set, each of the top 10 scoring species and 11 out of the top 12 scoring species were correctly identified. In all, 12 of the 15 species were identified at an estimated 1% false discovery rate. For these 12 species, the overlap in peptide-spectrum matches between the optimization method and the standard approach (identification using only the correct genomes) was 47-91%. For all 15 species, the weighted average of the overlap was a remarkable 79%. While this is less overlap than was seen in the data containing just three microbes discussed above (95% weighted average overlap), the performance was encouraging when the reasons for the decreases were examined. These are discussed are discussed next.

Three species were not identified - *Rhodopseudomonas Palustris CGA009, Saccharomyces cervisiae S288c,* and *Aspergillus carbonarius*. For *Aspergillus carbonarius*, 62 peptide-spectrum matches were found for a species from the same genus, *Aspergillus niger*. It is possible that determining the likelihood at the genus level instead of at the genome level would identify the *Aspergillus* genus. However, the number of significant peptide-spectrum matches found for each of these species was also quite low when a standard approach (only the correct protein sequences) was used. For instance, for *Aspergillus carbonarius* only 49 and 103 significant peptide-spectrum matches were found at 5% and 10% false discovery rates. The most likely cause is that the protein samples that were prepared for these species were too dilute. For *Aspergillus carbonarius*, a confounding problem is that the protein sequences are only available from the initial draft genome of the organism.

Species Target Species Related Species Other Species	Number of Identifications Optimization	Standard run on known DBs 5% FDR	 10% FDR	LR	Posterior Error Probability
Deinococcus radiodurans	1168	1048 83%	1191 90%	165.92	1.79E-05
Shewanella oneidensis	713	627 85%	714 91%	76.94	1.32E-04
Salmonella typhimurium LT2	482	495 83%	554 90%	50.86	3.86E-04
Synechococcus PCC 7002	247	230 84%	268 90%	41.70	6.47E-04
Arthrobacter FB24	323	299 85%	347 89%	31.58	1.33E-03
Heliobacterium modesticaldum Ice1	239	225 79%	269 87%	30.74	1.43E-03
Chloroflexus aurantiacus J 10 fl	226	208 80%	259 89%	26.81	2.04E-03
Desulfovibrio desulfuricans G20	177	153 77%	188 86%	20.19	4.27E-03
Anaeromyxobacter dehalogenans 2CP-C	265	243 66%	325 76%	15.77	8.11E-03
Geobacter uraniumreducens Rf	92	91 70%	119 79%	7.37	5.83E-02
Brucella melitensis bv	39			4.99	1.59E-01
Candidatus Korarchaeum cryptofilum	21			4.50	2.08E-01
Clostridium thermocellum ATCC 27405	40	41 67%	54 76%	4.03	2.77E-01
Beutenbergia cavernae DSM 12333	33			3.50	3.97E-01
Rhodobacter sphaeroides KD131	38	46 47%	79 52%	3.49	3.99E-01
Burkholderia cenocepacia J2315	59			3.06	5.60E-01
Acidovorax avenae citrulli AAC00-1	38			3.02	5.78E-01
Frankia alni ACN14a	53			3.00	5.91E-01
Desulfurivibrio alkaliphilus AHT2	21			2.88	6.51E-01
Rhodococcus erythropolis PR4	36			2.82	6.83E-01
Xylella fastidiosa	15			2.79	7.06E-01
Below 1% q-value cutoff:					
Rhodopseudomonas palustris BisB18	27			2.05	1.00E+00
Aspergillus niger CBS 513 88	62			1.31	1.00E+00
Rhodopseudomonas palustris BisB5	14			1.18	1.00E+00
Rhodopseudomonas palustris BisA53	12			0.91	1.00E+00
Rhodopseudomonas Palustris CGA009	10	23 5%	43 10%	0.79	1.00E+00
Saccharomyces cervisiae S288c	15	17 21%	42 36%	0.61	1.00E+00
Aspergillus carbonarius	18	49 7%	103 25%	0.48	1.00E+00

Table II. Optimization results for model microbial community containing 15 known microbes.

Likewise, *Rhodopseudomonas Palustris* CGA009 was also not identified at a statistically significant level. In this case, 10 peptide-spectrum matches were found in the optimization, while 23 and 43 matches were found when using a standard search of the known genomes only. This indicates that the sample was relatively dilute in peptides from this organism, but interestingly 43 peptide-spectrum matches were also assigned to closely related *Rhodopseudomonas* species. In total, 53 matches were found to the Rhodopseudomonas species, which is more than found in the standard search. This may indicate that the species grown in culture is no longer clonal with respect to the sequenced species *Rhodopseudomonas Palustris* CGA009. In this regard, the optimization approach may be useful for analyzing batch monocultures also against closely related genomes and would effectively allow for sequence variability in the peptide-spectrum matching process.

The third species that was not identified was *Saccharomyces cervisiae* S288c. In this case, 15 peptide-spectrum matches were found in the optimization, while only 17 and 42 matches were found at 5% and 10% FDR in a standard search of the known genome. However, even if the same number of matches were found in the optimization as was found at 10% FDR in the standard search against the known genome, this would not have been enough matches to be statistically significant. Again, the most likely explanation for this was that the sample simply did not contain enough proteins from this organism.

The three other data sets varied in the number of species in the sample from 6 to 12. The trends in these three data sets mirrored that seen for the other two data sets. In two of the data sets, the related species *Salmonella enterica serovar Typhimurium* was identified instead of *Salmonella typhimurium LT2*. *Aspergillus carbonarius* was consistently missed in each data set while the related *Aspergillus niger* was positively identified in one of the three data sets. While *Aspergillus niger* was not identified in the other two data sets, it did consistently have more peptide-spectrum matches (> 4-fold) than *Aspergillus carbonarius*. It is reasonable to expect that the number of *Aspergillus* peptide-spectrum matches could be significant at the genus level.

3.2 Scaling

As the processing time for a single processor job, T(1), takes longer to run on a single node than the job policy allowed at the time of these runs, we instead generate a weak scaling curve replacing T(1) by 128*T(128) shown in Figure 1. This amounts to taking the efficiency at 128 processors to be 1 for comparison purposes. We note that T1/(128*T(128)) < 127/128 (0.992), as the master processor does no work. The fall off in efficiency shown in Figure 2 at 1024 processors is an indicator that we are hitting inefficiency in the MPI infrastructure layer at scale, most likely due to too many messages being passed. This could also be also due to input/output bottlenecks if many processors are simultaneously writing output files. However, runtime

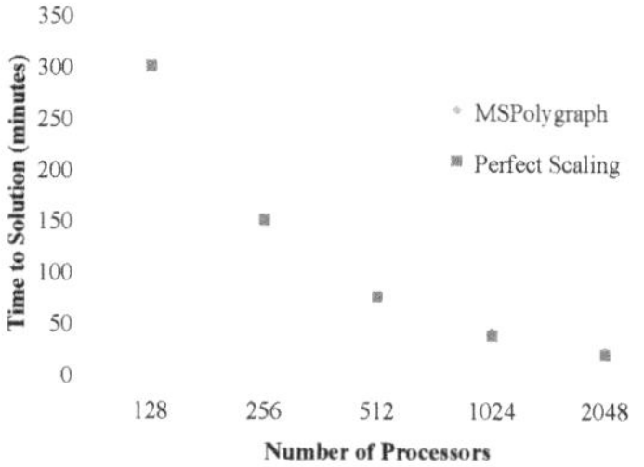

Figure 1. Scaling of MSPolygraph. The time to solution was determined for running 1258 fully sequenced genomes against 18,929 spectra on the Chinook supercomputer at EMSL.

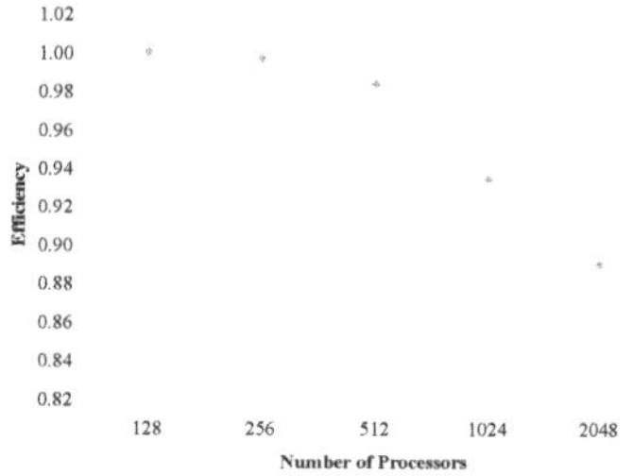

Figure 2. Parallel efficiency of MSPolygraph on data in Figure 1.

monitoring doesn't indicate that the code is any where close to the I/O bandwidth limits on the machine, making it unlikely that an I/O bottleneck causes the efficiency loss.

4. Discussion

The method discussed here allows us to use high performance computing to potential identify proteins from complex environmental samples without necessarily having sequenced the microbes in the sample. Despite analyzing over 1000 genomes against a spectral dataset, the method resulted in a 79-95% overlap with identifications made using only the correct protein sequences. There are several caveats to this approach that must be kept in mind. First, a high scoring genome does not necessarily imply that the microbe is present, only that the genome of the microbe is appropriate for analyzing the data. That being said, the presence of multiple high scoring genomes from the same genus does imply the presence of a microbe from that genus. Second, the method will actually become more powerful as more genomes become sequenced. In this regard, projects such as the Genomic Encyclopedia of Bacteria and Archaea [14], which seek to provide greater breadth of our knowledge of microbial genomes, are invaluable. At this time, there are approximately 1500 fully sequenced genomes of microbes available, and this number is increasing rapidly. At the same time, advances in instrumentation are resulting in a rapid increase in the number of spectra that are derived from a single sample. Currently, one sample may result in 10-30,000 spectra, but that number will continue to grow rapidly as the instrumentation technology develops. In summary, the statistical method described above combined with high performance computing offers a potentially significant break-through in the analysis of environmental samples.

Acknowledgements

This work was supported under contracts 57271 and 54976 from the Department of Energy's Office of Advanced Scientific Computing Research (OASCR) and Office of Biological and Environmental Research (BER) to develop new approaches for computational biology in areas of national interests. We would like to thank researchers in the laboratory of Richard Smith at PNNL who generated the datasets herein, as well as Dr. Smith, for providing the data. The calculations were performed on the Molecular Science Computing supercomputer, Chinook in EMSL, a national scientific user facility sponsored by the Department of Energy's Office of Biological and Environmental Research and located at the Pacific Northwest National Laboratory and at the National Energy Research Scientific Computing Center at Lawrence Berkeley National Laboratory. PNNL is operated by Battelle for the U.S. Department of Energy under Contract DE-AC06-76RLO 1830.

References

[1] K. Eng, *et al.*, "An approach to correlate tandem mass spectral data of peptides with amino acid sequences in a protein database," *Journal of the American Society of Mass Spectrometry*, vol. 5, pp. 976-989, 1994.

[2] R. Craig and R. C. Beavis, "TANDEM: matching proteins with tandem mass spectra," *Bioinformatics*, vol. 20, pp. 1466-7, Jun 12 2004.

[3] C. Y. Park, *et al.*, "Rapid and accurate peptide identification from tandem mass spectra," *J Proteome Res*, vol. 7, pp. 3022-7, Jul 2008.

[4] D. N. Perkins, *et al.*, "Probability-based protein identification by searching sequence databases using mass spectrometry data," *Electrophoresis*, vol. 20, pp. 3551-3567, DEC 1999.

[5] J. Colinge, *et al.*, "OLAV: towards high-throughput tandem mass spectrometry data identification," *Proteomics*, vol. 3, pp. 1454-63, Aug 2003.

[6] S. Tanner, *et al.*, "InsPecT: identification of posttranslationally modified peptides from tandem mass spectra," *Anal Chem*, vol. 77, pp. 4626-39, Jul 15 2005.

[7] W. R. Cannon, *et al.*, "Comparison of probability and likelihood models for peptide identification from tandem mass spectrometry data," *J Proteome Res,* vol. 4, pp. 1687-98, Sep-Oct 2005.
[8] W. R. Cannon and M. M. Rawlins, "Physicochemical/Thermodynamic Framework for the Interpretation of Peptide Tandem Mass Spectra," *Journal of Physical Chemistry C,* vol. 114, pp. 5360-5366, Apr 1 2010.
[9] W. R. Cannon, *et al.*, "Large Improvements in MS/MS-Based Peptide Identification Rates using a Hybrid Analysis," *J Proteome Res,* Mar 30 2011.
[10] J. F. Banfield, *et al.*, "Proteogenomic approaches for the molecular characterization of natural microbial communities," *OMICS,* vol. 9, pp. 301-33, Winter 2005.
[11] S. M. Sowell, *et al.*, "Proteomic analysis of stationary phase in the marine bacterium "Candidatus Pelagibacter ubique"," *Appl Environ Microbiol,* vol. 74, pp. 4091-100, Jul 2008.
[12] K. E. Burnum, *et al.*, "Proteome insights into the symbiotic relationship between a captive colony of Nasutitermes corniger and its hindgut microbiome," *ISME J,* vol. 5, pp. 161-4, Jan 2011.
[13] L. Kall, *et al.*, "QVALITY: non-parametric estimation of q-values and posterior error probabilities," *Bioinformatics,* vol. 25, pp. 964-6, Apr 1 2009.
[14] D. Wu, *et al.*, "A phylogeny-driven genomic encyclopaedia of Bacteria and Archaea," *Nature,* vol. 462, pp. 1056-60, Dec 24 2009.

PHYLOSEQ: A BIOCONDUCTOR PACKAGE FOR HANDLING AND ANALYSIS OF HIGH-THROUGHPUT PHYLOGENETIC SEQUENCE DATA

PAUL J. McMURDIE AND SUSAN HOLMES*

Statistics Department, Stanford University,
Stanford, CA 94305, USA
**E-mail: susan@stat.stanford.edu*
www-stat.stanford.edu/~susan/

We present a detailed description of a new Bioconductor package, *phyloseq*, for integrated data and analysis of taxonomically-clustered phylogenetic sequencing data in conjunction with related data types. The *phyloseq* package integrates abundance data, phylogenetic information and covariates so that exploratory transformations, plots, and confirmatory testing and diagnostic plots can be carried out seamlessly. The package is built following the S4 object-oriented framework of the R language so that once the data have been input the user can easily transform, plot and analyze the data. We present some examples that highlight the methods and the ease with which we can leverage existing packages.

Keywords: pyrosequencing; QIIME; microbiome; microbial ecology; vegan; adephylo; phylobase; OTUbase.

1. Introduction

High-throughput (HT) DNA sequencing[1] is allowing major advances in microbial ecology studies,[2] where our understanding of the presence and abundance of microbial species relies heavily on the observation of their nucleic acids in a "culture independent" manner.[3] Typically this is achieved by PCR amplification of a small (100–500 bp) fragment of a conserved gene (phylogenetic marker) for which there are taxonomically-informative reference sequences available. The most commonly-used phylogenetic marker gene is the small subunit ribosomal RNA (16S rRNA) gene,[3] for which there are also convenient tools[4] and large reference databases.[5–7]

The first step in interpreting this phylogenetic sequencing data is usually to define a species-equivalent *operational taxonomic unit* (OTU). Species clustering methods are a separate area of research[8,9] beyond the scope of this article, but there are several packages/pipelines currently available to perform processing of raw HT phylogenetic sequencing data such that sequences are both clustered and the clusters are classified taxonomically. These include QIIME,[10] *mothur*,[11] the RDP pipeline[6] and PANGEA.[12] [a] In some cases these applications are also able to perform downstream analyses, but the options are limited. While there may be some anecdotal advantages to having a downstream ecological analysis coupled with the application that defined the OTUs, we expect that they are outweighed by the advantages of a separate set of flexible, open-source analyses that can be applied consistently across experiments, independent of the choice of OTU clustering method. There are already several ecology

[a]Throughout this article we use regular or *italics* font for packages/applications with names that are capitalized or uncapitalized, respectively. We further use a `courier` style font for R code, including function and class names.

and phylogenetic packages available in R, including the *adephylo*,[13] *vegan*,[14] *ade4*,[15] *picante*,[16] *ape*,[17] *phangorn*,[18] *phylobase*,[19] and *OTUbase*[20] packages, and these can already take advantage of many of the powerful statistical and graphics tools available in R.[21] However, at present a user must devise their own methods for parsing the output of their favorite OTU clustering application, and, as a consequence, there is also no standard within *Bioconductor*[22] (or R generally) for storing or sharing the suite of related data objects that describe a phylogenetic sequencing project.

To address these issues, we have created a new package for Bioconductor, called *phyloseq*, that provides a related set of *S4* classes[23] that internally manage the handling tasks associated with organizing, linking, and storing phylogenetic sequencing data. The user is able to store all their relevant data types in a single *phyloseq* object. This approach has several advantages. First, it is easier to return to a previously-analyzed dataset, because the data is organized by design. Secondly, this data structure and accompanied methods make it is easier to share/compare data from separate experiments, as well as apply a consistent set of analysis methods to multiple experiments. For development, new method extensions can be created that recognize exactly the data types that are present in a particular *phyloseq* class. To instantiate an object of the appropriate *phyloseq* class, the user calls the initialization method (`phyloseq(...)`) with available core data types as input. Alternatively, the *phyloseq* package contains data input methods for each of the four main OTU clustering applications described above, allowing the user to import their data and check its compatibility in one function call.

The *phyloseq* package also allows the modification and subsetting of *phyloseq* objects, such that the component data types remain compatible at every step (e.g. contain exactly the same samples and taxa) and component data is easily accessed. Furthermore, the *phyloseq* package contains convenient wrapper methods for executing common analysis pipelines and creating useful graphics. Also provided are parallelized methods (using *Rmpi*[24]) for calculating the UniFrac[25] and weighted-UniFrac[26] distances — common methods for calculating the relative dissimilarities of the microbial communities of different samples. Parallelization of these two methods is especially important because they are computationally intensive calculations for experiments with a large number of diverse samples. If alternatively a user wants to off-load the UniFrac calculation to the UniFrac server,[27] *phyloseq* provides an export method that creates the required *environment* and NEXUS files directly from the abundance table and phylogenetic tree, respectively. Finally, we expect that the *S4* data structure and core object handling methods in the *phyloseq* package will facilitate development of novel methods and classes for phylogenetic sequencing analysis.

A related Bioconductor package, *OTUbase*,[20] currently allows for importing output from the *mothur* pipeline and retaining metadata associated with the raw HT sequencing, for example the read labels and sequence quality of individual reads. The *phyloseq* package provides some useful importing methods to support users that have already imported *mothur* datasets using *OTUbase*.

2. Class Structure

The class structure in the *phyloseq* package follows the inheritance diagram shown in Fig. 1. The *phyloseq* package contains multiple inherited classes with incremental complexity so that methods can be extended to handle exactly the data types that are present in a particular object. Currently, *phyloseq* uses 4 core data classes. They are the taxonomic abundance table (`otuTable`), a table of sample data (`sampleMap`), a table of taxonomic descriptors (`taxonomyTable`), and a phylogenetic tree (`phylo`) which is directly borrowed from the *phylobase* and *ape* packages. The `otuTable` class can be considered the central data type, as it directly represents the number and type of sequences observed in each sample. `otuTable` extends the numeric matrix class in the R base, and has a few additonal feature slots. The most important of these feature slots is the `speciesAreRows` slot, which holds a single logical that indicates whether the table is oriented with taxa as rows (as in the *genefilter* package in Bioconductor[22]) or with taxa as columns (as in *vegan* and *picante* packages). In *phyloseq* methods, as well as its extensions of methods in other packages, the `speciesAreRows` value is checked to ensure proper orientation of the otuTable. A *phyloseq* user is only required to specify the `otuTable` orientation during initialization, following which all handling is internal.

The `sampleMap` class directly inherits R's `data.frame` class, and thus effectively stores both categorical and numerical data about each sample. The orientation of a `data.frame` in this context requires that samples/trials are rows, and variables are columns (consistent with *vegan* and other packages). The `taxonomyTable` class directly inherits the `matrix` class, and is oriented such that rows are taxa (e.g. species) and columns are taxonomic ranks (e.g. phylum).

We use the term "higher-order classes" for those that contain two or more of the previously-described core data classes. We assume that *phyloseq* users will be interested in analyses that utilize their abundance counts derived from the phylogenetic sequencing data, and so all higher-order classes contain an `otuTable` slot. There are a number of common methods that require either an `otuTable` and `sampleMap` combination, or an `otuTable` and phylogenetic tree (the `phylo` or `phylo4` class) combination. These methods can operate on instances of the `otuSam` or `otuTree` classes, respectively, or their subclasses. Because of the inheritance structure of *phyloseq* classes, a method often only needs to be defined for a single class to work properly for all other relevant classes as well.

3. Input and initialization

An important feature of the *phyloseq* package is methods for input of phylogenetic sequencing data from common taxonomic clustering pipelines. These methods take file pathnames as input, read and parse those files, and return an object of the appropriate class. Initialization of higher-order objects can be achieved manually from core data objects using the initialization method (`phyloseq(...)`). An instance of a higher-order class has its component objects trimmed during initialization, such that the taxa and sample dimensions contain only their intersecting set of sample and taxa indices, respectively. This ensures compatibility between component data objects at instantiation of the higher-order object, as well as at subsequent subsetting operations because the initialization methods are used by the subsetting methods internally.

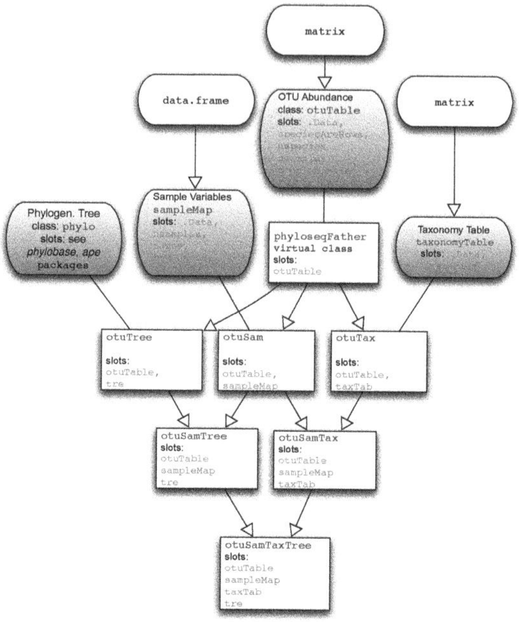

Fig. 1: Classes and inheritance in the *phyloseq* package. Core data classes are shown with grey fill and rounded corners. The class name and its slots are shown with red- or blue-shaded text, respectively. Inheritance is indicated graphically by arrows. Lines without arrows indicate that a higher-order object contains a slot with the associated class as one of its components.

The following line of code creates a `otuSamTaxTree` object from files created by the QIIME pipeline, a type of input method that is not previously available in R packages despite QIIME's popularity.

```
> otufilename = "../data/ex1_otutable.txt"
> mapfilename = "../data/ex1_samplemap.txt"
> trefilename = "../data/ex1_tree.tre"

> ex1 <- readQiime(otufilename, mapfilename, trefilename)
```

4. Graphical summary using default plot methods

The *phyloseq* package contains extensions for the default plot methods tailored to the data types present in each *phyloseq* class. For example, `otuSamTax` objects can be plotted as a series of stacked bars that represent the relative abundance of phylum-level groups in the combined community, as well as the relative abundance of individual taxa within each phylum (Fig. 2).

```
> taxaplot(ex1)
```

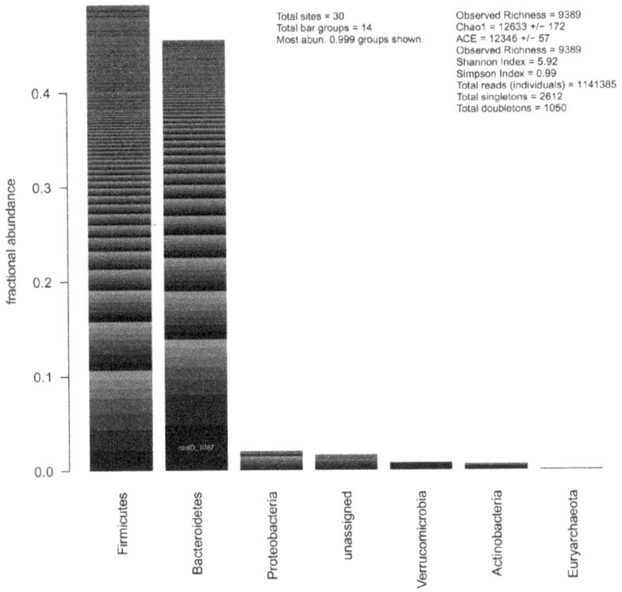

Fig. 2: Example of a default plot method for summarizing an object of class otuSamTax. Each *phyloseq* class has a specialized plot method for summarizing its data. In this case, relative abundance is shown quantitatively in a stacked barplot by phylum. Different taxa within a stack are differentiated by an alternating series of grayscale. The OTU identifier of taxa comprising a large enough fraction of the total community, 5% in this case, is labeled on the corresponding bar segment. Several diversity/richness indices are also shown.

5. Subsetting and filtering

In this particular example, we know that there were two separate Roche-454 sequencing trials included in the QIIME output, but the first sequencing trial ("Run 1") was of poor quality and should be removed from the dataset. We remove all of the Run 1 samples from `ex1` through a simple one-line combination of accessor and replacement methods for the `sampleMap` slot. On the right-hand side of the assignment operator (`<-`) we subset the `sampleMap`, and then assign it to the `sampleMap` slot of `ex1` on the left-hand side. This assignment replaces the original sampleMap in ex1, and also re-trims the other components of ex1 (e.g. the otuTable) such that only Run 2 samples remain.

```
> ex1 <- subset_samples(ex1, SeqRun == "R2")
```

An alternative approach would be to define a character vector with the sample names of Run 2, and then use the assignment operator: `sample.names(ex1)<-`. Similarly, we can use `species.names()<-` assignment to create a new `otuSamTaxTree` object in which all but the relatively abundant taxa have been trimmed, through a method borrowed from the *genefilter* package in *Bioconductor*.[22] In this case, we arbitrarily create a filter function that returns `TRUE` only for the most abundant 30 taxa from a sample, and then apply it to each sample in the `otuTable` of `ex1`, with the added parameter that a species must be `TRUE` in at least four samples (be among the most abundant 30 taxa in four or more samples). Like *genefilter*, the output of `genefilterSample()` is a logical vector with the same length as the number of taxa presently in `ex1`.

```
> fun1 <- filterfunSample(topk(30))
> wh1 <- genefilterSample(ex1, fun1, A = 4)
```

The taxa names are then accessed and subsetted based on the value in `wh1`, and this trimmed vector of taxa names (a total of 52 in this case) is assigned back to our *phyloseq* object, causing phyloseq's internal methods to trim the component objects in `ex1` to just these taxa. Using a similar approach, we also remove those samples that now contain unacceptably small (< 100, for instance) total numbers of individuals (total reads in this case).

```
> species.names(ex1) <- species.names(ex1)[wh1]
> sample.names(ex1) <- sample.names(ex1)[sampleSums(ex1) > 100]
```

6. UniFrac distances

UniFrac is a useful metric to summarize the difference between pairs of ecological communities, and can be used to create a distance matrix for samples in an experiment. An unweighted UniFrac distance matrix only considers the presence/absence of taxa, while weighted UniFrac accounts for the relative abundance of taxa as well as their phylogenetic distance. Presently, only one non-parallelized implementation of the unweighted UniFrac distance is available in R packages (`picante::unifrac`). In the *phyloseq* package we provide optionally-parallelized methods for calculating both UniFrac and weighted-UniFrac, as well a few key UniFrac variants, all of which return a sample-wise distance matrix from any *phyloseq* object that contains both a phylogenetic tree and an `otuTable` (`otuTree` and its subclasses).

Here is an example of the weighted UniFrac calculation using a dataset provided in the *picante* package. We first build an `otuTree` class object using the `phyloseq()` constructor, then use the `wUniFrac()` method to calculate and return the weighted UniFrac distance for all sample pair combinations.

```
> data(phylocom)
> tree <- phylocom$phylo
> OTU <- phylocom$sample
> ex3 <- phyloseq(otuTable(OTU, speciesAreRows = FALSE), tree)
> wUniFrac(ex3)
```

Extending our previous microbiome example, we now calculate the weighted-UniFrac distance matrix for `ex1`, then perform multi-dimensional scaling on the distance matrix and generate an annotated plot. This can be achieved with the wrapper function `wunifracMDS`, which also takes advantage of the enhanced plotting facilities provided by the *ggplot2* package[28] (Fig. 3).

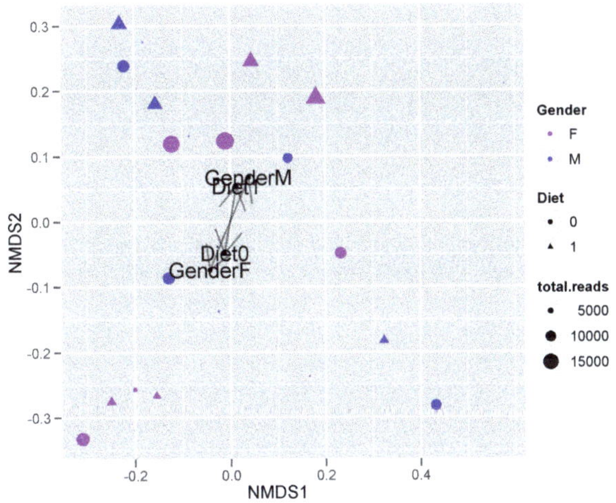

Fig. 3: NMDS ordination graphic generated by `wunifracMDS`. The NMDS coordinates are generated by `metaMDS()`, with the weighted-UniFrac distance matrix as argument, and 2-dimensions specified by default. A separate analysis was done using `adonis()`, which also did not find a compelling association between the weighted UniFrac distances and the gender ($p = 0.29$) or diet ($p = 0.9$) of subjects in the study.

7. Robust multiple-table methods

The NMDS of weighted-UniFrac distances did not reveal any clustering of gut microbiome samples according to a subject's diet type or gender. However, it may be that these differences are subtly confined to a subset of species that is not easily detectable with a community-wide summary statistic, like UniFrac. This effect is compounded by the observation that abundance counts as output from high-throughput phylogenetic sequencing have high variabilities,[29] leading to a need for robust methods[30] that can reveal subtle features of the data.

Many decompositions of inertia[31] — such as constrained correspondence analysis (`cca()`), analysis of diversity (`adonis()`), redundancy analysis (`rda()`) — use sums of squares. The simplest transformations take the ranked abundances and replace the values by their order statistic. Here we show an example in which we further threshold the ranks so that all the low-abundance taxa at noise-level (where presence/absence of a taxa is as much due to chance as biology) are given the same value.

We can apply this rank threshold transformation to our original data using a special enclosure form of the `threshrank()` function in *phyloseq*, and taking the threshold to be 500, for instance. We use a custom method for abundance table transformation in the *phyloseq* package (`transformsamplecounts`) that applies one or more transformation functions, in order, to each sample of the `otuTable` of the first argument. Although it operates only on the `otuTable` component, `transformsamplecounts` returns an object with the same class and components as its input.

```
> ex4 <- transformsamplecounts(ex4, threshrankfun(500))
```

In Fig. 4 we provide the results of a redundancy analysis on the ranked-thresholded abundance table, as well as a constrained correspondence analysis on the original abundance table. These plots were produced using the `calcplot()` convenience wrapper that — in addition to producing effective graphics with `ggplot2` — also provides the opportunity for further graphics manipulation using multiple layers and a mutable graphics description.[28] As anticipated, these two methods both appear to effectively cluster the samples by diet and gender, as specified. In addition, these ordination methods indicate which taxa best account for the clustering, which is highly valuable during exploratory analysis.

When analyzing a data abundance table with complementary covariates we use generalizations of PCA and Correspondence Analysis that attempt to account for the explanatory variables in the description of the overall variability. Redundancy Analysis (RDA) is a version of PCA with regards to instrumental variables[31] that tries to predict the multivariate ranked data by the covariates gender and diet. The left side of Fig. 4 shows the graphical output from such an analysis with a biplot representing both the explanatory factors, diet and gender, as well as the OTU distributions as dots. The OTUs actually have a structure formation along the axes of a grid; this is due to the ranking and thresholding of the data. We have magnified this in the left part of Fig. 5 to illustrate the graphical capabilities available through the combination of *phyloseq* with *ggplot2*.

On the right of Fig. 4 a different type of analysis was used directly on the abundance data, where we performed a constrained correspondence analysis (CCA)[32] and projected the output onto the first two constrained coordinates. We have also plotted the covariate groupings and the species. We can see in the magnification of the species plot on the right of Fig. 5 that the grid is replaced by the rhombus of extreme points. These are the OTUs with small abundances which are more strongly weighted in CCA, thus becoming the boundary of the projection region.

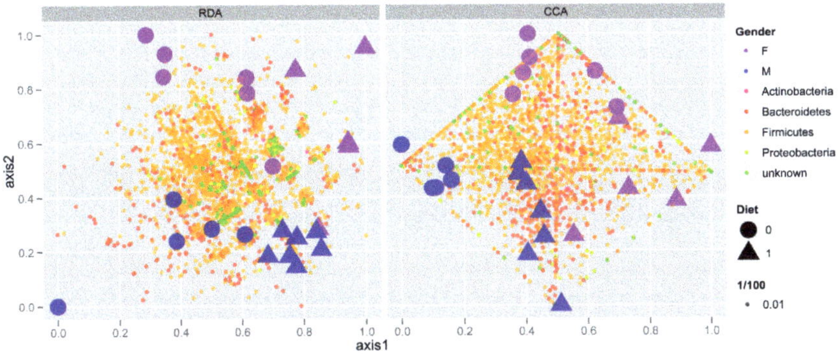

Fig. 4: Redundancy analysis and Constrained Correspondence Analysis. (Left) Redundancy analysis applied to a thresholded, ranked-transformed abundance table that had been trimmed such that only the phyla accounting for the top 99% of taxa are included. (Right) Original trimmed abundance table (no transformation nor threshold) subjected to Constrained Correspondence Analysis (CCA), constrained on a subject's diet and gender.

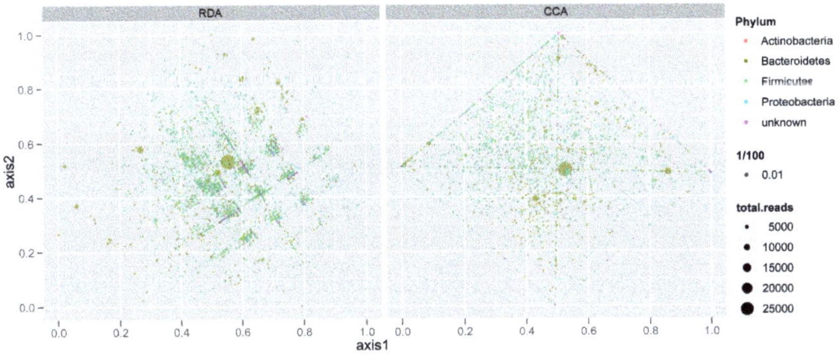

Fig. 5: Enlarged RDA and CCA plots emphasizing the taxa (species) coordinates. Graphics were produced with `calcplotrda()` or `calcplotcca()` convenience wrappers in *phyloseq*, which utilize analysis and graphics tools from the *vegan* and *ggplot2* packages, respectively. Only the phyla accounting for the top 99% of taxa are included.

8. Functional genes and non-standard phylogenetic markers

Environmental datasets also utilize novel markers or functional genes for which there is not a large curated database for comparison, nor clustering pipelines carefully tuned to define species-level taxonomic clusters. However, it is common and useful for researchers to calculate a phylogenetic tree from the sequence data, and the source sample from which the individual sequences were derived is generally known. For this situation *phyloseq* contains a method, `tipglom()`, that takes as input a phylogenetic tree and sequence-source hash table, and returns a higher-order object, including an `otuTable`. The `tipglom()` method uses a default greedy clustering based on a branch-length distance threshold — or an alternative user-provided tree-based method — to agglomerate closely related sequences as one taxa. Each merging step is carried out by a user-accessible method called `mergespecies()`, which ensures consistency between tree and species abundance tables during each merge. In Fig. 6 we provide an example of an unprocessed tree calculated from more than 400 sequences of the same functional gene, derived from multiple environmental samples. We also show the same tree after processing by the `tipglom` function (species branch-length cutoff of 0.1), and plotted using a *phyloseq* method for displaying `otuTree` objects. In addition to consolidating similar sequences for legibility, this new tree reveals information about particular taxa that appear in multiple samples, as well as portions of the tree that originate from only one sample type.

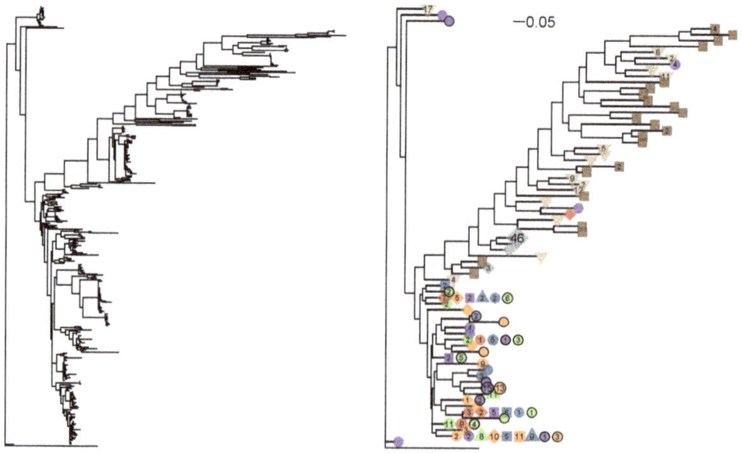

Fig. 6: Example of phylogenetic sequence data before and after basic clustering with `tipglom()` function. (Left) Standard phylogram produced using default plotting function and no OTU clustering. (Right) Annotated phylogram after OTU clustering with `tipglom()`. Different symbols next to each tip indicate different samples in which the OTU was observed. The number inside each symbol indicates the respective number of individuals of a given OTU were observed in each sample.

9. Conclusions

We describe a new Bioconductor package, *phyloseq*, for handling, filtering, and analysing high-throughput phylogenetic sequencing data after it has been processed in a sequence clustering pipeline. The *phyloseq* package provides extensions for leveraging analysis from other ecology-related packages, such as *adephylo*, *vegan*, *picante*, as well as other packages that we have found useful for data of this type. *phyloseq* also provides useful wrappers for key analysis pipelines that should help to streamline statistical analysis of data from phylogenetic sequencing projects. We hope that this package provides a useful class structure, methods, and methods extensions that will streamline the input of phylogenetic sequencing data, its QC/QA processing and trimming, and especially its analysis and graphical representation by this and other Bioconductor/R packages.

10. Acknowledgements

We would like to thank the developers of all the open source packages we used, in particular Rob Knight and his lab for QIIME, Hadley Wickham for *ggplot2* and the Bioconductor and R teams. Julia Fukuyama, Sam Pimentel, Kris Sarkan and Dustin Janatpour provided feedback on *phyloseq*. David Relman, Les Dethlefsen, and other members of the Relman lab, as well as Samantha Ying provided relevant data examples and data analysis questions. Alfred Spormann, Tyrrell Nelson and Tim Meyer also provided early versions of an illustrative data set. This work was also supported by grant NIH-R01GM086884.

References

1. M. L. Metzker, *Nature Reviews Genetics* **11**, 31 (2010).
2. M. Hamady, J. J. Walker, J. K. Harris, N. J. Gold and R. Knight, *Nature Methods* **5**, 235 (2008).
3. N. R. Pace, *Science* **276**, 734 (1997).
4. T. Z. DeSantis, P. Hugenholtz, K. Keller, E. L. Brodie, N. Larsen, Y. M. Piceno, R. Phan and G. L. Andersen, *Nucleic Acids Research* **34**, W394 (2006).
5. T. Z. DeSantis, P. Hugenholtz, N. Larsen, M. Rojas, E. L. Brodie, K. Keller, T. Huber, D. Dalevi, P. Hu and G. L. Andersen, *Applied and Environmental Microbiology* **72**, 5069 (2006).
6. J. R. Cole, Q. Wang, E. Cardenas, J. Fish, B. Chai, R. J. Farris, A. S. Kulam-Syed-Mohideen, D. M. McGarrell, T. Marsh, G. M. Garrity and J. M. Tiedje, *Nucleic Acids Research* **37**, D141 (2009).
7. E. Pruesse, C. Quast, K. Knittel, B. M. Fuchs, W. Ludwig, J. Peplies and F. O. Glöckner, *Nucleic Acids Research* **35**, 7188 (2007).
8. W. Li and A. Godzik, *Bioinformatics* **22**, 1658 (2006).
9. Y. Huang, B. Niu, Y. Gao, L. Fu and W. Li, *Bioinformatics* **26**, 680 (2010).
10. J. G. Caporaso, J. Kuczynski, J. Stombaugh, K. Bittinger, F. D. Bushman, E. K. Costello, N. Fierer, A. G. Peña, J. K. Goodrich, J. I. Gordon, G. A. Huttley, S. T. Kelley, D. Knights, J. E. Koenig, R. E. Ley, C. A. Lozupone, D. McDonald, B. D. Muegge, M. Pirrung, J. Reeder, J. R. Sevinsky, P. J. Turnbaugh, W. A. Walters, J. Widmann, T. Yatsunenko, J. Zaneveld and R. Knight, *Nature Methods* **7**, 335 (2010).
11. P. D. Schloss, S. L. Westcott, T. Ryabin, J. R. Hall, M. Hartmann, E. B. Hollister, R. A. Lesniewski, B. B. Oakley, D. H. Parks, C. J. Robinson, J. W. Sahl, B. Stres, G. G. Thallinger, D. J. Van Horn and C. F. Weber, *Applied and Environmental Microbiology* **75**, 7537 (2009).

12. A. Giongo, D. B. Crabb, A. G. Davis-Richardson, D. Chauliac, J. M. Mobberley, K. A. Gano, N. Mukherjee, G. Casella, L. F. W. Roesch, B. Walts, A. Riva, G. King and E. W. Triplett, *The ISME Journal* **4**, 852 (2010).
13. T. Jombart, F. Balloux and S. Dray, *Bioinformatics* **26**, 1907 (2010).
14. J. Oksanen, F. G. Blanchet, R. Kindt, P. Legendre, R. B. O'Hara, G. L. Simpson, P. Solymos, M. H. H. Stevens and H. Wagner, *vegan: Community Ecology Package*, (2011). R package version 1.17-10.
15. S. Dray, *Journal of Statistical Software* (2007).
16. S. W. Kembel, P. D. Cowan, M. R. Helmus, W. K. Cornwell, H. Morlon, D. D. Ackerly, S. P. Blomberg and C. O. Webb, *Bioinformatics* **26**, 1463 (2010).
17. E. Paradis, J. Claude and K. Strimmer, *Bioinformatics* **20**, 289 (2004).
18. K. P. Schliep, *Bioinformatics* **27**, 592 (2011).
19. `R Hackathon team`. (alphabetically: Ben Bolker, M. Butler, P. Cowan, D. de Vienne, D. Eddelbuettel, M. Holder, T. Jombart, S. Kembel, F. Michonneau, D. Orme, B. O'Meara, E. Paradis, J. Regetz and D. Zwickl), *phylobase: Base package for phylogenetic structures and comparative data*, (2011). R package version 0.6.3.
20. D. Beck, M. Settles and J. A. Foster, *Bioinformatics* (2011).
21. R Development Core Team, *R: A Language and Environment for Statistical Computing*. R Foundation for Statistical Computing, Vienna, Austria, (2011). ISBN 3-900051-07-0.
22. R. C. Gentleman, V. J. Carey, D. M. Bates and others, *Genome Biology* **5**, p. R80 (2004).
23. J. M. Chambers, *Software for data analysis* (Springer Verlag, 2008).
24. H. Yu, *Rmpi: Interface (Wrapper) to MPI (Message-Passing Interface)*, (2010). R package version 0.5-9.
25. C. Lozupone and R. Knight, *Applied and Environmental Microbiology* **71**, 8228 (2005).
26. C. A. Lozupone, M. Hamady, S. T. Kelley and R. Knight, *Applied and Environmental Microbiology* **73**, 1576 (2007).
27. C. Lozupone, M. Hamady and R. Knight, *BMC Bioinformatics* **7** (2006).
28. H. Wickham, *ggplot2: elegant graphics for data analysis* (Springer New York, 2009).
29. J. Zhou, L. Wu, Y. Deng, X. Zhi, Y.-H. Jiang, Q. Tu, J. Xie, J. D. van Nostrand, Z. He and Y. Yang, *The ISME Journal* (2011).
30. S. Holmes, A. Alekseyenko, A. Timme, T. Nelson, P. J. Pasricha and A. Spormann, *Pacific Symposium on Biocomputing* , 142 (2011).
31. S. Holmes, *Multivariate Analysis: The French Way* (Institute of Mathematical Statistics, Beachwood, Ohio, USA, 2008).
32. C. J. F. Ter Braak, *Plant Ecology* **69**, 69 (1987).

SEPP: SATé-Enabled Phylogenetic Placement

S. MIRARAB, N. NGUYEN, AND T. WARNOW*

Department of Computer Science
University of Texas at Austin
Austin, TX 78712, USA
**tandy@cs.utexas.edu*
www.cs.utexas.edu.edu

We address the problem of Phylogenetic Placement, in which the objective is to insert short molecular sequences (called query sequences) into an existing phylogenetic tree and alignment on full-length sequences for the same gene. Phylogenetic placement has the potential to provide information beyond pure "species identification" (i.e., the association of metagenomic reads to existing species), because it can also give information about the evolutionary relationships between these query sequences and to known species. Approaches for phylogenetic placement have been developed that operate in two steps: first, an alignment is estimated for each query sequence to the alignment of the full-length sequences, and then that alignment is used to find the optimal location in the phylogenetic tree for the query sequence. Recent methods of this type include HMMALIGN+EPA, HMMALIGN+pplacer, and PaPaRa+EPA. We report on a study evaluating phylogenetic placement methods on biological and simulated data. This study shows that these methods have extremely good accuracy and computational tractability under conditions where the input contains a highly accurate alignment and tree for the full-length sequences, and the set of full-length sequences is sufficiently small and not too evolutionarily diverse; however, we also show that under other conditions accuracy declines and the computational requirements for memory and time exceed acceptable limits. We present SEPP, a general "boosting" technique to improve the accuracy and/or speed of phylogenetic placement techniques. The key algorithmic aspect of this booster is a dataset decomposition technique in SATé, a method that utilizes an iterative divide-and-conquer technique to co-estimate alignments and trees on large molecular sequence datasets. We show that SATé-boosting improves HMMALIGN+pplacer, placing short sequences more accurately when the set of input sequences has a large evolutionary diameter and produces placements of comparable accuracy in a fraction of the time for easier cases. SEPP software and the datasets used in this study are all available for free at `http://www.cs.utexas.edu/users/phylo/software/sepp/submission`.

Keywords: Phylogenetic placement, metagenomic analysis

1. Introduction

Metagenomic datasets contain thousands to millions of short sequences, many from different species and for different genes. Determining the species present within a metagenomic dataset and their relative abundances are two of the main objectives in metagenomic analysis. However, these problems do not address other issues, such as the discovery of new species and the inference of evolutionary relationships between the species in the sample. Under the assumption that all short reads are from the same gene and that a tree and alignment for a large number of full-length sequences for that gene are available, each short read can be placed into a phylogenetic tree, thereby enabling species identification for these reads, the inference of evolutionary relationships between the reads, and potentially also the identification of reads coming from unknown species. This is called the "phylogenetic placement" problem, formally stated as follows:

Phylogenetic Placement Problem.

- Input: the *backbone* tree T and alignment A on set S of full-length sequences, and query sequence s.

- Output: tree T' containing s obtained by adding s as a leaf to T.

Several methods have been developed for this problem using the following two steps:

- Step 1: insert s into alignment A to produce the *extended alignment* A'
- Step 2: add s into T using A', optimizing some criterion

Methods for the first step include HMMALIGN[1] and the recently introduced PaPaRa[2] method. Methods for the second step include EPA[3] and pplacer,[4] which seek to optimize maximum likelihood (pplacer also provides a Bayesian approach). Methods for phylogenetic placement can therefore be described by how they handle each step. Three such methods include PaPaRa+EPA,[2] HMMALIGN+EPA,[3] and HMMALIGN+pplacer.[4] EPA and pplacer are comparably fast and have almost identical placement accuracy, but have somewhat different memory usage and algorithmic features;[4] hence the differences between HMMALIGN+EPA and HMMALIGN+pplacer do not impact the placement accuracy, and have a minor impact on running time and memory usage. The two techniques for computing the extended alignment, PaPaRa and HMMALIGN, are very different. HMMALIGN computes a HMM to represent the full-length alignment, and then aligns each query sequence to that HMM. In contrast, PaPaRa uses RAxML to estimate ancestral state vectors for each branch in the tree, aligns the query sequence to every ancestral state vector, selects the alignment that had the best score and uses it to extend alignment A to include s. Consequently, PaPaRa is computationally more expensive than HMMALIGN,[2] but EPA placements of query sequences based upon PaPaRa extended alignments can be more accurate than EPA placements based upon HMMALIGN extended alignments.

In this study, we introduce *SEPP*, a new basic algorithmic strategy for boosting the performance of methods for Phylogenetic Placement. We use a dataset decomposition within SATé-2[5] to decompose the set of full-length sequences into disjoint sets based upon the input tree and alignment, so that each set contains a small number of closely related sequences. We then compute the extended alignment for each query sequence and the placement of the query sequences into the input tree by running HMMALIGN and pplacer on the subsets instead of on the full tree. Depending upon the decomposition sizes, this approach can result in much faster and more accurate placements, as we will show. The parameters of the divide-and-conquer technique thus allow the user to trade-off running time and accuracy.

We report on a study comparing SEPP to HMMALIGN+pplacer and PaPaRa+pplacer on a collection of simulated datasets (with 500-taxon backbone trees and alignments) and one large biological dataset with a curated alignment of its 13,822 full-length sequences. Our study shows that SEPP is more accurate than PaPaRa+pplacer and HMMALIGN+pplacer when the full-length sequence dataset are evolutionary distant, and yields placements of comparable accuracy but much more efficiently when the sequences are more closely related.

We finish by defining some terms we will use throughout this paper.

- Reference alignment: We use the true alignment for the simulated datasets and the curated alignment for the biological dataset.
- Reference tree: For the simulated datasets, we use the true (model) tree with all zero-event branches collapsed. For the biological dataset, we use the RAxML[6] bootstrap tree computed

on the curated alignment, with all branches having less than 75% bootstrap support contracted. For a given set of full-length sequences and one query sequence, we restrict the reference tree on the full set of sequences to the subset, in order to define the reference topology for that subset.

- Query sequence: the sequences that are not part of the reference alignment, and which will be inserted into the reference tree.
- Backbone tree and alignment: this is the tree and alignment on the set S of full-length sequences, provided as input (along with the query sequences) to the phylogenetic placement problem.
- Extended alignment: The alignment on $S \cup \{s\}$ produced by inserting s into the backbone alignment on S.

Note: The terms "reference alignment" and "reference tree" are used differently in other papers,[2-4] where these refer to the input alignment and the ML tree estimated on that alignment. Because our study evaluates accuracy with respect to the true tree (known for the simulated data) or to the curated tree (for the biological dataset), we reserve the term "reference tree" for these objects, instead of for trees estimated on estimated alignments.

2. SEPP: SATé-Enabled Phylogenetic Placement

SEPP is a meta-method that works with existing methods for the two steps of phylogenetic placement (computing the extended alignment, and placing query sequences into a backbone tree). SEPP utilizes the dataset decomposition technique in SATé-2, a method that co-estimates sequence alignments and phylogenies. This technique takes as input a tree T and a target size K, and it partitions the leaf set of T into smaller subsets, as follows. SATé-2 removes a centroid edge (an edge that splits the taxon set into two approximately equally sized subsets) from the input tree, thus dividing the tree into two subtrees, and repeats the process until each subset has at most K leaves (taxa). Thus, the taxa within any single subset are close together within the tree and densely sampled within that subset.

The input to SEPP consists of

- the backbone tree T and alignment A for the full-length sequences and a set of query sequences,
- positive integers a and p, with $p \geq a$,
- a technique for aligning the query sequence to a multiple sequence alignment of full-length sequences, and
- a technique for inserting the query sequence into a backbone tree, given the extended alignment that includes the query sequence.

The output of SEPP is the placement of each query sequence into the backbone tree. In this study we explore SEPP where we use HMMALIGN to produce extended alignments and pplacer to insert query sequences into backbone trees.

We now show how SEPP uses the parameters a and p to compute the extended alignment and placement of a set of query sequences into the tree. For the sake of simplicity of exposition, we describe this for a single query sequence.

- We use the SATé-2 dataset decomposition strategy to divide the set of taxa in the tree T into disjoint subsets of size at most p. These subsets are called the "taxon-insertion-subsets."
- We further divide each taxon-insertion subset into smaller subsets of size a. These subsets are the "alignment–subsets". Thus, each alignment-subset is a subset of exactly one taxon-insertion-subset.
- We compute the HMM profile using HMMER for each of the alignment-subsets, and we find the alignment-subset that the query sequence s has the best match to. We use HMMALIGN to produce an alignment of s to the backbone alignment on the alignment-subset, and then use that alignment to produce the extended alignment for $S \cup \{s\}$.
- We find the taxon-insertion-subset that contains the selected alignment-subset, and we use pplacer to locate the query sequence s into the subtree of the backbone tree induced by the taxon-insertion-subset. Finally, we use the location of s in the subtree to insert s into the backbone tree on the entire set of taxa.

Thus, the two parameters a and p control the behavior of SEPP. We let a range from 10 to 250 for the simulated datasets and from 10 to 2500 for the biological dataset. We let p range from 10 to 500 for the simulated datasets and from 100 to 13,822 for the biological dataset.

3. Study Design

We evaluate phylogenetic placement methods on both empirical and simulated datasets. We include HMMALIGN+pplacer and PaPaRa+pplacer as representatives of currently available methods for phylogenetic placement. We also include SEPP used with HMMALIGN to produce extended alignments and pplacer to place the query sequences.

We studied performance of these phylogenetic placement methods on 61 sequence datasets (available at `http://www.cs.utexas.edu/users/phylo/software/sepp/submission`). We included 20 simulated 1000-taxon datasets that have evolved with substitutions and indels from each of three different model conditions (M2, M3, and M4), each with the "medium" gap length distribution (see Liu et al.[7] for these data). The three model conditions are chosen such that one dataset is hard, one is moderate, and one is easy. Because these are simulated datasets, the true alignment and true tree are known for each datasets.

We also used a large bacterial dataset, 16S.B.ALL, with 27,643 16S rRNA sequences, originally taken from the Gutell Comparative Ribosomonal Website (CRW),[8] and also studied by Liu et al.[5] This dataset has a curated alignment based upon confirmed secondary (and higher-order) structures, which are highly reliable. We use a ML bootstrap tree as the curated tree for this dataset, retaining only those branches with bootstrap support above 75%.[5] Thus, the 16S.B.ALL dataset has a curated tree and alignment as well.

Each dataset was randomly divided into two subsets of equal size, with one subset (S) used to define the backbone alignment and tree, and the other subset (R) used to produce the query sequences. These query sequences are created by taking substrings of normally-distributed lengths (from two distributions, described below), and with the start positions chosen uniformly at random.

Two categories of reads are generated for each sequence in the M2, M3, and M4 datasets: "long" reads, with a mean length of 250 and a standard deviation of 60, and "short" reads, with a mean length of 100 and a standard deviation of 20. A total of 10 fragmentary sequences are generated for each sequence, with half long and half short. Since these datasets each include 500 reference and 500 non-reference sequences, this process yields 2500 short and 2500 long reads per dataset. In summary, each M2, M3, and M4 dataset has a reference tree and alignment with 500 taxa and a total of 5000 fragmentary sequences, of which half are "short" and half are "long".

For the 16S.B.ALL biological dataset, we create two categories of reads, with length distributions identical to those of simulated datasets. This dataset contains 27,643 taxa, of which we use 13,822 sequences for the backbone tree, leaving us with 13,821 sequences for creating fragmentary reads. For each of these 13,821 sequences, we generated one fragmentary sequence, randomly choosing between the long and short distributions. Thus, for this dataset the backbone tree and alignment has 13,822 taxa, and there are 13,821 fragmentary sequences.

The sequences in S are used to create two backbone alignments and trees, as follows. For sets S that are produced by simulating sequence evolution, we have the true alignment and the true tree. We restrict each of these (which have 1000 taxa) to the subset of 500 full-length sequences, and then run RAxML on the resultant tree/alignment pair in order to optimize the branch lengths and GTR+Gamma parameters. This produces the first alignment/tree backbone. The second backbone alignment/tree pair is produced by running SATé on the set of full-length sequences.

For the 16S.B.ALL dataset, we run RAxML on the curated alignment to produce a binary tree. We then restrict the tree to the subset of 13,822 sequences, and optimize the branch lengths and GTR+Gamma parameters on the tree using RAxML. This produces the first backbone alignment/tree pair. We use SATé on the subset of 13,822 full-length sequences to produce the second.

We used SATé to produce these estimated alignment/tree pairs because SATé produces more accurate alignments and trees than any two-phase method (where an alignment is first estimated and then a tree computed on that alignment) for these datasets.[5] We used SATé-2, the new algorithm design for SATé; this produces an alignment and an ML tree on the alignment estimated using RAxML. For the 16S.B.ALL dataset, we used FastTree[9] within SATé-2 in each iteration, and finished with RAxML in order to produce optimized GTR+Gamma parameters on the final tree.

We classify each query sequence for its likely difficulty in phylogenetic placement as follows. We use HMMER to produce a HMM profile for the reference alignment, and then to classify the query sequences with respect to the HMM profile. The fragmentary reads are classified as easy to align ("easy") if the obtained E-value is less than 10^{-5}, and as "hard" otherwise. Among the hard reads, there are some reads for which HMMER does not report any E-value due to default filtering settings of HMMER. We classify such reads as "very hard" reads. In earlier phylogenetic placement studies, the hard fragments are excluded;[4] however, our study does not automatically eliminate hard fragments. Many very hard reads are able to be placed by SEPP, because the reads will receive E-values with respect to the smaller alignment subsets. Those that fail to be placed at all by SEPP are removed from the experimental study; this process removes 9 from all the simulated datasets together and 5 from the biological dataset. Thus, results for each simulated model condition are for 99997-100000 retained query sequences (20 replicas, each with 5000 query sequences) and 13,819-13,821 retained query sequences for the two 16S.B.ALL datasets.

Table 1 shows various statistics for the true or curated alignment of the datasets included in our study. The p-distance is the fraction of sites within an alignment in which two sequences are different and "% gaps" is the percentage of gaps within the alignment. The empirical statistics show that the

Dataset	Type	Size backbone	Num generated query seqs	Avg p-dist	Max p-dist	% gap
M2	sim	500	5000	0.68	0.76	67
M3	sim	500	5000	0.66	0.74	53
M4	sim	500	5000	0.50	0.60	51
16S.B.ALL	emp	13,822	13,821	0.21	0.52	74

Table 1: Dataset statistics: We present statistics for the true alignments for the simulated datasets (M2, M3, M4) and statistics for the curated alignment on the biological dataset, 16S.B.ALL. However, a small number of query sequences is deleted from some of the runs.

datasets vary substantially in terms of evolutionary distances, with datasets from model M2 having the largest evolutionary distances and 16S.B.ALL having the smallest.

Measurements. We measure placement accuracy (averaged over all the query sequences), running time, and peak memory usage, for each method on each dataset. For the simulated datasets we report averages for these measurements over the 20 replicates in each model condition.

Placement accuracy: All our phylogenetic placement methods use pplacer to position each query sequence into the backbone tree, and we use the most likely placement computed by pplacer (multiple possible placements for each read, along with the likelihood of each placement, are provided by pplacer.) We then compare the tree that is created to the reference tree, and compute the number of missing branches. This number in isolation is hard to interpret, for at least two reasons. In the case where SATé alignment/tree is the input, the backbone tree itself contains error. The error of the initial backbone tree is a lower bound on the tree error after placement of reads (in fact it is a rather liberal lower bound, as the optimal placement of fragments can still have errors higher than the initial tree). In the case where true or curated alignment/tree is the input, the initial tree has no error, but we can still establish useful lower bounds of the tree error. This can be done by using the reference alignment of query sequences to be the backbone alignment as input to the pplacer. The resulting placement of query sequences is the best we can realistically hope for.

To account for the lower bounds described above, we also define and report the "delta error" for each technique, as follows. For each read s placed on the SATé backbone tree, we report the difference between the number of missing branches of the initial backbone tree and the number of missing branches after placement of s. When the backbone tree is the reference (true or curated) tree, we report the difference between the number of missing branches of the tree produced by placement of s according to the reference alignment of s to S and the number of missing branches of the tree after placement of s. In all cases, the number of missing branches in each tree is defined with respect to the reference tree for the taxa in the given tree. Thus, the number of missing branches in the backbone trees is defined by the reference tree on the set S of backbone taxa, and the number of missing branches in the tree produced by placing the query sequence into the backbone tree is defined by the reference tree on the set $S \cup \{s\}$.

Note that in the case where the backbone tree is the reference tree, the number of missing branches is equal to the node distance between the correct placement of reads and actual placements, the error used in the literature.[2–4] However, this edge distance is not as meaningful as the number of missing branches with respect to either the true or curated tree, since estimated trees will generally have error.

Computational aspects: We report the running times and peak memory usage, each measured separately for the computation of extended alignments and placement of query sequences. These reported values are for placement of all query sequences in each set. Due to memory requirements of PaPaRa and pplacer, 16S.B.ALL experiments are run on a Linux machine with 16 cores and 256GB of main memory. The results for simulated datasets are obtained on a heterogeneous condor cluster.

4. Results

Algorithm design experiments.

Figure 1 shows results where we vary the two algorithmic parameters a and p, with peak memory usage indicated by the size of the circle. Note that decreasing a to 50 (and sometimes to 10) and increasing p tends to improve the placement accuracy, but at a running time cost. Also, bigger improvements in accuracy are obtained by decreasing a than by increasing p. However, for most conditions, there is a wide range of parameter settings in which the differences in placement error are quite small (often less than half an edge), and within this collection there can be substantial differences in running time.

To set the default parameters, we sought a setting that worked reasonably well with respect to both running time and placement accuracy. Setting $a = p = 1000$ for the 16S.B.ALL datasets and $a = p = 50$ for the simulated datasets produced good results. These settings correspond to setting the subset sizes to about 10% of the number of taxa in the backbone tree. Note, however, that setting $a = p = 50$ is by no means optimal for the M2 model condition (four other settings, have less error and complete faster). Similarly, setting $a = p = 1000$ is the fastest for the 16S.B.ALL datasets, but more accurate results can be obtained with other settings (with a below 1000) for a running time cost.

Comparisons using the default setting for SEPP.

We present results for PaPaRa+pplacer, HMMALIGN+pplacer, and the default setting for SEPP where we set $a = p$ to approximately 10% of the number of taxa in the backbone tree. This yields parameters 50/50 for the simulated datasets (backbone trees have 500 taxa) and 1000/1000 for the 16S.B.ALL dataset (backbone trees have 13,822 taxa).

Results on simulated datasets. The simulated datasets have backbone trees with 500 sequences and fairly high rates of evolution, with M2 having the highest rate and M4 having the lowest rate (Table 1). Placement error rates were impacted by the model, so that the missing branch rate for all methods is higher on model M2 than on model M3, and higher on model M3 than on model M4 (Table 2). Not surprisingly, absolute error rates are lower with the true alignment and tree than with the SATé alignment and tree. These trends also held for PaPaRa and SEPP.

Figure 2 and Table 2 show results for PaPaRa+pplacer, HMMALIGN+pplacer, and SEPP(50,50) (i.e., SEPP ran with the default setting on this model of $a = p = 50$). Note that SEPP(50,50) has the lowest delta-error of the three methods by far, followed by HMMALIGN+pplacer, and then by PaPaRa+pplacer. Furthermore, the differences are substantial. The methods are clearly also distinguished by running time and peak memory usage. HMMALIGN+pplacer is the fastest, SEPP(50,50) is somewhat slower, and PaPaRa uses much more time. Both PaPaRa+pplacer and

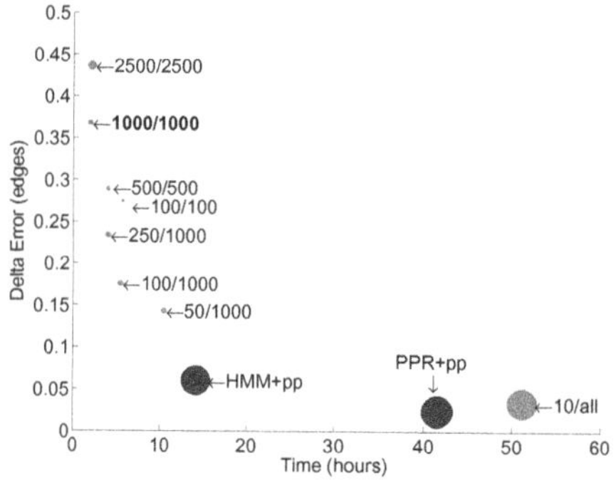

(a) 16S.B.ALL, Curated backbone

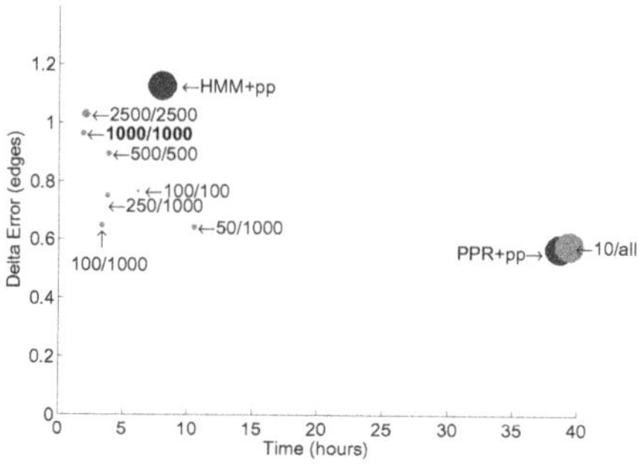

(b) 16S.B.ALL, SATé backbone

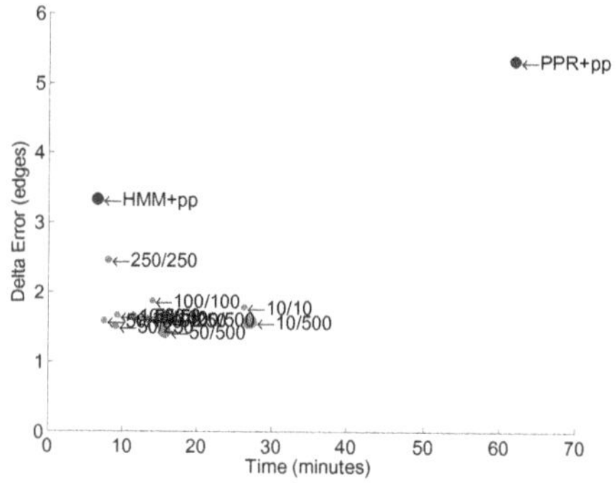

(c) M2, SATé backbone

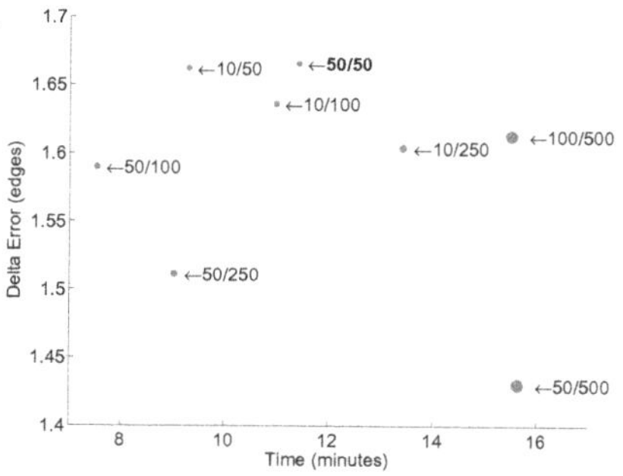

(d) M2, SATé backbone - most accurate settings

Fig. 1: Scatter plot of delta error (x) versus time (y) versus memory (circle diameters). The symbol "x/y" refers to SEPP(x,y). The default setting is 50/50 for M2 and 1000/1000 for 16S.B.ALL; these points are bold-faced. Note that the default setting for SEPP is far from optimal, with other settings providing better accuracy (and in some cases also better speed).

HMMALIGN+pplacer use more memory than our method.

Results for M3 (see Table 2) are quite similar to M2, with HMMALIGN+pplacer was much more accurate than PaPaRa+pplacer and SEPP(50/50) produced more accurate placements than HMMALIGN+pplacer. However, the gap between SEPP(50,50) and HMMALIGN+pplacer was reduced to only half an edge. On M4 (see Table 2), however, the relative performance between SEPP(50,50) and HMMALIGN+pplacer depended on the backbone tree. For the SATé alignment/tree, SEPP(50,50) was more accurate but slightly slower than HMMALIGN+pplacer. For the true alignment/tree, HMMALIGN+pplacer was somewhat more accurate and took less time. However, the difference in placement accuracy between SEPP(50,50) and HMMALIGN+pplacer was extremely small - less

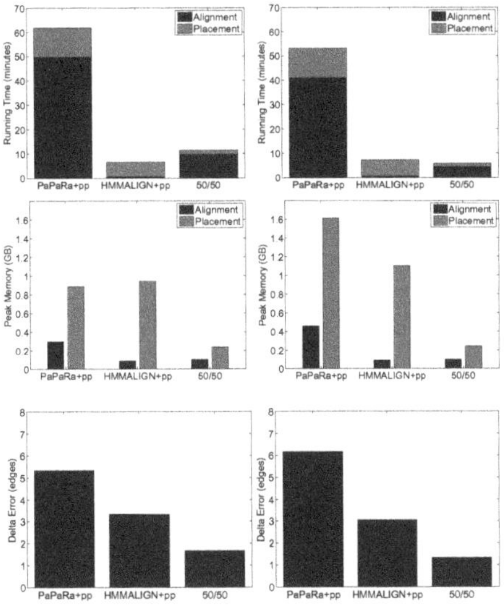

(a) SATé backbone (b) True backbone

Fig. 2: Results on simulated datasets for model M2. We show running time (top), peak memory usage (middle), and average number of additional missing branches per query sequence (bottom). Results for the SATé backbone alignment and tree are on the left, and results for the true backbone alignment and tree are on the right. The SATé backbone tree has 12.1% missing branch rate and the backbone tree based upon the true alignment has 0.09% missing branch rate. The number of additional missing branches shown (bottom) is the increment above that amount.

than one-ninth of an edge for both backbones.

16S.B.ALL. The datasets based upon 16S.B.ALL presented a different kind of challenge. Each dataset had 13,821 query sequences and a backbone tree with 13,822 sequences. Thus, backbone trees were much larger, but the backbone trees and alignments reflected lower rates of evolution.

Figure 3 shows results for both backbones, in which we ran SEPP(1000,1000) (i.e., the default setting). Note that PaPaRa+pplacer provides a small improvement in placement accuracy (slightly more than half an edge) in comparison to the other methods. However, PaPaRa+pplacer is enormously computationally intensive, using 40 hours to analyze these data, much longer than either other method. Also, HMMALIGN+pplacer and PaPaRa+pplacer have very large peak memory usage, near 60GB on both backbone trees. Thus, PaPaRa+pplacer is computationally extremely intensive, but both HMMALIGN+pplacer and PaPaRa+pplacer have very large peak memory usage.

A comparison of SEPP(1000,1000) to HMMALIGN+pplacer shows that both have extremely good placement accuracy, with delta-error approximately one edge for both methods on the SATé backbone tree and well under half an edge on the curated backbone tree. HMMALIGN+pplacer pro-

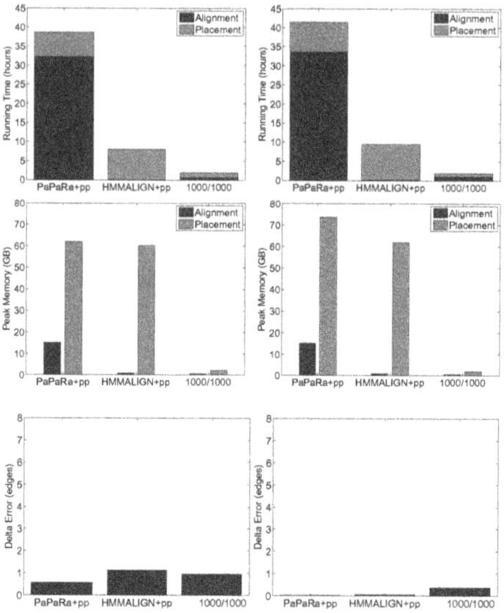

(a) SATé backbone (b) Curated backbone

Fig. 3: Results on 16S.B.ALL. We show running time (top), peak memory usage (middle), and average number of additional missing branches per query sequence (bottom). Results for the SATé backbone alignment and tree are on the left, and results for the curated backbone alignment and tree are on the right. The SATé missing branch rate is 7.64%, and the missing branch rate for the backbone tree defined by the true alignment is 1.835%. The number of additional missing branches shown (bottom) is the increment above that amount.

duces more accurate placements than our method for the curated backbone and SEPP(1000,1000) produces more accurate placements for the SATé backbone, but the differences between the two methods are small in both cases (less than a third of an edge). The methods are, however, distinguished by their computational requirements, as HMMALIGN+pplacer is much slower (at least 4 times as much time) and uses dramatically more memory (60GB as compared to about 2GB).

Comparing methods on query sequences of different levels of difficulty.

Table 2 compares methods in terms of their placement accuracy as a function of the level of difficulty in placing the query sequence, as predicted by HMMER (see the discussion in Section 3). Note that error increases as the reads become more difficult, as HMMER predicts. We see that SEPP run in default mode performs very well in general (as observed earlier) in comparison to HMMALIGN+pplacer and PaPaRa+pplacer, but has a particularly strong advantage on the hard and very hard reads. Interestingly, PaPaRa+pplacer does well on hard and very hard reads for 16S.B.ALL but not on simulated datasets.

	All reads				Hard reads				Very hard reads			
	bio.	M2	M3	M4	bio.	M2	M3	M4	bio.	M2	M3	M4
	SATé Backbone											
count	13819	99998	99999	99999	104	79510	58924	3989	21	63613	40495	844
HMM	1.1	3.4	1.4	**0.3**	2.4	4.2	2.3	**0.7**	3.9	5.2	3.2	1.5
PPR	**0.6**	5.4	3.6	0.4	**0.8**	5.8	4.4	1.0	**0.9**	6.2	4.9	1.5
SEPP	1.0	**1.7**	**1.0**	0.4	2.4	**2.0**	**1.4**	0.8	3.8	**2.4**	**1.8**	**1.0**
	True or Curated Backbone											
count	13818	99997	99999	99999	104	79511	58924	3989	21	63614	40495	844
HMM	**0.0**	3.2	1.2	**0.0**	0.5	4.0	2.0	**0.2**	2.2	5.0	2.9	0.9
PPR	**0.0**	6.2	3.8	0.2	**0.1**	6.7	4.7	0.5	**0.0**	7.1	5.2	0.9
SEPP	0.4	**1.4**	**0.7**	0.1	1.1	**1.7**	**1.1**	0.4	1.5	**2.0**	**1.4**	**0.5**

Table 2: Mean delta-error for different categories of query sequences. We show the mean delta-error for each method on each model condition, as a function of the level of difficulty for the query sequence, as estimated by HMMER. Count refers to the number of query sequences processed and placed, HMM refers to HMMALIGN+pplacer, PPR refers to PaPaRa+pplacer, and SEPP refers to SEPP run in default mode.

5. Discussion

There are several observations we can make. First, the methods we evaluated for phylogenetic placement–PaPaRa+pplacer, HMMALIGN+pplacer, and SEPP methods–often produce placements that are extremely accurate, increasing the topological error over the input backbone tree by at most an edge (often much less than an edge) on average. Furthermore, while these methods do sometimes have differences in placement accuracy that go beyond an edge, these differences are sometimes still small enough to be relatively unimportant, compared to the computational cost required to obtain the improved placement accuracy.

However, we did observe conditions in which the differences in placement accuracy were quite large, suggesting that increased effort in placing query sequences correctly was merited. For example, we see big differences in placement accuracy on model M2, resulting in several edges improvement produced by SEPP(50,50) over HMMALIGN+pplacer. The conditions under which accuracy differences are substantial are characterized by large evolutionary distances between some pairs of full-length sequences. We conjecture that in such conditions, the HMMs produced by HMMER on the full set of taxa may not be sufficient to produce highly accurate alignments for the query sequences, and will result in degraded placement accuracy. The technique we introduce here avoids this problem by using HMMER to produce HMMs only on smaller, less diverse, subsets of the taxa. As a result, the HMMs may produce more accurate alignments to the query sequences, and result in improved phylogenetic placement.

We note also the interesting differences between HMMALIGN+pplacer and PaPaRa+pplacer. The only dataset on which PaPaRa+pplacer produces more accurate placements than HMMALIGN+pplacer is 16S.B.ALL, while PaPaRa+pplacer has substantially less accurate placements for the faster evolving datasets. We conjecture that the reason for this difference is that 16S.B.ALL has a low evolutionary diameter, while the simulated datasets have larger evolutionary diameters, and the estimation of transition state matrices on each edge may only be highly accurate when the backbone tree has a low evolutionary diameter.

Furthermore, these methods differ dramatically with respect to running time, with PaPaRa+pplacer much more computationally intensive than HMMALIGN+pplacer and default SEPP, thus suggesting that PaPaRa+pplacer is unlikely to be useful in largescale metagenomic analyses.

The comparison between HMMALIGN+pplacer and SEPP is more complex, because SEPP is parameterized by the two algorithmic parameters a and p. Here we see that some very simple settings for these parameters ($a = p$, both set to about 10% of the number of taxa in the backbone tree) produces very fast results with generally very good accuracy, coming close to the accuracy obtained by the best methods (or improving on them), but in a fraction of the time. Other settings for the parameters can improve the placement accuracy but require greater running time and memory usage.

It is also helpful to put the running time results in context. New sequencing technologies have enabled gathering millions of short reads from microbial communities, and a question of great practical importance is whether analyzing a million reads is feasible in a reasonable time frame. Since the placement problem treats reads independently, the running time of tools developed for the placement problem increases linearly with the number of reads (in fact most tools have pre-processing steps, the cost of which is constant with regards to the number of reads). Therefore by a simple interpolation one can estimate how long it would take to place a milion reads given our reported results for 13,000 reads. If we were to place 1 milion reads on a tree of about 13,000 reference sequences, PaPaRa+pplacer would take about 133 days and HMMALIGN+pplacer about 30 days. SEPP on the other hand would take just about 6 days, which should be feasible for many applications. SEPP therefore can enable the analysis of very large data sets in relatively short time frames.

6. Conclusions and Future Work

The method we have presented, SEPP, is a general technique for boosting the accuracy and/or speed of a phylogenetic placement method. In this study, we explored its performance when coupled with HMMALIGN and pplacer, and showed improvements in placement accuracy and/or running time. Given the plans to analyze millions of reads, the speed-ups that SEPP provides could be essential to providing scalability for phylogenetic placement methods. Improved accuracy, furthermore, might be obtained by coupling SEPP with PaPaRa for those cases where the backbone tree and alignment has a small enough evolutionary diameter to produce highly accurate extended alignments.

References

1. S. R. Eddy, *Bioinformatics* **14**, 755 (1998).
2. S. A. Berger and A. Stamatakis, *Bioinformatics* (2011).
3. S. A. Berger, D. Krompass and A. Stamatakis, *Systematic Biology* **60**, 291 (May 2011).
4. F. Matsen, R. Kodner and E. V. Armbrust, *BMC Bioinformatics* **11**, 538+ (October 2010).
5. K. Liu, T. Warnow, M. Holder, S. Nelesen, J. Yu, A. Stamatakis and C. Linder, *Systematic Biology* (2011), in press.
6. A. Stamatakis, *Bioinformatics* **22**, 2688 (2006).
7. K. Liu, S. Raghavan, S. Nelesen, C. R. Linder and T. Warnow, *Science* **324**, 1561 (June 2009).
8. J. Cannone, S. Subramanian, M. Schnare, J. Collett, L. D'Souza, Y. Du, B. Feng, N. Lin, L. Madabusi, K. Muller, N. Pande, Z. Shang, N. Yu and R. Gutell, *BMC Bioinformatics* **3**, p. 2 (2002).
9. M. N. Price, P. S. Dehal and A. P. Arkin, *PLoS ONE* **5**, p. e9490 (03 2010).

ARTIFICIAL FUNCTIONAL DIFFERENCE BETWEEN MICROBIAL COMMUNITIES CAUSED BY LENGTH DIFFERENCE OF SEQUENCING READS*

QUAN ZHANG

School of Informatics and Computing, Indiana University
150 S. Woodlawn Ave, Bloomington, IN 47405, USA
Email: quzhang@indiana.edu

THOMAS G. DOAK

Biology Department, Indiana University
1001 E. 3rd Street, Bloomington, IN 47405, USA
Email: tdoak@indiana.edu

YUZHEN YE

School of Informatics and Computing, Indiana University
150 S. Woodlawn Ave, Bloomington, IN 47405, USA
Email: yye@indiana.edu

Homology-based approaches are often used for the annotation of microbial communities, providing functional profiles that are used to characterize and compare the content and the functionality of microbial communities. Metagenomic reads are the starting data for these studies, however considerable differences are observed between the functional profiles—built from sequencing reads produced by different sequencing techniques—for even the same microbial community. Using simulation experiments, we show that such functional differences are likely to be caused by the actual difference in read lengths, and are not the results of a sampling bias of the sequencing techniques. Furthermore, the functional differences derived from different sequencing techniques cannot be fully explained by the *read-count bias*, *i.e.* 1) the higher fraction of unannotated shorter reads (*i.e.*, "read length matters"), and 2) the different lengths of proteins in different functional categories. Instead, we show here that specific functional categories are under-annotated, because similarity-search-based functional annotation tools tend to miss more reads from functional categories that contain less conserved genes/proteins. In addition, the accuracy of functional annotation of short reads for different functions varies, further skewing the functional profiles. To address these issues, we present a simple yet efficient method to improve the frequency estimates of different functional categories in the functional profiles of metagenomes, based on the functional annotation of simulated reads from complete microbial genomes.

* This work is supported by NSF grant DBI-0845685 and NIH grant 1R01HG004908.

1. Introduction

By enabling the direct analysis of microbial communities (containing many unculturable and often unknown microbes), metagenomics is revolutionizing the field of microbiology, and has excited researchers in the many disciplines that benefit from the study of environmental microbes, including those in ecology, environmental sciences, and biomedicine [1-6]. Functional analysis of a metagenomic dataset provides valuable insights into the functionality of the corresponding environmental microbial community. Also, it is very interesting to compare microbial communities in terms of their functionalities. One way of comparing microbial communities is to build a *functional profile* for each community and then compare the profiles across multiple communities, known as *functional metagenomics* [7, 8]. To predict the function of sequencing reads derived in metagenomics studies, homology-based methods are generally used. Several fast and efficient similarity search tools, such as BLAST [9] and HMMER [10], were introduced in the early 90's, and are still used for this kind of analyses, to classify metagenomic sequences into functional categories, such as COG protein families [11, 12] and KEGG pathways [13]. Subsequently, the "read-count" approach is often used to derive functional profiles of metagenomic datasets, simply equating the frequency of a functional category to the proportion of reads that have this particular functional annotation (*e.g.*, a COG family, or a K number as used in KEGG pathways). However, the reads-count approach has been shown to have a "read-count bias": annotations of short reads will be missed [14], and the reads-count approach favors functional categories (*e.g.* the protein families) composed of longer genes. As a result, the "reads-count" approach sometimes overestimates the frequency of functional categories containing longer genes and underestimates the frequency of functional categories containing shorter genes [15]. It is one of the various kinds of artifacts present in metagenomics data, caused by the limitations of the experimental protocols and/or inadequate data analysis procedures, which can lead to incorrect conclusions about microbial communities [16, 17].

An example of these artifacts is found in the comparison of the distal gut microbiota of genetically obese mice and their lean littermates. Obesity is associated with changes in the relative abundance of Bacteroidetes and Firmicutes species, and Gordon and colleges [3] demonstrated through metagenomic and biochemical analyses that this affected the metabolic potential of the mouse gut microbiota. Interestingly, they sequenced samples from a lean and an obese mouse using two different sequencing techniques—Sanger sequencing and pyrosequencing (which produced reads of ~100 nt at the time, equivalent to current Illumina read lengths; current 454 sequences can produce much longer reads)—and compared the functional profiles of the resulting datasets (Figure 1b in [3]). Their results indicated that different metagenomic datasets sequenced with the same technique share higher functional similarity than do metagenomic datasets sequenced from the same biological sample but using different sequencing techniques. The authors proposed the hypothesis that different sequencing techniques may favor different functional categories, which, if true, would pose a significant challenge for the shotgun next-generation sequencing of metagenomes adopted by most current metagnomics projects.

We used simulation studies to explore the possible explanations for the observed functional differences between annotations of the same microbial communities built from different sequencing techniques. In our first study, we simulated short reads (of ~100 nt) from the Sanger reads for the mice microbiomes [3]. We then applied the same functional annotation pipeline used for annotating the Sanger reads to the annotation of the simulated short reads, to examine and compare the functional profile of the simulated dataset with that of the original Sanger dataset. In the second study, we simulated short reads from 7 complete genomes (so we know what functions we expect in the short reads) to compare the real annotations of short reads with the expectation. Our simulation studies revealed that there are two confounding factors that bias the functional profiles of a metagenome: 1) an unequal fraction of reads tend to be missed in functional annotation for different functional categories (*i.e.* protein families); and 2) different functional categories tend to have different proportions of wrong annotations. We report below the details of our simulation studies and the results. Based on these results, we present a simple yet efficient method to improve the frequency estimates of different functional categories in metagenomes, utilizing the functional annotations of simulated short reads from complete genomes.

2. Methods

2.1. *Datasets*

2.1.1. *Obese- and lean-mouse metagenomic sequences*

The Sanger and pyrosequences (of ~100 nt) of obese mice and lean mice [3] were obtained from the MG-RAST server (http://metagenomics.anl.gov/).

2.1.2. *Complete genomes*

Table 1. A list of complete genomes for short reads simulation.

Genome id	Species name
NC_002737	*Streptococcus pyogenes* M1 GAS
NC_002927	*Bordetella bronchiseptica* RB50
NC_002937	*Desulfovibrio vulgaris* subsp. vulgaris str. Hildenborough
NC_003902	*Xanthomonas campestris* pv. campestris str. ATCC 33913
NC_006905	*Salmonella enterica* subsp. enterica serovar Choleraesuis str. SC-B67
NC_006932	*Brucella abortus* biovar 1 str. 9-941 chromosome I
NC_007795	*Staphylococcus aureus* subsp. aureus NCTC 8325

We selected 7 genomes (see Table 1) to simulate short reads from, to estimate the annotation accuracy of different protein families. The genomes were downloaded from the NCBI ftp site and the NCBI annotations of the protein-coding genes were used in this study.

2.1.3. *Simulated short reads*

1. Two datasets of short reads of ~100 nt (denoted as "Short-simu", short for "short simulated reads") were simulated from the lean-mouse and obese-mouse Sanger reads, respectively. See details of the simulated datasets in Table 2.

Table 2. Statistics of the real and simulated metagenomic datasets.

Reads type	Lean mouse		Obese mouse	
	Average length of reads (bp)	Total number of reads	Average length of reads (bp)	Total number of reads
Sanger	780	11,321	766	11,381
Short	108	1,045,701	110	675,880
Short-simu	107	1,000,000	110	670,000

"Sanger" is for Sanger sequences; "Short" for pyrosequences; and "Short-simu" for the short reads simulated from the Sanger dataset.

2. Short reads of ~100 nt were simulated from the 7 selected complete genomes (at 10X coverage) using MetaSim [18] (Version 0.9.1). For simplicity, no sequencing errors were introduced in the simulation.

2.2. *Functional Profiling of Metagenomes*

Here we used COG categories to create functional profiles for metagenomes: the extended COG definitions [11, 12] consist of 4873 COG families, which can be grouped into 25 broad functional categories. The functional profiles of metagenomes can then be built at the COG family level (*i.e.* a functional profile is represented by a vector of 4873 dimensions), or at the level of broad functional categories (*i.e.* a functional profile is represented by a vector of 25 dimensions). To compute the functional profile of a metagenomic dataset using the broad functional categories, the frequency of a category is estimated by the total reads that can be assigned to the families belonging to that category, normalized by the total number of sequencing reads in the dataset.

2.3. *Functional Annotation*

We used similarity-search-based methods for the functional annotation of long reads, or the simulated short reads. We tested both BLAST and HMMER searches (HMMER3) [19]. For BLAST searches, we used an E-value cutoff of 10^{-3} (a typical threshold used in functional annotations of metagenomes [3, 20]), and for HMMER-search-based annotation (by hmmscan from the HMMER3 package), we used an E-value cutoff of 10^{-2}. For both methods, we retained the best hit from non-overlapping regions, so that if a gene (or a read) contains multiple domains each with a distinct function, all the functions will be reported.

For HMMER searches, we built Hidden Markov Models (HMMs) for all COG families, using the sequences annotated in the eggNOG database (version 1.0, http://eggnog.embl.de/) [21]. MUSCLE [22] was used to generate a multiple alignment for each protein family, and the HMM builder from the HMMER3 package was then applied to build a HMM for each COG.

2.4. *"Perfect" Annotation of Simulated Short Reads*

For simulated short reads, we "assign" their functional annotations based on the functional annotation of the longer reads, or the complete genes from which the short reads were sampled (*i.e.* the *parent* sequence). Considering that similarity searches will miss annotations of extremely short reads, we only transfer functions of the parent sequence to the simulated reads if the simulated reads and the parent sequence overlap by at least 60 nt (*i.e.* encoding 20 aa). In this way, the simulated short reads "inherit" functions from their parent long reads or complete genes, achieving 'perfect' annotations, which could not in reality be achieved by similarity-search-based methods.

3. Results

3.1. *Read Length Differences can Cause Artificial Functional Differences between Metagenomes without Real Functional Differences*

The first question we asked is *if* different read lengths can cause artificial functional differences between metagenomes. To address this question, we simulated short reads (of ~100 nt) from actual experimental Sanger reads (long reads): we do not expect to observe real biological difference between the simulated short-reads dataset and the original Sanger dataset (referred as the *parent* datasets). We then built functional profiles (based on the broad COG categories) based on HMMER searches against COG proteins, for the simulated datasets and for the parent Sanger datasets for further comparison.

We show that if the functional annotations were perfect (by simply transferring the function annotations from the Sanger reads to simulated short reads), there would be no significant difference between the functional profiles of the simulated dataset and its parent dataset, although small differences are expected due to the randomness of read sampling (see Figure 1A). However, when we applied actual annotation to the simulated reads, we observed a significant difference between the functional profile of the simulated short-read dataset and that of the parent Sanger dataset (Figure 1B), similar to the difference observed between the profiles of the real pyrosequence dataset and the Sanger dataset (see Figure 3B below). As a result, the profiles of the datasets appear to be more similar if the reads are of similar length, regardless of the sample resources of the datasets (*e.g.*, obese versus lean mice), similar to what had been reported in [3].

To examine the effect of different similarity search tools on building the functional profiles, we also tried BLAST searches (similar methods based on BLAST searches were used for the annotation in [3]), and observed similar results: the functional profiles of the datasets appear to be more similar if the reads are of similar length, regardless of the sample resources of the datasets (data not shown). Also, correcting for the protein lengths of different families did not reduce the differences (data not shown). Taking advantage of the fast speed of HMMER3 and the high sensitivity and significance evaluation provided, we used HMMER searches as the functional annotation method in our subsequent studies.

In summary, our simulation experiments indicated that the observed functional difference reported in [3] could be simply caused by the length difference of the reads, even when there was no real biological difference involved. We show that differences in read lengths (not sequencing method) result in biased functional profiles of metagenomic samples, if the length difference is not considered properly.

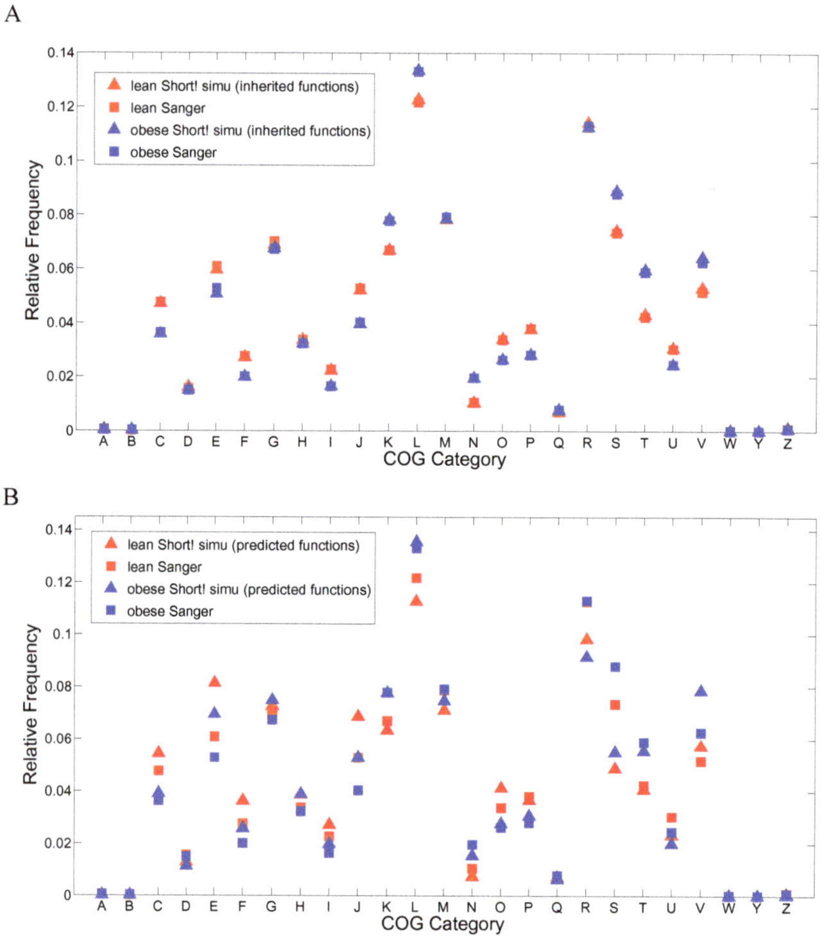

Fig. 1. Comparison of functional profiles of lean- (shown in red) and obese-mouse fecal microbiomes (shown in blue) built from Sanger reads (shown in squares) and simulated short reads (triangles), (A) providing 'perfect' annotation for the short reads, and (B) applying real annotations for the short reads based on HMMER search results.

3.2. *Read Length Matters Differentially to Different Protein Families*

We then asked *how* different read lengths cause artificial functional differences. To address this question, we simulated short reads from 7 selected complete genomes (see Table 1), so that we can compare annotations based on the short reads to the annotations of the complete genes. It has been shown that read length matters and for all protein families, there is a significant sensitivity loss in the similarity searches for short reads of < 200 nt (homology finding is limited by the read length) [14]; we further show that 1) unequal fractions of reads tend to be missed in annotations for different functional categories (*i.e.* protein families); and 2) different functional categories tend to have different proportions of wrong annotations. These two confounding factors together can bias functional profiling of metagenomes, causing artificial functional differences between metagenomes—if the lengths of reads are not carefully controlled for.

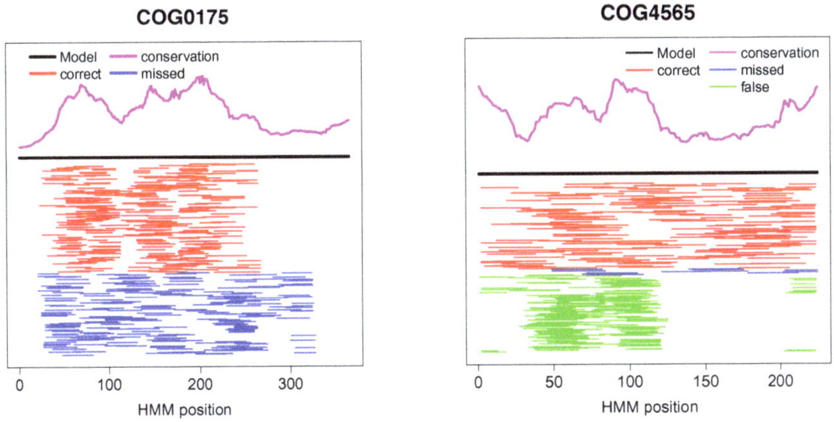

Fig. 2 Demonstration of functional annotation of short reads assigned to COG0175 (A) and COG4565 (B) The black line represents the HMM of the corresponding COG family. Red lines represent the reads that have the correct annotation. Blue lines represent the reads that are of that function, but are missed by the annotation pipeline. Green lines represent the reads that are incorrectly assigned to the family. The magenta curve above the HMM line shows the conservation score of the family. The higher the value is, the more conserved the position is.

For each family, we calculated the *precision* of the annotations, defined as the proportion of correctly annotated reads among all the reads assigned to the family, and the *recall*, defined as the number of reads correctly annotated over the actual number of reads with the function. Note that in the simulation experiment, we can assign any read based on the functional annotation of the complete gene, and thus the number of reads within a functional category can be precisely counted. Both the precision and recall of annotations differ among different protein families. Among the 4077 COG families present in the 7 genomes, 2146 families have a precision >90%, and 674 families have a precision <= 20%; 880 families have a recall > 90% and 804 families have a recall <=20%. Figure 2 shows two examples of COG families with different levels of false annotations and missed annotations: COG0175 has significant missed annotations (293 reads are correctly assigned to this function, while 223 reads of this function are missed); by contrast, COG4565 has a large number of false annotations (158 reads are incorrectly assigned to this function) but very few missed annotations (12 reads). In the figure, we indicate the positional conservation for each COG (estimated based on the "hitting probability" [23]), which helps us understand the patterns of missed and false annotations of each COG family (see Discussion).

Some COG families share sequence similarity, and we observed incorrect functional assignments of short reads among these families. For example, COG3829 (transcriptional regulator containing PAS, AAA-type ATPase, and DNA-binding domains) and COG2204 (response regulator containing CheY-like receiver, AAA-type ATPase, and DNA-binding domains) both belong to the functional category T (signal transduction mechanisms); COG3829 also belongs to category K (transcription). 85 reads that are sampled from genes of function COG3829 were incorrectly assigned to COG2204, while 520 reads that are of function COG2204 were assigned to COG3829. Another pair is COG4664 (TRAP-type mannitol/chloroaromatic compound transport system, large permease component) and COG1593 (TRAP-type C4-dicarboxylate transport system, large permease component); these two families belong to different functional categories: COG4664 belongs to Q (secondary metabolites biosynthesis, transport and catabolism) and COG1593 belongs to G (carbohydrate transport and metabolism). 480 reads that are sampled from genes of function COG1593 were incorrectly assigned to COG4664, while no reads that are of function COG4664 were assigned to COG1593. An extreme case is function COG4565 (response regulator of citrate/malate metabolism): short reads simulated from genes of seven different functions (including COG3290, COG0745, COG2197, COG2204, COG3279, COG1629, COG0784) are assigned to this function.

Table 3. Annotations of the reads simulated from gene YP_012234.1.

Gene/reads	Positions	COG3604	COG2197	COG3437	COG3829	COG4565	COG2204
YP_012234.1	1–1419	6.80e-114	2.90e-19	4.80e-27	3.50e-115	1.70e-23	***6.40e-145***
r231087.1	49–162	-	-	***7.90e-05***	-	-	-
r173226.1	57–167	-	-	***9.80e-06***	-	0.036	0.016
r4681.1	140–240	-	7.40e-05	4.60e-06	-	***2.90e-07***	6.00e-05
r122743.1	160–268	-	0.00012	0.0021	-	***2.30e-05***	0.00021
r111158.1	178–286	-	***0.00016***	0.0053	-	0.013	0.0032
r191730.1	470–585	1.10e-15	-	-	***2.10e-16***	-	3.80e-14
r196288.1	547–662	***8.10e-18***	-	-	4.80e-16	-	1.40e-15

This table lists the E-values of selected reads sampled from gene YP_012234.1 (with sampled regions shown in the 'Positions' column) that have different best matches than the complete gene. The best COG family assignment (with lowest E-value) for each read was highlighted with the E-value shown in bold and italic, with '-' indicating an insignificant match between the read and corresponding COG family.

We show that false annotations of short reads among related families may be explained by the local sequence variation of the involved families and the fact that different families may share local similarities [24, 25]. Table 3 lists the annotations of short reads sampled from gene YP_012234.1. Gene YP_012234.1 can be assigned to function COG2204, although it also shares significant similarity with several other COG families, including COG3604 and COG3284. 66% of the short reads sampled from this gene can be assigned to a function (with E-value smaller than the cutoff), among which a mere 23% are assigned to the same function (COG2204) as the complete gene, using the best hit strategy. Among the short reads that were assigned to different COGs, read 'r173226.1' (which is sampled between 57 and 167 bp of this gene) can be assigned to COG3437 with an E-value of 9.80e-06; by contrast, the match between this read and the "correct" function COG2204 is much less significant, with an E-value of 0.016 (see more examples in Table 3).

3.3. *Artificial Bias of Functional Profiles can be Adjusted by Considering Read Lengths*

Having shown that different read lengths can cause artificial functional differences because of the differential impact of short reads on different protein family, we then asked: *can* the frequency estimates be adjusted, to allow functional profiling of a metagenome independent of read length? We proposed to adjust the functional profile of a metagenome—built with reads of a certain length—using the recall and the precision of functional annotations of different families we learned from simulated reads of the same length:

$$Hit(f) = \frac{Hit_0(f)Precision(f)}{Recall(f)Len(f)} \tag{1}$$

where $Hit_0(f)$ and $Hit(f)$ are the original and adjusted number of query sequencing reads that are assigned to function f, respectively. $Precision(f)$ and $Recall(f)$ are the precision and recall of functional annotation for functional category f based on similarity searches of simulated reads of the same length as short reads. $Len(f)$ is the average length of the protein sequences that are of the function f (calculated from the protein sequences that are assigned to functional category f collected in the eggNOG database). These family-specific values of precision and recall can be applied to any metagenomic project, once calculated for applicable read-lengths.

We used Eq. (1) to adjust the frequency of different COG families for the simulated twin-mouse short-read datasets, resulting in improvements of the frequency estimates (comparing to those estimated from the Sanger reads) for many functional categories, including E (amino acid transport and metabolism), F (nucleotide transport and metabolism), J (translation, ribosomal structure and biogenesis), N (cell motility), and S (function unknown) (see Figure 3A). Here we show the details of the comparison for functional category E. Its frequencies estimated from both obese and lean datasets (shown in red and blue unfilled triangles in Figure 3A, respectively) were significantly overestimated; as a result, they differed greatly from those estimated from the Sanger datasets (shown in squares in Figure 3A), even though the short reads were simulated from the Sanger datasets. The adjustment decreased the artificial functional difference caused by the read length difference, revealing a clearer abundance difference of this functional category between the obese and lean metagenomes. We believe that such adjustment is important for interpreting real metagenomes derived from different sequencing techniques, as shown in the twin-mouse datasets (see Figure 3B). For most functional categories (except for functional category V, defense mechanisms), the frequency adjustment helped to decrease the differences between the frequencies estimated from the short-read datasets and the Sanger datasets of the same microbial community, revealing consistent functional differences (*e.g.*, for functional category L, replication, recombination and repair) between obese and lean metagenomes (functional category L is more abundant in the obese metagenomes).

4. Discussion

Using simulation studies, we show that differences in read lengths can result in biased functional profiles of metagenomic samples, if the length difference is not treated properly. This is an important observation especially for metagenomic research, which utilize different kinds and generations of next-generation sequencing (NGS) techniques that produce reads of various lengths. It indicates that conclusions need to be drawn carefully when 1) characterizing the functional content (capacity) of a metagenome derived by NGS (as the functional profile built from short reads may not faithfully reflect the actual functional content of the microbiome), and when 2) comparing different metagenomes that are produced by different NGS techniques (*e.g.*, 454 versus Illumina). While Gordon *et al.* [3] suggest that different sequencing techniques will produce reads in a biased way (*e.g.*, pyrosequencing may under-sample reads from functional category A) to explain the functional difference observed between metagenomes containing reads of various lengths, our conclusion is that biased-annotation of short reads can explain the observed difference, indicating sequencing is not biased, rather annotation is biased. We expect that such functional annotation bias due to read length differences exists in other projects as well, as long as the projects involve short reads (such as those from current Illumina sequencers).

It is important to see how read length impacts other functional classifications: we plan to extend our study to include more functional categories, including the FIG families [26] and the KEGG pathways. We will try more sophisticated methods for functional profile adjustments, aiming to achieve better improvements. And more systematic tests are needed. One direction is to analyze the individual families (*e.g.*, their overall evolutionary conservation and positional-conservation patterns, as shown in Figure 2) and their relationships (*e.g.*, the local similarity between different families) in more details, and utilize this information when constructing functional profiles. Another direction is to consider the read length explicitly when adjusting the functional profiles, so that we do not need a specific learning procedure for each read length.

We have presented a simple method to adjust functional profiles, based on the read length, and we demonstrate that it is possible to alleviate the "read-length-bias" problem. We feel that this approach is immediately applicable to many past and on-going metagenomic studies, including the Human Microbiome Project [27], impacting not only the functional annotation of metagenomes (as shown in this paper), but also other aspects, such as the enterotype-study of human gut microbiomes [1], which is based on read mapping to reference genomes for quantification of the abundances of different genera, and where obvious differences can be observed in the results drawn from metagenomes sequenced by Illumina or Sanger sequencing [1].

5. Acknowledgements

The authors thank Drs. Mina Rho and Haixu Tang for helpful discussions.

References

1. M. Arumugam, J. Raes, E. Pelletier *et al*, *Nature*, **473**:174-180 (2011).
2. J. Qin, R. Li, J. Raes *et al*, *Nature*, **464**(7285):59-65 (2010).
3. P. J. Turnbaugh, R. E. Ley, M. A. Mahowald *et al*, *Nature*, **444**(7122):1027-1031 (2006).
4. G. W. Tyson, J. Chapman, P. Hugenholtz *et al*, *Nature*, **428**(6978):37-43 (2004).
5. J. C. Venter, K. Remington, J. F. Heidelberg *et al*, *Science*, **304**(5667):66-74 (2004).
6. C. S. Riesenfeld, P. D. Schloss and J. Handelsman, *Annu Rev Genet*, **38**:525-552 (2004).
7. S. G. Tringe, C. von Mering, A. Kobayashi *et al*, *Science*, **308**(5721):554-557 (2005).
8. W. Xie, F. Wang, L. Guo *et al*, *ISME J*, **5**(3):414-426 (2011).
9. S. F. Altschul, W. Gish, W. Miller *et al*, *J Mol Biol*, **215**(3):403-410 (1990).
10. A. Krogh, M. Brown, I. S. Mian *et al*, *J Mol Biol*, **235**(5):1501-1531 (1994).
11. R. L. Tatusov, N. D. Fedorova, J. D. Jackson *et al*, *BMC Bioinformatics*, **4**:41 (2003).
12. R. L. Tatusov, M. Y. Galperin, D. A. Natale *et al*, *Nucleic Acids Res*, **28**(1):33-36 (2000).
13. M. Kanehisa, M. Araki, S. Goto *et al*, *Nucleic Acids Res*, **36**(Database issue):D480-484 (2008).
14. K. E. Wommack, J. Bhavsar and J. Ravel, *Appl Environ Microbiol*, **74**(5):1453-1463 (2008).
15. I. Sharon, A. Pati, V. M. Markowitz *et al*, *A statistical framework for the functional analysis of metagenomes*, in *RECOMB 2009; (*Tucson, AZ, 2009), pp 496-511.
16. J. C. Wooley and Y. Ye, *J Comput Sci Technol*, **25**(1):71-81 (2009).
17. N. Shah, H. Tang, T. G. Doak *et al*, *Comparing bacterial communities inferred from 16S rRNA gene sequencing and shorgun metagenomics*, in *Pac Symp Biocomput; (*The Big Island of Hawaii, 2011), pp 165-176.
18. D. C. Richter, F. Ott, A. F. Auch *et al*, *PLoS One*, **3**(10):e3373 (2008).
19. S. R. Eddy, *Genome Inform*, **23**(1):205-211 (2009).
20. E. A. Dinsdale, R. A. Edwards, D. Hall *et al*, *Nature*, **452**(7187):629-632 (2008).
21. L. J. Jensen, P. Julien, M. Kuhn *et al*, *Nucleic Acids Res*, **36**(Database issue):D250-254 (2008).
22. R. C. Edgar, *Nucleic Acids Res*, **32**(5):1792-1797 (2004).
23. B. Schuster-Bockler, J. Schultz and S. Rahmann, *BMC Bioinformatics*, **5**:7 (2004).
24. G. Agarwal, S. Mahajan, N. Srinivasan *et al*, *PLoS One*, **6**(3):e17826 (2011).
25. B. Morgenstern, K. Frech, A. Dress *et al*, *Bioinformatics*, **14**(3):290-294 (1998).
26. R. Overbeek, T. Begley, R. M. Butler *et al*, *Nucleic Acids Res*, **33**(17):5691-5702 (2005).
27. J. Peterson, S. Garges, M. Giovanni *et al*, *Genome Res*, **19**(12):2317-2323 (2009).

A

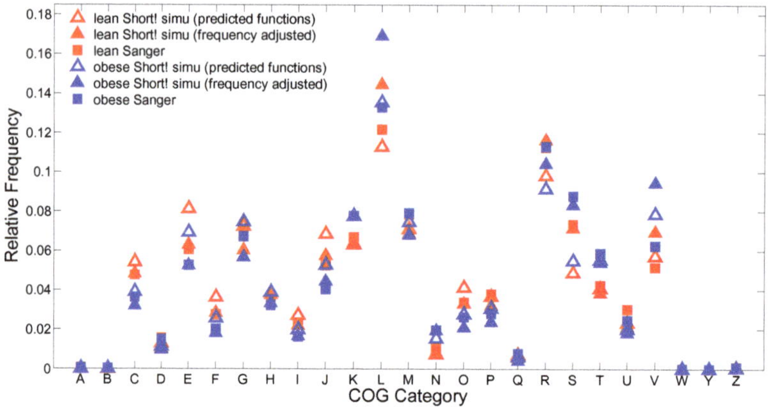

B

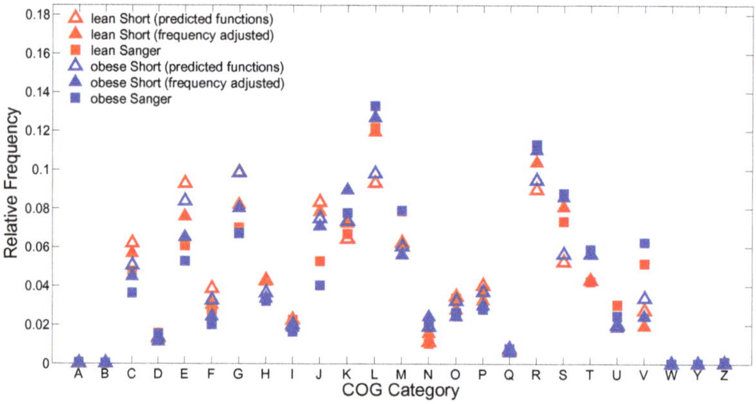

Fig. 3 Comparison of functional profiles of lean- (shown in red) and obese-mouse fecal microbiomes (shown in blue) built from Sanger reads (shown in squares) and short reads (triangles). Both original functional profiles (shown in unfilled triangles) and adjusted ones by applying Eq. 1 (shown in filled triangles) are shown in the figures: (A) for simulated short reads, and (B) for actual pyrosequence datasets.

METADOMAIN: A PROFILE HMM-BASED PROTEIN DOMAIN CLASSIFICATION TOOL FOR SHORT SEQUENCES

YUAN ZHANG and YANNI SUN

Dept. of Computer Science and Engineering, Michigan State University,
East Lansing, MI 48824, U.S.A.
E-mails: zhangy72@msu.edu
yannisun@msu.edu

Protein homology search provides basis for functional profiling in metagenomic annotation. Profile HMM-based methods classify reads into annotated protein domain families and can achieve better sensitivity for remote protein homology search than pairwise sequence alignment. However, their sensitivity deteriorates with the decrease of read length. As a result, a large number of short reads cannot be classified into their native domain families. In this work, we introduce MetaDomain, a protein domain classification tool designed for short reads generated by next-generation sequencing technologies. MetaDomain uses relaxed position-specific score thresholds to align more reads to a profile HMM while using the distribution of alignment positions as an additional constraint to control false positive matches. In this work MetaDomain is applied to the transcriptomic data of a bacterial genome and a soil metagenomic data set. The experimental results show that it can achieve better sensitivity than the state-of-the-art profile HMM alignment tool in identifying encoded domains from short sequences. The source codes of MetaDomain are available at http://sourceforge.net/projects/metadomain/.

Keywords: Protein domain classification; metagenomics; short reads; profile HMM.

1. Introduction

With the advent of next-generation sequencing and culture-independent methods, an enormous amount of metagenomic data have been sequenced from microbial communities from different habitats. In order to understand the phylogenetic complexity and biological functions of microbial communities, as well as their interactions with the host, automatic annotation tools such as CAMERA,[1] MG-RAST,[2] and MEGAN[3] are being used for annotating metagenomic data sets. As an important component of these metagenomic annotation tools, protein homology search provides basis for identifying putative genes and assigning those genes to annotated functional categories (e.g. protein domain families). There are two major methods for protein homology search. The first method is based on pairwise sequence alignment. Putative genes can be identified by comparing metagenomic reads against annotated protein databases using BLASTX.[4] Although BLAST is one of the most efficient protein homology search tools, probabilistic model-based methods have better sensitivity for remote protein homology recognition. In particular, using profile hidden Markov models (HMMs) to represent a protein family greatly improves homology search sensitivity between highly diverged sequences.[5] Thus it is desirable to conduct protein domain classification using profile HMM-based tools such as HMMER.[6] In conjunction with a fast-growing protein domain family database Pfam,[7] HMMER is able to classify sequences into different domain families with high accuracy. In addition, the latest implementation of profile HMM-based domain classification tool HMMER 3.0[6] has achieved comparable speed to BLAST, making it suitable for large-scale protein compositional

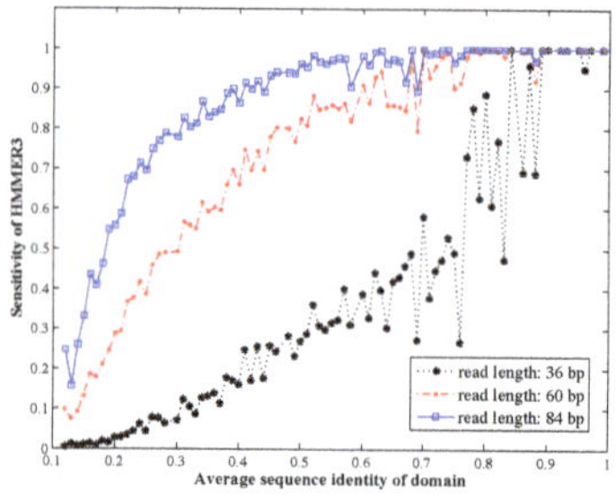

Fig. 1. Change of the read classification sensitivity of HMMER over read length and the average sequence identity of domain families

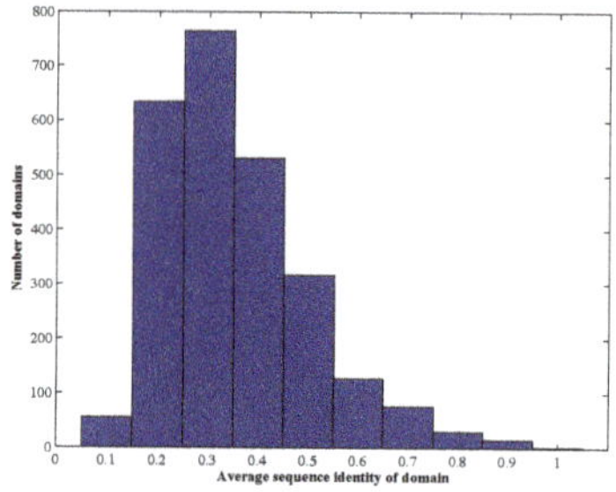

Fig. 2. Histogram of the average pairwise sequence identity for 2558 domains

analysis. For the convenience of discussion, we use HMMER to refer to HMMER 3.0 hereafter unless otherwise specified.

Low sensitivity of HMMER on classifying short reads. Because of the high sensitivity of remote homology recognition, HMMER has been successfully applied to genome-wide domain analysis. However, its sensitivity is significantly limited by the short reads of metagenomic data sets and poorly conserved domains. In order to investigate how read length and domain identity affect the sensitivity of HMMER, we randomly sampled 200 peptides with lengths of 12, 20, and 28 amino acids from the seed sequences of each of the 2,558 Pfam domains, which contain the word "Bacteria" in their descriptions. The peptides were aligned with the domain families using HMMER. We used the E-value cutoff 1000 in order to boost the sensitivity. For each domain, the read classification sensitivity of HMMER is measured as the ratio of the number of aligned reads to the total number of sampled reads. We sort all data points by domain identity in ascending order and plot them in Figure 1. For domains with the same identity, their average sensitivity is reported.

Figure 1 shows that the sensitivity of HMMER deteriorates with the decrease of the query sequence length and domain identity. The sensitivity is decreased from 90% to 65-70% when the lengths of reads change from 28 residues (i.e., 84 bp for corresponding DNA reads) to 20 residues (i.e., 60 bp for DNA reads) for domains with identity around 40%. Although next-generation sequencing technologies are producing longer reads and assembly tools may be available to assemble short reads into longer contigs, there is still a need for a protein domain analysis tool for short reads. First, many finished or on-going metagenomic sequencing projects contain reads with lengths from 35 to around 400 bp depending on the chosen sequencing technologies. In addition, peptide sequences encoded in individual metagenomic sequence reads may share only small overlaps with existing domain families. Thus, a sizable portion of many available data still contains short reads. Second, the sheer amount of data and the complexity of many metagenomic data sets pose a great challenge for assembly tools.[8] A large portion of short reads cannot be correctly assembled into longer contigs. Third, many domain families exhibit low average sequence identity, which poses a challenge for short and medium-sized reads. Figure 2 shows the histogram of pairwise sequence identity for domains related to bacteria. Of 2558 domains, there are about 43% domains with average identity no greater than 0.3. For these domains, the sensitivity of HMMER is between 0.7 and 0.8 for reads of length 84 bp, between 0.4 and 0.6 for reads of length 60 bp, and smaller than 0.1 for reads of length 36 bp. As a result, although a large number of reads are sequenced from genes, which are highly compact in microbial genomes, only a small percentage of the short reads can be classified into their native domains using existing tools.

In this work, we introduce *MetaDomain*, a protein domain classification tool designed for short reads in metagenomic data sets. MetaDomain provides a complementary protein analysis tool to HMMER on assigning short reads into their native families.

2. Related Work

Profile HMM-based protein homology search is widely used for mining microbial genomes. For example, Ellrott et al.[9] investigated the distribution of protein families in the available human gut genomic and metagenomic data. As the data set contains assembled contigs, using HMMER is expected to achieve high sensitivity. Schlüter et al.[10] used HMMER to understand the genetic diversity and composition of a plasmid metagenome from a wastewater treatment plant. The reads have an average length of 104 bp, which is also adequate for HMMER to achieve high sensitivity.

Besides providing a basis for functional profiling, profile HMM-based homology search was also used for phylogenetic complexity analysis in metagenomic data. The phylogenetic algorithm CARMA[11] uses all Pfam domain and protein families as phylogenetic markers to identify the source organisms of environmental DNA fragments as short as 80 bp. As we show in Figure 1, profile HMM-based tools have sensitivity of at least 0.9 in classifying reads of 80 bp into domains with average sequence identity above 40%. However, for poorly-conserved domains, a significant number of reads might be missed. A similar but faster tool Treephyler[12] conducted community profiling in metagenomics and metatranscriptomics based on Pfam domain assignments. Treephyler was applied to a data set with average read length of 200 bp. It is unclear how shorter reads affect its performance.

Our previous work designed a tool HMM-FRAME,[13] which can identify and correct frameshift errors in pyrosequencing reads during protein domain classification using profile HMM-based alignment. However, it was not specifically designed to handle short reads.

Finally, we note that the method used in MetaDomain shares a similar rationale to the recent work by Weng et al.[14] Weng et al. reported that taxonomic binning tools for metagenomes discard 30-40% of Sanger sequencing data due to the stringency of BLAST cut-offs. Thus, they re-analyzed the discarded reads using less stringent cut-offs. In order to control the false positive matches introduced by the relaxed cut-offs, they used the evolutionary conservation of adjacency between neighboring genes as an additional criterion.

3. Method

HMMER uses E-values as the discrimination threshold to determine the membership of a query sequence. However, short reads may only generate low alignment scores and thus in-

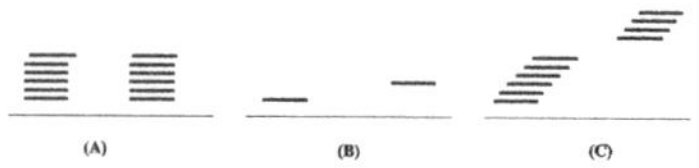

Fig. 3. Three types of alignment distributions

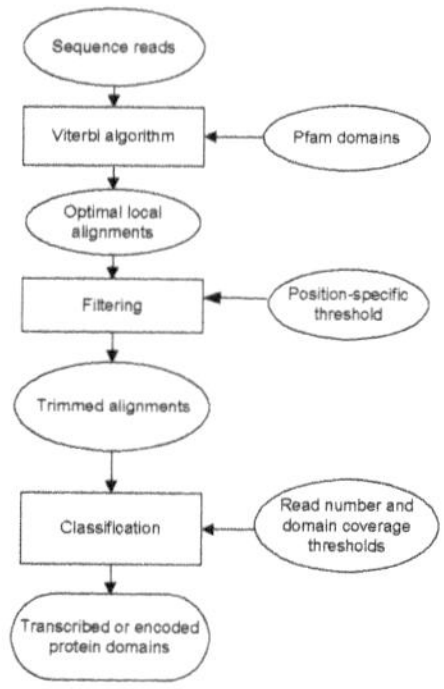

Fig. 4. Pipeline of MetaDomain

significant E-values. In particular, the conservation across the entire length of a domain family can be highly variable, posing a great challenge for classifying reads sequenced from poorly conserved sub-regions. In order to increase the sensitivity of aligning remotely-related short reads, we propose position-specific score cutoffs, by which poorly conserved regions allow more relaxed discrimination thresholds than well-conserved regions. However, the low thresholds can easily incur random matches. In order to control the false positive rate, we examine the position distribution of read alignments. The position distribution of read alignments on a truly encoded domain is expected to be more uniform than a domain that incurs random read alignments.[15,16] Figure 3 shows the schematic representations of three types of distributions of read alignments along a domain. The alignments in (A) and (B) are more likely to be random. Thus the domains may not be encoded in the data set. The alignment distribution in (C) exhibits a much more uniform distribution, providing strong evidence for the existence of the underlying domain in the data set. Thus, by using relaxed position-specific score cutoffs and inspecting the distribution of alignments, we expect to classify more short reads into the correct domain families while not falsely reporting domains that are not characterized in the data.

3.1. *Pipeline of MetaDomain*

The input to MetaDomain includes sequence reads and a list of protein domains. The output is a list of domains encoded in the underlying data set and the number of aligned reads. Figure 4 shows a schematic representation of the pipeline of MetaDomain.

MetaDomain consists of three main stages: short read alignment, filtering, and classification. In the alignment stage, we use the Viterbi algorithm[5] to search for the best local alignment between a query sequence and a profile HMM-represented domain family. In the filtering stage, we first apply a position-specific score threshold to eliminate insignificant alignments. Then we remove stacked alignments with the same alignment positions inside a poorly conserved region. In the final stage, we use the number of aligned reads and the distribution of alignment positions to determine whether a domain is encoded.

3.2. *The Viterbi algorithm*

The Viterbi algorithm aligns a query sequence to a profile HMM by searching for the most probable state path in the model. Unlike HMMER, MetaDomain directly aligns a DNA sequence to a profile HMM. To do so, we implicitly align translated peptides under different reading frames with a profile HMM. Let π be a state path in a profile HMM M and let x be a query DNA sequence. The Viterbi algorithm searches for the most probable path π^* such that $\pi^* = argmax_\pi(x, \pi)$. The output of the Viterbi algorithm includes the optimal alignment and its score. As Viterbi is a standard algorithm designed for HMMs, we refer readers to Durbin et al.[5] for a detailed illustration of the dynamic programming equations for finding π^*. The major difference between our implementation and the standard Viterbi algorithm includes : 1) our implementation accepts a DNA rather than a peptide sequence as input; 2) a local alignment can start and end with any state without incurring insertion or deletion penalties.

3.3. *Alignment Filtering*

MetaDomain employs two filtering mechanisms to increase its sensitivity in aligning short reads while maintaining a low false positive rate: position-specific thresholds (PSTs) and trimming.

3.3.1. *Position specific threshold*

PST allows different alignment thresholds for well conserved and poorly conserved regions. Let the length of a query DNA sequence be L (in bp). Denote the profile HMM as M. Let $M_{i,j}$ be a sub-model formed by all consecutive states from the ith match state M_i to the jth match state M_j. The upper bound of the alignment score against $M_{i,j}$ is the maximum score that can be generated by aligning any input sequence of length $j - i + 1$ with $M_{i,j}$. Let $a_{i,j}$ denote the transition probability from state M_i to state M_j. Let $e_i(a)$ denote the probability of state M_i emitting amino acid a. Then the upper bound $U_{i,j}$ for sub-model $M_{i,j}$ is calculated as follows:

$$U_{i,j} = \prod_{k=i}^{j} a_{k,k+1} \times max(e_k(a))$$

where $a_{j,j+1}$ is set to 1 because j is the ending state of the sub-model.

We define PST for the submodel $M_{i,j}$ as:

$$PST_{i,j} = \gamma U_{i,j}$$

where the coefficient γ is a user-specified parameter in the range of [0,1]. It can be flexibly adjusted to control the trade-off between sensitivity and false positive rate of MetaDomain. The default value is 0.6, which is used in our experiments.

3.3.2. *Alignment trimming*

Alignment with scores larger than their corresponding PSTs will pass the first filtering stage. As each domain has various conservation along the entire length of the model, well-conserved sub-regions have high PSTs while poorly-conserved sub-regions yield low PSTs. Thus, random sequences tend to be aligned to poorly-conserved regions by MetaDomain, incurring a high FP rate. Our empirical experiments show that dozens of reads that are not sequenced from the underlying domain can be aligned to the same position in a poorly-conserved subregion. In order to minimize the effects of noise, we discard stacked alignments that have the same alignment positions.

3.4. *Protein domain classification*

In this stage we extract two features from the collected read alignments for each domain: the number of aligned reads and the domain coverage. The domain coverage is the fraction of positions covered by at least one read alignment in a domain. MetaDomain then applies a simple decision tree to classify all the target domains into two classes: encoded domains and non-encoded domains. If both features of a domain are equal to or bigger than their corresponding thresholds, this domain will be classified as encoded. Otherwise it is not encoded in the sample. By default, the cutoff for domain coverage is 30%. Ideally, the cutoff for the number of aligned read should be determined based on the properties of data such as sequencing depth. If users do not specify this value, we use 20 by default.

4. Experimental Results

In order to evaluate the performance of MetaDomain on real data generated by next-generation sequencing technologies, we applied MetaDomain to protein domain analysis in two data sets. The first one is the transcriptome generated using RNA-seq for *Burkholderia cenocepacia*. As both the reference genome and its domain annotations are available, we can quantify the sensitivity and false positive (FP) rate of MetaDomain. The second one is metagenome data sequenced from soil. We applied MetaDomain to identify domains encoded in the underlying data. In addition, we compared HMMER and MetaDomain in both applications.

4.1. *Identifying transcribed protein domains in transcriptome*

In this experiment, we conducted transcribed domain analysis in the transcriptome from one strain of *B. cenocepacia* named AU1054.[17] By using Illumina RNA-seq, the authors generated multiple samples for AU1054 in two growth media. We used one replicate of cDNA sample of AU1054 in the growth medium cystic fibrosis. In total, 3,361,008 reads of a length of 41 bp were downloaded from the website provided by the authors. We evaluated the performance of read classification and domain identification of MetaDomain and HMMER.

4.1.1. *Performance of read classification*

The performance of read classification is quantified using both read classification sensitivity and FP (false positive) rate. In this experiment, the read classification performance is computed on reads that can be mapped to annotated domains. Below we sketch the main steps to obtain mapped reads for a domain using the reference genome and the domain annotations. First, we downloaded the genome of AU1054 and the annotated genes and domains from the IMG website.[18] There are 2,181 annotated Pfam domains. Second, the reads were mapped to the reference genome using Bowtie[19] with two mismatches allowed. Third, we compared the positions of read mapping and annotated domains. For a domain, all reads that fall into it are defined as "mapped" reads. Denote the set of mapped reads as M. All other (unmapped) reads constitute set U. For a domain classification tool, let the set of aligned reads for a domain be A. Thus, the sensitivity and FP rate of read classification for a domain are $\frac{A \cap M}{M}$ and $\frac{A-M}{U}$, respectively. Sensitivity of 100% indicates that all mapped reads can be aligned. A zero FP rate indicates that only mapped reads can be aligned to a domain.

Of the 2,181 annotated families, we evaluated the performance of HMMER and MetaDomain on 1406 families which have at least one mapped read. Of the 1406 tested domains, HMMER could not align any read to 1150 domains, resulting in zero sensitivity and FP rate. For the rest 256 domains, all aligned reads by HMMER are non-mappable reads, resulting in zero sensitivity and a positive FP rate. The comparison between HMMER and MetaDomain is summarized using a bubble chart in Figure 5. The biggest bubble indicates that HMMER has zero sensitivity and zero FP rate for 1150 domains. This experiment shows that it is highly difficult for HMMER to correctly align reads as short as 41 bp. There are two possible reasons for the low sensitivity of HMMER on short reads. First, the parameter training in E-value calculation of HMMER is based on much longer reads (100 amino acids). Thus, the small alignment scores generated by the short reads yield large E-values and cannot pass the E-value threshold. Second, the small alignment scores of short reads may not pass the filtration stage of HMMER.

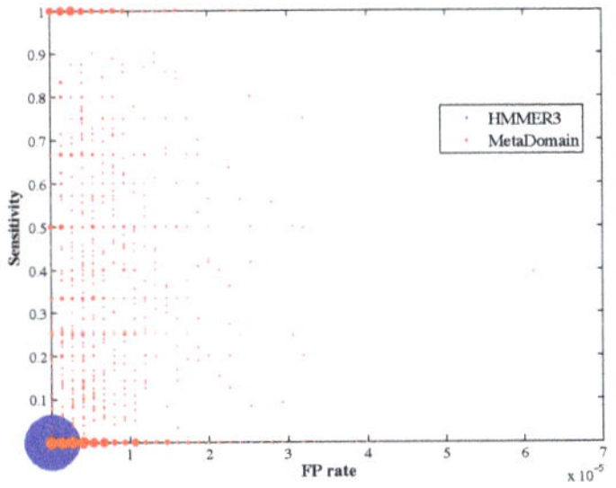

Fig. 5. Read classification sensitivity and FP rate of HMMER and MetaDomain. The size of each bubble represents the number of data points (i.e., domains) with the same sensitivity and FP rate.

4.1.2. *Identifying transcribed domains in the transcriptome*

Figure 5 only shows the read classification performance. MetaDomain uses both aligned read number and domain coverage as thresholds for domain identification. We expect that the additional constraint will reduce the false positive rate in domain identification. Because of the low read classification sensitivity, we speculate that HMMER will have low sensitivity in identifying transcribed domains.

Positive and negative test sets In order to quantify the performance of domain identification, we need to build positive and negative test sets, which include transcribed and non-transcribed domains based on mapped reads. There is no commonly accepted criterion to define transcribed genes using the number of mapped reads. Various expression scores such as an average coverage depth across the entire length of each gene[20] and reads per kilobase of exon model per million mapped reads (RPKM) are used to quantify transcriptional level. In addition, the cutoffs of defining highly transcribed, lowly transcribed, or non-transcribed genes are variable in different applications.[21] In this work, we define transcribed domains based on the rationale that a truly transcribed domain should be mapped by a number of reads at different positions. Correspondingly, we use the following criteria to determine whether a domain inside a gene is transcribed: 1) at least **N** reads are mapped to a domain; 2) at least 30% of positions in a domain are mapped by reads. A domain is labeled "non-transcribed" if the number of mapped read is zero. For domains that fall between the criteria for transcribed and non-transcribed domains, they are labeled "unknown" and are excluded from the test sets. Table 1 shows the size change of the positive and negative test sets over the cutoff N. Intuitively, bigger N creates an easier case for domain classification than smaller N.

Table 1. Number of transcribed and non-transcribed domains using different cutoffs (N) for the number of mapped reads

N	transcribed	unknown	none-transcribed
10	318	1317	546
15	262	1373	546
20	226	1409	546
25	195	1440	546
30	169	1466	546

Domain analysis using MetaDomain and HMMER We align all reads to the transcribed and non-transcribed domains using MetaDomain and HMMER. The "unknown" domains are removed due to their ambiguity. For HMMER, we first translated the short reads into peptide sequences using 6-frame translations. We then aligned the domains with the translated sequences using 1000 as the E-value threshold, which is chosen to maximize the sensitivity. For MetaDomain we directly aligned the short reads with the domains. The pipeline in Figure 4 was used to output a list of transcribed domains for MetaDomain. Let D^+ and D^- be the number of transcribed and non-transcribed domains identified using the read mapping results in Section 4.1.2. Let M^+ and M^- be the predicted number of transcribed and non-transcribed domains by MetaDomain or HMMER. The sensitivity and FP rate of domain classification tools are defined using the following equations:

$$\text{Sensitivity} = \frac{D^+ \cap M^+}{D^+}$$
$$\text{FP rate} = \frac{D^- \cap M^+}{D^-}$$

The values of D and M are affected by several options. First, D^+ and D^- can change over the cutoff N as shown in Table 1. Second, we used both the domain coverage and the number of aligned reads to determine whether a domain is encoded or transcribed. In this experiment, the cutoff for domain coverage is 30%, which we found reasonable across different experiments. Thus, M^+ and M^- mainly change over the required number of aligned reads to a domain. For simplicity, we denote the cutoff as τ. Increasing τ implies a more stringent constraint for defining transcribed domains, and thus might result in lower sensitivity and a smaller FP rate. Decreasing τ is likely to increase the sensitivity while incurring a higher FP rate. In order to compare the performance of MetaDomain and HMMER under different τ, we plotted the ROC curves by changing τ from 1 to N for N=10, 20, and 30 in Figure 6.

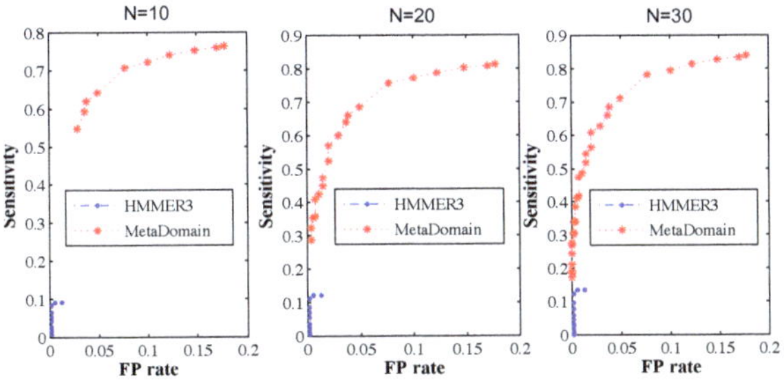

Fig. 6. ROC curves of HMMER and MetaDomain

Figure 6 shows that HMMER is highly specific (FP rate $\leq$ 1.3%). However, as we speculated, its sensitivity is low, with the highest sensitivity being only 0.135. HMMER misses a large portion of short reads that can be mapped to protein domains even when we use a very relaxed E-value cutoff. When both tools incur an FP rate of 0.02, the sensitivity of MetaDomain is 0.53 vs. 0.13 for HMMER. When N decreases from 30 to 10, the size of the positive test set D^+ becomes larger and the sensitivity of both HMMER and MetaDomain decreases. Note that the sensitivity and FP rate of HMMER keep the same for many different thresholds (i.e., τ), resulting in compact ROC curves. Overall, the ROC curves show that MetaDomain can achieve higher sensitivity while keeping a similar FP rate as HMMER for domain classification in this experiment. In addition, Figure 6 provides guidance on determining appropriate τ for MetaDomain in order to achieve desired sensitivity and FP rate.

On average, it took MetaDomain 280 seconds to align 752,156 reads with one domain on a 2.2GHz dual-core AMD Opteron machine. There are 2181 domains in this experiment.

4.2. *Protein domain analysis in a soil metagenomic data set*

In the first experiment, we demonstrated the accuracy of MetaDomain in classifying short RNA-seq reads into their native domain families. In this section, we present the utility of MetaDomain in identifying encoded protein domains in a complicated metagenomic data set, which is sequenced from the microbes dwelling in the soil from a long term cultivated corn site at Iowa using Illumina HiSeq platform [a]. There are 520,346,510 sequence reads of various lengths, ranging from 31 bp to 114 bp. The average length of the reads is $\sim$ 73 bp. Figure 7

[a]Sequenced by James Tiedje Lab at Michigan State University. Unpublished yet.

shows the distribution of the read lengths in this data set. The sheer amount of data and the complexity of this data set pose a great challenge for read assembly programs. Thus, we directly apply MetaDomain and HMMER to unassembled reads.

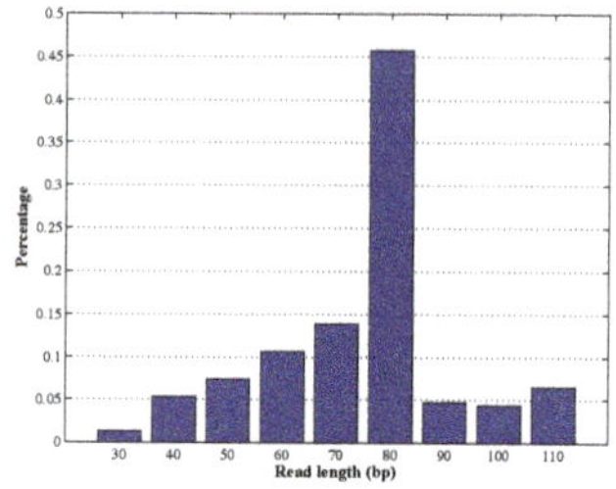

Fig. 7. Read length distribution in the soil data set.

We first used HMMER to align all the reads against the 2558 Pfam domains that contain the word "Bacteria" in their descriptions. Using E-value 1000, HMMER classified 34,602,784 reads into 2558 Pfam domains, accounting for 6.65% of all reads in the data set. The classifiable reads have an average length of $\sim$ 80 bp. A large number of reads shorter than 60 bp were not classified. By conducting a complementary domain analysis on short reads using MetaDomain, we expect to classify more reads to their native families. As many domains have been aligned with a large number of reads by HMMER, it is highly likely that they are encoded by the bacterial species in the soil data set. We thus excluded them from further screening. There are 80 domains with less than 20 reads aligned or with a smaller domain coverage than 30%. The left panel of Figure 9 shows an example of such domain. The small number of aligned reads and their biased distribution do not support the representation of this domain in this data set. We thus applied MetaDomain to the 80 domains and investigated whether they are encoded. On average, it took MetaDomain about 31 CPU minutes to align 11,194,176 sequences that were not classified by HMMER and are shorter than 60 bp against one domain family.

Figure 8 presents the number of aligned reads and the domain coverage output by HMMER and MetaDomain. Note that MetaDomain was only applied to reads that were not classifiable by HMMER. Thus, the total number of reads that can be classified into each domain should be the sum of the output of HMMER and MetaDomain. This figure shows that significantly more reads can be classified into the corresponding domains. We need to specify thresholds for domain coverage and the number of aligned reads in order to define an encoded domain. Similar to previous experiments, the domain coverage cutoff is 30%. According to Figure 6, the sensitivity and FP rate of MetaDomain are 0.34 and 0.004 when τ is 20. We thus choose 20 as the cutoff for the number of aligned reads. Out of the 80 Pfam domains, 24 have less than 20 reads aligned or a domain coverage no greater than 30%. So these 24 domains are not likely to be encoded in this data set. For the other 56 domains, their average domain coverage by MetaDomain alone is 97.25%. The number of aligned reads by MetaDomain is 169.52 versus 15.27 by HMMER. This provides strong evidence that these protein domains are actually

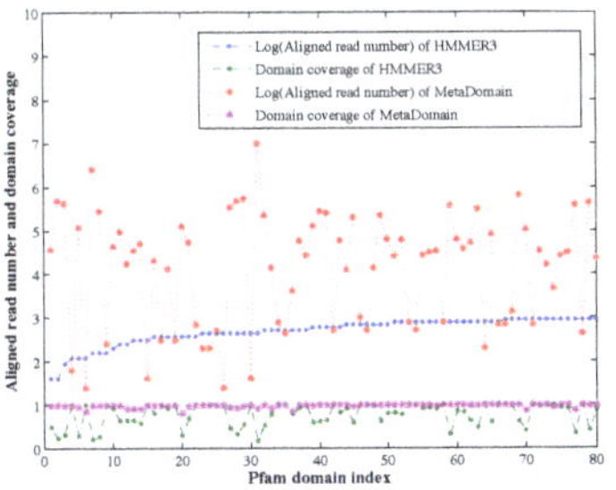

Fig. 8. Reads aligned by HMMER and MetaDomain.

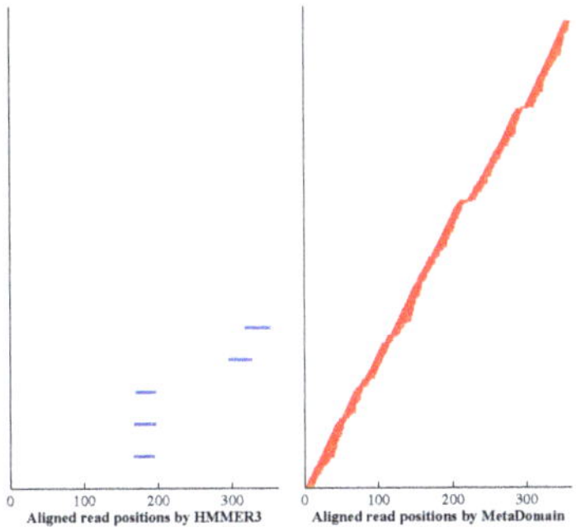

Fig. 9. The distributions of aligned reads for PF09703 by HMMER and MetaDomain.

encoded. The average read length aligned by MetaDomain is 38 bp. As an example, Figure 9 shows the distribution of aligned reads by HMMER and MetaDomain for the domain PF09703. In summary, by using MetaDomain, we are able to identify 56 more domains encoded in this data set. 130,930 (0.025%) more reads are classified into these domain families.

Among these 56 protein domains, 21 have unknown functions. 6 domains are CRISPR-associated domains. These kinds of domains are found in the genomes of approximately 40% of bacteria and 90% of archaea. More detailed analysis is needed to understand whether the functions of these domains are important to the specific habitat.

5. Conclusion and future work

In this work, we introduce MetaDomain, a protein domain classification tool for short reads produced by next-generation sequencing technologies. It provides a complementary domain classification tool to HMMER on classifying short reads into domain families with low sequence identity. Our experimental results show that it can achieve a better tradeoff between sensitivity and FP rate than HMMER in classifying short sequences. Its current version is based on a faithful implementation of Viterbi and is slow when applied to thousands of millions of reads and the whole Pfam database. We plan to improve its efficiency by using filtration strategies such as ungapped alignment and parallel programming. In addition, we plan to improve the method of designing position-specific score thresholds in order to achieve a better discrimination power.

6. Acknowledgements

We would like to thank Dr. James Tiedje and Dr. James Cole for providing us the soil metagenomic data set from the cultivated corn site. The sequencing efforts were supported by JGI and DOE.

References

1. "CAMERA: Community cyberinfrastructure for advanced microbial ecology research and analysis." `http://camera.calit2.net/`.

2. F. Meyer, D. Paarmann, M. D'Souza, R. Olson, E. Glass, M. Kubal, T. Paczian, A. Rodriguez, R. Stevens, A. Wilke, J. Wilkening, and R. Edwards, "The metagenomics RAST server - a public resource for the automatic phylogenetic and functional analysis of metagenomes," *BMC Bioinformatics*, vol. 9, no. 1, p. 386, 2008.
3. D. H. Huson, A. F. Auch, J. Qi, and S. C. Schuster, "MEGAN analysis of metagenomic data," *Genome Research*, vol. 17, no. 3, pp. 377–386, 2007.
4. S. F. Altschul, G. Warren, W. Miller, E. W. Myers3, and D. J. Lipmanl, "Basic local alignment search tool," *J. Mol. Biol.*, vol. 215, pp. 403–410, 1990.
5. R. Durbin, S. Eddy, A. Krogh, and G. Mitchison, *Biological Sequence Analysis Probabilistic Models of Proteins and Nucleic Acids*. UK: Cambridge University Press, 1998.
6. "HMMER3: a new generation of sequence homology search software." `http://hmmer.janelia.org/`.
7. R. D. Finn, J. Mistry, J. Tate, P. Coggill, A. Heger, J. E. Pollington, O. L. Gavin, P. Gunasekaran, G. Ceric, K. Forslund, L. Holm, E. L. L. Sonnhammer, S. R. Eddy, and A. Bateman, "The Pfam protein families database," *Nucleic Acids Research*, vol. 38, no. suppl 1, pp. D211–D222, 2010.
8. S. C. Schuster, "Next-generation sequencing transforms today's biology," *Nat Meth*, vol. 5, no. 1, pp. 16–18, 2008.
9. K. Ellrott, L. Jaroszewski, W. Li, J. C. Wooley, and A. Godzik, "Expansion of the protein repertoire in newly explored environments: Human gut microbiome specific protein families," *PLoS Comput Biol*, vol. 6, no. 6, p. e1000798, 2010.
10. A. Schlüter, L. Krause, R. Szczepanowski, A. Goesmann, and A. Pühler, "Genetic diversity and composition of a plasmid metagenome from a wastewater treatment plant," *Journal of Biotechnology*, vol. 136, no. 1-2, pp. 65–76, 2008.
11. L. Krause, N. N. Diaz, A. Goesmann, S. Kelley, T. W. Nattkemper, F. Rohwer, R. A. Edwards, and J. Stoye, "Phylogenetic classification of short environmental DNA fragments," *Nucleic Acids Research*, vol. 36, no. 7, pp. 2230–2239, 2008.
12. F. Schreiber, P. Gumrich, R. Daniel, and P. Meinicke, "Treephyler: fast taxonomic profiling of metagenomes," *Bioinformatics*, vol. 26, no. 7, pp. 960–961, 2010.
13. Y. Zhang and Y. Sun, "HMM-FRAME: accurate protein domain classification for metagenomic sequences containing frameshift errors," *BMC Bioinformatics*, vol. 12, no. 1, p. 198, 2011.
14. F. Weng, C. H. Su, M. T. Hsu, T. Y. Wang, H. K. Tsai, and D. Wang, "Reanalyze unassigned reads in sanger based metagenomic data using conserved gene adjacency," *BMC Bioinformatics*, vol. 11, no. 1, p. 565, 2010.
15. J. C. Dohm, C. Lottaz, T. Borodina, and H. Himmelbauer, "Substantial biases in ultra-short read data sets from high-throughput DNA sequencing," *Nucleic Acids Research*, vol. 36, no. 16, p. e105, 2008.
16. K. D. Hansen, S. E. Brenner, and S. Dudoit, "Biases in Illumina transcriptome sequencing caused by random hexamer priming," *Nucleic Acids Research*, 2010.
17. D. R. Yoder Himes, P. S. G. Chain, Y. Zhu, O. Wurtzel, E. M. Rubin, J. M. Tiedje, and R. Sorek, "Mapping the Burkholderia cenocepacia niche response via high-throughput sequencing," *Proceedings of the National Academy of Sciences*, 2009.
18. "IMG: Integrated microbial genomes." `http://img.jgi.doe.gov/`.
19. B. Langmead, C. Trapnell, M. Pop, and S. L. Salzberg, "Ultrafast and memory-efficient alignment of short DNA sequences to the human genome," *Genome Biology*, vol. 10, p. R25, 2009.
20. K. D. Passalacqua, A. Varadarajan, B. D. Ondov, D. T. Okou, M. E. Zwick, and N. H. Bergman, "Structure and complexity of a Bacterial transcriptome," *J. Bacteriol.*, vol. 191, no. 10, pp. 3203–3211, 2009.
21. B. A. Williams, B. A. Williams, K. McCue, L. Schaeffer, and B. Wold, "Mapping and quantifying mammalian transcriptomes by RNA-Seq," *Nat Meth*, vol. 5, no. 7, pp. 621–628, 2008.

MODELING HOST-PATHOGEN INTERACTIONS: COMPUTATIONAL BIOLOGY AND BIOINFORMATICS FOR INFECTIOUS DISEASE RESEARCH

JASON E. MCDERMOTT[1,2]
Pacific Northwest National Laboratory, Computational Biology and Bioinformatics Group. Richland, WA 99352 USA. Jason.McDermott@pnnl.gov

PASCAL BRAUN
Center for Cancer Systems Biology (CCSB) and Dana-Farber Cancer Center (DFCI), Boston, 450 Brookline Avenue, MA 02215; and Harvard Medical School, Department of Genetics, 240 Longwood Avenue, Boston, MA 02115

RICHARD BONNEAU
Departments of Biology and Computer Science, New York University. New York, NY.

DANIEL R. HYDUKE[1]
Bioengineering, University of California San Diego, San Diego, CA

Introduction

Infections are a major cause of both human disease and loss of crop yields and animal stocks and thus cause immense damage to the worldwide economy. The significance of infectious diseases is expected to increase in an ever more connected warming world, in which new viral, bacterial and fungal pathogens can find novel hosts and ecologic niches. At the same time, the complex and sophisticated mechanisms by which diverse pathogenic agents evade defense mechanisms and subvert their hosts networks to suit their lifestyle needs is still very incompletely understood especially from a systems perspective [1]. Thus, understanding host-pathogen interactions is an important and timely topic.

Recently technological advances have afforded the opportunity to investigate host-pathogen interactions on a level of detail and scope that offers immense computational and analytical possibilities. Genome sequencing was pioneered, in part, with pathogens; and, the number of strains and variants of pathogens sequenced to date vastly outnumbers the number of host genomes available. At the same time, for both plant and human hosts more and more data on population level genomic variation becomes available and offers a rich field for analysis into the genetic interactions between host and pathogen.

---

1 Work partially supported by NIH/NIAID interagency agreement Y1-AI-8401
2 Work partially supported by NIH/NIAID Contract No. HHSN272200800060C

On a molecular level deep-sequencing [2], increasingly sophisticated proteomic tools [3] and recently interactome analysis [4-7] have emerged as powerful tools to dissect pathogen virulence programs and to investigate how pathogens rewire cellular transcriptional and protein networks [8]. However, very difficult problems are posed by these interactions in terms of data analysis, making useful predictions, and modeling. The rapid evolution of pathogens, coupled with the strong selection exerted by the host defense system means that pathogens can be exquisitely adapted to their environments. The exploitation of the host by the pathogen can prove difficult to extract from experimental observation of these interactions [9, 10]. Knowledge of these interactions has direct applications in terms of improved therapeutic strategies. Historically, the molecular and structural differences pathogens and their hosts have been exploited as targets and led to some of the most successful treatments of disease ever developed, for example vaccines, antibiotics and antivirals. However, a major problem confronting the medical and research communities currently is that of the development and rapid growth of antibiotic resistance in a large number of important pathogens for a range of traditional antibiotics. Accordingly, developing methods to study these pathogens with the eventual goal of developing novel therapies is of utmost importance.

This session is focused on discussing computational approaches to studying the interactions between pathogens and their hosts. The session is composed of an invited talk, three selected papers, and a discussion panel.

The invited speaker, Dr. Ram Samudrala Department of Microbiology University of Washington, is a pioneer in the area of structural prediction. His recent work employs modeling approaches to aid in drug discovery, and he is a recipient of a prestigious NIH Pioneer award to identify novel drugs that target various pathogens using computational approaches.

Interactions between a pathogen and its host are dynamic with the pathogen executing a complicated virulence program during infection and the host activating a number of orchestrated defense responses to counter the pathogen. Gordan, et al. employ comparative genomics, promoter analysis and transcriptional data from the fungal pathogen Candida albicans to identify a set of genes that may have a role in the activating or maintaining the virulent hyphal state.

Franzosa, et al. and Nouretdinov, et al. describe methods for predicting protein-protein interactions (PPIs) between pathogen effector proteins and their host targets. These interactions form the interface between host and pathogen and are targets for further experimental investigation and potentially therapeutic drug development. Franzosa, et al. describe a careful evaluation of their method for predicting PPIs based on homology modeling of protein structures and corresponding interactions. Nouretdinov, et al. use a novel statistical approach, the conformal method, to predict PPIs between HIV proteins and the human host and provide a corresponding confidence measure.

In the discussion panel we will discuss issues related to the computational analysis and modeling of host-pathogen interactions, outline major barriers to current work, and identify upcoming challenges given the state of the field today. We believe this session will provide a solid basis for further discussion in this important area.

Acknowledgements

We would like to thank the authors who submitted their papers for consideration in this session as well as the reviewers for their time and thoughtful feedback.

1. Aderem, A., et al., A systems biology approach to infectious disease research: innovating the pathogen-host research paradigm. MBio, 2011. 2(1).
2. Peng, X., et al., Unique signatures of long noncoding RNA expression in response to virus infection and altered innate immune signaling. MBio, 2010. 1(5).
3. Niemann, G.S., et al., Discovery of novel secreted virulence factors from Salmonella enterica serovar Typhimurium by proteomic analysis of culture supernatants. Infect Immun, 2011. 79(1): p. 33-43.
4. de Chassey, B., et al., Hepatitis C virus infection protein network. Mol Syst Biol, 2008. 4: p. 230.
5. Shapira, S.D., et al., A physical and regulatory map of host-influenza interactions reveals pathways in H1N1 infection. Cell, 2009. 139(7): p. 1255-67.
6. Mukhtar, M.S., et al., Independently evolved virulence effectors converge onto hubs in a plant immune system network. Science, 2011. 333(6042): p. 596-601.
7. Das, S. and G.V. Kalpana, Reverse two-hybrid screening to analyze protein-protein interaction of HIV-1 viral and cellular proteins. Methods Mol Biol, 2009. 485: p. 271-93.
8. McDermott, J.E., et al., Technologies and approaches to elucidate and model the virulence program of salmonella. Front Microbiol, 2011. 2: p. 121.
9. Shi, L., et al., Proteomic analysis of Salmonella enterica serovar typhimurium isolated from RAW 264.7 macrophages: identification of a novel protein that contributes to the replication of serovar typhimurium inside macrophages. J Biol Chem, 2006. 281(39): p. 29131-40.

10. Eriksson, S., et al., Unravelling the biology of macrophage infection by gene expression profiling of intracellular Salmonella enterica. Mol Microbiol, 2003. 47(1): p. 103-18.

STRUCTURAL MODELS FOR HOST-PATHOGEN PROTEIN-PROTEIN INTERACTIONS: ASSESSING COVERAGE AND BIAS

ERIC A. FRANZOSA

Bioinformatics Program
Boston University, 24 Cummington Street
Boston, MA 02215, USA
Email: franzosa@bu.edu

YU XIA

Bioinformatics Program, Department of Chemistry, Department of Biomedical Engineering
Boston University, 24 Cummington Street
Boston, MA 02215, USA
Email: yuxia@bu.edu

Recently, we applied structural systems biology to host-pathogen interaction and constructed the human-virus structural interaction network (SIN) based on a combination of solved structures and homology models. Subsequent analysis of the human-virus SIN revealed significant differences between antagonistic human-virus and cooperative within-human protein-protein interactions (PPIs). Although the SIN approach is advantageous due to the complementary nature of 3D structure and network data, integration of these data sources is associated with two potential issues: reduced coverage of the full human-virus PPI network, and the introduction of specific biases from structure determination. In this work, we evaluate the impact of these issues by comparing the growth and properties of human-virus and within-human PPI networks with and without structural models. We find that although the human-virus SIN is small in size, it is largely depleted for false positives, which are common in the full network. In addition, the SIN shows potential for major growth in the near future. Furthermore, compared to the full network, the coverage of viral species in the human-virus SIN is large relative to its size, suggesting that it may be a less biased sampling of the universe of human virus PPIs. Next, we systematically compare structural versus full networks of human-virus and within-human PPIs in terms of functional, physicochemical, and network properties. We find that although there exist biases inherent to the structural approach, such biases tend to affect both human-virus and within-human PPIs equally. As a result, the significant differences between structured human-virus and within-human PPI networks are never contradicted by the full networks. Collectively, these results suggest that a structural approach to host-pathogen systems biology is not only justified, but also highly complementary to previous approaches. In particular, conclusions drawn from direct comparisons of host-virus and host-host PPIs within the SIN are minimally confounded by the inherent biases of the structural approach.

1. Introduction

Systems biology approaches have been recently applied to immunology [1] and pathogen research [2] following significant progress in the genome-wide mapping of host-pathogen protein-protein interactions (PPIs) for selected pathogens [3-8]. This work has been successful in revealing systematic trends in host-pathogen interaction networks, e.g., that viruses tend to target host protein interaction hubs involved in characteristic cellular processes, such as cell cycle regulation [2]. Concurrently, a structural approach to general problems in systems biology has started to gain momentum [9-11]. Structural systems biology focuses on the analysis of structural interaction networks (SINs), a class of PPI network in which every interaction is supported by a solved 3D structure of the interacting proteins. These two data types are highly complementary: network information provides a holistic but low-resolution view of cellular events, and 3D structure information provides a mechanistic and high-resolution view of interfaces between specific biomolecules. Indeed, the structural systems biology approach has already been applied to answer interesting questions in network biology, such as the physical nature of party- and date-type protein hubs [12], and the role of native disorder in protein-protein interaction [13].

In a recent work we combined these two trends, applying a structural systems biology approach to the analysis of host-pathogen interaction networks [14]. We combined 3D structure and traditional network data for human-virus (exogenous) and within-human (endogenous) PPIs to construct a combined human-virus SIN. Our analysis of the human-virus SIN revealed that viral proteins tend to bind to and mimic existing endogenous interfaces, and that they often do so without structural homology to an endogenous binding partner. Furthermore, we found that the endogenous interfaces targeted by viruses tend to be otherwise occupied by multiple, transiently bound ("date-like") endogenous regulators, and tend to evolve significantly faster than generic protein surface sites. Interface similarity in the absence of structural homology, date-like binding, and fast evolutionary rate are comparatively rare among cooperative endogenous interactions and interfaces, and hence our discoveries highlighted the distinct consequences of antagonistic host-virus interaction.

While a SIN provides many advantages over a traditional PPI network, it also associated with several known or potential drawbacks. For example, by enforcing that all PPIs in our human-virus SIN be supported by a solved 3D structure, our coverage of the universe of known human-virus and within-human PPI is dramatically reduced. In addition to incompleteness, experimental and investigator bias remain as potential limitations for the structural systems biology approach given the inherent difficulties associated with protein structure determination. For example, the human-virus SIN will tend to under-represent interactions involving highly disordered proteins, and may be biased toward well-studied proteins. Given that our comparisons and contrasts between exogenous and endogenous interactions are carried out within the SIN, the hope is that our conclusions are minimally confounded by the inherent biases from the structural approach. In this work, we provide evidence in support of this assertion, and present additional support for a structural approach to the analysis of host-pathogen interaction networks.

We begin by analyzing the size and rate of growth of the currently known universe of structured and binary human-virus and within-human PPIs. In this work, the term *binary* interaction refers to a physical association between two proteins that is supported by experimental evidence [15]. A *structured* interaction refers to a direct, physical association between two proteins that is supported by a 3D model. Naturally, structured interactions are a subset of known binary interactions. We demonstrate that, while small, the set of structured PPIs constituting the human-virus SIN is more accurate and more representative of the true universe of human-virus interaction than the full binary network. We then compare sequence, structural, and functional properties within these networks, revealing that, although biases from the structural approach exist, they tend to affect exogenous and endogenous interactions equally, and hence do not have a confounding effect on trends interpreted from the SIN. With the number of structured exogenous interactions increasing dramatically, structural systems biology is rapidly becoming a key route for exploring principles of host-pathogen interactions.

2. Constructing Binary and Structural Human-Virus Interaction Networks

We constructed structural and binary networks of human-virus (exogenous) and within-human (endogenous) protein-protein interactions (PPIs) for analysis and comparison. Endogenous PPI data were assembled from the IntAct database [15]; there are a total of 35,303 endogenous interactions. Human-virus PPI data were taken principally from VirusMINT [16] and supplemented with additional reports from IntAct; this produced an initial set of 1,814 exogenous interactions. Exogenous interactions were filtered to remove redundancy (e.g., orthologous viral proteins from different strains with the same human protein target). Based on sequence data sequence from the UniProt database [17], two exogenous interactions were said to be redundant if the human proteins involved were the same or highly similar (BLAST [18] E-value $< 10^{-5}$) and the virus proteins involved were the same or highly similar. Among redundant exogenous interactions, only the interaction with the largest number of independent reports was considered. Endogenous interactions are considered to be non-redundant by definition.

Structural models were assigned to interactions following previous work [14, 19]. Briefly, a 3D structure of interacting proteins from the Protein Databank [20] acts as a model for an exogenous interaction if (i) the structured proteins are a human protein and a virus protein, (ii) the structured proteins are a human ortholog and a virus protein, or (iii) the structured proteins have very strong sequence homology (BLAST E-value $< 10^{-10}$) to human and virus proteins that are independently known to interact. Models satisfying the latter two conditions are referred to as *homology models*. Structural models of exogenous interactions were filtered for redundancy following the procedure described above; there are a total of 53 structured exogenous interactions. Structural models of endogenous interactions were assigned following conditions (i) and (iii) for pairs of interacting human proteins; there are a total of 3,039 structured endogenous interactions.

3. Sizes and Growth Rates of Human-Virus Interaction Networks

From the period of 1989 to 2009, the number of experimentally determined interactions in each of the four classes (binary endogenous, structured endogenous, binary exogenous, and structured exogenous) has increased by orders of magnitude (Figure 1A). As a result of large-scale functional genomics techniques such as yeast two-hybrid and affinity capture, it is now possible to rapidly enumerate binary protein-protein interactions within a species. The number of experimentally determined human endogenous interactions (27,646) dwarfs all other counts by the end of 2009, and appears to be growing dramatically (Figure 1B). It is worth noting, however, that although the number of *reported* endogenous interactions is large, the vast majority of these (95.2% as of the end of 2009) have never been *confirmed* by a second experiment, and hence their accuracy remains largely unknown.

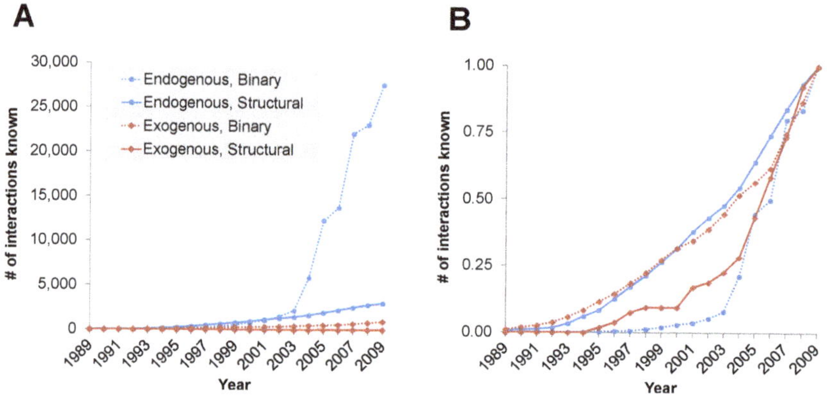

Figure 1. Growth of exogenous (human-virus) and endogenous (within-human) interaction networks, binary and structural. (**A**) Absolute numbers of interactions known, 1989-2009. (**B**) Relative numbers of interactions known, compared to end-of-2009 totals, 1989-2009.

Structured endogenous interactions are the second most abundant class, but their number (3,039) remains far below that of their binary counterparts (Figure 1A). Unlike binary endogenous interactions, the techniques for generating structures of protein complexes remain mostly low-throughput and labor-intensive, and so these interactions have been accumulating at a slower, more linear pace (Figure 1B). Roughly 10% of the structured endogenous class is composed of homology models—i.e., pairs of proteins that are known to interact and have strong sequence similarity to a solved structure (which serves as a 3D "scaffold" for their interaction). In order to maintain a high degree of accuracy, we employed a stringent sequence similarity cutoff when assigning homology models to proteins (BLAST E-value $< 10^{-10}$; 80% average alignment coverage). Relaxing this cutoff or employing alternative homology modeling techniques could increase the number of models substantially, and the size of the structured endogenous class by extension. Compared with binary interactions derived from noisy, high-throughput experiments, structured interactions—including homology models—will be dominated by true positives.

Exogenous interactions are the next-most abundant class, numbering 976 at the end of 2009 (Figure 1A). Compared with endogenous interactions, the number of experimentally determined exogenous interactions is growing at a more linear pace (Figure 1B), and a larger (but still small) fraction of these interactions (17.5%) have been confirmed by multiple experiments. Only 53 interactions fall in the structured exogenous class (Figure 1A), of which 11 were inferred from homology. As with structured endogenous interactions, the structured exogenous list could be expanded using additional homology modeling techniques or relaxed homology criteria. For example, alternative methods could be used to build models of viral proteins which present short linear motifs to human linear motif-binding domains [21]. Curiously, although determining the structure of exogenous interactions is subject to the same challenges and time constraints of endogenous structure determination, their relative numbers have recently begun increasing dramatically (Figure 1B). This suggests to us that structural biologists are rapidly turning their attention toward problems in host-pathogen interaction. We believe that the growth of the exogenous and endogenous networks will further accelerate as structural genomics projects naturally shift their focus from single proteins to protein-protein interactions in the near future.

4. Viral Diversity in Human-Virus Interaction Networks

The size of the total universe of exogenous interactions remains unknown. We assert here that pooled human-virus interactions from the VirusMINT and IntAct databases represent a thorough catalog of the *known* universe of binary human-virus interactions, although these are probably only a small sample of the total universe. Even if we discard the approximately half of these interactions which are redundant, the fraction of the remaining exogenous interactions having known structures is small (5.3%; Figure 2A). We therefore conclude that the human-virus SIN represents a very small sample of the total universe of exogenous interactions. This does not imply that it is inappropriate to make statistical inferences based on the structured sample, as the statistical significance of a trend depends upon its effect size (weak or strong) and the sample's absolute size, but not its relative coverage of the population.

On the other hand, the structured exogenous interactions could be a *biased* sample of the universe of human-virus PPIs, which could influence measured effect sizes. However, the structured exogenous interactions cover 21.7% of the viral proteins in the non-redundant exogenous dataset (Figure 2B), and 50.0% of the viral species (as determined by NCBI taxonomic ID; Figure 2C). These differences in interactions-to-protein and interactions-to-species ratios between structured and non-redundant exogenous interactions result from the latter's overrepresentation of interactions from viruses which have been the focus of high-throughput surveys, such as HIV and Epstein-Barr virus [5, 7]. As a result, with respect to viral diversity, the structured exogenous network may represent a less biased sample of the total universe of exogenous PPIs than the binary exogenous PPI network, and could therefore be a more appropriate sample to consider for the purposes of statistical hypothesis testing. This finding also demonstrates that, while we are principally concerned here with biases inherent to the structural interaction network, traditional networks are also vulnerable to experimental and investigator biases.

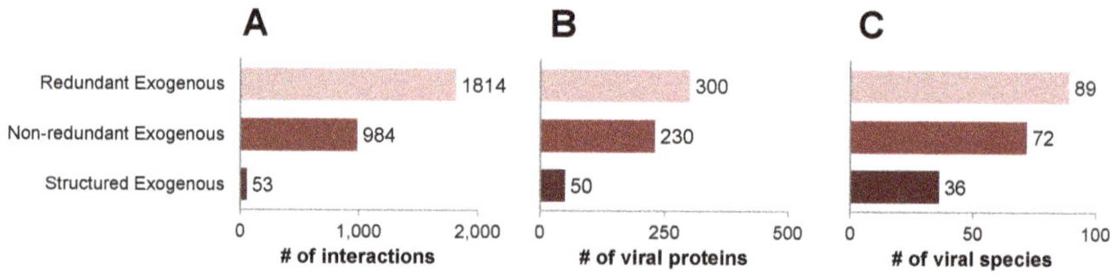

Figure 2. Properties of exogenous interaction networks. (**A**) Number of exogenous interactions represented in the redundant, non-redundant, and structured human-virus interaction networks. (**B**) Number of viral proteins represented in these networks. (**C**) Number of viral species represented in these networks (based on NCBI taxonomic ID). The structural interaction network covers 5% of known non-redundant interactions but 50% of their source species.

5. Evaluating Potential Biases from the Structural Systems Biology Approach

Although we assert that the human-virus SIN includes a relatively unbiased sampling of viral diversity, it may still be biased in other ways. As discussed in the Introduction, structured interactions will tend to reflect the experimental constraints of structure determination. For example, proteins with large disordered regions or extensive transmembrane domains are difficult to crystallize, and hence their structures may elude discovery by X-ray crystallographic methods. Given the targeted rather than high-throughput nature of structured interaction determination, investigator bias may also play a role in skewing the sample of interactions in a structural interaction network. For example, having determined the structure of a given protein in complex with one of its binding partners, an investigator may tend to continue to study the structural interactions of this protein (or similar proteins), rather than moving on to a completely different system.

Such biases from the structural approach are likely to exist in the human-virus SIN. What remains to be shown is whether or not a measured difference between the exogenous and endogenous interactions in the human-virus SIN can be *explained* by such biases. Compared to within-host PPI interfaces, we observed in our previous work that host-virus PPI interfaces tend to be more transient, targeted by more host proteins, more regulatory in function, faster evolving, and more reliant on convergent evolution to achieve interface mimicry. If we checked each of these trends within the total binary exogenous and endogenous interaction networks, several scenarios are possible (Figure 3). The first possibility is that the binary networks follow the same trend observed in the SIN (Figure 3A). In this case, there is no bias from the structural approach, and the absolute and relative level of a certain property for exogenous and endogenous interactions does not depend on whether or not the interactions are structured. The second possibility is that there is a bias introduced by the structural approach, but that it affects the exogenous and endogenous interactions uniformly (Figure 3B). In this case, the absolute level of a certain property for exogenous and endogenous interactions *does* depend on whether or not the interactions are structured, but the relative difference between them does not. In both of these cases (Figure 3 A and B), the exogenous levels are uniformly different from the endogenous levels, indicating that the trend is biologically relevant. This may not always be the case. In Figure 3C, a trend difference between the structured exogenous and structured endogenous interactions does *not* recur between the total binary exogenous and total binary endogenous interactions, and in Figure 3D the trend is reversed. In these cases, the trend from the structural interaction network appears to be driven solely by biases from the structural approach, and is not biologically relevant.

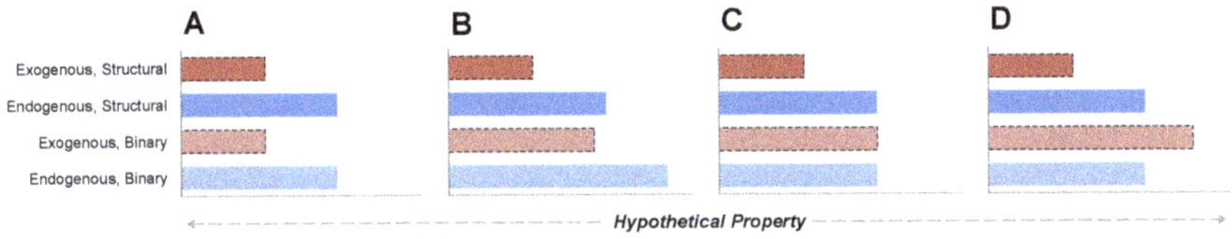

Figure 3. Illustration: The influence of structural bias on trends from the human-virus SIN. In each scenario, there is strong trend between structured exogenous and structured endogenous interactions with respect to some hypothetical property (x-axis); assume all differences are significant. (**A**) An example with no structural bias: the structured and full binary human-virus networks show the same trend. (**B**) An example with uniform structural bias: the structured and full binary human-virus networks show the same trend. (**C** & **D**) Examples with confounding structural bias: a trend observed in the human-virus SIN is not reflected (C) or is reversed (D) in the full binary network.

Many of the properties in which we are interested can only be measured in a structural interaction network—an issue which motivates the structural approach. For example, we cannot determine from binary interactions alone the average degree of interface overlap between exogenous and endogenous interfaces (interface mimicry), a trend we know to be significant from the structural network [14]. In order to demonstrate that trends from the human-virus structural interaction network hold for the larger human-virus binary interaction network, we must employ properties that can be measured with high coverage in both networks. We overcame this obstacle by focusing on the human proteins in these networks (enumerated in Figure 4A), and by selecting properties that were (i) available as high coverage annotations, e.g., major Gene Ontology (GO) Slim Process terms [22]; (ii) based on large-scale experiments, e.g., coexpression with binding partners [23]; or (iii) directly predicted from sequence, e.g., hydropathy index [24]. The complete set of properties that we considered is listed in Table 1. Alpha helix, beta sheet, and disorder propensities were based on previously published amino acid scales [25, 26]; domains and predicted transmembrane helices were assigned using the *Pfam* database [27] and *TMHMM* server [28], respectively; the remaining properties were directly calculated from our previously described network and sequence data. Recall that human proteins involved in the exogenous networks are viral target proteins.

GO Slim Process Terms	Structured vs. Binary	Binary Exo. vs. Binary Endo.	Struct. Exo. vs. Struct. Endo.
Biosynthesis	>	>	
Catabolism	>>	>	
Cell differentiation	>	>	
Cellular process	>	>	
Development	>	>	
Macromolecule metabolism	>	>	
Metabolism	>	>	
Nucleic acid metabolism	>	>	
Regulation of biological process*	>	>	>
Response to stimulus*	>>	>>	>>
Transport	>	>	

Other Properties	Structured vs. Binary	Binary Exo. vs. Binary Endo.	Struct. Exo. vs. Struct. Endo.
Alpha helix propensity	>		
Beta sheet propensity	<		
Disorder propensity	<		
Extent of *Pfam* domain coverage	>	>	
Hydropathy index	<	>	
Predicted transmembrane helices*	<<	<<	
Size	<	<	
Clustering coefficient		<	
Coexpression w/ binding partners*	>>	>>	
Degree in endogenous network*	>>	>>	
Extent of gene duplication	>		

KEY:
Sample is...

3/2 fold enriched	>>
significantly enriched	>
not significantly different	
significantly depleted	<
2/3 fold depleted	<<

Table 1. Properties of human proteins in structured and binary human-virus PPI networks. We explored enrichments for 22 protein properties in three sets of comparisons: structured endogenous vs. binary endogenous, binary exogenous vs. binary endogenous, and structured exogenous vs. structured endogenous. Statistical significance was determined from 10^4 rounds of random sampling and subjected to FDR correction with $\alpha = 0.1$ following the Benjamini-Hochberg method. See the inset key to interpret the significance, sign, and magnitude of the enrichments in each comparison. Starred (*) comparisons are highlighted in Figure 4.

Table 1 further summarizes the results of the comparisons made within and between the structured and binary networks with respect to our chosen protein properties. Statistical significance was determined from 10^4 rounds of random sampling and subjected to FDR correction with $\alpha = 0.1$ following the Benjamini-Hochberg method to account for multiple hypothesis testing. Many of the properties surveyed differ significantly between human proteins found in structured interactions versus those found in binary interactions, and in five cases the differences are very strong (Table 1). In most cases the effect size is quite small, and the statistical significance is mainly due to large sample size. This confirms our intuition that the challenges of protein structure determination impose experimental and investigator biases on the structural interaction network. For example, structured proteins are strongly depleted for predicted transmembrane helices (0.5-fold depletion, $p < 0.0001$), a form of experimental bias, and are consistently enriched for all major GO Slim Process terms, which may be a sign of investigator bias toward well-studied proteins.

Two of the 22 properties are significantly enriched among the structured exogenous group relative to the structured endogenous group: the GO Slim Process terms "regulation of biological process" (1.3-fold enrichment, $p = 0.0074$; Figure 4B) and "response to stimulus" (1.8-fold enrichment, $p < 0.0001$; Figure 4C). These are examples of significant structural exogenous-endogenous trends as illustrated in Figure 3. In both cases we observe a bias from the structural approach: structured endogenous interactors are more enriched for these GO terms than generic human interacting proteins (Figure 4 B and C). However, as in the example from Figure 3B, this bias appears to be uniform: i.e., structured viral targets are also more enriched for regulatory activity and stimulus response than generic viral targets (Figure 4 B and C). As a result, these significant trends in the human-virus SIN also appear in the binary human-virus interaction network. The idea that viral targets are enriched for regulatory activity, in particular regulation of the cell cycle, has been previously suggested [2]. Enrichment for the broad term "response to stimulus" is not surprising given that this term is inclusive of the more specific "immune response" term.

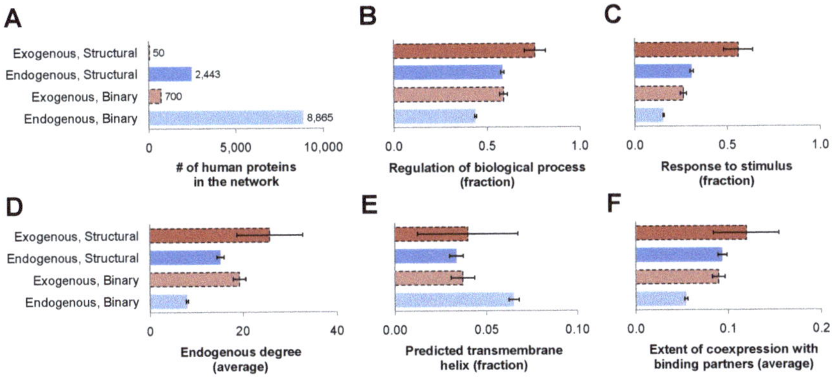

Figure 4. Significant trends in human-virus PPI networks. (**A**) The number of human proteins in each of the four networks; human proteins involved in the exogenous networks are viral target proteins. (**B** & **C**) For these properties we observe a significant exogenous-endogenous trend in the structural networks. In both cases the trend is supported by the binary interaction networks, indicating that it cannot be explained by structural biases, although they are present. (**D**, **E**, & **F**) For these properties we observe a significant exogenous-endogenous trend in the binary networks, but the structural trend is borderline significant (D) or not significant (E & F). Such discrepancies may result from reduced sample size in the structural networks (A), but do not undermine the structural approach. Error bars reflect 95% bootstrap confidence intervals.

There are other properties which differ in a strong, significant manner between the binary exogenous and binary endogenous networks, but not between their structured counterparts (Table 1). For example, as previously demonstrated by Dyer et al. [2], generic viral targets tend to have higher average endogenous degree than generic human proteins, and are said to be more hub-like (2.4-fold increase, $p < 0.0001$; Figure 4D). This trend achieves borderline significance within the human-virus SIN (1.7-fold increase, $p = 0.0560$; Figure 4D). Generic viral targets also tend to be depleted for predicted transmembrane helices (0.6-fold decrease, $p = 0.0013$; Figure 4E) and are more likely to be coexpressed with their endogenous binding partners (1.6-fold increase,

$p < 0.0001$; Figure 4F). Notably, in none of these instances does a significant difference between generic viral targets and human interacting proteins *contradict* a significant difference between their structured counterparts, as postulated in Figure 3D (Table 1). Although there are cases where a significant difference is *absent* in the structural interaction network, in both cases where a significant difference is present, a complementary significant difference is also observed in the binary networks (Table 1; Figure 4 B and C).

The fact that some differences are only observed as significant in the full network speaks to one of the strengths of binary host-pathogen interaction networks: scale. Due to the small size of the human-virus SIN (Figure 4A), and in particular the small number of structured exogenous interactions, a structural systems biology approach may lack the power to detect a weak exogenous-endogenous interaction trend. To understand the ramifications of such a trend in the structural interaction network, including its precise physical basis, will require the generation of a larger number of structured exogenous interactions. Fortunately, interest in determining the structures of host-virus interactions appears to be growing rapidly (Figure 1B). It is also worth noting that a significant trend observed only in the binary network could be explained by experimental and investigator biases that are not found in the human-virus SIN, such as over-representation of certain viruses that have been targeted by large-scale screens.

6. Conclusions

There are many advantages to a structural approach to studying the systems biology of host-pathogen interaction. By previously applying such an approach [14], we demonstrated that, compared to within-host protein-protein interaction (PPI) interfaces, host-virus PPI interfaces tend to be more transient, targeted by more host proteins, more regulatory in function, faster evolving, and more reliant on convergent evolution to achieve interface mimicry. These types of high resolution trends would be difficult or impossible to ascertain using traditional systems biology techniques that rely on binary interaction data.

In this work, we addressed two potential drawbacks inherent to a structural approach to systems biology: reduced coverage, and the introduction of biases from structure determination. With respect to coverage, we argued that—while the human-virus SIN is indeed small compared to current binary networks—the number of exogenous interactions of known structure is growing rapidly, and that these numbers can be further bolstered by the application of homology modeling techniques. Moreover, we emphasized that relative coverage of the total universe of human-virus PPIs is not critical to statistical hypothesis testing, provided that the sample size is sufficiently large to detect a trend of a given effect size. In addition, compared with binary interaction networks, interactions within the human-virus SIN are highly accurate (contain few false positives) and provide a less biased representation of viral diversity, both of which facilitate statistical hypothesis testing.

By analyzing a variety of properties that can be calculated for both structural and binary human-virus interaction networks, we demonstrated that, although strong structural biases exist, they tend to uniformly affect viral target proteins and generic human interacting proteins. In other words, if there is a significant exogenous-endogenous trend in the human-virus structural interaction network, the trend will also be significant in the binary human-virus interaction network. For example, we confirmed here that viral target proteins (both structured and non-structured) tend to be enriched for regulatory activities [2]. This finding supports our original assertion that conclusions based on comparisons and contrasts made within the SIN ought to be minimally confounded by inherent biases from the structural approach. Conversely, it is sometimes possible to observe a significant trend in the binary human-virus interaction network but not in the human-virus SIN. For example, we confirmed here that viral target proteins in the binary network tend to be hubs [2], and showed further that they tend to be co-expressed with their binding partners. In these cases, a discrepancy between the structured and traditional approach is acceptable: due to the reduced size of the SIN relative to the full binary network, the structural approach will sometimes lack the statistical power to detect weaker trends.

Previously, we demonstrated that the interfaces mimicked by viruses tend to be otherwise used by multiple, transiently bound (date-like) human interactors [14]. However, in this work, we found that viral target proteins in the binary interaction network tend to be coexpressed with their endogenous binding partners ($p < 0.0001$), a hallmark of constitutive, "party-like" protein-protein interaction—the opposite of transient, date-like behavior. These two results seem contradictory, but are in fact compatible. In the structural systems biology approach, a protein is a 3D structure, and we are able to make statements about its individual interfaces. In the traditional systems biology approach, a protein is a node, and so we can only speak of the protein's average behavior (or the behavior of an "average interface" of that protein). Hence, while the average endogenous interaction involving a generic viral target protein tends to involve coexpression, the specific interactions and interfaces targeted by viruses may still tend to be transient. In other words, party-like hubs may contain date-like interfaces, and these interfaces may be preferentially targeted by viruses (perhaps due to regulatory function). Such contrasts point to the fact that traditional systems biology research is highly complementary to the structural approach, and we expect both methods to continue to contribute to our understanding of host-pathogen interactions in the near future.

References

1. Gardy, J.L., et al. *Trends Immunol*, **30**:249-262 (2009).
2. Dyer, M.D., T.M. Murali, and B.W. Sobral. *PLoS Pathog*, **4**:e32 (2008).
3. Maxwell, K.L. and L. Frappier. *Microbiol Mol Biol Rev*, **71**:398-411 (2007).
4. Tan, S.L., et al. *Nat Biotechnol*, **25**:1383-9 (2007).
5. Calderwood, M.A., et al. *Proc Natl Acad Sci U S A*, **104**:7606-11 (2007).
6. von Schwedler, U.K., et al. *Cell*, **114**:701-13 (2003).
7. Brass, A.L., et al. *Science*, **319**:921-6 (2008).
8. Uetz, P., et al. *Science*, **311**:239-42 (2006).
9. Aloy, P. and R.B. Russell. *Nat Rev Mol Cell Biol*, **7**:188-97 (2006).
10. Beltrao, P., C. Kiel, and L. Serrano. *Curr Opin Struct Biol*, **17**:378-84 (2007).
11. Kiel, C., P. Beltrao, and L. Serrano. *Annu Rev Biochem*, **77**:415-441 (2008).

12. Kim, P.M., et al. *Science*, **314**:1938-41 (2006).
13. Kim, P.M., et al. *Mol Syst Biol*, **4**:179 (2008).
14. Franzosa, E.A. and Y. Xia. *Proc Natl Acad Sci U S A*, **108**:10538-43 (2011).
15. Aranda, B., et al. *Nucleic Acids Res*, **38**:D525-31 (2010).
16. Chatr-aryamontri, A., et al. *Nucleic Acids Res*, **37**:D669-73 (2009).
17. The UniProt Consortium. *Nucleic Acids Res*, **38**:D142-8 (2010).
18. Altschul, S.F., et al. *Nucleic Acids Res*, **25**:3389-402 (1997).
19. Franzosa, E.A. and Y. Xia. *Mol Biol Evol*, **26**:2387-95 (2009).
20. Berman, H.M., et al. *Nucleic Acids Res*, **28**:235-42 (2000).
21. Davey, N.E., G. Trave, and T.J. Gibson. *Trends Biochem Sci*, **36**:159-69 (2011).
22. Barrell, D., et al. *Nucleic Acids Res*, **37**:D396-403 (2009).
23. Lee, H.K., et al. *Genome Res*, **14**:1085-94 (2004).
24. Kyte, J. and R.F. Doolittle. *J Mol Biol*, **157**:105-32 (1982).
25. Koehl, P. and M. Levitt. *Proc Natl Acad Sci U S A*, **96**:12524-9 (1999).
26. Campen, A., et al. *Protein Pept Lett*, **15**:956-63 (2008).
27. Finn, R.D., et al. *Nucleic Acids Res*, **36**:D281-8 (2008).
28. Krogh, A., et al. *J Mol Biol*, **305**:567-80 (2001).

IDENTIFICATION OF CELL CYCLE-REGULATED, PUTATIVE HYPHAL GENES IN *CANDIDA ALBICANS*

RALUCA GORDÂN†

Division of Genetics, Department of Medicine,
Brigham & Women's Hospital and Harvard Medical School,
Boston, MA 02115, USA
Current address: Department of Biostatistics and Bioinformatics,
Institute for Genome Sciences and Policy,
Duke University
Email: raluca.gordan@duke.edu

SAUMYADIPTA PYNE†

Department of Medical Oncology,
Dana-Farber Cancer Institute, Harvard Medical School,
Boston, MA 02115, USA
Broad Institute of MIT and Harvard, Cambridge, MA 02142, USA
Email: spyne@broad.mit.org

MARTHA L. BULYK†

Division of Genetics, Department of Medicine, Department of Pathology,
Brigham & Women's Hospital and Harvard Medical School, Boston, MA 02115, USA
Harvard-MIT Division of Health Sciences & Technology (HST)
Harvard Medical School, Boston, MA 02115, USA
Email: mlbulyk@receptor.med.harvard.edu

†Correspondence can be addressed to R.G., S.P. or M.L.B.

Candida albicans, a major fungal pathogen in human, can grow in a variety of morphological forms ranging from budding yeast to pseudohyphae and hyphae, and its ability to transition to true hyphae is critical for virulence in various types of *C. albicans* infections. Here, we identify 17 putative hyphal genes whose expression peaks during the S/G2 transition of the cell cycle in *C. albicans*. These genes are *Candida*-specific (*i.e.*, they do not have orthologs in *S. cerevisiae*, a related fungal species that does not exhibit hyphal growth and is primarily non-pathogenic), and their promoters are enriched for the DNA binding site motifs of Tec1 and Rfg1, two transcription factors (TFs) known to play important roles in hyphal growth and virulence. For 5 of the 17 genes we found strong evidence in the literature that confirms our hypothesis that these genes are involved in hyphal growth and/or virulence, for 5 additional genes we found suggestive (albeit weak) evidence, while the other genes remain to be tested. It will be interesting to determine in future studies whether these 17 putative hyphal genes, whose expression peaks during the S/G2 transition, are part of a mechanism for this pathogenic fungus to 'turn on' hyphal growth late during the cell cycle, or if these genes are used to sustain hyphal growth and ensure that the cell does not transition back to yeast growth. In either case, the involvement of these genes in hyphal growth makes them putative targets for new antifungal drugs aimed at inhibiting hyphae formation in *C. albicans*.

1. Introduction

Candida albicans is a major human pathogen and the number one cause of fungal infection in human. Unlike other pathogens, it can be found in skin and the gastrointestinal tract as a harmless commensal organism, producing serious disease in people with weakened immune systems [1]. *C. albicans* is a truly polymorphic organism: it has the ability to undergo morphological changes between the yeast form (with rounded cells and daughter buds that physically separate from the mother cell), the pseudohyphal form (which consists of chains of cells with various degrees of elongation that still show constrictions between adjacent cells), and the true hyphal form (which consists of long tubes with parallel sides and no constrictions) [2]. Yeast cells disseminate more easily in the bloodstream, while hyphae are invasive and can penetrate host tissues during the early stages of infections [2]. Furthermore, switching of *C. albicans* to the hyphal form in the host has long been considered to be important for pathogenesis, since mutants defective in hyphal growth are known to be less virulent [3]. Thus, identification of genes involved in the yeast-to-hyphae transition is important for the development of new antifungal agents.

Here, we identify 17 putative hyphal genes that have a particular characteristic: they are periodically expressed during the *C. albicans* cell cycle, with their expression peaking during the S/G2 transition. We analyzed the gene expression data of Côte *et al.* [4], who examined the periodic expression of genes through the cell cycle in cultures of *C. albicans* synchronized by mating pheromone treatment. Côte *et al.* reported a set of 494 genes that are periodically expressed during the cell cycle, 100 of which do not have homologs in *S. cerevisiae*, a related fungal species that does not exhibit hyphal growth and is primarily non-pathogenic [5]. We henceforth refer to these 100 genes as "*Candida*-specific", and we anticipate that at least some of these genes may be necessary for hyphal growth or pathogenicity.

We investigated the transcriptional regulation of the *Candida*-specific genes, in an attempt to find possible clues about *C. albicans* hyphal growth and its connection to the cell cycle. We analyzed the promoter regions of periodically expressed genes that peak during different cell cycle transitions: G1/S, S/G2, G2/M, and M/G1. For each cell cycle transition we performed two motif enrichment analyses to identify: 1) DNA motifs enriched upstream of genes that peak at that particular transition, and 2) DNA motifs enriched upstream of *Candida*-specific genes that peak at that particular transition. Since high-resolution DNA binding site motifs, such as motifs derived from protein binding microarray (PBM) [6], SELEX-seq [7], or MITOMI [8] data, are not available for *C. albicans* TFs, we used as a proxy 139 high-resolution motifs of *S. cerevisiae* TFs [9-11] (see Section 2.2). To find significantly enriched motifs in the promoters of *C. albicans* cell cycle-regulated genes, we use a method that we developed recently to compute the enrichment of TF DNA binding motifs in genome-scale chromatin immunoprecipitation data on *in vivo* TF occupancy (ChIP-chip) [11]. Previously, we used this method successfully to distinguish between direct and indirect TF-DNA interactions in the yeast *S. cerevisiae*. Here, we apply a similar enrichment analysis to the sets of promoters of *C. albicans* cell cycle-regulated genes.

We find that the DNA motifs of Tec1 and Rfg1, two known regulators of hyphal growth and virulence [12], are significantly enriched upstream of *Candida*-specific genes that peak during the S/G2 transition, and are not enriched in general upstream of genes that peak at this stage. Since these 17 genes are regulated by hyphal TFs and do not have orthologs in *S. cerevisiae* (which is non-pathogenic and does not form true hyphae), we hypothesize that the 17 genes may be involved in hyphal growth and/or virulence. To test this hypothesis, we performed a literature search to see whether these genes are overexpressed during hyphal growth or whether there is any evidence of a role in virulence. Most of the 17 genes have unknown functions [13] and are not well represented in *C. albicans* gene expression data. Despite this fact, our literature search revealed strong evidence that 5 of the 17 genes are indeed involved in hyphal growth. For 5 additional genes we found suggestive (albeit weak) evidence, while the other 7 genes remain to be tested (see Table 1).

Since the expression of these 17 *Candida*-specific genes peaks at late stages of the cell cycle (more precisely, the S/G2 transition), and since it has been shown previously that *C. albicans* can be induced to start hyphal growth not only in G1 but also in later stages [14], our results suggest that these genes may be part of a mechanism for this pathogenic fungus to 'turn on' hyphal growth late during the cell cycle. Alternatively, the genes may be important for sustaining hyphal development throughout the cell cycle and preventing the cell from transitioning back to yeast growth.

2. Data and Methods

2.1. *Candida albicans gene expression data*

We used the cell cycle gene expression data of Côte *et al.* [3] to build sets of periodically expressed genes that are potentially regulated by a shared set of TFs. Côte *et al.* examined the periodic expression of genes through the cell cycle in cultures of *C. albicans* synchronized by mating pheromone treatment, and found 494 genes that are periodically expressed and peak at different cell cycle transitions: G1/S, S/G2, G2/M, or M/G1. The samples were collected at 0, 30, 60, 90, 120, 150, and 180 minutes after pheromone treatment, and each time point was manually assigned to a cell cycle transition [3]. For each cell cycle transition we built a "foreground" set of sequences that contains the 1-kb regions upstream of the transcription start sites of all *C. albicans* genes whose expression peaks at that particular transition, as shown in Figure 1 for the S/G2 transition. Next, for each foreground set (e.g., $F_{S/G2}$) we built a corresponding "background" set (e.g., $B_{S/G2}$) that contains the 1-kb promoter regions of the remaining *C. albicans* genes. Having constructed the foreground and background sequence sets, we next computed the enrichment of each query DNA motif in the foreground sequences as compared to the background sequences, for each cell cycle transition.

Of the 494 genes that are periodically expressed during the *C. albicans* cell cycle, 100 genes do not have orthologs in the non-pathogenic yeast *S. cerevisiae* [4]. We used the promoters of these *Candida*-specific genes to construct new sets of foreground (e.g., $FC_{S/G2}$) and background sequences for each cell cycle transition, to search for significantly enriched motifs upstream of the cell cycle-regulated *Candida*-specific genes. We were particularly interested to see whether there are TF motifs significantly enriched upstream of *Candida*-specific genes but not enriched upstream of all genes that peak at a particular time during the cell cycle.

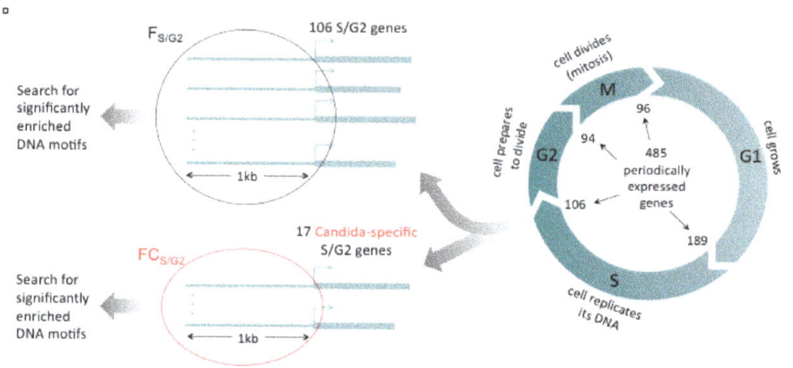

Figure 1. Building the sets of foreground sequences. We use the promoter regions of genes that peak at each cell cycle transition to construct sets of foreground sequences that are then searched for significantly enriched motifs.

2.2. *TF binding site motif data*

Ideally we would perform the motif enrichment analysis using *C. albicans* high-resolution TF DNA binding site motifs, such as motifs derived from PBMs [6], SELEX-Seq [7], or MITOMI [8] data. However, since such motifs are not available for *C. albicans* TFs, in our analyses we used *S. cerevisiae* TF motifs as a proxy. We tested a previously assembled collection of 139 high-resolution motifs derived from universal PBM data [9-11]. For the TF binding site motifs significantly enriched in particular *C. albicans* data sets (including Tec1 and Rox1/Rfg1 – see Section 3.1), we compared the DNA binding domain (DBD) of the *S. cerevisiae* TF against the DBD of the *C. albicans* ortholog to ensure that the DBDs of the two proteins are similar, implying that the *C. albicans* TF likely has highly similar, if not essentially the same, DNA-binding specificity as its *S. cerevisiae* ortholog. To verify the similarity between an *S. cerevisiae* TF and its *C. albicans* ortholog, we performed a BLASTP search using the DBD of the *S. cerevisiae* TF (as defined in UniProt) or of the entire protein if the DBD was not well defined, and required that the *C. albicans* ortholog be recovered at an E-value < 1e-10. The DNA binding specificity of each TF is represented as a position weight matrix (PWM) [15].

2.3. *Method for computing enrichment of DNA motifs*

We recently developed a novel method for computing the enrichment of a TF DNA binding site motif in a set of foreground DNA sequences compared to a set of background sequences (derived from ChIP-chip data), and we used this method to successfully distinguish between direct and indirect TF-DNA interactions in *S. cerevisiae* [11]. Here, we apply a similar enrichment method to the sets of promoter regions of *C. albicans* cell cycle-regulated genes [4]. Previously [11], we used our method to compare the enrichment of several TFs in each set of foreground sequences; in this work, we compare the enrichment of each TF across several foreground sets (see below). Formally, the method can be described as follows.

Let $\boldsymbol{F}$ and $\boldsymbol{B}$ denote the sets of foreground and background sequences, respectively (e.g., $F_{S/G2}$ and $B_{S/G2}$, as described in Section 2.1). Let $\boldsymbol{T}$ denote a TF, and $\boldsymbol{M}$ denote the PWM describing the DNA binding specificity motif of $\boldsymbol{T}$: $\boldsymbol{M}(b, j)$ = the probability of finding base b at location j within the binding site (b = A, C, G, or T, and $1 \le j \le k$, where k is the motif width). Let $\boldsymbol{Q}$ denote the background nucleotide frequencies. Given a DNA site $\boldsymbol{S} = S_1S_2\ldots S_k$, we score it according to the PWM and background models, and use the ratio of the two scores to approximate the dissociation constant:

$$K_d(\boldsymbol{T},\boldsymbol{S}) = \prod_{j=1}^{k} \left(Q(S_j)/\boldsymbol{M}(S_j,j)\right) \tag{1}$$

Next, we write the probability that $\boldsymbol{T}$ binds $\boldsymbol{S}$ as:

$$P(\boldsymbol{T}\text{ binds } S) = \frac{[T\cdot S]}{[T\cdot S]+[S]} = \frac{[\boldsymbol{T}]}{[\boldsymbol{T}]+K_d(\boldsymbol{T},\boldsymbol{S})} = 1\Big/\left(1+\frac{1}{[\boldsymbol{T}]}\times\prod_{j=1}^{k}\frac{Q(S_j)}{\boldsymbol{M}(S_j,j)}\right) \tag{2}$$

where the concentration of free TF, [$\boldsymbol{T}$], is set to the dissociation constant for the site with the optimal PWM score, as in the GOMER [16] model (this is equivalent to setting the TF occupancy of the optimal binding site to 50%). For a DNA sequence $\boldsymbol{X}$ longer than the motif width k, the probability that $\boldsymbol{T}$ binds $\boldsymbol{X}$ is:

$$P(\boldsymbol{T}\text{ binds } \boldsymbol{X}) = P(\boldsymbol{T}\text{ binds any } X_{i\ldots i+k-1}) = 1-\prod_{i=1}^{n-k+1}\left(1-1\Big/\left(1+\frac{1}{[\boldsymbol{T}]}\times\prod_{j=i}^{i+k-1}\frac{Q(X_j)}{\boldsymbol{M}(X_j,j-i+1)}\right)\right) \tag{3}$$

Previously [11], we also considered the nucleosome occupancy when computing P($\boldsymbol{T}$ binds $\boldsymbol{S}$) (Eq. (3)). However, since here we are analyzing cell cycle expression data and nucleosome occupancy data for *C. albicans* cells in particular stages of the cell cycle are not yet available, here we do not incorporate DNA accessibility into the score.

After scoring all foreground and background sequences using Eq. (3), we use the computed scores to construct a receiver operating characteristic (ROC) curve and calculate the area under the ROC curve (AUC) as a measure of enrichment of the PWM in the foreground set as compared to the background set. An AUC value of 1 corresponds to perfect enrichment (*i.e.*, all foreground sequences have higher scores than the background sequences), while an AUC of 0.5 corresponds to lack of any enrichment (or of depletion), such as would be obtained using a random motif.

To compute the statistical significance of an AUC enrichment value (which corresponds to the results obtained for a particular PWM and particular sets of foreground and background sequences), we randomly permute the PWM 1,000 times as previously described [11], and for each random PWM we compute its AUC enrichment in the foreground versus the background sequences. We use the 1,000 AUC values of random motifs to compute an empirical p-value for the PWM of interest. We consider the PWM significantly enriched if it has an AUC >= 0.6 and an associated p-value <= 0.005 (i.e., for at most 2 out of 1,000 random motifs we obtain an AUC greater than or equal to the AUC of the real motif). The Perl scripts used to conduct the analyses are available online at http://thebrain.bwh.harvard.edu/psb2012.

3. Results

3.1. *DNA motifs significantly enriched upstream of periodically expressed genes*

When searching for DNA motifs significantly enriched upstream of genes that peak during specific cell cycle transitions in *C. albicans*, we recover known master regulators of the cell cycle: Mbp1, Swi4, Mcm1, Fkh2, etc. (Figure 2).

When we restrict the analysis to *Candida*-specific genes, we again recover the motifs of the master regulators at the various cell cycle transitions. In addition, we find that the DNA binding site motifs of *S. cerevisiae* TFs Tec1 and Rox1 are significantly enriched upstream of S/G2 genes unique to *C. albicans*, and are not significantly enriched in the set of all S/G2 genes. The orthologs of these two *S. cerevisiae* TFs in *C. albicans* are Tec1 and Rfg1, respectively, and they are both known regulators of hyphal growth and virulence in *C. albicans* [12]. Thus, we hypothesize that the 17 *Candida*-specific genes whose expression peaks at the S/G2 transition may also play a role in hyphal growth and/or virulence (see below).

Since we used *S. cerevisiae* motifs, and not *C. albicans* motifs, in our motif enrichment analysis, we checked to make sure that the proteins are well conserved over their DNA binding domains (TEA/ATTS domain for Tec1, and HMG box domain for Rox1/Rfg1), which implies that the *C. albicans* TFs likely have highly similar DNA binding specificities as their *S. cerevisiae* orthologs. Conservation over the DBDs was high, with BLASTP E-values as follows for the *C. albicans* orthologs recovered for the TFs shown in Figures 2 and 3: Fkh2-Fkh2 (1e-37), Mcm1-Mcm1 (1e-25), Hcm1-Hcm1 (1e-26), Swi4-Swi4 (1e-66), Mbp1-Mbp1 (1e-11), Azf1-Azf1 (1e-74), Ace2-Ace2 (1e-38), Tec1-Tec1 (1e-17), Rox1-Rfg1 (1e-21).

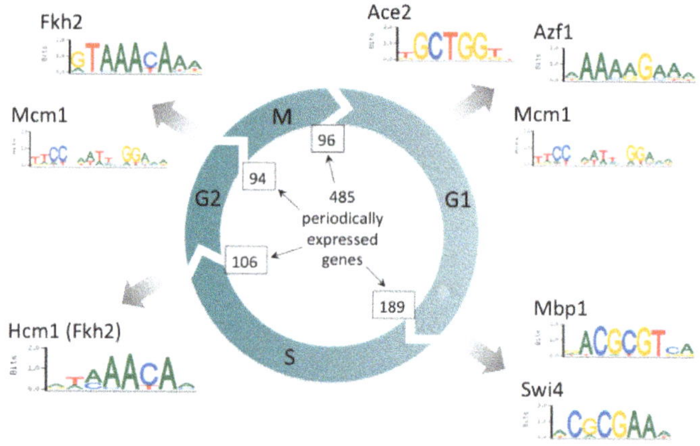

Figure 2. Enriched DNA motifs. The DNA binding site motifs of master regulators of the cell cycle are significantly enriched upstream of genes periodically expressed during the *C. albicans* cell cycle.

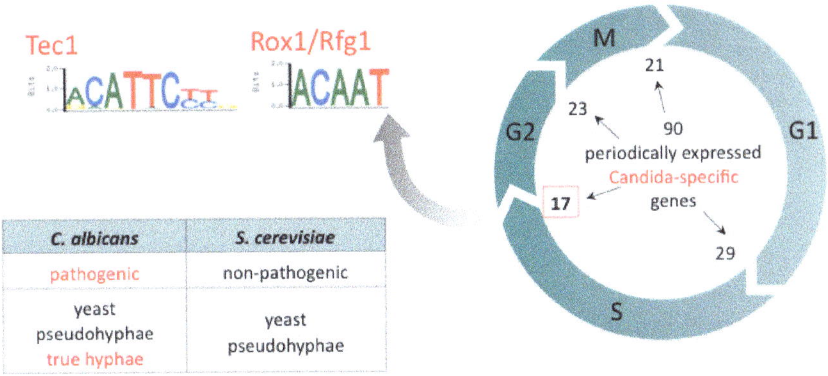

Figure 3. Motifs specifically enriched upstream of *Candida*-specific genes. We find that the DNA binding site motifs of *S. cerevisiae* TFs Tec1 and Rox1 (orthologs of *C. albicans* TFs Tec1 and Rfg1, respectively) are significantly enriched upstream of *Candida*-specific S/G2 genes.

3.2. *Tec1 and Rfg1 – regulators of hyphal growth and virulence*

Tec1 is a well-studied TEA/ATTS TF conserved across many fungal species, including *S. cerevisiae* and *C. albicans*. In *S. cerevisiae*, this factor is known to be required for haploid invasive and diploid pseudohyphal growth [17]. In *C. albicans*, Tec1 is involved in regulation of hypha-specific genes and it is required for wild-type biofilm formation [13]. *C. albicans* Tec1 acts downstream of Efg1, a protein that plays a critical role in hyphal morphogenesis [18]. Efg1 is required for expression of all hyphal-specific genes, and Tec1 overexpression has been shown to restore filamentous growth in an *efg1/efg1* mutant [12]. Furthermore, Tec1 is also involved in pathogenesis, as *TEC1* homozygous null mutants show decreased or absent hyphal growth and virulence [13].

The *C. albicans* TF Rfg1 is the ortholog of *S. cerevisiae* Rox1. However, unlike Rox1 in *S. cerevisiae*, *C. albicans* Rfg1 is not responsible for hypoxic repression [19]. Instead, it regulates filamentous growth and hyphal genes, acting in both Tup1p-dependent and -independent pathways [13]. Similarly to Tec1, Rfg1 controls both filamentous growth and virulence [19], and *RFG1* homozygous null mutants show decreased filamentous growth and are not virulent [13].

3.3. *Prior literature support for the hypothesis that the 17 S/G2 Candida-specific genes are putative hyphal genes*

Unlike *S. cerevisiae*, *C. albicans* is truly dimorphic: it has the ability to undergo morphological changes between the yeast, pseudohyphal, and the true hyphal forms. The fact that the 17 S/G2 *Candida*-specific genes do not have homologs in *S. cerevisiae* and seem to be regulated by Tec1 and Rfg1, two known regulators of hyphal genes, suggests that these 17 genes are also involved in hyphal growth and they may provide a connection between cell cycle and polymorphism in *C. albicans*. These genes may be involved in a mechanism that acts late during the *C. albicans* cell cycle and provides a last chance for the cell to commit to hyphal growth before entering a new cycle.

Testing putative hyphal genes in the laboratory is not trivial. Genetic studies in *C. albicans* are especially challenging because of its diploid genome, chromosomal instability, and incomplete sexual cycle [14, 20]. Nevertheless, several genes involved in hyphal growth have been identified thus far, so we searched the literature for evidence that supports our hypothesis that the 17 *Candida*-specific genes are part of the hyphal morphogenesis program. Except for orf19.3430 (*BUD21*) and orf19.6877 (*PNG2*), the remaining 15 genes are annotated in the *Candida* Genome Database (CGD) [13] as "Uncharacterized ORFs", and thus there is little information about them in the literature. Still, for two of the 17 genes (orf19.5848 and orf19.4905) Nantel *et al.* [21] have reported a significant increase in gene expression during yeast-to-hyphal transition. For one additional gene (*BDA1* or orf19.376), the homozygous deletion mutant shows substantial morphology defects [20], while the null mutant of orf19.4905 (a putative MFS* transporter) shows abnormal infectivity in a mouse infection model [20]. For two additional genes (orf19.3516 and orf19.3430), invasive growth is decreased in a heterozygous null mutant [5]. Relevant information about all 17 genes is summarized in Table 1. Most of these genes are well conserved across related fungi, but their orthologs are also uncharacterized.

We also note that although for some of these genes we have not found strong evidence of their direct involvement in hyphal growth, there is weak evidence supporting our hypothesis (see Table 1). For example, orf19.5549 encodes a protein that is very rich in Ser residues, a characteristic of cell surface proteins (which play an important role in the pathogenic process). Six of the 17 genes encode proteins that contain predicted transmembrane domains [13] (orf19.5549, orf19.5848, orf19.4905, orf19.3430, orf19.1350, orf19.6877), and 4 were predicted to contain signal peptides (orf19.5549, orf19.5848, orf19.7606, orf19.876), which are important for directing proteins to the cell wall and for secretion into the extracellular matrix. The gene *PNG2* was predicted to contain three PRICHEXTENSN domains characteristic of proline-rich extensins (plant cell wall proteins with functions in cell wall strengthening [23]). Another gene, orf19.876, codes for a putative GPI-anchored protein, and many hypha-specific genes encode GPI-anchored cell surface proteins [12].

4. Discussion

In this work, we identify 17 putative hyphal genes in the pathogenic yeast *C. albicans*, which have the specific characteristic that their expression is cell cycle-regulated and peaks at the S/G2 transition. These 17 genes are *Candida*-specific and seem to be regulated by Tec1 and Rfg1, two master regulators of hyphal growth, and thus we hypothesize that the 17 genes are also involved in hyphal growth. A literature search revealed evidence that supports our hypothesis for 10 of the 17 genes, while the other seven remain to be verified in the laboratory.

---

* Major facilitator superfamily (MFS) is one of the two major superfamilies of plasma membrane efflux proteins involved in antifungal drug resistance [22].

Table 1. *Candida*-specific genes that peak at the S/G2 transition.

ORF (gene name)	Evidence strength	Gene information, protein information, and/or evidence of role in hyphal/invasive growth
orf19.5549	weak	Proteins rich in Ser residues
orf19.5848	strong	Late-stage biofilm-induced gene [24] Upregulated during yeast-to-hyphal transition [21]
orf19.6238	weak	CGD molecular function: oxidoreductase activity PSI-BLAST: potential FAD-dependent oxidoreductase, similarity to Ser/Thr-protein kinase Chk2 in *P. pastoris*
orf19.376 *(BDA1)*	strong	Mutant shows substantial morphology defects; [20]
orf19.4905	strong	Mutant shows abnormal infectivity; putative MFS transporter [20] Upregulated during yeast-to-hyphal transition [21] CGD biological process: transmembrane transport
orf19.7606	no evidence	
orf19.836	no evidence	CGD description: Protein likely to be essential for growth, based on an insertional mutagenesis strategy
orf19.3516	strong	Invasive growth is decreased in a heterozygous null mutant [5] CGD molecular function: carbonate dehydratase activity
orf19.389	no evidence	CGD description: Hap43p-induced gene
orf19.3430 *(BUD21)*	strong	Invasive growth is decreased in a heterozygous null mutant [5] CGD description: plasma membrane-associated protein
orf19.1350	weak	Included in the "opaque-induced" transcriptional module [25] CGD molecular function: electron carrier activity; protein disulfide oxidoreductase activity PSI-BLAST: Thioredoxin_like superfamily
orf19.3245	no evidence	
orf19.876 *(PGA33)*	weak	CGD description: putative GPI-anchored protein CGD cellular component: cell surface
orf19.1050	no evidence	
orf19.6579	no evidence	
orf19.6877 *(PNG2)*	weak	CGD: transcription upregulated by treatment with caspofungin, ciclopirox olamine, ketoconazole or hypoxia; gene of core caspofungin response; CGD biological process: protein deglycosylation CGD cellular component: plasma membrane CGD conserved domains: 2 PNGaseA domains and 3 PRICHEXTENSN (proline-rich extensin signature) domains.
orf19.3871 *(DAD3)*	no evidence	Subunit of the Dam1 (DASH) complex, which acts in chromosome segregation by coupling kinetochores to spindle microtubules [13]

For all the analyses presented here we used *S. cerevisiae* TF DNA binding site motifs because high-resolution motifs are not yet available for *C. albicans*. However, some *C. albicans* TFs may have slightly different DNA binding specificities as compared to their *S. cerevisiae* orthologs. Furthermore, some TFs do not have orthologs in *S. cerevisiae*. Thus, once high-resolution DNA binding site motifs become available for *C. albicans* TFs, it will be interesting to repeat the analyses presented here to determine whether additional DNA motifs are significantly enriched upstream of *Candida*-specific genes. In addition, future analyses that combine *C. albicans* TF DNA binding data with cell cycle gene expression data and also gene expression data obtained during the yeast-to-hyphal transition could help us to understand how hyphal growth is initiated and maintained during the cell cycle, and what are the roles of these *Candida*-specific genes.

When computing the enrichment of a DNA motif in a set of foreground sequences compared to a set of background sequences, a commonly used approach is to search for good matches to the motif in both the foreground and background sets, and then use Fisher's exact test (*i.e.*, the hypergeometric p-value) to determine whether the motif is overrepresented in the foreground sequences [26, 27]. This approach is sensitive to the size of the two sets and the cutoff used to determine "matches" to the DNA motif, which is why we used an alternative, AUC-based approach (see Section 2.3.) that allowed us to directly compare the enrichment of different DNA motifs in different sets of sequences. Using the AUC-based approach we were able to identify the two motifs (Tec1 and Rox1/Rfg1) that are enriched upstream of *Candida*-specific S/G2 genes and not enriched in general upstream of S/G2 genes. For comparison, we also performed an enrichment analysis using Fisher's exact test (see Supplementary Figure 1, available online at http://thebrain.bwh.harvard.edu/psb2012). The results were inconclusive: we did not find a significant enrichment for either Tec1 or Rox1/Rfg1, and it was unclear whether the motifs were more enriched in the *Candida*-specific promoters or all S/G2 promoters. Furthermore, as expected, the computed p-values for the Tec1 and Rox1/Rfg1 motifs varied widely depending on the motif cutoff. Our AUC-based approach alleviates the need to choose a motif cutoff by taking into account all possible binding sites in a given sequence, weighted according to how well they match the motif of interest. Our approach is not limited to *Candida*-specific genes or to cell cycle data, but rather can be used for any organism to identify regulators of organism-specific genes that exhibit a particular characteristic.

Our finding that the DNA binding site motifs of hyphal TFs Tec1 and Rox1 are enriched upstream of the 17 *Candida*-specific S/G2 genes is intriguing. We had expected to find the DNA binding site motifs of master hyphal growth regulators enriched upstream of genes whose expression peaks early in the cell cycle, most likely in the G1 phase, since it has been proposed that there is a point of phenotypic commitment to hyphal growth in G1 [14, 28]. However, the 17 *Candida*-specific, putative hyphal genes we identified peak at the S/G2 transition, which suggests that they may be involved in the initiation of hyphae formation (or hyphal evagination, given that some genes may encode cell surface proteins) late during the cell cycle. We note that whether hyhae can be induced from all cell cycle stages in *C. albicans* is a matter of some controversy [29]. Still, some studies have shown that *C. albicans* can be induced to start hyphal growth not only in G1, but also in later stages (definitely S and G2, and possible M also) [14], so the 17 *Candida*-specific S/G2 genes could potentially be involved in hyphal initiation.

The ability of *C. albicans* to switch between yeast and hyphal growth has been linked to its virulence [3], so involvement of the 17 *Candida*-specific genes in hyphal growth suggests that they may also be important for virulence/pathogenesis†. We note that the connection between morphological switching and virulence is controversial in the *Candida* literature, as it has recently been demonstrated that the yeast-to-hyphae transition is not necessarily required for infectivity in a mouse model of disseminated candidiasis and infection of the kidney [20]. However, it is still generally accepted that hyphal growth is critical for virulence in many types of *C. albicans* infections [31].

The hypothesis that the 17 *Candida*-specific genes may be important for virulence is supported by the fact that the TFs Tec1 and Rfg1, whose DNA binding site motifs are enriched upstream of the 17 genes, play important roles in both hyphal growth and virulence [13]. Reduced virulence has been observed in both *TEC1* and *RFG1* mutants, and also in *EFG1* mutants [12] (the TF Efg1 acts upstream of Tec1 in several signal transduction pathways that regulate the yeast-to-hyphae transition [18]). Efg1 is known to be important for *C. albicans* pathogenesis in several *in vivo* models of infection [12], and one of pathways known to regulate yeast-to-hyphae transition via Efg1 (a cAMP-dependent protein kinase pathway) plays a major role in both hyphal development and pathogenesis [12]. Furthermore, at least one of the 17 *Candida*-specific genes (orf19.4905) has been shown to play a role in pathogenesis, as a homozygous null mutant showed abnormal infectivity in a mouse model [20]. Thus, a number of the other 16 putative hyphal genes we identified may also be important for *C. albicans* pathogenesis. We note that acquiring new genes is not the only way to develop virulence/pathogenicity. Some protein-coding genes present in both *S. cerevisiae* and *C. albicans* may have acquired coding mutations that allow them to be more virulent. Some genes conserved at the sequence level may have diverged in their gene expression due to changes in non-coding *cis*-regulatory elements. Such genes should also be included in future analyses aimed at deciphering regulatory networks of *C. albicans* virulence.

Identifying genes responsible for hyphal growth and virulence in *C. albicans* is undoubtedly important for understanding how this fungus invades human cells and how it switches from a harmless organism to a pathogen. The work presented here provides a useful resource for future studies of *Candida*-specific hyphal and virulence genes. Such studies may aid in understanding the biology of this important human pathogen.

Acknowledgments

We thank Gerry Fink and James Konopka for helpful discussion. This work was supported by NIH/NHGRI grant # R01 HG003985 (M.L.B.). R.G. was funded in part by an American Heart Association postdoctoral fellowship #10POST3650060.

† The terms virulence and pathogenesis are closely related. The former is viewed from the point of view of the fungus, while the latter from its effect on the host [30].

References

1. T. E. Zaoutis, J. Argon, J. Chu, J.A. Berlin, T. J. Walsh and C. Feudtner, *Clin. Infect. Dis.* **41**, 1232-9 (2005)
2. P. Sudbery, N. Gow and J. Berman, *Trends Microbiol.* **12**, 317-24 (2004)
3. H. J. Lo, J. R. Kohler, B. DiDomenico, D. Loebenberg, A. Cacciapuoti and G. R. Fink, *Cell* **90**, 939-49 (1997)
4. P. Côte, H. Hogues and M. Whiteway, *Mol. Biol. Cell.* **20**, 3363-73 (2009)
5. J. Oh, E. Fung, U. Schlecht, R. W. Davies, G. Giaever, R. P. St Onge, A. Deutschbauer and C. Nislow, *PLoS Pathog.* **6**, e1001140 (2010)
6. M. F. Berger, A. A. Philippakis, A. M. Qureshi, *et al.*, *Nat. Biotechnol.* **24**, 1429-35 (2006)
7. A. Jolma, T. Kivioja, J. Toivonen, L. Cheng, G. Wei, M. Enge, M. Taipale, J. M. Vaquerizas, J. Yan, M. J. Sillanpää, M. Bonke, K. Palin, *et al.*, *Genome Res.* **20**, 861-73 (2010)
8. S. J. Maerkl and S. R. Quake, *Science* **315**, 233-7 (2007)
9. C. Zhu, K. Byers, R. McCord, Z. Shi, M. Berger, D. E. Newburger, K. Saulrieta, Z. Smith, M. V. Shah, M. Radhakrishnan, A. A. Philippakis, *et al.*, *Genome Res.* **19**, 556-66 (2009).
10. G. Badis, E.T. Chan, H. van Bakel, L. Peña-Castillo, *et al.*, *Mol. Cell* **32**, 878-87 (2008)
11. R. Gordân, A. J. Hartemink and M. L. Bulyk, *Genome Res.* **19**, 2090-100 (2009).
12. H. Liu, *Int. J. Med. Microbiol.* **292**, 299-311 (2002)
13. M. S. Skrzypek, M. B. Arnaud, M. C. Costanzo, D. O. Inglis, P. Shah, G. Binkley, S. R. Miyasato and G. Sherlock, *Nucleic Acids Res.* **38**, D428-32 (2010)
14. I. Hazan, M. Sepulveda-Becerra and H. Liu, *Mol. Biol. Cell* **13**, 134-45 (2002)
15. G. D. Stormo, *Bioinformatics* **16**, 16-23 (2000)
16. J. A. Granek and N. D. Clarke, *Genome Biol.* **6**, R87 (2005).
17. T. Köhler, S. Wesche, N. Taheri, G. H. Braus, H. U. Mösch, *Eukaryot. Cell* **1**, 673-86 (2002)
18. J. Berman and P. E. Sudbery, *Nat. Rev. Genet.* **3**, 918-30 (2002)
19. D. Kadosh and A. D. Johnson, *Mol. Cell. Biol.* **21**, 2496-505 (2001)
20. S. M. Noble, S. French, L. A. Kohn, V. Chen and A. D. Johnson, *Nat. Genet.* **42**, 590-8 (2009)
21. A. Nantel, D. Dignard, C. Bachewich, D. Harcus, A. Marcil, A.-P. Bouin, C. W. Sensen, H. Hogues, M. van het Hoog, P. Gordon, T. Rigby, *et al.*, *Mol. Biol. Cell* **13**, 3452-65 (2002)
22. R. Cannon, E. Lamping, A. Holmes, K. Niimi, *et al.*, *Clin. Microbiol. Rev.* **22**, 291-321 (2009)
23. V. Stiefel, L. Perez-Grau, F. Albericio, E. Giralt, L. Ruiz-Avila, M. D. Ludevid and P. Puigdomenech, *Plant Mol. Biol.* **11**, 483-93 (1988)
24. J. Bonhomme, M. Chauvel, S. Goyard, P. Roux, T. Rossignol and C. d'Enfert, *Mol. Microbiol.* **80**, 995-1013 (2011)
25. J. Ihmels, S. Bergmann, J. Berman and N. Barkai, *PLoS Genet.* **1**, e39 (2005)
26. S. Tavazoie, J. Hughes, M. J. Campbell, R. Cho and G. Church, *Nat. Genet.* **22**, 281-5 (1999)
27. J. D. Hughes, P. W. Estep, S. Tavazoie and G. M. Church, *J. Mol. Biol.* **296**, 1205-14 (2000)
28. L. H. Mitchell and D. R. Soll, *Exp. Cell Res.* **120**, 167-79 (1979)
29. J. Berman, *Curr. Opin. Microbiol.* **9**, 595-601 (2006)
30. R.A. Weiss, *Trends Microbiol.* **10**, 314-17 (2002)
31. J. Shareck, A. Nantel and P. Belhumeur, *Eukaryot. Cell* **10**, 565–77 (2011)

DETERMINING CONFIDENCE OF PREDICTED INTERACTIONS BETWEEN HIV-1 AND HUMAN PROTEINS USING CONFORMAL METHOD

ILIA NOURETDINOV*, ALEX GAMMERMAN*, YANJUN QI**, JUDITH KLEIN-SEETHARAMAN***

**Computer Learning Research Centre, Royal Holloway University of London, UK*
***Machine Learning Department, NEC Labs America, Princeton, NJ 08540, USA ***Research Centre Júlich, 52425 Júlich, Germany / University of Pittsburgh, Pittsburgh, PA 15260, USA*
{ilia,alex}@cs.rhul.ac.uk, yanjun@nec-labs.com, jks33@pitt.edu,

Identifying protein-protein interactions (PPI's) is critical for understanding virtually all cellular molecular mechanisms. Previously, predicting PPI's was treated as a binary classification task and has commonly been solved in a supervised setting which requires a positive labeled set of known PPI's and a negative labeled set of non-interacting protein pairs. In those methods, the learner provides the likelihood of the predicted interaction, but without a confidence level associated with each prediction. Here, we apply a conformal prediction framework *to make predictions and estimate confidence of the predictions.* The conformal predictor uses a function measuring relative 'strangeness' *interacting pairs* to check whether prediction of *a new example* added to the sequence of already known PPI's would conform to the 'exchangeability' assumption: *distribution of interacting pairs is invariant with any permutations of the pairs. In fact, this is the only assumption we make about the data. Another advantage is that the user can control a number of errors by providing a desirable confidence level. This feature of CP is very useful for a ranking list of possible interactive pairs. In this paper, the conformal method has been developed to deal with just one class - class interactive proteins - while there is not clearly defined of 'non-interactive' pairs.* The confidence level helps the biologist in the interpretation of the results, and better assists the choices of pairs for experimental validation. We apply the proposed conformal framework to improve the identification of interacting pairs between HIV-1 and human proteins.

Keywords: protein-protein interaction, suspected interactions, confident prediction

1. Introduction

1.1. *Protein-protein interactions*

Protein-protein interactions (PPI's)[1,2] are fundamental building blocks of communication within and across cells: cells responding to signals in the environment and pathogens interacting with host cells do so via complex networks of PPI's. Identification of the full sets of PPI's (interactomes), for individual species is an important general problem in biology. Experimental, computational and combined efforts are underway[1,4,5] to identify such interactomes in the form of lists of pairs of proteins that are labeled as one of two classes: interacting and not interacting. However, this binary label is a simplification of the problem, as for most PPI labels we are not absolutely certain if the proteins interact or not. This is because experimental data is noisy and the confidence placed in a label vastly depends on the type and number of experiments carried out that support a particular interaction. Furthermore, there are no negative labels at all. Negative results in experiments designed to test for PPI do not provide ultimate proof that two proteins don't interact, because not all of the criteria necessary for an interaction to occur may have been fulfilled. These parameters will vary for different proteins and their interactions. Therefore, we avoid using terms sensitivity/specificity and precision/recall as they are related to the case when there are both positives and negatives for validation. The identification of protein interactomes has been focused on the positive (interacting) pairs. Prediction output may be presented as a sorted list of protein pairs. Pairs in its beginning are considered more likely to be interacting. Thus, a reasonable choice which pair to experimentally validate by a biological experiment, is to choose pairs that are ranked high in the list. Typically, biological intuition or interest is further used in selecting pairs from this top-ranking list for experimental validation. This interest varies with the potential application area for studying an interaction. For example, a structural biologist is interested in those pairs that are well characterized, from a biological as well as biophysical perspective. Biological characterization of a protein pair (for example co-crystallization trials) can require a large amount of time and effort. Thus it is useful to have an estimate for likelihood of success to decide whether structural studies are meaningful. On the other hand, a biomedical researcher might be more interested in identification of truly novel interactions, for which no prior evidence was reported in the literature. Such pairs could provide novel drug targets or biomarkers of disease. This divergence in biological interest in PPI's is an important consideration for experimentalists to use computational predictions, but it has not been addressed previously.

One biomedical application area for PPI prediction is the study of host-pathogen interactions such as between human and human immunodeficiency virus type 1 (HIV-1) proteins. HIV-1 is the causative agent for the AIDS epidemic that has affected more than 33 million people across the world. Currently, for treating HIV-1 infection therapeutic agents use reverse transcriptase, protease, and envelope. However, drug resistance, poor compliance and side effects pose problems in efficient antiviral therapy justifying the search for alternative approaches to viral inhibition. One avenue is to target interactions between HIV-1 and host proteins. HIV-1 depends on its host for virtually every aspect of its life cycle, and the communication between virus and host is via PPI's. Targeting such PPI's therefore has great promise for anti-HIV-1 drug design. However, more than 2500 putative interactions involving some 1440 human proteins have been reported.[2] Selecting a pair from this large list to pursue for further studies is not trivial, while at the same time the list is likely not complete. The analysis of such data would greatly benefit from being able to quantify a required level of confidence in the data.

In this paper we approach this current gap in PPI research by incorporating a pre-determined level of confidence rather than relying on empirical performance such as precision/recall curves given by prediction scores. This allows the biological user to specify their required level of confidence in an interaction and the need to recover known interactions in the list of predicted interactions. This approach is based on a machine learning method called conformal predictors,[3,8] a new technique for 'hedging' predictions. We 'hedge' our predictions by quantitative measure of the accuracy and reliability of predictions. In particular, each individual prediction is supplied by its confidence measure. The problem of hedged prediction is connected with the problem of testing randomness. The hedged predictions include quantitative measures of the confidence and reliability of the predictions. These measures are provably valid under the assumption of exchangeability. One of the major advantages is that the method allows us to control the reliability of prediction by selecting a suitable confidence level.

The utility of this new approach is demonstrated for the HIV-1, human PPI case. We used the protein interaction dataset studied previously.[6] This dataset contains labels of interaction derived from the NIAID database[2] by defining critera for interaction related to physical contact between the proteins and 35 features collected from various databases, used to predict potential PPI's between the 17 HIV-1 and the full human proteome consisting of 20873 human proteins. Description of these features and other details can be found in supplementary materials[7] to our previous work.[6] There are 1063 interactions in this dataset. The rest of the 353778 pairs are to be labeled as interacting or not. In the previous work,[6] we randomly sampled from this pool examples taken as negative pairs for training of the classifier. In this work, we do not define a negative set, but rather consider their labels to be unknown.

The aim here is to look for interactions between these HIV-1 and human proteins. Some of the interactions (1063 ones) are already known and the tasks are (1) to find additional new pairs that are also likely to interact and (2) to report how confident we are in the predicted interaction. Here, we will demonstrate the application of conformal predictors to these tasks. In the evaluation, we will use two performance measures to check our framework: *validity* that guarantees that a true interaction is covered by a list of predictions with certain confidence, and *efficiency* is characterized by how often interaction hypotheses are rejected (*i.e.* how many known interactions are not included in the list of interactions).

1.2. *Background: conformal prediction*

Originally, the conformal prediction approach[3,8] was introduced for the supervised machine learning problem: to predict a label for a new example from a given training set *and estimate confidence of the prediction*.

The advantage of the conformal predictor is that it has a proven validity: *if we set up a confidence level, say, 95%, then we can guarantee that in the long run our prediction would contain the true label with probability at least 95%. To prove this property of validity, the only assumption we make is the* i.i.d. assumption (data examples are generated from identical distribution independently on each other) or weaker exchangeability one (data examples appear in random order).

The main idea is to check each possible hypothesis about the label of a new example, and to the label of new example would conform the assumption of exchangeability, or with which label the example 'fits well' into the training set?

There are *two* ways to present the results. Either we can provide the prediction of a new label together with measures of its individual *confidence*. Alternatively, if we set up a pre-determined *confidence level* γ (or significance level $1-\gamma$) we present a prediction region: a list of classifications which meet this confidence requirement. The correspondence between two types of output is following: individual confidence is the highest confidence level at which prediction region consists of only one value. In terms of p-values assigned to different labels, it is a complement to 1 of the second highest p-value. As we said before the only requirement for this approach to be applicable is that the data follows the exchangeability assumption, no matter what probability distribution the examples follow and no matter what nonconformity measure is used to construct the conformal prediction region. Supposing the examples in the data sequence $z_1, \ldots, z_l, \ldots, z_N$ where $z_i = (x_i, y_i)$ are generated in random order, if the $N!$ different orderings of this sequence are equally likely, then we say that $z_1, \ldots, z_l, \ldots, z_N$ are exchangeable.

The calculations of prediction regions are based on a special function called *non-conformity measure (NCM)* that basically reflects how strange an example is with respest to others. The validity property implies that the probability of error (in the sense of a true value being outside the prediction region) is at most $1 - \gamma$ whenever the exchangeability (or i.i.d.) assumption is true.

1.3. *Problem setting: prediction of interactions between proteins*

Suppose, there is a pool of examples (protein pairs); in this pool some examples (a subset) have a specific property: they are known to *interact*. Our task is to select more examples that are likely to have this property.

Let us introduce the following notation.

- $x_1, \ldots, x_l$ are known to interact;
- x_{l+1} is a new example from the pool being classified;
- $U = \{x_{l+2}, \ldots, x_{l+s+1}\}$ is a random selection from other examples in the pool ('background' set);
- the labels are used to denote examples having property of interaction: $y_i = 1$ for interactions $y_i = 0$ for non-interactions; so $y_i = 1$ for $i = 1, \ldots, l$ and $y_i \in \{0, 1\}$ for $i > l$.

The conformal prediction method needs some modifications in order to be applied to the PPI prediction task. In particular, there is the unique characteristic in PPI's that just one class is clearly defined, namely the class of interactions; there are no clear examples of non-interactions. Thus, our aim becomes to test the hypothesis that the example is an interaction and see if we can reject this hypothesis. In general, conformal prediction deals with 2 or more classes, with second largest p-value as individual confidence. However, here we deal with one clearly defined class and our p-values are related to the hypothesis that the label is positive. So small p-value assigned to an example means that this example is unlikely to be an interaction. This may be opposite to usual use of p-values when a null hypothesis is rejected if p-value is small.

The 'background' set contains unlabelled examples and we can expect that at least some of them represent protein pairs that do not interact, although we do not know which of them.

We shall make the reasonable assumption ('exchangeability') about the mechanism of discovery: any of the undiscovered interactions has equal chance to be discovered next.

2. Conformal approach for protein interaction search

2.1. *The algorithm*

The conformal approach to the protein interaction search presented here is given by Algorithm 2.1.

X is the set of possible objects. In the PPI problem, each protein pair is presented as a vector

$$x_i = (x_i(1), \ldots, x_i(k))$$

of k numeric attributes measuring within-pair similarities and other properties of the pair, so X is a k-dimensional linear space, in the considered dataset $k = 35$.[6]

The paramater γ corresponds to the confidence level γ. Due to the validity property, if this level is pre-selected, then a true interaction will be predicted at this level as interactions with probability at least γ. Usually in statistics a selected confidence level is close to 1, such as $\gamma = 95\%$ or $\gamma = 99\%$.

Algorithm 2.1 Unlabelled Conformal Predictor in search of interactions

Input: training examples (known interactions) $x_1, x_2, \ldots, x_l \in X$
Input: a new example $x_{l+1} \in X$
Input: unlabelled examples $U = \{x_{l+2}, x_{r+2}, \ldots, x_{l+s+1}\} \subset X$
Input: a confidence level γ
Input: a NCM A
for j in $1, 2, \ldots, l+1$ **do**
  $\alpha_j = A(x_j, \{x_1, \ldots, x_l, x_{l+1}\}, U)$
end for
$p = \frac{\#\{j=1,\ldots,l+1:\alpha_j \geq \alpha_{l+1}\}}{l+1}$
Output: x_{l+1} is predicted as a suspected interaction if $p > 1 - \gamma$.

2.2. *Non-conformity measure*

As stated, our aim is to test the hypothesis that a given example is an interaction. For a new example x_{l+1}, the conformal predictor checks whether the hypothesis that $x_1, \ldots, x_l, x_{l+1}$ is generated by an exchangeable distribution can be rejected, and assigns an individual p-value to x_{l+1}. Note that individual p-value does not depend on the selected confidence level γ. So if the program outputs individual p-value for each pair, it allows users to select appropriate confidence level later, and cut the list, leaving only pairs with individual p-values at least $1 - \gamma$.

This is being done by an NCM that is a function of an element $x \in X$ and two subsets $T \subset X$ and $U \subset X$, defined whenever $x \in T$. Its aim is to measure how strange is X relatively to other elements of T; U plays an auxiliary role. By defintion, NCM can be any function of this type, but performance depence on its choice and relevence for a specific problem. Usually in supervised learning the NCM is constructed on the basis of a prediction algorithm. We can use an underlying method such as SVM or k-Nearest-Neighbours for separation of two or several classes. However the PPI problem really requires a one-class NCM applied to unlabelled objects.

2.3. *Example of one-class NCM*

The described conformal prediction framework requires that we first choose an NCM. In the supervised classification case, NCMs are based on underlying algorithms, and applied to sets of example and label pairs (x_i, y_i), where y_i is usually taken from a finite set. However, in this work, the NCM will be defined on sets of unlabelled examples x_i instead of pairs, and may optionally depend on an additional 'background' set U - a random selection of examples other than examples known to interact.

Algorithm 2.2 formalizes the following concepts of non-conformity ('strangeness'):

- a pair is 'strange' with respect to the interaction set if one of its features has a value either too large or too small; it is even 'stranger' if there are several such features;
- since we expect that the background set mainly consists of non-interactions, additional value is added to NCM if the example is not 'strange' with respect to the background set.

We shall call this approach a feature-wise NCM. Although the last point is indirectly using the second class (in the form of the background set), the NCM itself is still a one-class one, as its scores are finally assigned only to $x_1, \ldots, x_{l+1}$.

Algorithm 2.2 Feature-wise NCM

Input: data examples $T = \{x_1, \ldots, x_{l+1}\} \subset X = R^k$
Input: $x_j \in T$ (one of examples above)
Input: Background set $U = \{x_{l+2}, \ldots, x_{l+s+1}\} \subset X$.
for i in $1, 2, \ldots, k$ **do**
let a_i be percentage of $x_1(i), \ldots, x_{l+1}(i)$ larger or equal than $x_j(i)$, or percentage of $x_1(i), \ldots, x_{l+1}(i)$ smaller or equal than $x_j(i)$, (the smallest one of two, taking values from $1/(l+1)$ to $1/2$);
let a'_i be percentage of $x_{l+2}(i), \ldots, x_{l+s+1}(i); x_j(i)$ larger or equal than $x_j(i)$, or percentage of $x_{l+2}(i), \ldots, x_{l+s+1}(i)$ smaller or equal than $x_j(i)$, (the smallest one of two);
end for
Output: a NCM (ranged is from $-k\log(l+1)$ to $+k\log(l+1)$)

$$\alpha_j = -\sum_{i=1}^{k} \log(a_i/a'_i)$$

---

3. Results

We applied our approach to the HIV-1, human PPI prediction task. The full list of individual predictions sorted by p-value will be presented in the Supplementary website `http://www.clrc.rhul.ac.uk/people/alex/HIVpsb12/`.

Each protein pair (for example: POL-PROTEASE and MYH7) is assigned an individual p-value (0.100564 in this example). A small p-value means the interaction hypothesis can be rejected.

Table 1 summarizes the results for the whole dataset in order to assess the overall quality of prediction with the selected NCM. The first column lists confidence level γ. As mentioned above, γ determines the confidence that a true interaction will be predicted as interaction The second and third columns show how many examples are predicted as interactions at different confidence levels γ. An example is predicted as interaction if its p-value is at least $1-\gamma$. We ask the question how many examples have such p-values. Since there were two types of examples, namely the 1063 known interacting protein pairs and the remaining 353778 pairs with unknown labels, we consider them separately. Column 2 is related to individual p-values assigned to pairs already known to interact. These figures illustrate validity: 90% of these interactions are really reported as interactions at confidence level 90% when the algorithm is applied to them in leave-one-out mode.

As mentioned, we assume that all the other examples can be either true interactions or not, so we need to deal with all of them. Column 3 shows how many of these examples are predicted as possible interactions.

Consider a yet unknown interacation. To be confident at level 90% that the prediction list covers it, we need to include into it all predictions with individual p-value at least 10%. Table 1 shows that in our experiment 34.2% of all pairs do have such p-values.

Table 1. Size of prediction lists

confidence level	pred.interactions (amongst known interactions)	pred.interactions (amongst other pairs)
0%	1/1063(0.1%)	4/353778(0%)
1%	11/1063(1%)	37/353778(0%)
5%	54/1063(5.1%)	241/353778(0.1%)
10%	107/1063(10.1%)	604/353778(0.2%)
20%	213/1063(20%)	2185/353778(0.6%)
30%	320/1063(30.1%)	5446/353778(1.5%)
40%	426/1063(40.1%)	10380/353778(2.9%)
50%	533/1063(50.1%)	18988/353778(5.4%)
60%	639/1063(60.1%)	31412/353778(8.9%)
70%	745/1063(70.1%)	49019/353778(13.9%)
80%	852/1063(80.2%)	72743/353778(20.6%)
90%	957/1063(90%)	121091/353778(34.2%)
95%	1010/1063(95%)	150711/353778(42.6%)
99%	1053/1063(99.1%)	189432/353778(53.5%)
100%	1063/1063(100%)	353778/353778(100%)

This is related to efficiency: the smaller this percentage is, the more interaction hypotheses are rejected and the more efficient is a conformal predictor with selected NCM.

4. Biological interpretation

The full list of rank-ordered interactions sorted by p-value of the conformal predictor are provided in the supplement, together with additional information, including the random forest (RF) score of the previous classifier.[6] Fig. 1 shows a comparison of the predictions made by the RF classifier and the conformal predictor with the known interactions. One can see that both methods agree on many of the top-ranked predicted interactions. The known interactions are given a particularly high RF score, especially when the p-value of the conformal predictor is high. The trend is towards lower RF scores as the p-value decreases.

Because of the guaranteed number of known interactions at a given confidence level, an informative way to compare the previous RF classifier with the conformal prediction is by plotting the number of predicted interactions at a given p-value for each prediction method, shown in Fig. 2.

Of the 354,848 possible interactions, there is a total of 1063 interacting pairs in the NIAID dataset. Thus, as expected at a confidence level of 0.5 there are 532 known interactions, corresponding to 50% of the known interactions. At this confidence level there are 19,460 interactions suspected by the conformal predictor. Only 2376 of these have RF scores greater than zero. For 7424 of these suspected interactions there is some additional evidence of interaction (see Section 4.4, below). 1678 of these have RF scores greater than 0 (see Supplement). The comparison shows that the RF classifier may miss many potential interactions, but there are a larger number of interactions suspected by the conformal predictor, indicating the chance for false interactions. Because the false interaction rate is not known, this is to be expected, but does mean that potentially more experiments need to be carried out to find novel interactions, especially when looking at low p-values.

Below, we discuss some of the specific characteristics of the predictions made.

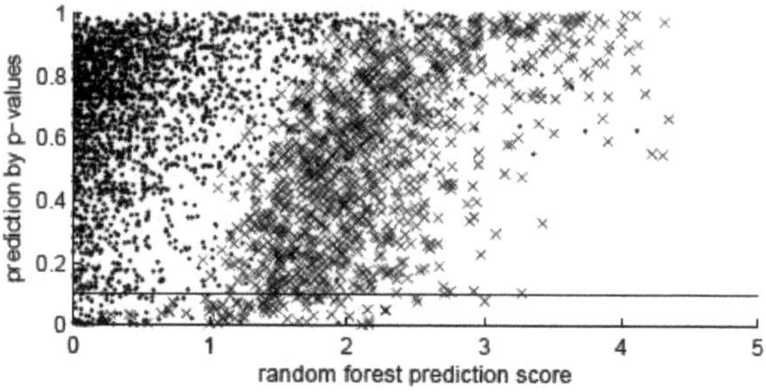

Fig. 1. Random Forest (RF) prediction scores[6] (where they are positive) and p-values obtained by the conformal predictor for the same pairs. Crosses (interactions) are known interactions, other examples are represented as points. Examples with RF prediction score less than zero are not included. The horizontal line corresponds to 10% cut-off for p-value (significance level), or confidence level 90% (for a true interaction, we are 90% sure that it is above this line).

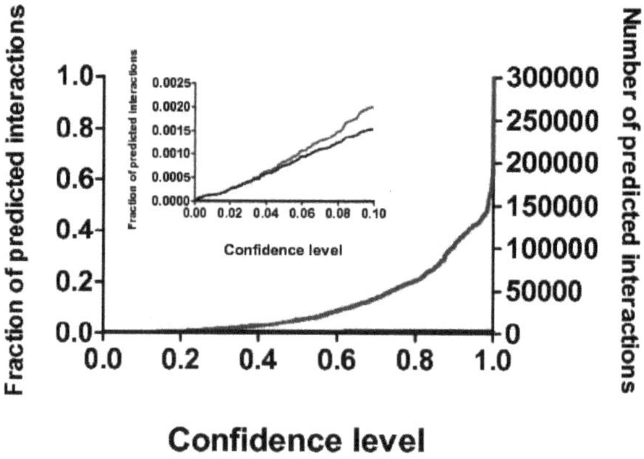

Fig. 2. Dependence of the size of conformal prediction list on the confidence level. Lower line corresponds to the percentage of pairs predicted as interactions both by conformal prediction and by positive random forest score.

4.1. *Top-ranked predictions*

When resources are limited or the interaction to be pursued should be very likely (e.g. as a candidate for structural biology), one would look at the top-group of predicted interactions. For example, for p-values greater than 0.95, there are 295 interactions, with 54 interactions being already known. 267 have RF scores greater than 0. As we will see in section 4.4, more than half (117) of the 242 novel predictions have some evidence from 'siRNA', 'in virion' or other interactors supporting that these interactions may be true.

When looking at the top-ranked predictions, the first disagreement is ranked interaction 105 between tat and thyroid hormone receptor interactor 13, TRIP13. This interaction is not predicted by the RF classifier, but is ranked with a p-value of 0.98. Similarly ranked and not predicted by the classifier are interactions of tat with CHD4, EZH2, DYNLL1, Stat4 and Stat5B. All of these proteins except DYNLL1 are involved in regulation of transcription and may well interact with tat because the function of tat is to increase the level of transcription of the HIV-1 dsRNA. In addition, HIV-1 tat targets microtubules to induce apoptosis,[9] thus making the predicted interaction with DYNLL1 plausible. In many of the suspected interactions, there is additional evidence supporting the interactions. For example, ranked at a p-value of 0.97 is the interaction between tat and IRS1, insulin receptor substrate 1. There are several publications that implicate the interaction between early insulin signaling, including phosphorylation of IRS1 with HIV-1 infection and treatment, discussion and links to them can be found in the previous work.[10]

4.2. *Intermediate and low p-values*

Statistically, we would normally reject the hypothesis that two proteins interact if their p-value is less than 10%. However, many of the known interactions will be found at such lower values and it may be of interest to a biologist to look at putative pairs amongst lower values, in particular when novelty in the new pairs is a criterion. Looking at the extreme - the interaction with the lowest p-value - is actually a known interaction, gp160 with PRMT6, protein arginine methyltransferase 6. This interaction is also not predicted by the RF. However, the evidence is rather strong for this interaction to be true: rev, tat and gp160 are all methylated by this protein.[11–13] Just above are interactions of almost every single HIV-1 protein predicted by the RF with CLCA3, chloride channel, calcium activated, family member 3, which is known to interact with gp120.[14] The actual interaction between CLCA3 and gp120 has an RF score of 1.00 and a p-value of 0.0056. Although the conformal predictor uses the same features, it appears to be better able to differentiate between such non-specifically predicted HIV-1 interactions.

4.3. *Tat predictions*

Because of the nature of the dataset, many of the predicted interactions are with tat. This is simply because the dataset from which both methods learned the features characterizing true interactions contained tat interactors in excess over the other HIV-1 proteins. However, because of this relatively unspecific spectrum of tat interactors already known, newly predicted interactions are not biologically as interesting as those involving other HIV-1 proteins. In the case of the conformal predictor, there are more non-tat predictions ranked highly. This becomes particularly apparent, when going to slightly lower p-values. For example, there are 712 interactions with p-value greater than 0.9, of these 646 are for tat (91%). Amongst these are 107 known interactions and 549 have RF scores greater than 0. In contrast, when looking at p-values between 0.8 and 0.9, there are 1687 interactions, including 106 known interactions. Here, 608 have RF scores greater than 0. Of these only, 726 are tat (43%). 255 of these tat interactors have RF scores greater than 0. Thus, in addition to the observation that the top-ranked predicted interactions not previously known make sense and would be very reasonable to validate experimentally (Section 4.1 above), there may be an additional advantage of the conformal predictor over the RF classifier, in being less affected by artifacts produced through an imbalanced representation of some HIV-1 proteins (especially tat) in the dataset.

4.4. *Comparison to in virion and siRNA human gene lists*

As can be seen for individual predictions in the Supplement, for many predicted human binding partners there is evidence for a relationship with HIV-1 function, even when a direct interaction has not yet been experimentally established. We quantified this observation by quantifying how many of the top ranked predicted HIV-1, human pairs contain human proteins that are also part of the following two human gene list: 'in virion', representing 316 human proteins that are highjacked by HIV-1 in its virion[16] and 'siRNA', 282 human genes that have been reported in the siRNA screen to have an effect on HIV-1 infection upon silencing.[15] These lists can be accessed at `www.cs.cmu.edu/~qyj/HIVsemi/validateHumanGeneLists/` and are summarized in Table 2.

The 'virion' and 'siRNA' datasets are not directly evidence about virus-protein interactions. They are functional related evidence though. Thus the comparison of our predictions to them is not a direct 'validation' experiment, but a 'functional validation'.

In this table, we compare how many of the human genes involved in top-ranked novel interactions, listed in the last column of Table 1, are also present in the siRNA and in virion lists. Note that these lists contain only human proteins, so we consider every predicted pair a hit, if the human protein is predicted to be involved in an interaction, regardless of the identity of the HIV-1 protein. This results in a very high overlap: for example, at confidence level 90% our prediction list involves 264 of 316 (83.5%) human genes from the in virion list and 165 of 282 (58.5%) human genes from the siRNA list. This high overlap further supports the functional relevance of the predictions made.

Table 2. Overlap with in virion and siRNA lists.

Conf.level	novel interactions	overlap with in virion	overlap with siRNA.
10%	604	38	11
20%	2185	78	26
30%	5446	111	57
40%	10380	140	68
50%	18988	173	85
60%	31412	187	103
70%	49019	217	124
80%	72743	239	142
90%	121091	264	165

5. Conclusion and discussion

In this paper, we have developed a new framework for the prediction and quantitative ranking of suspected and known PPI's, referred to as conformal predictors. In contrast to the typical classification approach integrating multiple feature evidences or the commonly used simpler predictions based on single features such as sequence homology, this approach allows the user to specify the required confidence in predicted interactions. Inspection of the predictions made by the conformal predictor for HIV-1, human interactions indicates that the predicted novel pairs are excellent candidates for experimental validation; for many of them, there is indirect evidence from various databases and publications supporting the interaction, which was not used in the prediction and thus independently validates the predictions made.

Comparison with the previously developed RF based classifier on the same data indicates that the conformal prediction may be less prone to artifacts related to some of the unspecific aspects of the features used in the prediction (*e.g.* if one HIV-1 protein interacts with a human protein, the classifier tends to predict a large number of additional HIV-1 proteins to interact with the same protein) or the relative abundance of HIV-1 proteins in pairs used for training, such as tat. On the other hand, the number of suspected interactions is larger compared to those obtained by the RF, potentially affecting the false interaction rate.

This is of particular note in the PPI prediction problem because proving experimentally a negative interaction is not possible. Another major advantage of the conformal predictor approach is the quantitative treatment of the ranking of suspected pairs. The biologist can use the p-value as a guide to inspect the predictions because it guarantees a given number of known interactions to be present amongst the predictions. Thus, biologists with different interests can decide on the range of interactions to study because novel interactions are ranked quantitatively with respect to known interactions. Thus, a structural biologist interested in high-confidence interactions would chose a different range of p-values than a virologist interested in discovery of novel interactions that may represent new, yet potentially risky drug targets.

The conformal predictor approach described here presents the first application of its kind to a problem in computational biology and is general. Thus, it is likely that the good results achieved here with the HIV-1, human PPI task would translate to other PPI predictions and potentially other tasks in computational biology.

A possible direction of research to improve on the proposed framework is to find other NCMs with better performance. On the current dataset, the NCM based on SVM shows slightly less efficiency as compared to feature-wise NCM. This may change with improvements in the methodology. However, in the PPI prediction task we assumed true percentage of interactions between all pairs to be unknown; thus, we cannot say what will be the 'ideal' result. Especially interesting extensions of the approach may be based on support vector machines. This method can be kernelized, equivalent to mapping the original vector into a high-dimensional space.

In this work we use exchangeability assumption as a basic model, although in fact some kinds of interactions may be not presented in the training set. Testing of exchangeability and its possible replacement with another assumption is also a subject for our future work.

6. Acknowledgements

We would like to thank Oznur Tastan for sharing the data, and her valuable input and discussions, and Dmitry Adamskiy for discussion. This work was mainly supported by EraSysBio+ grant funds from the European Union, BBSRC and BMBF 'Living with uninvited guests: comparing plant and animal responses to endocytic invasions' to the Salmonella Host Interactions Project European Consortium, SHIPREC. It was also supported by MRC grant G0802594 (Application of conformal predictors to Functional magnetic resonance imaging research); MRC grant G0301107 (Proteomic analysis of the human serum proteome); Veterinary Laboratories Agency of Department for Environment, Food and Rural Affairs on Machine learning algorithms for analysis of large veterinary datasets; EU FP7 grant O-PTM-Biomarkers; grant 'Development of New Venn Prediction Methods for Osteoporosis Risk Assessment' from the Cyprus Research Promotion Foundation.

References

1. J.Espadaler, O.Romero-Isart, R.M.Jackson, and B.Oliva. Prediction of protein-protein interactions using distant conservation of sequence patterns and structure relationships. *Bioinformatics*, 21(16):3360–3368, Aug 2005.
2. W.Fu, B.E.Sanders-Beer, K.S.Katz, D.R.Maglott, K.D.Pruitt, and R.G.Ptak. Human immunodeficiency virus type 1, human protein interaction database at ncbi. *Nucleic Acids Res*, 37(Database issue):D417–D422, Jan 2009.
3. A.Gammerman and V.Vovk. Hedging predictions in machine learning. *Comput. J*, 50:164–172, 2007.
4. B.A.Shoemaker and A.R.Panchenko. Deciphering protein-protein interactions. part i. experimental techniques and databases. *PLoS Comput Biol*, 3(3): e42, 2007.
5. B.A.Shoemaker and A.R.Panchenko. Deciphering protein-protein interactions. part ii. computational methods to predict protein and domain interaction partners. *PLoS Comput Biol*, 3(4):e43, 2007.
6. O.Tastan, Y.Qi, J.G.Carbonell, and J.Klein-Seetharaman. Prediction of interactions between HIV-1 and human proteins. *Pacific Symposium on Biocomputing*, 516, 2009.
7. O.Tastan, Y.Qi, J.G.Carbonell, and J.Klein-Seetharaman. Supporting on-line material for prediction of interactions between HIV-1 and human proteins by information integration. *Available on-line at* `http://www.cs.cmu.edu/~oznur/hiv/data/Tastan_PSB09_Supplement.pdf`
8. V.Vovk, A.Gammerman, and G.Shafer. *Algorithmic Learning in a Random World.* Springer-Verlag New York, Inc., Secaucus, NJ, USA, 2005.
9. D.Chen, M.Wang, S.Zhou, and Q.Zhou. HIV-1 Tat targets microtubules to induce apoptosis, a process promoted by the pro-apoptotic Bcl-2 relative Bim. *EMBO J., 21(24): 68016810, 2002.*
10. M.Schütt, M.Meier, M.Meyer, J.Klein, S.P.Aries and H.H.Klein. The HIV-1 protease inhibitor indinavir impairs insulin signalling in HepG2 hepatoma cells. *Diabetologia*, 43(9): 1145-1148.
11. C.F.Invernizzi, B.Xie, S.Richard, M.A.Wainberg. PRMT6 diminishes HIV-1 Rev binding to and export of viral RNA. *Retrovirology*, 3:93, 2006 Dec. 18.
12. M.C.Boulanger, C.Liang, R.S.Russell, R.Lin, M.T.Bedford, M.A.Wainberg, S.Richard. Methylation of Tat by PRMT6 regulates human immunodeficiency virus type 1 gene expression. *J. Virol.*, 79(1): 124-31, 2005 Jan.
13. N.M.Willemsen, E.M.Hitchen, T.J.Bodetti, A.Apolloni, D.Warrilow, S.C.Piller, D.Harrich. Protein methylation is required to maintain optimal HIV-1 infectivity. *Retrovirology*, 3:92, 2006 Dec. 15.
14. Q.H.Liu, D.A.Williams, C.McManus, F.Baribaud, R.W.Doms, D.Schols, E.De Clercq, M.I.Kotlikoff, R.G.Collman, B.D.Freedman. HIV-1 gp120 and chemokines activate ion channels in primary macrophages through CCR5 and CXCR4 stimulation. *Proc. Natl. Acad. Sci. USA.*, 97(9):4832-7., 2000.
15. A.L.Brass, D.M.Dykxhoorn, Y.Benita, N.Yan, A.Engelman, R.J.Xavier, J.Lieberman, S.J.Elledge. Identification of Host Proteins Required for HIV Infection Through a Functional Genomic Screen. *Science*, 2008.
16. D.E.Ott. Cellular proteins detected in hiv-1. *Rev. Med. Virol.*, 18(3):159175, 2008.

PERSONALIZED MEDICINE: FROM GENOTYPES AND MOLECULAR PHENOTYPES TOWARDS COMPUTED THERAPY

OLIVER STEGLE

Max Planck Institutes Tübingen, 72076 Tübingen, Germany
Email: oliver.stegle@tuebingen.mpg.de

FREDERICK P. ROTH

University of Toronto, Donnelly Centre, 160 College Street, Toronto, ON M5S 3E1, Canada
Email: fritz.roth@utoronto.ca

QUAID MORRIS

University of Toronto, Donnelly Centre,160 College Street, Toronto, ON M5S 3E1, Canada
Email: quaid.morris@utoronto.ca

JENNIFER LISTGARTEN

Microsoft Research, 110 Glendon Avenue, Suite PH1, Los Angeles, CA
Email: jennl@microsoft.com

Joint genotyping and large-scale phenotyping of molecular traits are currently available for a number of important patient study cohorts and will soon become feasible in routine medical practice. These data are one component of several that are setting the stage for the development of personalized medicine, promising to yield better disease classification, enabling more specific treatment, and also allowing for improved preventive medical screening. This conference session explores statistical challenges and new opportunities that arise from application of genome-scale experimentation for personalized genomics and medicine.

1. Overview of major statistical and computational challenges

Realizing the promises of personalized medicine requires robust analysis approaches, handling the breadth of data types and addressing key statistical challenges. Examples of these challenges include hidden structure within the data that can confound analysis results and lead to loss of power; missing or incomplete information; data heterogeneity; and the burden of multiple testing.

While these statistical challenges are not new, *per se*, the scale of genomic datasets comes along with additional difficulties but also opportunities for methodological innovation. For example, genome-wide association studies (GWAS) generate millions of hypotheses; it requires special consideration to reduce the burden of multiple testing so that the rate of false discoveries can be controlled[1] while retaining sufficient statistical power to detect true genetic associations, for example with single nucleotide polymorphisms (SNPs). One can begin to tackle these issues by incorporation of prior information (e.g. Lee et al.[2] and Sun et al.[3]), or using multivariate modeling[4]. Tied in with these techniques are also methods that combine groups of candidate features (*e.g.*, SNPs) in such a way as to obtain higher power, thereby attributing larger effect sizes, and uncovering a more complete picture of the underlying sources of heritability (e.g. Yang et al.[5] and Tatonetti et al.[4]).

Very large-scale datasets also support analysis strategies not available on smaller datasets. These include the ability to deduce and model hidden confounders from high-dimensional measurements, by way of Principal Components Analysis (e.g. Eigenstrat[6]), Factor Analysis[7,8], and Linear Mixed Models[9,10,11], for example. All of these approaches leverage on high data dimensionality, assuming that confounders act similarly on a large fraction of SNPs or phenotypes, which allows these factors to be reconstructed solely from the observed data.

Another challenge is the development of high quality software for the community, so that resources and expertise can be appropriately leveraged and shared. Use of such software exposes weaknesses that can then be addressed by further developments. For example, Linear Mixed Models software for GWAS has been increasingly made faster and faster, as the speed and memory constraints of each new software release becomes the bottleneck for larger and larger data sets[9,10,11].

2. Open statistical challenges

Statistical genomics is further complicated by the fact that, in real world settings, multiple confounders with intertwined impacts affect data, and as such, heterogeneous data need to be analyzed together rather than independently. Thus, tools are needed that tackle these statistical challenges in a joint fashion.

For example, when relating genotype to phenotype in a GWAS, population structure and family relatedness can reduce power to detect true associations and cause spurious associations[6]. Most molecular phenotypes, such as gene expression, are additionally contaminated with experimental artifacts or environmental influences. Such confounding factors, sometimes termed *expression heterogeneity*, have been shown to severely corrupt results of naïve analyses[7,8,12]. When seeking the genetic underpinnings of gene expression, such as in an expression quantitative trait loci analysis, problems of population structure, family relatedness and expression heterogeneity can be jointly present, and therefore models that address all of them simultaneously are required[12]. Additionally, individual readings of high-dimensional cellular phenotypes cannot be considered as independent, and thus hypothesizing and learning hidden regulatory causes of co-expression, such as cell type or transcription factor activity, has been shown to shed light on otherwise incomprehensible expression patterns[13]. However, further work in this area is still needed.

3. Open and usable software tools

Ultimately, personalized medicine needs to make its way into the clinic and the results of statistical inference need to be communicated to both clinicians and patients. How much statistics, molecular genetics and machine learning do users need to know to be able to interpret the results? Should software come with user-friendly tutorials on overfitting, multiple testing issues, p-values, false discovery rates and the 'winner's curse'? Although physicians and patients may be interested in inferences about health and disease, what they most require is assistance in acting on these inferences, i.e., making medical and lifestyle decisions that maximize expected benefit to the patient. So, is there another way to communicate an intuition about what it was about the primary data that led to the inference so that users can place their results in the context of current knowledge and evolving expertise? On top of addressing these difficult requirements, software must also safeguard patient privacy.

4. Session contribution

Our session explores these statistical and software development challenges within the context of personalized medicine.

In this session, **Karczewski et al.** propose new software, which allows the lay user to view and draw understanding from their genome sequence. It allows users to browse individual genome information in an interface that pays heed to security and privacy concerns. Of particular interest to computational biologists and statisticians engaged in tools development for personal genomics is a plugin interface that promises to ease the transition from ideas to implementation to wide use by physicians, patients and curious consumers.

Other directions of development addressed in the session are uncovering relationships between genotype and phenotype. Established modeling approaches are predominantly based on a linear model to predict the phenotypic readout from genotype, whereas heterogeneous disorders such as diabetes group distinct genetic diseases together. **Warde-Farley et al.** investigate using a mixture model of phenotypes and genotypes to map the age of diagnosis of type 2 diabetes. Using this simple, non-linear mapping, allows them to fit a set of simple genotype models that collectively predict phenotype. Perhaps many complex disorders with related phenotypes can be decomposed into multiple simple genetic disorders. Complementary to this work, **Curtis et al.** propose a new method to identify important transcriptome and phenome relations in the GWAS data by incorporating relational information on the gene and trait level. This new framework is based on a two-step procedure, where the first step identifies the important transcriptome relations (from SNPs to genes) and the second step identifies gene-to-trait relationships, thereby simplifying the problem of mapping from SNPs to traits.

Finally, our session explores computational methods to predict the sensitivity of tumor disease systems to external drug therapy. **Pal et al.** propose a novel approach that derives an abstract representation of cancer pathways from data, yielding Boolean drug kinase inhibition maps, which can then be used to predict sensitivity of systems with respect to a previously unseen drug treatment.

References

1. JD. Storey, R. Tibshirani, *Proc Nat Acad Sci* **16**, 9440 (2003).
2. S.-I. Lee, A.M. Dudley, D. Drubin, P.A. Silver, N.J. Krogan, D. Pe'erand D. Koller, *PLoS Genet* **5**, e1000358 (2009).
3. L. Sun, RV. Craiu, AD. Paterson, SB. Bull, *Genet Epidemiol* **6,** 519-30 (2006).
4. NP. Tatonetti, JT. Dudley, H. Sagreiya, AJ. Butte, RB. Altman, *BMC Bioinf* **11**, S9 (2010).
5. J. Yang, B. Benyamin, BP. McEvoy, S. Gordon *et al.*, *Nat Genet* **42**, 565–569 (2010).
6. AL. Price, NA. Zaitlen, D. Reich, N. Patterson, *Nat Rev Genet* **11**, 459–463 (2010).
7. JT. Leek, JD. Storey JD., PLoS Genet **3**, e161 (2007).
8. O. Stegle, L. Parts, R. Durbin, J. Winn, PLoS Comp Biology **6**, e1000770 (2010).
9. HM. Kang, J.H. Sul *et al.*, *Nat Genet* **42**, 348–354 (2010).
10. Z. Zhang, Z. Ersoz, CQ. Lai, *et al., Nat Genet* **42**, 355–360 (2010).
11. C. Lippert, J. Listgarten, Y. Liu, C. Kadie, R. Davidson, D. Heckerman, *Nat Methods* **8**, 833-835 (2011).
12. J. Listgarten, C. Kadie, EE. Schadt, D. Heckerman D., *Proc Nat Acad Sci* **107**, 16465 (2010).
13. L. Parts, O. Stegle, J. Winn, R. Durbin, *PLoS Genet* **7**, e1001276 (2011).

FINDING GENOME-TRANSCRIPTOME-PHENOME ASSOCIATION WITH STRUCTURED ASSOCIATION MAPPING AND VISUALIZATION IN GENAMAP

ROSS E CURTIS[1,2], JUNMING YIN[2], PETER KINNAIRD[3], AND ERIC P XING[4]

[1]*Joint Carnegie Mellon-University of Pittsburgh PhD Program in Computational Biology,* [2]*Lane Center for Computational Biology,* [3]*Human Computer Interaction Institute,* [4] *Machine Learning Department*
Carnegie Mellon University, Pittsburgh, PA 15213, USA
Email: {rcurtis, junmingy, kinnaird, epxing}@cs.cmu.edu

Despite the success of genome-wide association studies in detecting novel disease variants, we are still far from a complete understanding of the mechanisms through which variants cause disease. Most of previous studies have considered only genome-phenome associations. However, the integration of transcriptome data may help further elucidate the mechanisms through which genetic mutations lead to disease and uncover potential pathways to target for treatment. We present a novel structured association mapping strategy for finding genome-transcriptome-phenome associations when SNP, gene-expression, and phenotype data are available for the same cohort. We do so via a two-step procedure where genome-transcriptome associations are identified by GFlasso, a sparse regression technique presented previously. Transcriptome-phenome associations are then found by a novel proposed method called gGFlasso, which leverages structure inherent in the genes and phenotypic traits. Due to the complex nature of three-way association results, visualization tools can aid in the discovery of causal SNPs and regulatory mechanisms affecting diseases. Using well-grounded visualization techniques, we have designed new visualizations that filter through large three-way association results to detect interesting SNPs and associated genes and traits. The two-step GFlasso-gGFlasso algorithmic approach and new visualizations are integrated into GenAMap, a visual analytics system for structured association mapping. Results on simulated datasets show that our approach has the potential to increase the sensitivity and specificity of association studies, compared to existing procedures that do not exploit the full structural information of the data. We report results from an analysis on a publically available mouse dataset, showing that identified SNP-gene-trait associations are compatible with known biology.

1. Introduction

Understanding the genetic basis of human diseases has the potential for improved treatment and prevention. Genome-wide association study (GWAS) is a popular approach that uncovers genetic variants associated with complex traits. Typically, GWAS uses machine learning or statistical approaches to search for genetic polymorphisms, often single-nucleotide polymorphisms (SNPs), which are associated with a disease. These studies find mutations in the genome that play a role in disease development and progression. Successful studies over the last few years have uncovered susceptibility loci for many diseases such as Crohn's disease[1], asthma[2], and heart disease[3]. GWAS is commonly regarded as a powerful tool for uncovering the genetic basis of human diseases[4].

Despite its recent success and popularity, fundamental limitations of GWAS impede its ability to explain disease variation and lead to clinical treatments[5]. For example, while many studies have identified SNPs associated with different diseases, these SNPs often explain only a fraction of the heritability of the disease, suggesting that other genetic factors remain undiscovered[4].

Additionally, a majority of identified SNPs do not affect protein sequence; these SNPs may have a yet unknown regulatory role in the cell[6]. Hence, we are far from a complete understanding of how discovered SNPs actually regulate cellular pathways leading to disease[7]. Furthermore, without adequate understanding of how SNPs operate in a biological context, it is often impossible for a study's results to lead to clinical applications[5]. Overcoming these barriers is a significant problem facing the genetics community today. A proposed solution to these barriers is the combination of additional data types to the traditional GWAS data (SNP genotyping and trait measurements)[5,8].

Recently, GWAS analyses have incorporated gene expression data[6-7,9]. Given gene expression data for a given cell type or tissue, SNPs can be mapped to the gene expression levels. SNP-gene associations can provide insight into regulatory patterns in the genome and help elucidate the mechanism by which a SNP leads to disease. Moreover, methods leveraging gene-network structure have been developed to enhance the discovery of associations from SNPs to cellular pathways[10-11]. The integration of expression data into GWAS analyses has led to the identification of new disease genes[12-14], demonstrating the potential of this approach.

To date, the general strategy has been to integrate gene expression data into a GWAS study *after* the main analysis and from individuals *unrelated* to the original study[6-7]. However, in the increasingly common scenario where expression, phenotype, and genomic data are available from the same cohort, expression data can be incorporated directly into the primary analysis. In this case, all the data guides the discovery of genome-transcriptome-phenome associations. This unified three-way association framework has the potential to reveal the functional relationships between associated genomic variations and physical phenotypes, via intermediate phenotypes. This has a direct impact on personalized medicine as different patients may have different regulatory mechanisms by which disease arises, but traditional SNP-trait association studies are insufficient to uncover the hidden mechanisms. Despite the promise of unified three-way association analysis, work done on this problem is limited[15-16].

In this paper, we present a new algorithmic and visualization strategy for three-way association analysis with GWAS and gene expression data. We use *structured association mapping* to find genome-transcriptome-phenome associations. In the framework of structured association mapping, the structure of the data guides the algorithm to find true associations and eliminate false positives. While incorporating gene expression data and its structure into GWAS can enhance discovery statistically, the additional data also increases the difficulty of interpreting the results. The results, with the trait and genetic structures, are a large, complex sea of data that must be explored, suggesting the applicability of a visualization approach[17]. Available visualization tools for general GWAS[18-19] are not built for this type of data, thus new visualization tools are needed to use the structure of genome-transcriptome-phenome association data to guide the analyst to discover the biological mechanisms driving genome-phenome associations.

In Section 2, we present a novel structured association mapping strategy that incorporates genome-wide SNP data, expression data, and phenotype data in a unified three-way association framework. We find genome-transcriptome-phenome associations through a two-step process. First, we find the association of SNPs to gene expression data by leveraging the genetic network structure among genes by applying GFlasso[10]. In the second step, we find associations from genes to phenotypes using a new proposed method called gGFlasso, which incorporates structure information from both the predictors (genes) and responses (traits).

In Section 3 we present novel visualizations we have designed to guide geneticists through the exploration of three-way association results, guided by the structure of the data. We implement these tools in GenAMap, visual analytics software for association analysis[20].

Finally, we demonstrate our approach by simulation (Section 4) and on a NIH heterogeneous stock mice dataset[21] (Section 5). Results on simulated datasets show that our strategy outperforms existing methods in terms of sensitivity and specificity. Using the mouse data, we have uncovered the associations of eight genes in the mouse H2 complex on Chr 17 with mouse immunology traits, compatible with current knowledge.

2. Statistical Model

Let $\boldsymbol{X}$ be an $N \times P$ genotype matrix for N individuals where each row represents the allele states of an individual at P loci; let $\boldsymbol{Y}$ be an $N \times J$ gene expression matrix where expression levels of J genes are measured for the same set of individuals; finally, let $\boldsymbol{Z}$ be an $N \times K$ phenotype matrix where each row records K phenotypic traits of an individual. We assume that there are two *undirected weighted* relevance graphs $G_G = (V_G, E_G)$ and $G_T = (V_T, E_T)$ available for J genes and K traits, respectively. Each node in G_G represents a gene and each node in G_T represents a trait, hence $|V_G| = J$, $|V_T| = K$. There is a vast literature on network construction algorithms, and any strategy can be applied to create a network where a connection between nodes u and v represents a relationship between nodes. Each edge $\{u, v\} \in E_G$ or E_T is associated with a weighted connection between nodes. The three-way association study is then performed in a two-step procedure: first, we find genome-transcriptome associations by applying GFlasso to $\boldsymbol{X}$ and $\boldsymbol{Y}$; next, we search for transcriptome-phenome associations between $\boldsymbol{Y}$ and $\boldsymbol{Z}$ using a new structured sparse regression method called graph-graph-guided fused lasso (gGFlasso).

2.1. *Finding genome-transcriptome associations using GFlasso*

To find associations between SNPs and gene expression levels, we use GFlasso[10], a structured association mapping approach that estimates the association strengths of predictors jointly for multiple correlated responses by a sparse regression strategy. In our case, SNPs are treated as predictors and gene expression levels are treated as responses. GFlasso assumes that a relatively small number of markers are associated with each gene and that highly correlated genes tend to be influenced by a common subset of SNPs. This assumption is explicitly expressed as two regularization terms in a linear regression model (Eq. 1). Input to GFlasso includes $\boldsymbol{X}$, $\boldsymbol{Y}$, and G_G; the output is a regression coefficient matrix $\boldsymbol{B_1}$, where B_{pv} denotes the strength of pth SNP associated with vth gene. $\boldsymbol{B_1}$ is estimated by solving the following optimization problem[22]:

$$B_1 = \underset{B \in \mathbb{R}^{P \times J}}{argmin} \|Y - XB\|_F^2 + \lambda \sum_j \sum_p |B_{pj}| + \gamma \sum_{\{u,v\} \in E_G} \sum_p |B_{pu} - sign(\rho_{uv}) B_{pv}|, \quad (1)$$

where $\|.\|_F$ is the Frobenius norm of the matrix. The second term is the L_1 lasso penalty[23], which has the property of shrinking the strengths of irrelevant SNPs towards zero; the last term is the *graph-guided fused lasso* penalty, which encourages highly correlated genes (connected by an edge in E_G) to be associated with the same SNPs. $\text{sign}(\rho_{uv})$ controls the pattern of fusion applied to the association strengths: if two genes are negatively correlated, their corresponding association strengths are encouraged to have different sign. λ and γ are the regularization parameters and can be estimated by cross-validation on a validation set. For each gene, the set of associated SNPs are those with non-zero association strengths in the estimated coefficient matrix B_1.

2.2. *Finding transcriptome-phenome associations using gGFlasso*

To find associations between genes and phenotypic traits, we develop a new method by extending GFlasso, called gGFlasso, which incorporates the correlation structure of genes G_G as well as the dependency graph of traits G_T. In this setting, expression levels are treated as predictors and traits are regarded as responses. The GFlasso framework only uses the dependency structure on responses and assumes that variations of correlated responses (traits) are likely to be explained by a common set of predictors (genes). However, predictors can also be correlated, such as co-expressed genes, and it seems natural to also exploit the predictor's dependency graph. Given G_G, gGFlasso assumes that correlated (connected) genes tend to influence the same subsets of traits. This assumption is encoded as an additional fusion penalty in the linear regression model:

$$B_2 = \underset{B \in \mathbb{R}^{J \times K}}{argmin} \|Z - YB\|_F^2 + \lambda \sum_j \sum_k |B_{jk}| + \gamma_1 \sum_{\{u,v\} \in E_G} \sum_k |B_{uk} - sign(\rho_{uv}) B_{vk}|$$

$$+ \gamma_2 \sum_{\{m,l\} \in E_T} \sum_j |B_{jm} - sign(\rho_{ml}) B_{jl}|. \quad (2)$$

As in Eq. 1, λ, γ_1, and γ_2 are regularization parameters and B_{jk} is the association strength of *j*th gene with *k*th trait. Note that there are two graph-guided fused lasso penalty terms in Eq. 2: if two genes u and v are connected in G_G, the fusion penalty encourages their influence on each trait to be similar; if two traits m and l are connected by an edge in E_T, the association strength B_{jm} and B_{jl} for each gene j is encouraged to have the same absolute value. As we will demonstrate in our experiments, this joint framework that accounts for the information in the correlation structure of predictors (gene expressions) and responses (traits) has the potential to increase the sensitivity and specificity of association studies. Combined with the L_1 lasso penalty, the estimated coefficient matrix B_2 has a large fraction of zero entries and the remaining non-zero entries tend to show a block structure (see Figure 4).

2.3. *Implementation notes*

We have automated GFlasso-gGFlasso in GenAMap. We use a coordinate descent approach and the "η-trick"[24] to optimize Eq. 2. The update equations and a C++ implementation of gGFlasso are available for download from sailing.cs.cmu.edu/genamap. GenAMap uses a linear search strategy and hold-out validation to find values for the regularization parameters λ, γ_1, and γ_2. For our simulated data with 100 SNPs, 500 genes, and 20 traits, the full GFlasso-gGFlasso runs in less than one day. Using GenAMap to run the GFlasso-gGFlasso for 2500 SNPs with associations to 2000 genes would return results in less than one week.

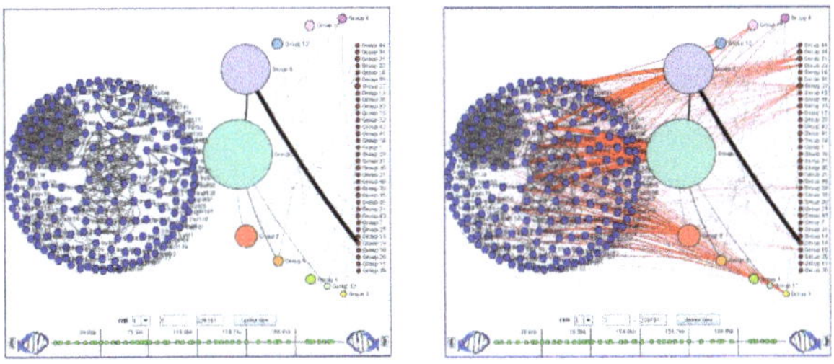

(a) (b)

Fig. 1. Overview of three-way association visualization using the KK-layout[a]. (a) The structure (edges) of the traits (blue hexagons) and gene groups (circles) are displayed without showing the association edges. (b) The red association edges from gene groups to traits are shown. SNPs are represented at the bottom of the display as green circles in GenAMap's genome browser.

3. Visualization Strategies in GenAMap

The GFlasso-gGFlasso analysis of a genome-transcriptome-phenome dataset leaves the genetics analyst with a large, complex sea of data to interpret. In addition to the gene-gene and trait-trait relationships, potentially hundreds of SNP-to-gene and gene-to-trait associations are identified. The analyst must explore this data to pinpoint the associations that lead to insight into disease.

The analyst explores the data with different strategies, depending on what questions he is interested in. In one scenario, the analyst comes into the study with questions about specific traits. The analyst explores the network relationships between these traits and looks for associated gene networks. He then examines the gene networks for genomic associations, leading to the discovery of SNPs that perturb the gene networks associated with the phenotypes of interest. Alternatively,

[a] Large, high resolution figures from this paper are available at http://sailing.cs.cmu.edu/genamap/threeway.html

the analyst starts with a genomic region of interest. In this case, the first considers the genomic region and its associations to genetic pathways. He then identifies the traits associated with the discovered genes. In either scenario, analysts have to filter through the dataset in an exploratory fashion. As visualization methods are particularly adept at guiding the identification of interesting data through exploratory analysis[17], visualization can be a powerful tool in either scenario.

To design a visualization system that would facilitate the exploration of three-way association results with gene and trait structure we follow Shneiderman's mantra[25]: 1) *overview first*, 2) *zoom and filter*, and 3) *details on demand*. In this section, we discuss our design of the visualizations, which are implemented in GenAMap using JAVA and the JUNG toolkit[26]. A video overview of our visualization tools is available online at http://sailing.cs.cmu.edu/genamap/threeway.html.

3.1. *Overview first: an introduction to the visualization*

When the analyst first examines three-way association results in GenAMap, he is presented with an overview of the data (Figure 1, 2). Traits are represented as blue hexagons and edges in the trait network are displayed as weighted (by strength of correlation) gray lines. Genes are grouped according to GFlasso results: genes associated with common SNPs form a group. Visually, gene groups are represented as circles where the size of the circle represents the size of the group. Edges between genes groups are shown as black lines, the thickness of the line representing the number of edges between the genes in the two groups.

GenAMap enables the analyst to explore the three-way results through two different layouts. The KK-layout is designed to present trait-trait and gene-gene structures separately (Figure 1). The positions of traits are determined by the KK-layout algorithm[27] and gene groups are plotted in a half-circle with the ten largest groups placed in an arc. Association edges are shown in red. When the analyst hovers over a gene group, a tool-tip reports the group's gene count and significant gene ontology (GO) enrichment.

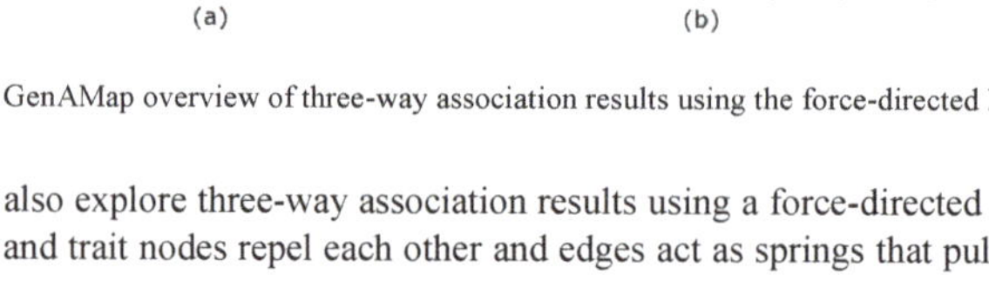

Fig. 2. GenAMap overview of three-way association results using the force-directed layout.

Analysts can also explore three-way association results using a force-directed layout (FDL). In this layout, gene and trait nodes repel each other and edges act as springs that pull nodes together.

The analyst can adjust the repulsion and attraction parameters to adjust the display. In Figure 2, we show the data from Figure 1 now represented in the FDL. In Figure 2a, the association and correlation edge spring tension is high, causing connected nodes to pull into a tightly clustered group. As the spring tension is relaxed (Figure 2b), the structure of the connected gene-trait clusters is visible. In both the KK-layout and FDL, the analyst can customize the display by turning labels on and off, adjusting parameters, or manually repositioning nodes.

GenAMap enables the analyst to simultaneously explore SNP-gene and gene-trait associations. The SNP-gene association strengths are visualized using color-encoding. For example, the analyst selects SNPs of interest and clicks to color all genes by association. The gene view updates so that the brightness of the color of each group represents its strongest association to the selected SNPs. Alternatively, the analyst selects gene groups of interest and colors the SNPs by association. SNPs are then colored: white represents a strong association and black represents no association. In Figure 3e we demonstrate both of these encodings: SNPs are colored by association to the teal gene group, and groups are colored by association to selected SNPs.

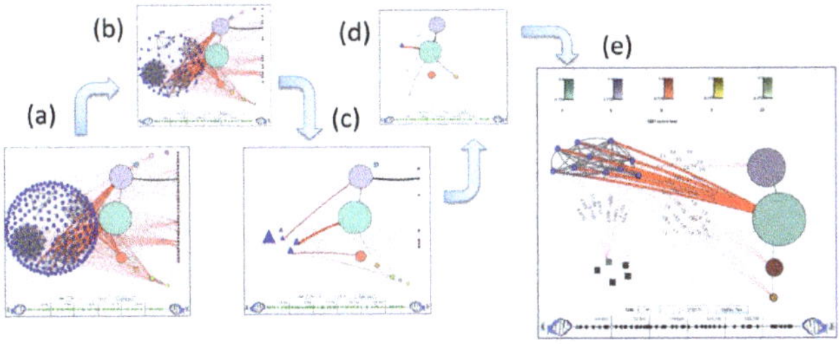

Fig. 3. Demonstration of zoom and filter tools in GenAMap. (a) The analyst starts with all data visible and filters data based on association resulting in (b). He adjusts the association edge threshold and groups traits into trait groups (c). Trait groups and gene groups not connected/associated with a trait group are removed (d). A trait group and gene group of interest are expanded and the analyst considers gene-transcriptome-phenome association simultaneously (e).

3.2. *Zoom and filter: identifying interesting signals in the data*

GenAMap is designed to enhance the analyst's ability to filter through the dataset to identify genome-transcriptome-phenome associations. We demonstrate GenAMap's filter and zoom tools through a series of steps shown in Figure 3. In Figure 3a, the data is loaded in the KK-layout. The analyst uses a filter to removes all genes and traits without associations (Figure 3b). Next, the analyst adjusts the association edge threshold to remove weak associations. Interested in the largest gene group (Group 2, teal color), he removes all groups that do not have a network edge to Group 2. Each connected component in the trait graph is then collapsed into a *trait group* (represented as a triangle) to simplify the display (Figure 3c). To explore only the trait group with the most associations to Group 2 (thickest edge), all trait groups and gene groups not connected to this trait group are manually removed (Figure 3d). Finally, the analyst expands the trait group and a strongly-associated gene group. He colors the SNPs by association to the now-visible genes (squares) in this gene group. After identifying associated SNPs, the analyst colors all genes by association to the identified SNPs (Figure 3e). In summary, GenAMap allows the analyst to filter and explore based on network connectivity, association, edge thresholds, and grouping strategies. These strategies can be employed in any three-way analysis, starting from traits, SNPs, or genes.

3.3. *Details on demand: resources for further exploration*

Once the analyst has found interesting gene-trait associations, GenAMap directly links the analyst to more information. The analyst can directly link to the UniProt database[28], Google search, or to dbSNP[29]. GO information for genes is available through GenAMap's integration with BiNGO[30].

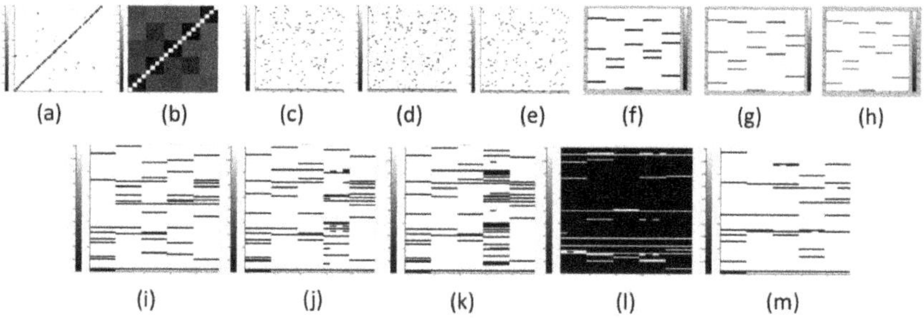

Fig. 4. Results of a simulation study on a dataset simulated with $\sigma_1^2 = \sigma_2^2 = 1$. Dark pixels indicate large values. (a) Correlation matrix of genes thresholded at 0.5 (b) Correlation matrix of traits thresholded at 0.5 (c) True $\boldsymbol{B_1}$ matrix (d) Recovered $\boldsymbol{B_1}$ matrix by lasso (e) Recovered $\boldsymbol{B_1}$ matrix by GFlasso. (f) True $\boldsymbol{B_2}$ matrix (g) Recovered $\boldsymbol{B_2}$ matrix by gGFlasso. (h) Recovered $\boldsymbol{B_2}$ matrix by GFlasso (i) True SNP-trait association matrix $\boldsymbol{B_3}$ (j) Recovered $\boldsymbol{B_3}$ by GFlasso-gGFlasso strategy. (k) Recovered $\boldsymbol{B_3}$ by GFlasso-GFlasso strategy (l) Recovered $\boldsymbol{B_3}$ by lasso (m) Recovered $\boldsymbol{B_3}$ by plink, values plotted are $-\log 10(p\text{-value})$.

4. Simulated Data Analysis

We performed a simulation experiment to validate our GFlasso-gGFlasso strategy for three-way association analysis, comparing the results with a GFlasso-GFlasso strategy, a lasso-lasso strategy and PLINK[31], a popular pairwise GWAS tool. We randomly selected $N = 250$ individuals and $P = 100$ SNPs from the mice dataset[21] to create a genotype matrix $\boldsymbol{X}$. The number of genes J was set to 500; we assumed 50 correlated gene groups of equal size. The true $P \times J$ association matrix $\boldsymbol{B_1}$

was created as follows: for each group, we randomly selected three causal SNPs; one (random) housekeeping SNP was assumed to be associated with all genes. Strengths of association for selected SNP-gene pairs were set to 1 (Figure 4c). Given $\boldsymbol{X}$ and the association matrix $\boldsymbol{B}_1$, the gene expression matrix $\boldsymbol{Y}$ was simulated as $\boldsymbol{Y} = \boldsymbol{XB}_1 + \boldsymbol{\varepsilon}_1$, where $\boldsymbol{\varepsilon}_1$ is an $N \times J$ matrix with entries independently generated from $N(0, \sigma_1^2)$. The gene correlation matrix, simulated with noise level $\sigma_1^2 = 1$ and thresholded at 0.5, reveals the inherent structure of the genes (Figure 4a). We set K = 20 traits. To generate the trait matrix $\boldsymbol{Z}$, we first created a $J \times K$ association matrix $\boldsymbol{B}_2$ similar to the procedure for obtaining $\boldsymbol{B}_1$: we assumed five groups of correlated traits of equal size each with three causal gene groups (Figure 4f). Then, $\boldsymbol{Z} = \boldsymbol{YB}_2 + \boldsymbol{\varepsilon}_2$, where entries of $\boldsymbol{\varepsilon}_2$ were independently generated from $N(0, \sigma_2^2)$. We also calculated the true SNP-trait association matrix, $\boldsymbol{B}_3$ (Figure 4i), by assuming that if a SNP is associated with a gene and that gene influences a trait, then there is an association between the SNP and the trait.

Table 1. True positive rates (TPR) and false positive rates (FPR) of different methods in recovering the association matrices at different noise levels.

	$\sigma_1^2 = \sigma_2^2 = 1/4$		$\sigma_1^2 = \sigma_2^2 = 1$		$\sigma_1^2 = 4, \sigma_2^2 = 16$	
	TPR	FPR	TPR	FPR	TPR	FPR
$\boldsymbol{B}_1$ by GFLasso	0.9454	0.0060	0.8965	0.0057	0.7884	0.0092
$\boldsymbol{B}_1$ by Lasso	0.9758	0.7632	0.9535	0.0763	0.9081	0.7528
$\boldsymbol{B}_2$ by gGFLasso	1.0000	0.0000	0.9333	0.0016	0.7067	0.0205
$\boldsymbol{B}_2$ by GFLasso	0.9800	0.0004	0.9233	0.0039	0.7000	0.0207
$\boldsymbol{B}_3$ by GFlasso-gGFlasso	0.9200	0.0266	0.8600	0.0305	0.8400	0.2450
$\boldsymbol{B}_3$ by GFlasso-GFlasso	0.9200	0.0266	0.8600	0.0615	0.8400	0.2500
$\boldsymbol{B}_3$ by Lasso-Lasso	1.0000	0.8803	1.0000	0.9030	1.0000	0.8559
$\boldsymbol{B}_3$ by PLINK	0.6300	0.0150	0.5357	0.0234	0.5000	0.0294

In Table 1, we present a summary of performance results evaluated by the true positive rate (TPR) and false positive rate (FPR). As the table and Figure 4e show, in a low noise setting, GFlasso is able to recover SNP-gene pairs with true association with a tiny FPR. We compared the performance of GFlasso and gGFLasso in recovering $\boldsymbol{B}_2$. Although both methods successfully recover true gene-trait pairs, GFlasso tends to produce a number of false positive pairs with weak associations (Figure 4h) and hence has a higher FPR. Both GFlasso-GFlasso and GFlasso-gGFlasso have a much lower FPR than the lasso-lasso strategy in terms of recovering $\boldsymbol{B}_3$ (Figure 4l). We were interested in whether the methods that do not incorporate the gene expression data would similarly recover $\boldsymbol{B}_3$, and so we ran the default association algorithm in PLINK to find SNP-trait associations (Figure 4m). We found that although there is a low FPR, the power of detecting true associations decreases significantly (Figure 4m).

5. Mouse Data Analysis

We ran GFlasso-gGFlasso on the NIH heterogeneous stock mice dataset[21] using GenAMap to demonstrate that: 1) our method scales to real-world problems, 2) the visualization tools can filter and zoom to find association signals in real data, and 3) the method finds genome-transcriptome-phenome associations compatible with established biological knowledge.

The NIH heterogeneous stock mice dataset has data for 259 mice with 12545 SNPs, gene expression data (from the liver) for 46147 genes, and 178 phenotypic traits. We excluded genes that did not have signals in at least 50 mice, leaving 9742 genes, and normalized the expression values using lumi[32]. All traits with more than 30% of the mice having missing values were excluded, leaving 161 traits. Missing values were imputed using *k*-nearest neighbor imputation[33]. The network was created using a soft-threshold method for finding scale-free networks[34].

We compared the results from the GFlasso-gGFlasso strategy to results found from PLINK on the SNP-trait data, where all SNP-trait pairs with a *p*-value < 1*e*-4 were considered significant associations. We found that 1385 SNPs had associations to 131 traits in the PLINK results. In the two-step results, we found that 943 SNPs were associated with 746 genes and that 412 genes were associated with 133 traits. We found 604 SNPs that were associated to a trait in the PLINK analysis and were associated to a gene using GFlasso-gGFLasso. Of the 604 SNPs with an association to a trait using PLINK and to a gene using GFlasso, we found that 546 also had at least one associated gene that was associated with the same trait in the gGFlasso results. These findings suggest that the GFlasso-gGFlasso has found SNP-gene-trait associations that may help to explain associations found in the data by the PLINK analysis, in addition to the new signals that were not found by PLINK.

We wanted to explore this large results dataset to find biologically interesting results. We explored the association results using GenAMap's three-way visualization tools (see supplementary website for video). In our analysis, we found a group of 15 genes that was enriched for the GO process *antigen processing* (*p*-value = 4.54e-6). Using GenAMap, we identified eight genes in the group (*H2-T23, H2-Q1, H2-Ab1, Bat5, Tmem63b, C4a, Psmb9, NM_010396*) that were associated with a sub-network of seven immunology traits (%CD8+/CD3+, CD4+/CD8+, %CD4+/CD3+, %CD8+, %B220+, %CD4+, %CD3+). We found that these genes were associated to SNPs on Chr 17 (Figure 5), with the strongest signals to *rs8242408*.

rs8242408 is located on Chr 17 in the intron region of *Tap1* and 6kb upstream from *Psmb9*. All of the identified SNPs are part of the mouse H2, the major histocompatibility complex (MHC). The eight genes we identified are also located in the H2 region; three (*H2-T23, H2-Q1, H2-Ab1*) are known H2 genes and *Psmb9* is known to be involved in antigen processing to create class I binding peptides (http://www.phosphosite.org). The H2 region is the mouse ortholog to the human HLA region and encodes genes involved in the mouse immune response[35]. Two classes of H2 genes are presented to immune cells for identification. Some H2 genes (class I) are expressed in virtually all cells and display "self" antigens, while others (class II) are expressed only in antigen-presenting cells[36]. CD8+, CD4+, etc., refer to proteins on the surface of immune response cells that bind to the antigens on the surface of the other cells in the organism.

Despite the known association of H2 genes and immunology traits, the precise mechanism of how the identified SNPs influence the expression of these associated H2 genes is yet unknown. It is likely that the discovered SNPs act as *cis*-regulatory elements of these genes, which in turn have effects on the immunology traits. Although this investigation merits further study, the discovered SNP-gene-trait associations are compatible with current biological knowledge and could lead to further insight into how SNPs in the H2 region affect the immune system. Of note is the strong association to *rs8242408* in the promoter region of *Psmb9*, which suggests expression levels of *Psmb9* could have a regulatory effect on other identified genes.

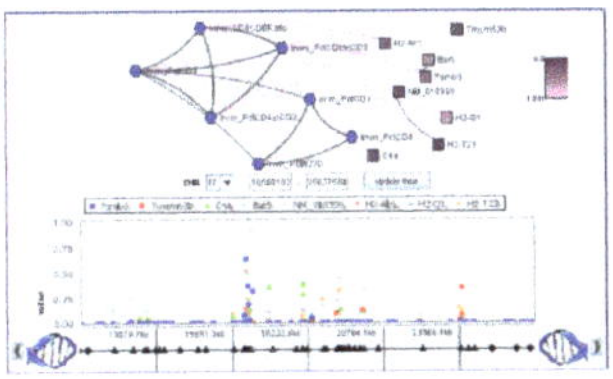

Fig. 5. Results of mouse analysis. In our mouse three-way association analysis we found eight genes associated in *cis* to the mouse H2 region and are associated with immune response phenotypes.

6. Discussion

High-throughput sequencing technology has advanced rapidly over the past few years and it will soon become routine to obtain personal genome information. Personal genomic data will possibly allow us to provide personalized treatment for each individual. While a number of exciting findings have emerged from genome-wide association studies, many results haven't translated to the clinic for improved patient care. In the post-GWAS era, it has been suggested that the GWAS data could be more insightful when integrated with other data types. In this paper we considered the problem of combining GWAS and gene expression data to find genome-transcriptome-phenome associations from an algorithmic and a visualization perspective. We have proposed a novel two-step strategy that employs structured association mapping to find associations from genome to phenome by leveraging gene expression data and its network structure. Using simulated and mouse datasets, we demonstrated that integrating this intermediate data type into a unified framework has the potential to increase the accuracy of studies and offer more insights on the functional relationships between identified GWAS loci and associated phenotypic traits. Although it is challenging to have a full understanding of such relationships and translate them into medical practice, we believe we have taken an important step forward.

We are currently undertaking further steps to demonstrate our approach with biological analysis using the NIH heterogeneous stock mice. Additionally, more sophisticated optimization algorithms could be developed to increase the computational efficiency of gGFlasso.

Acknowledgments

This work is supported by the Defense Advanced Research Projects Agency [Z931302]; National Science Foundation [DBI-0640543, IIS-0713379]; National Institutes of Health [1R01GM087694,

1R01GM093156, 0015699]; an Alfred P. Sloan Research Fellowship to EPX; and a Ray and Stephenie Lane Research Fellowship to JY.

References

1. G. Lettre and J. D. Rioux, *Hum. Mol. Genet.* **17**(R2), R116-R121 (2008).
2. D. S. Postma and G. H. Koppelman, *Proceedings of the American Thoracic Society* **6**, 283-287 (2009).
3. R. McPherson et al., *Science* **316**(5830), 1488-91 (2007).
4. R. A. Manolio et al., *Nature* **461**, 747-753 (2009).
5. E. E. Schadt, *Nature* **461**, 218-223 (2009).
6. E. E. Schadt et al., *PLoS Biol* **6**(5), e107 (2008).
7. M. I. McCarthy and J. N. Hirschorn, *Hum. Mol. Genet.* **17**(R2), R156-R165 (2008).
8. A. Califano, A. Butte, S. Friend, T. Ideker, and E. E. Schadt, *Nature Precedings* **5732**(1), doi:10.1038 (2011).
9. Y. Gilad, S. A. Rifkin, and J. K. Pritchard, *Treds Genet* **24**(8), 408-145 (2008).
10. S. Kim and E. P. Xing, *PLoS Genet* **5**(8), e1000587 (2009).
11. J. Zhu et al., *Nature Genet* **40**(7), 854-861 (2008).
12. W. Cookson, L. Liang, G. Abecasis, M. Moffatt, and M. Lanthrop, *Nature Reviews Genetics* **10**, 184-194 (2009).
13. A. C. Silveira et al., *Vision Research* **50**(7), 698-715 (2010).
14. Y. Hsu et al., *PLoS Genetics* **6**(6), e1000977 (2010).
15. Y. Chen et al., *Nature* **452**, 429-435 (2008).
16. V. Emilsson et al., *Nature* **452**(27), 423-430 (2008).
17. J. D. Fekete, J. J. vanWijk, J. T. Stasko, and C. North, *LNCS* **4950**, 1-18 (2008).
18. R. J. Pruim et al., *Bioinformatics* **26**(18), 2336-2337 (2010).
19. D. Ge et al., *Genome Res* **18**(4), 640-3 (2008).
20. R. E. Curtis and E. P. Xing, in *Proc. ISMB*, Technology Track (2010).
21. M. Johannesson et al., *Genome Res* **19**(1), 150-8 (2009).
22. X. Chen, S. Kim, Q. Lin, J. G. Carbonell, and E. P. Xing, *CoRR* (2010).
23. R. Tibshirani, *Royal Statist Soc B* **58**(1), 267-288 (1996).
24. A. Rakotomamonjy, F. R. Bach, S. Canu, and Y. Grandvalet, *JMLR* **9**, 2491-2521 (2008).
25. B. Shneiderman, *IEEE Software* **11**(6), 70-77 (1994).
26. J.O. Madahain., D. Fisher. P. Smyth. S. White. and Y.B. Boey, *J Stat Softw* **VV**(II) (2005).
27. T. Kamada and S. Kawai, *Information Processing Letters* **31**(1), 7-15 (1989).
28. The UniProt Consortium, *Nucleic Acids Res.* **39**, D214-D219 (2011).
29. S. T. Sherry et al., *Nucleic Acids Res.* **29**(1), 308-11 (2001).
30. S. Maere, K. Heymans, and M. Kuiper, *Bioinformatics* **21**, 3448-3449 (2005).
31. S. Purcell, et al., *American Journal of Human Genetics*, 81 (2007).
32. P. Du, W.A. Kibbe, S.M. Lin. *Bioinformatics* **24**(13),1547-8 (2008).
33. O. Troyanskaya, et al., *Bioinformatics* **17**(6):520-525.
34. B. Zhang and S. Horvath. *Stat Appl Genet Molec Bio* 4(1): Article 17 (2005).
35. P.M. Stuart, *eLS* doi:10.1002/9780470015902.a0000921.pub3 (2010).
36. A. Kumanovics, T. Takada, K. F. Lindahl. *Immunology* 21:629-657 (2003).

INTERPRETOME: A FREELY AVAILABLE, MODULAR, AND SECURE PERSONAL GENOME INTERPRETATION ENGINE

KONRAD J. KARCZEWSKI[1,2,†,*], ROBERT P. TIRRELL[1,*], PABLO CORDERO[1], NICHOLAS P. TATONETTI[1,2], JOEL T. DUDLEY[1], KEYAN SALARI[2], MICHAEL SNYDER[2], RUSS B. ALTMAN[2], STUART K. KIM[2,3,†]

[1]Training Program in Biomedical Informatics,
[2]Department of Genetics,
[3]Department of Developmental Biology
Stanford University School of Medicine, Stanford, CA 94305, USA
** denotes equal contribution*

The decreasing cost of genotyping and genome sequencing has ushered in an era of genomic personalized medicine. More than 100,000 individuals have been genotyped by direct-to-consumer genetic testing services, which offer a glimpse into the interpretation and exploration of a personal genome. However, these interpretations, which require extensive manual curation, are subject to the preferences of the company and are not customizable by the individual. Academic institutions teaching personalized medicine, as well as genetic hobbyists, may prefer to customize their analysis and have full control over the content and method of interpretation. We present the Interpretome, a system for private genome interpretation, which contains all genotype information in client-side interpretation scripts, supported by server-side databases. We provide state-of-the-art analyses for teaching clinical implications of personal genomics, including disease risk assessment and pharmacogenomics. Additionally, we have implemented client-side algorithms for ancestry inference, demonstrating the power of these methods without excessive computation. Finally, the modular nature of the system allows for plugin capabilities for custom analyses. This system will allow for personal genome exploration without compromising privacy, facilitating hands-on courses in genomics and personalized medicine.

1. Background and Significance

The rapid decrease in the price of genotyping and sequencing technologies, with the race to the $1,000 genome, has brought forth an age of genomic personalized medicine. The market of direct-to-consumer (DTC) genotyping, with the emergence of companies such as 23andMe, Navigenics, and Lumigenix, has put personal genotype information in the hands of patients and health care providers, based around the central idea that individuals are the owners of their genotype data. However, the problem has now shifted from the generation of accurate genotype data to tackling the problem of the "$1,000,000 interpretation." Without the proper tools, both patients and physicians will find it difficult to interpret and analyze the extraordinary amount of data, effectively rendering it useless.

DTC genetic testing companies normally provide some data analysis, but such an approach has a number of drawbacks. First, DTC genetic testing companies may sometimes use proprietary algorithms that remain undisclosed, or use genetic data that are private and not available to the public. Hence, their analysis is not always transparent and the user may not understand how the analysis was done or be able to independently replicate the results. Second, the analysis can only be modified, expanded, or tweaked by the genotyping service itself, disallowing the application of

† Corresponding authors: konradjkarczewski@gmail.com, stuartkm@stanford.edu

other analysis by a third party. Finally, and perhaps most importantly, the consumer's information is necessarily stored in a company's server, to which people other than the user may have access.

In addition, the age of genomic personalized medicine has brought the use of genetic data into the clinical setting. However, the pace of medical education has not kept up with this demand and patients are beginning to enter clinics seeking guidance in interpreting their personal genomes. Stanford University has introduced a pioneering course in Personalized Medicine and Genomics, aimed at medical and graduate students interested in interpreting personal genomes. While interpretation is offered by these DTC genetic testing companies, medical schools and universities avoiding conflicts of interest may prefer to be independent and retain the ability to customize and expand their interpretations.

Various tools that have been developed for genomic analysis can be extended to interpret personal genotype information. For instance, genome-wide association studies (GWAS) have discovered the genetic factors related to various diseases and traits, which can be applied in reverse to personal genotypes to predict traits based on genetics. Additionally, approaches from population genetics that distinguish populations can be used to infer an individual's ancestry. Many such techniques already exist and more are being developed every day and a systematic evaluation of these methods is crucial to present a compact and informative report to the end users. Equally important is the way to present this report, with the necessary background to understand each analysis, including its accurate interpretation and limitations. Additionally, more knowledgeable users, such as physicians or bioinformaticians, may wish to fine-tune the parameters of these analyses to fully exploit the given data.

We have developed a web-based genome interpretation engine that addresses these needs by providing comprehensive, secure, and highly customizable framework to analyze personal genotype information. Leveraging modern browser technology, including HTML5, CSS3, and the document canvas, we have built a system to analyze whole-genome genotype data within the user's browser. The key feature of this approach is that the server is never sent any genotype data except when the user expressly requests to do so.

2. Methods

To accomplish these goals, we have built a client-side genome interpretation system, have implemented and developed advanced analyses for personal genomes, and built a framework for customization of annotations.

2.1. *Client-side system*

We leveraged several application and user interface (UI) frameworks for use on the client-side. We chose Backbone as an application framework, which separates client-resident code (Figure 1) into models (managing and manipulating data), views (responsible for the user interface of any particular section), and controllers (which route requests and manage application-level logic, e.g. session and history). In this terminology, the models correspond to a user, the views correspond to each analysis module, and we have a single application-level controller.

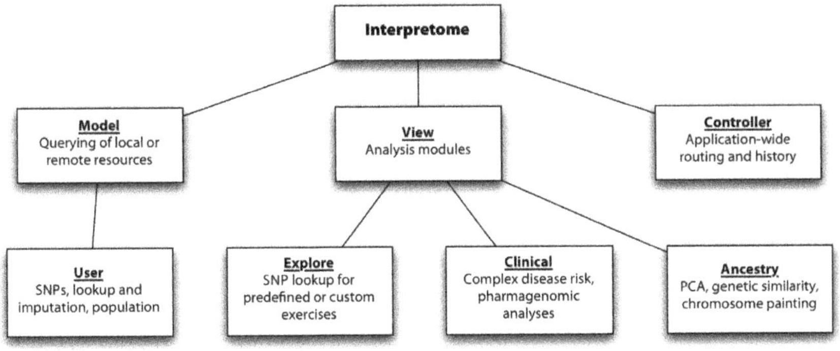

Figure 1: Interpretome is designed along the Model-View-Controller pattern, separating the application into distinct components corresponding to data, analysis, and navigation.

As the ultimate goal of this application is to communicate genetic information in a clear and concise manner, making informed decisions about the user interface and representation of data was critically important. We conducted a survey of health-related websites in order to gauge the 'state of the art' in this domain. All attempt to balance accessibility and information content - many erring on the side of data overload. We decided to maintain a sparser interface, employing widgets from the jQueryUI, Google charts, and Highcharts libraries.

Since the entire application is loaded dynamically, our Backbone views utilize jQuery and jQueryUI to update the interface in response to user interaction. The clear separation of interface and logic afforded by our design choices enabled us to preserve application state across different modules. As users navigate to new modules, the Javascript logic and HTML content corresponding to those modules are loaded dynamically.

2.1.1. *The user model*

Determining how to load user genomes and how to represent a user was one of the first challenges in building Interpretome. Even a year ago, it would not have been possible to load a file into the Javascript machine without using obscure developer versions of a web browser. Since then, the newest releases of Chrome and Firefox have added support for the FileReader API, a new standard developed to support reading of text and other files in Javascript. Notably, this API does not have access to the filesystem, only to files the user has selected explicitly.

The user supplies a tab-delimited file (with RSID, chromosome, position, genotype), a format utilized by many DTC companies; additionally, we provide conversion scripts on request for all major DTC vendors, as well as other formats, including VCF files for full genomes. We parse these files line-by-line and store each SNP as an object in a hash table associated with a newly created user instance, and a progress bar provides a visual cue of the process. Even with larger files (several million SNPs) and older computers, this takes no more than thirty seconds.

2.1.2. *Analysis views*

When a user runs an analysis, a function is dispatched that runs the main computation. In many cases, the result of the first function is a block of data received from the server, which defines parameters of the exercise (e.g., a set of SNPs). Specifically, when genotype-specific information is requested, data for all possible genotypes are typically retrieved, preventing the deduction of the individual's genotype by intercepting this query. After receiving the relevant data, the client queries the model for a user's genotype at these SNPs (which may be measured directly in the user's genotype or imputed using public data, as described below). Once the client receives the necessary data, the algorithm is run, without sending any genotype information to the server. Finally, the view updates the interface with the results and generates associated plots and figures.

2.1.3. *Scalability*

Delegating most of the computation into the browser has major advantages for scalability. Since our backend server is largely responsible for sending (as opposed to receiving) content, and database access is mostly limited to large cacheable chunks, scaling the application is relatively simple. We are able to increase site availability by simply adding more database servers and can ignore issues of synchronization across database replicates, which are huge challenges for other dynamic web applications.

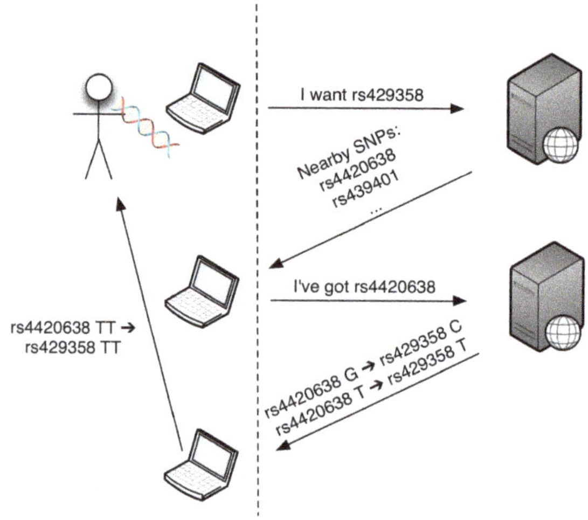

Figure 2: Imputation of a user's genotype is done directly in the browser. Allele data from public databases needed to impute a user's genotype can be obtained by just requesting the necessary SNPs through their rsids or genomic coordinates. No genotype information from the user is ever sent through the network.

2.2. *Analyses*

We have implemented a set of standard genome analysis modules for the Interpretome. These analyses utilize our client-side imputation method, which demonstrates the power and features of the private analysis system. Additionally, we have implemented clinical and ancestry analysis methods, as well as a number of exploratory tools, which are easily expandable.

2.2.1. *Imputation*

To expand the number of SNPs available for analysis, we first implemented a client-side imputation by proxy method. In this scheme, all the computation required for the task is performed on the client-side, with public information downloaded as required from the server (Figure 2). The user requests a number of SNPs not in the personal genotype file and a request is sent with RSID identifiers and a target population. The server responds by providing all SNPs in linkage disequilibrium with the requested SNP in the selected population (from Hapmap data). On the client side, the system determines which of these SNPs are contained in the personal genotype file, and thus, will be suitable for imputation. The client requests phase information for these SNPs from Hapmap genotype data from the server. These data are returned and the resulting SNPs are "imputed" from the returned phases in the browser.

2.2.2. *Clinical analysis*

We have implemented a number of analyses that demonstrate the methods available for clinical interpretation of a personal genome. First, we have implemented a disease risk calculation, as in

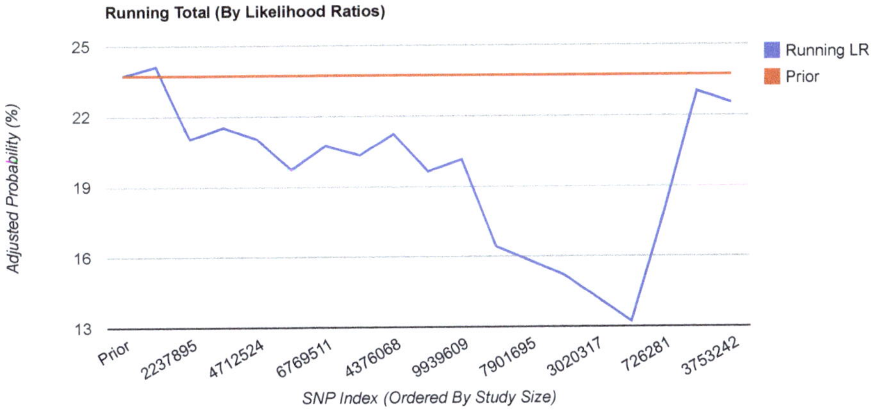

Figure 3: Diabetes risk calculator. Using likelihood ratios calculated from published association studies, the diabetes risk exercise computes a user's risk of developing Type 2 Diabetes. The estimate is based on a population and sex-specific prior for each user, adjusted by the user's genotypes.

the first clinical assessment of a personal genome [Ashley et al., 2010]. We have employed the risk calculation method of likelihood ratios and demonstrate how each variant affects an individual's risk of Type 2 Diabetes (Figure 3). Prior to the computation, the likelihood ratio for each genotype is downloaded from the server (with no genotype data sent to the server). The user inputs a population and sex, which define a prior probability for the disease. Then, the likelihood ratios are chained together (using the actual genotypes for the individual) as in [Morgan et al., 2010] to determine a posterior probability.

Additionally, we demonstrate the applications of a personal genome in pharmacogenomics, or the study of genetic variation related to drug response. For instance, warfarin, an anti-coagulant prescribed to millions of patients every year, has a high therapeutic range and genetic variants in genes such as VKORC1 and CYP2C9 have been implicated in this variation. These variants, along with clinical factors such as age and weight, can be used to predict an optimal predicted starting dose of warfarin [Consortium, 2009]. We have implemented this warfarin dose calculator, which predicts an optimal starting dose given a personal genotype and clinical parameters. In addition, we extend the pharmacogenomic application of personal genomes to other drugs, using a set of annotations from PharmGKB (www.pharmgkb.org; Michelle Carillo; personal communication).

Finally, we include a section for further exploration of rare pharmacogenomics variants (Table 1). This analysis searches for rare, non-synonymous variants in putative pharmacogenes (genes with drug-gene interaction data from DrugBank; www.drugbank.ca). The functional impact of these variants is predicted by PolyPhen2 [Adzhubei et al., 2010], which are pre-computed for all variants in dbSNP.

dbSNP	Genotype	MAF	Gene Name	Drug Name	PolyPhen Class	PolyPhen Score
rs16985442	CG	0.041	SLC12A5	Bumetanide	benign	0
rs10075302	AC	0.049	SLC25A2	L-Ornithine	benign	0.064
rs11548670	AG	0.022	NDUFS1	NADH	probably damaging	0.999
rs933135	CT	0.022	Plcd1	Acetate Ion	possibly damaging	0.822
rs9332608	AG	0.049	F5	Phenylmercury	benign	0.021
rs4252128	CT	0	PLG	Bicine	possibly damaging	0.418
rs363504	AG	0.022	GRIK1	Topiramate	benign	0
rs1801690	CG	0.046	APOH	Alpha-D-Mannose	probably damaging	0.938
rs1805321	AG	0	PMS2	Adenosine-5'-Diphosphate	benign	0.002

Table 1: Rare Pharmacogenomic Variants. Non-synonymous, rare variants (MAF < 5%) in genes predicted to interact with drugs from DrugBank are shown for a personal genome. The PolyPhen Class and Score predict whether a variant may be damaging to the function of the protein, which may affect an individual's drug response.

2.2.3. *Ancestry analysis*

As methods for population genetics can be applied to infer ancestry from personal genomes, we have implemented client-side methods for global ancestry similarity, individual similarity, and chromosome painting. First, we have enabled individuals to compare their personal genomes to a reference panel, using principal component analysis (PCA). Typically, to run such an analysis, an individual genotype would be added to a reference panel, such as the HGDP [Cann et al., 2002] or POPRES [Novembre et al., 2008] datasets, and principal components would be calculated for the combined dataset, which can take from 10 minutes to an hour for each dataset. In this method, we have instead pre-computed the eigenvectors and loadings for each SNP, as well as projections of the individuals in the reference panels. When the analysis is run, the client downloads these data and then projects the user's genotype onto the same dataset to compute the principal component coordinates, and the resulting projections are plotted using the Highcharts library (Figure 4). One limitation to this approach is that the user requires the same SNPs as those used to pre-compute the PCA results. We avoid this problem by providing multiple options for performing the projections, based on common platforms (Illumina Hap550+ and Illumina OmniExpress+) and this problem will be solved when full genomes are supported.

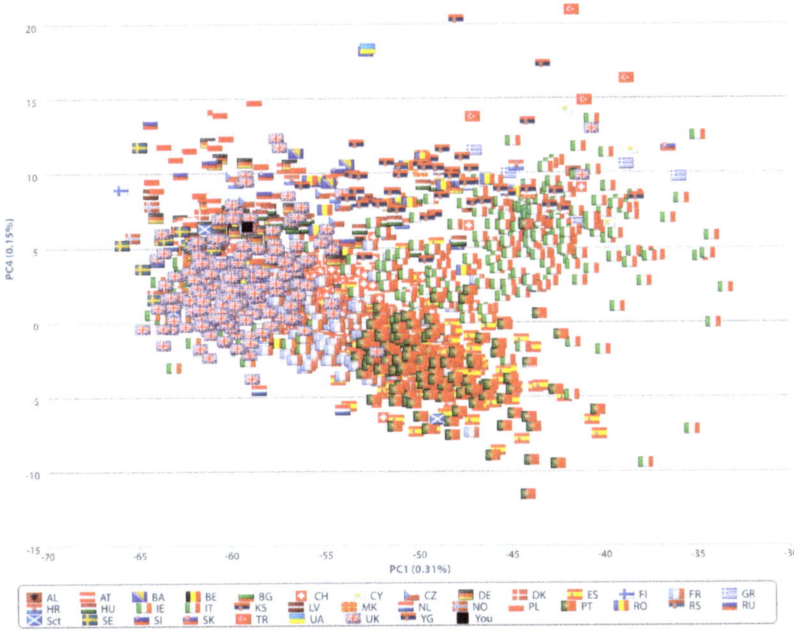

Figure 4: Ancestry analysis by PCA. Loadings for numerous population data sources are precomputed, allowing a user to project their data onto any one of those datasets. Here, an Eastern European individual is plotted in the upper-left quadrant among the POPRES European reference panel.

Additionally, we implemented a heuristic algorithm for chromosome painting. The state-of-the-art algorithms were not suitable for this task, as they require phased data and employ computationally expensive hidden Markov models (HMMs) to determine the most likely ancestry for each allele. Therefore, we designed a Monte Carlo simulation method to generate an approximation. First, we pre-computed the most informative population-differentiating SNPs and the client requests the allele frequencies for these SNPs in the selected reference panel. Then, for each "block" of the genome, we sample an allele from each genotype randomly (since we cannot determine phase) and use the allele frequency for that SNP to update a Bayesian model, which represents the likelihood of the block originating from each population given the data. For each iteration, the most likely population is chosen for each block, and this simulation is run multiple times to generate a number of votes for each block. These votes are then aggregated and ancestry is assigned: if the proportion of votes crosses a "heterozygosity threshold," both blocks are painted with the highest voted ancestry; otherwise, the highest and 2nd highest ancestries are chosen. The results are then smoothed and the results are plotted in Canvas (Figure 5).

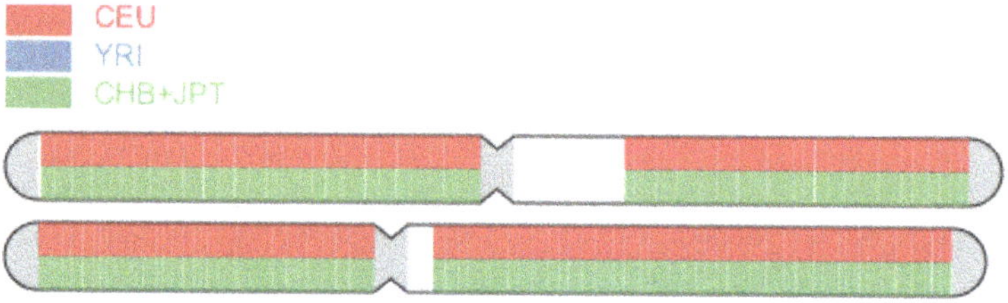

Figure 5: Chromosome Painting. The first two chromosomes from a half-European, half-Asian individual are shown. CEU, YRI, CHB, and JPT refer to European, African, Chinese, and Japanese Hapmap populations, respectively.

2.2.4. *Exploratory analysis*

Finally, we also implement a number of exploratory analyses and modules that were integrated with lectures of the Stanford course in Personalized Medicine and Genomics (Figure 6, left). For instance, we aggregated the SNPs associated with height from the GWAS catalog [Hindorff et al., 2009] and combined their effect sizes to create an approximate height prediction algorithm. Additionally, we created a widget to count the number of Neandertal-derived alleles [Green et al., 2010] in a personal genotype (Figure 6, right). Other exercises were developed to explore "SNPs of interest" that would integrate with a lecture, where students could optionally submit their allele information for real-time aggregation of allele frequencies.

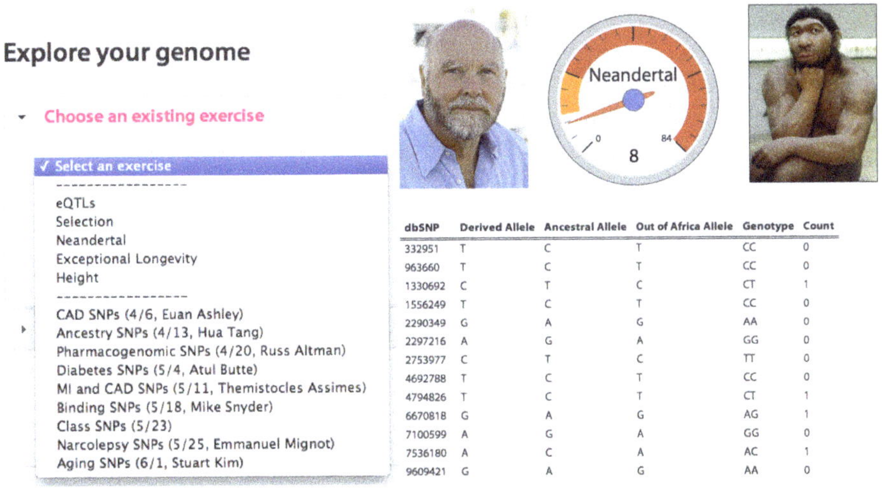

dbSNP	Derived Allele	Ancestral Allele	Out of Africa Allele	Genotype	Count
332951	T	C	T	CC	0
963660	T	C	T	CC	0
1330692	C	T	C	CT	1
1556249	T	C	T	CC	0
2290349	G	A	G	AA	0
2297216	A	G	A	GG	0
2753977	C	T	C	TT	0
4692788	T	C	T	CC	0
4794826	T	C	T	CT	1
6670818	G	A	G	AG	1
7100599	A	G	A	GG	0
7536180	A	C	A	AC	1
9609421	G	A	G	AA	0

Figure 6: Exploratory analysis. (Left): Numerous exercises are predefined, some with content from lectures of the Stanford Personalized Medicine course. Each of these is implemented independently, but all share a common data table format. (Right): One such analysis; Neandertal alleles in a personal genotype.

Through the development of these exercises, we observed that one major use case involved the counting of "risk" alleles (or alleles of some effect or significance), possible weighted by some "effect size" measure, such as odds ratios for traits, or centimeters for height. Therefore, we developed a customization framework for users to perform their own analyses.

2.3. *User Customization*

Although we wanted to provide curated datasets for standard analysis of a user's personal genome, we also wanted to allow the possibility of custom analyses. We therefore added functionality that allows the user to load custom annotated SNP lists. The user can then compare personal genotype information to this SNP list, as with the default exercises. For instance, a user may be interested in how many rare variants in a specific gene are found in a personal genotype and compare their results with those of colleagues or other personal genotypes.

The custom SNPs are loaded as a tab-delimited file, containing a header line (that correspond to the header of the output table) and the first column must indicate the SNP rsid(s) in question. An example custom annotation file snippet can be viewed by clicking on the 'Example Annotation File' link. As with the default exercises in the Explore tab, a table showing the user's genotype of his SNPs that were contained in the custom file, as well as its respective annotations, is presented to the user upon clicking the 'Lookup custom exercise' button.

In the course of development, we have noticed that one major use case involves reporting the allele count of the user's genotype against particular annotated columns. Therefore, we also allow the user to specify which columns should be used for allele counting by surrounding the column header with the count(.) syntax. It is worth noting that further functionality can be easily added to the custom exercise lookup; such as ethnicity specific SNP filtering or further aggregation, perhaps even SQL-like, functions. This could eventually allow researchers and developers to distribute custom annotation files and queries to expand the interpretive power of this system.

While a main focus of Interpretome is to maintain privacy of the user's genotype data, we are aware that users may want to share their results with others or even submit their genotype information to contribute to the enhancement of Interpretome. We have thus included both the option to share the exploration exercises results through a social network site and to submit their raw genotype information in an anonymous fashion. These two options give the user the possibility to explore a spectrum of privacy restrictions: from the default, most restrictive setting in which the user does not choose to share any of his information, to the other extreme of sharing both the results from the analysis and even genotype information. Sharing is an opt-in choice left to the user, and we have included a detailed description of the possible consequences of sharing in the Start page, as well as pop-up dialogs that ask the user to confirm all submissions of results or genotype information.

3. Results

We present the Interpretome at www.interpretome.com, a system for exploratory personal genome analysis, including guided explanations for clinical and ancestry analysis. The system is fast and easy-to-use and has been demonstrated in the Stanford course in Personalized Medicine and Genomics.

This system can load 1 million SNPs from a personal genotype into modern browsers (including Chrome and Firefox) in ~5-10 seconds. Further analyses require a server query, which range from ~1 KB to ~15 MB. These downloads typically take a few seconds to less than a minute for relatively local users (Northern California users with at least a cable modem connection). Once downloaded, the computational load on the client-side is very light for most applications (running in <5 seconds). A notable exception is the chromosome painting algorithm, which utilizes a Monte Carlo simulation to infer ancestry for specific chromosomal regions. However, even this analysis runs in ~15-20 seconds on a new laptop using the default parameters.

We have demonstrated the use of this system in the pioneering course on Personalized Medicine and Genomics at Stanford University. In this course, medical and graduate students learned about genomic personalized medicine through a hands-on analysis of their personal genotypes, for which we required an easy-to-use system that could accomplish sophisticated genotype interpretation tasks. The system was deployed for the Spring 2011 course and accomplished these goals. Overall, course students gave positive feedback on the system, expressing that its interactivity and ease of use enabled non-experts to extract meaning out of their genomic information. Particularly, they found the ability to instantly see their personal alleles for specific traits accompanied by relevant descriptions and annotations useful to interpret the results. Furthermore, advanced users liked having the option of tweaking the parameters for each module, as they found it useful to see how the methods performed with different values. These comments emphasize that a system of genomic interpretation must have both experts and non-experts in mind to both gain acceptability by the general public and convince experts of its usability.

The speed of the system and submission logic also allowed for further integration with lectures. Throughout the course, instructors were able to discuss SNPs for which there was an interesting association and students would have an option of submitting their genotypes for each SNP anonymously. The submitted genetic information was then aggregated and real allele frequencies were displayed to the instructor and the class, allowing for interactive participation in course material.

4. Discussion

In this work, we present the Interpretome, a system for private personal genotype interpretation and education. We believe that this approach will overcome a major hurdle to wider adoption of personal genotyping: the question of privacy and ownership. Users of Interpretome are assured privacy, as their data remains on their computer and in their browser. There exists no mechanism to track a user across uses of the website or to correlate data requests with client profile information (sex, population, etc.). However, genotyping services, such as direct-to-consumer companies, currently store the consumer's genomic information in their own servers. It may be preferable for service providers to provide users with an option for whether their genotype data should be stored at the company. Indeed, it would be ideal if the notion of privacy persisted through each step of the genotype pipeline, ensuring that only the consumer has exclusive access to their data.

The customizable nature of the Interpretome provides a platform for researchers to make their genomic annotations available to the general public. While we already enable the user to use their own SNP annotations, it would be straightforward to implement a web development framework, perhaps based on Javascript, for external modules that could be loaded at runtime. Such functionality would allow researchers to publish their methods as "Interpretome modules" for experts and non-experts to evaluate.

At present, we have included options for sharing of analysis results. While including these options may be considered controversial, it is our belief that enabling people to make informed choices about sharing their own genetic information will lead to an optimal trade-off between privacy and actionability of personal genomic data. The debate over privacy issues on genotype data is far from over. Thus, we believe that providing a genotype interpretation system that accommodates both extremes is essential to solving such conflicts.

Acknowledgments

We thank Euan Ashley, Atul J. Butte, Carlos D. Bustamante, Michelle Carillo, and Hua Tang for providing access to other data sources. We thank the students of the Stanford University course in Personalized Medicine and Genomics (Genetics 210, Spring 2011) for beta testing. KJK, JTD, and RPT were funded through training grant NIH LM007033, PC is funded through a Hewlett-Packard Stanford Graduate Fellowship. NPT is supported by DOE SCGF. RBA is a paid advisor to 23andme, Inc. MS and RBA are founders of Personalis, Inc.

KJK, RPT, NPT, and SKK conceived of the idea. KJK and RPT designed the site with significant input from NPT, PC, and SKK. KJK, RPT, NPT, JTD, and KS implemented the core analysis functionality; PC and KJK wrote the advanced graphing modules, user customizable framework, and data sharing schemes. SKK, RBA, and MS guided development and provided data and computational resources. KJK, RPT, PC, RBA, and SKK wrote the manuscript. All authors reviewed and approved the manuscript.

References

1. Ashley, E.A., et al. Clinical assessment incorporating a personal genome. *Lancet* **375**, 1525-1535 (2010).
2. Cann, H.M. et al. A human genome diversity cell line panel. *Science* **296**, 261–262 (2002).
3. International Warfarin Pharmacogenetics Consortium et al. Estimation of the warfarin dose with clinical and pharmacogenetic data. *N Engl J Med.* **360**, 753–764 (2009).
4. Green, R.E. et al. A Draft Sequence of the Neandertal Genome. *Science* **328**, 710–722 (2010).
5. Hindorff LA, Sethupathy P, Junkins HA, Ramos EM, Mehta JP, et al. Potential etiologic and functional implications of genome-wide association loci for human diseases and traits. *Proc Natl Acad Sci U S A.* **106**: 9362–9367 (2009).
6. Morgan, A.A., Chen, R. & Butte, A.J. Likelihood ratios for genome medicine. *Genome Med.* **2**, 30 (2010).
7. Adzhubei, I.A. et al. A method and server for predicting damaging missense mutations. *Nature Methods* 7(4), 248–249 (2010).
8. Novembre J, Johnson T, Bryc K et al. Genes mirror geography within Europe. *Nature.* **456**, 98–101 (2008).

A KINASE INHIBITION MAP APPROACH FOR TUMOR SENSITIVITY PREDICTION AND COMBINATION THERAPY DESIGN FOR TARGETED DRUGS

RANADIP PAL* NOAH BERLOW+

Department of Electrical and Computer Engineering, Texas Tech University, Lubbock, TX, 79409, USA
*E-mail: *ranadip.pal@ttu.edu +noah.berlow@ttu.edu*

Drugs targeting specific kinases are becoming common in cancer research and are a basis for personalized cancer therapy. Some of these drugs have the capacity to target multiple kinases. Promiscuous kinase inhibitors can be effective but the "off-target" effects can bring in toxicity for the patient. Thus the success of targeted cancer therapies with nominal harmful side effects is dependent on administering a single or multiple combinations of kinase inhibitors that targets the minimum number of kinases required to inhibit the tumor pathways. This requires a framework to predict the tumor sensitivities of a drug or drug combination based on the knowledge of the kinase inhibitors of a drug. In this article, we present a novel approach to predict the tumor sensitivities of a drug based on the generation of deterministic and stochastic Kinase Inhibition Maps. We build sensitivity maps or truth tables for a cell line from experimentally generated tumor sensitivities to kinase inhibitor drugs and use them to predict the sensitivity of a new drug or drug combinations based on known kinase inhibitor targets. We test our algorithms on a dataset of a dog osteosarcoma cell line with 317 possible kinase inhibitor targets after application of 36 targeted drugs. Our proposed algorithms are able to predict the sensitivities with high accuracy based on the given kinase inhibitor targets.

Keywords: Targeted therapy design, Drug sensitivity prediction

1. Introduction

Drugs that target specific kinases are becoming common in cancer research (see[1–5]). Cancer-related kinases are the paradigm of molecularly-targeted therapies, and a cornerstone of Personalized Cancer Therapy. The kinome consists of greater than 500 diverse tyrosine- or serine-threonine kinases ([6]). Some of these drugs have the capacity to target multiple kinases ([7]) and effects of multiple kinase inhibition are still to be properly understood. A prediction problem that we often encounter is predicting the sensitivity of a new drug for cell lines when the kinases that are inhibited by the drug are known. To address this, we formulate the problem as a boolean logic problem and build Kinase inhibiting maps for each cell line based on prior data. The Kinase inhibiting maps are used to answer the prediction problem.

The structure of kinases have been well studied for increasing the specificity of targeted drugs (see[8]) or high throughput kinase profiling has been used as a platform for drug discovery ([9]). However, approaches for high throughput analysis of multiple kinase inhibitor data to predict the behavior of new drugs are lacking in the literature. Prediction of tumor sensitivity to drugs have been approached earlier as a classification problem using gene expression profiles in *Staunton et al.*[10] . In *Staunton et al.*,[10] gene expression profiles are used to predict the binarized efficacy of a drug over a cell line with the accuracy of the designed classifiers ranging from 64% to 92%. In our proposed approach, we use the kinase inhibitor profiles of drugs as opposed to gene expression profiles of cell lines. The kinase inhibitor profiles of a drug are not dependent on a cell line and thus for a new cell line corresponding to a patient, we dont need to measure the gene expression profiles and just have to measure the tumor sensitivity after application of different drugs. Furthermore, our stochastic approach can predict the effectiveness of a drug on more than binary levels and produce higher accuracy results. In *Lee et al.*,[11] a co-expression extrapolation (COXEN) approach is used to predict the drug sensitivity in data points outside the training set with an accuracy of around 75% in predicting the binarized sensitivity. Our proposed approach is unique in not using gene expression biomarkers for prediction and using set theory and stochastic extensions to predict the non-binarized tumor sensitivity with high accuracy.

Drugs like Staurosporine have numerous kinase targets and are extremely effective in inhibiting the tumor but also have high toxicity ([7]). In fact, the toxicity of Staurosporine is similar to chemotherapy and thus working against the goal of targeted action on tumor cells without damaging normal cells and tissues. On the other hand, Imatinib (Gleevec) that has minimal kinase targets are being taken by leukemia patients for over a decade with no major side effects. Thus, it is extremely important that the off target effects of drugs are taken into consideration before prescribing a therapeutic regime for a tumor patient. Having a mathematical framework for the possible tumor pathways can allow us to generate the minimal set of kinase inhibition combinations that are required to block all the activated tumor pathways. A combination of drugs can be selected that just targets a minimal set of kinase inhibitors, with extremely few off target inhibitions. Majority of the current approaches for modeling genetic regulatory networks are not well suited for tackling this issue as the data requirements for model parameter estimation are significantly more in terms of number of samples and requirement of primarily time series data for estimation of the model parameters.[12] We propose a novel way to utilize steady-state experimental data on drug sensitivities over cell lines or animal models to generate robust maps for representing the tumor pathways. We exploit the fact that if a drug with inhibitor set S is effective, then any other drug with inhibitor set containing S will also be effective.

The paper is organized as follows: the mathematical formulation of the problem is provided in Section 2, algorithms to generate kinase inhibition maps are provided in Section 3, Section 4 covers the algorithm to predict the sensitivity of a new drug based on kinase inhibitor targets, Section 5 covers the probablistic maps; the results are presented in section 6 and finally section 7 considers the conclusions.

2. Mathematical Formulation

Let us a consider that we have drug sensitivity data for n cell lines after application of m drugs. The known multi kinase inhibiting sets for these drugs are denoted by S_1,S_2,...,S_m. The cardinality or number of elements of S_i is given by a_i for $i = 1, 2, \cdots, m$. Let S denote the union of $S_1, S_2, ..., S_m$ i.e. $S = \cup_{i=1}^{m} S_i$ and a denote the cardinality of S i.e. $a = |S|$. Thus there are a kinases that we are interested in and we will denote them as $k_1, k_2, \cdots, k_a$. The elements of sets S_i are denoted by $[e_{i,1}, e_{i,2}, \cdots, e_{i,a_i}]$ for $i = 1, 2, \cdots, m$. The $e_{i,j}$'s for $j = 1, .., a_i$ and $i = 1, 2, \cdots, m$ are elements from the set $[k_1, k_2, \cdots, k_a]$. For this scenario, the maximum possible number of distinct multiple kinase inhibitor activities exhibited by drugs are 2^a since each kinase can be either inhibited or not and there are a kinases, thus number of possibilities are $(2 \times 2 \times \cdots \times 2)_{a \text{ times}} = 2^a$. To solve our prediction problem, we will construct Kinase Inhibiting Maps (KIMs) for each cell line similar to Karnaugh maps ([13]) to simplfy boolean expressions.

As an example, let us consider the abstract representation of a biological pathway shown in Fig. 1. We will consider that a drug works if it inhibits all the paths between X and Y.

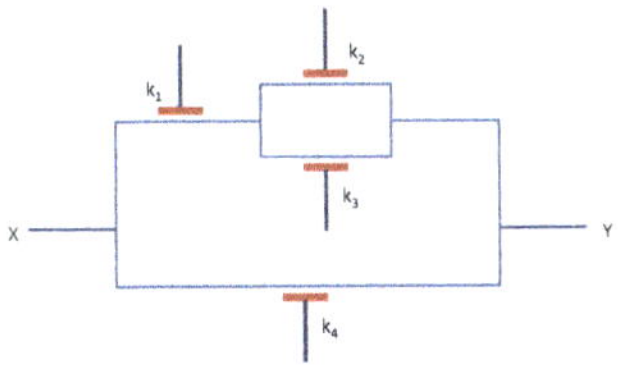

Fig. 1. Circuit 1

The circuit shown in Fig. 1 has three paths from X to Y through kinases $\{k_1, k_2\}$; $\{k_1, k_3\}$ and $\{k_4\}$. To block all the paths from X to Y, we have to inhibit all these parallel pathways. The first two pathways can be blocked by either inhibiting k_1 or inhibiting both k_2 and k_3. The third pathway has to be blocked by inhibiting k_4. If we represent this information as a map where a 1 indicates all pathways being blocked for that set of kinase inhibitions, then the resultant maping will be as shown in Figure 2. For example, a 1 in the 2nd row and 3rd column of the 4×4 matrix denotes that all the pathways for circuit 1 can be blocked by the combination $k_1 = 0, k_2 = 1, k_3 = 1, k_4 = 1$ i.e. any drug inhibiting k_2, k_3 and k_4 will have high sensitivity for circuit 1. Similarly, a 0 in the 2nd row and 2nd column denotes that inhibiting k_2 (i.e. k_2=1) and inhibiting k_4 (i.e. k_4=1) cannot produce high sensitivity for circuit 1.

$k_3\,k_4$ / $k_1\,k_2$	0 0	0 1	1 1	1 0
0 0	0	0	0	0
0 1	0	0	1	0
1 1	0	1	1	0
1 0	0	1	1	0

Fig. 2. Kinase Inhibiting Map (KIM) corresponding to Circuit 1

The map shown in Table 2 will be termed as Kinase Inhibiting Map (KIM) and can be constructed from Circuit 1 by checking the output for all the 2^4 combinations of kinase activities. The boolean logic for the circuit can be constructed based on the fact that each series connection is like Boolean **OR** function and each parallel connection is like boolean **AND** function, thus circuit 1 refers to (k_1 **OR** (k_2 **AND** k_3)) **AND** (k_4).

3. Constructing kinase inhibition maps from drug targets and sensitivity data

For the construction of the kinase inhibition maps, the following two sets of rules relevant to our problem will assist in filling the entries of the map.

Rule 1: If S_i is the inhibiting set of kinases for drug i and the drug is successful in inhibiting the circuit, then any set B containing the set S_i (i.e. B is a superset of S_i, $B \supset S_i$) will also be successful in inhibiting the circuit.

Rule 2: If S_i is the inhibiting set of kinases for drug i and the drug is unsuccessful in inhibiting the circuit, then any set B that is the subset of set S_i (i.e. $B \subset S_i$) will also be unsuccessful in inhibiting the circuit.

Rule 1 essentially says that if inhibiting a number of kinases has blocked all the paths, then inhibiting more kinases will not open any path. For instance, if we consider circuit 1 and our experiments denote that the set $\{k_1, k_4\}$ is able to inhibit the circuit, then any superset of $\{k_1, k_4\}$ such as $\{k_1, k_4, k_2\}$ will also inhibit the circuit. The number of possible supersets of a set S_i containing a_i elements among possible a elements is 2^{a-a_i}. For circuit 1, if $S_i = \{k_1, k_4\}$, the number of possible supersets of S_i is $2^{4-2} = 4$ and they are $\{k_1, k_4\}, \{k_1, k_4, k_2\}, \{k_1, k_4, k_3\}$ and $\{k_1, k_4, k_2, k_3\}$. If we consider the kinase inhibiting map in Fig. 2, then knowing the information that $\{k_1, k_4\}$ inhibits the circuit, we can fill the entries of its superset as 1. The entries corresponding to its superset are $[k_1k_2k_3k_4] = \{1001, 1101, 1011, 1111\}$ which fills the KIM(3,2), KIM(3,3), KIM(4,2) and KIM(4,3) entries of the inhibition map. Here KIM(i, j) denotes the ith row and jth column entry of the Inhibition map. The kinase inhibition map is a matrix of size $2^{p_1} \times 2^{p_2}$ where $p_1 + p_2 = a$ and $p_1 = \lfloor a/2 \rfloor$ and $p_2 = a - p_1$.

Rule 2 captures the fact that if a set of kinase inhibitors is unsuccessful in blocking the paths of a circuit, then any reduced number of kinase inhibitors among the inhibiting kinases cannot block all the paths. For instance, in circuit 1, if our experiments denote that the set $\{k_1, k_2, k_3\}$ of kinase inhibition is not sucessful in blocking all the paths of the circuit, then any subset of $\{k_1, k_2, k_3\}$ such as $\{k_1, k_2\}$ will also be unsuccessful in blocking the paths of the circuit. The number of possible subsets of a set S_i containing a_i elements is 2^{a_i}. For circuit 1, if $S_i = \{k_1, k_2, k_3\}$, the number of possible subsets of S_i is $2^3 = 8$ and they are $\{k_1, k_2, k_3\}, \{k_1, k_2\}, \{k_1, k_3\}, \{k_2, k_3\}, \{k_1\}, \{k_2\}, \{k_3\}$ and $\{\}$. In the boolean logic, the subsets will be as follows $[k_1k_2k_3k_4] = \{1110, 1100, 1010, 0110, 1000, 0100, 0010, 0000\}$. Thus, if a drug with set S_i of inhibitors is unsuccessful, then we can mark the states in the kinase inhibition map that has zeros for the $S - S_i$ kinases as zeros. Thus for our example, $S - S_i = \{4\}$ and any state having zero for k_4 will be marked zero in the kinase inhibition map. So the first and fourth column of the kinase inhibition map will be marked 0 based on the experimental piece of information that the set $\{k_1, k_2, k_3\}$ is unsuccessful. This approach fills up a large number of entries of the kinase inhibition map based on limited experimental knowledge. For our example, the two experimental results that $\{k_1, k_4\}$ is successful and $\{k_1, k_2, k_3\}$ is not successful fills in 12 out of 16 entries in the kinase inhibition map. Higher number of entries are filled when a large set of kinase inhibitors is unsuccessful or a small set of kinase inhibitors is successful.

The algorithm for constructing the kinase inhibitor map is shown in Algorithm 3.1.

Algorithm 3.1 Algorithm for constructing Kinase Inhibitor Map KIM

for $i = 1$ to m **do**
  for $j = 1$ to n **do**
    if Drug is successful **then**
      {Use Rule 1 to fill up entries in the kinase inhibition map for cell line j (KIMj)}
      for all R such that $R \supset S_i$ **do**
        KIMj (R) = 1;
      end for
    else
      {Use Rule 2 to fill up entries in the kinase inhibition map for cell line j (KIMj)}
      for all R such that $R \subset S_i$ **do**
        KIMj (R) = 0;
      end for
    end if
  end for
end for

4. Predicting sensitivity of a new drug based on its set of kinase inhibitors

If the set of kinase inhibitors is known for a new drug, then we can check the kinase inhibition map entry for that set of inhibitors for each cell line and predict the outcome of the drug when applied to that cell line.

Example 4.1.

Let us consider that we have two cell lines whose abstract circuit representation of the pathways are shown in Figures 3 and 4. Based on experimental data on these cell lines, we can construct the Kinase Inhibitors Maps KIM1 and KIM2 as shown in Figures 5 and 6. Here $S = \{k_1, k_2, k_3, k_4\}$ and k_4 is not directly involved in pathway 1 whereas k_1 is not directly involved in pathway 2. If a new drug D_{m+1} has the kinase inhibition set $S_{m+1} = \{k_2, k_3, k_4\} = \{0111\}$, then based on the 2nd row, 3rd column entries of the inhibition maps IM1 and IM2, the drug will be ineffective for CP1 and effective for CP2.

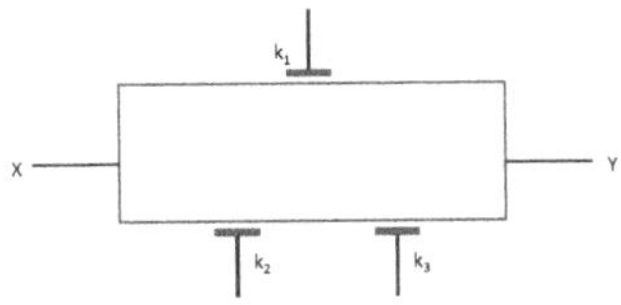

Fig. 3. CP1: Cellular pathway representation 1

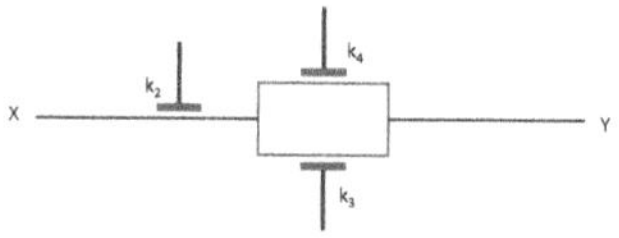

Fig. 4. CP2: Cellular pathway representation 2

$k_1 k_2$ \ $k_3 k_4$	0 0	0 1	1 1	1 0
0 0	0	0	0	0
0 1	0	0	0	0
1 1	1	1	1	1
1 0	0	0	1	1

Fig. 5. KIM1: Kinase Inhibition map corresponding to Cellular pathway representation 1 shown in Fig. 3

$k_1 k_2$ \ $k_3 k_4$	0 0	0 1	1 1	1 0
0 0	0	0	1	0
0 1	1	1	1	1
1 1	1	1	1	1
1 0	0	0	1	0

Fig. 6. KIM2: Kinase Inhibition map corresponding to Cellular pathway representation 2 shown in Fig. 4

5. Stochastic Extension

The approach described in the previous sections depend on the accuracy of the experimental data and accurate modeling assumptions. When there is a mismatch in the biological pathway and its abstract modeling and there are noises in the extraction of the data, we have to consider modifications of our algorithm to make it robust to uncertainties. Currently, the entries of the kinase inhibition map are updated based on the latest experimental data and in the process, the inferences of previous experimental data can be lost. For instance, let us consider the pathway representation of Figure 1 and our first experiment states that a drug combination of $\{k_1, k_2, k_3\}$ is ineffective, then the third row fourth column entry is zero. If now, an error in the data extraction of a new experiment states that $\{k_1, k_3\}$ is effective, then the third row fourth column entry will be erroneously changed to 1. Furthermore, the sensitivity to drugs will not necessarily be binary (effective or ineffective), thus our modified algorithm will also take into account the percentage of effectiveness of a drug. The effectiveness of a drug is usually measured by their IC_{50} values which denotes the amount of drug required to inhibit the activity of a process by half. So if the levels of drugs used for our experiment is from 1 unit to U units (say 1 to 10,000 nano-molar units), we will use the following approach to get the scaled sensitivity y of the drug with IC_{50} value x:

$$y = \frac{U - x}{U} \tag{1}$$

Note that equation 1 is one possible approach to map the IC_{50} values to numbers between 0 and 1 and there can be many other ways for this maping. To arrive at the new stochastic Kinase Inhibition Map, we will initially consider two matrices IM^1 and IM^0 denoting the most effective and ineffective combinations respectively. Then, the two matrices are combined in a proportional ratio to arrive at the final Probabilistic Kinase Inhibition Map ($PKIM$).

Algorithm 5.1 Algorithm for constructing Probabilistic Kinase Inhibitor Map PKIM

{Generate Kinase Inhibition Maps IM^1 and IM^0}
for $j = 1$ to n **do**
  Initialize $IM_j^1 = IM_j^0 = \mathbf{0}$
  for $i = 1$ to m **do**
    {Use Rule 1 to fill up entries in IM^1} for cell line j
    for all R such that $R \supset S_i$ **do**
      IM_j^1 (R) = IM_j^1 (R) + y_i;
    end for
    {Use Rule 2 to fill up entries in IM^0} for cell line j
    for all R such that $R \subset S_i$ **do**
      IM_j^0 (R) = IM_j^0 (R) + $1 - y_i$;
    end for
  end for
end for
{Generate Probabilistic Kinase Inhibition Map $PKIM$ }
for all $CL_j \in \{CL_1, \cdots, CL_n\}$ **do**
  Initialize $PKIM_j = \mathbf{0}$
  for all $v \in V$ **do**
    $PKIM_j$ (v) = $\frac{IM_j^1(v)}{IM_j^1(v)+IM_j^0(v)}$;
  end for
end for

return $PKIM_1, PKIM2, ..., PKIM_n$

6. Results

To illustrate the effectiveness of the proposed algorithms, we consider experimental data on a canine osteosarcoma cell line *CanOS1224* that was treated with 36 targeted cancer drugs and the tumor sensitivities measured. The experiments were conducted in the labs of Charles

Keller and Brian Druker in Oregon Health and Science University, Portland, USA. The cell line *CanOS1224* was derived from an actual canine with osteosarcoma who is being treated in OHSU as part of a clinical trial. Thus the data reflects a new cell line without much prior biological knowledge specific to that cell line and success of tumor sensitivity prediction on such a cell line holds promise for personalized therapy for a new patient coming to the clinic. The canine osteosarcoma primary cell culture was plated in 96 well plates at a seeding density of 2000 cells per well over graded concentrations of 36 small-molecule kinase inhibitors. Each inhibitor was plated individually at four concentrations predicted to bracket the IC_{50} for that drug. Cells were cultured in RPMI 1640 supplemented with 2mM glutamine, 2mM sodium pyruvate, 2mM HEPES, 1% penicillin streptomycin, and 10% fetal bovine serum for 72 hours. At the end of the 72 hour incubation, cell viability was assessed using the MTS assay. All values were normalized to the mean of seven wells on each plate containing no drug. The IC_{50} for each drug was then determined by identification of the two concentrations bracketing 50% cell viability and application of the following formula: [((A-50)/(A-B))*(Dose B - Dose A)]+ Dose A where cell viability value above 50% = A (drug dose for this value is Dose A) and cell viability value below 50% = B (drug dose for this value is Dose B).

The drug targets (kinases inhibited by a drug) are obtained from the supplementary tables of[7] and[14] based on experimental quantitative dissociation constant (k_d) values for each drug across 317 kinase assays. We considered a drug to be inhibiting the kinase if the k_d value is less than 10 percent of the maximum. The sensitivities (in terms of IC_{50} values) of the 36 drugs are experimentally generated and converted to numbers between 0 and 1 with 2 significant digits using Eq. 1 where U is 10000nM. Using all the 317 kinases to build the kinase inhibition map has the following problems: (a) the size of the Kinase Inhibition Map with $2^{317} = 2.67 \times 10^{95}$ entries will be computationally intractable; (b) data on 36 drugs is not rich enough to fill majority of the entries and keeping too many kinases will lead to over-fitting and (c) a number of kinases in the non-activated pathways of the cell line are not necessary to predict the tumor sensitivity to drugs. Thus, we used sequential feature selection to narrow down our number of Kinases.The cost function considered was mean absolute error in predicting the tumor sensitivities. We initially started with the kinase EGFR and sequentially added other kinases to it with 8 being the maximum number of kinases to be selected. For a set of kinases at any stage of feature selection, we used leave-one-out error estimation technique to calculate the mean absolute error. The 8 kinases selected through sequential selection approach are PIM1, PIK3CA, MRCKA, EPHA3, MAP4K5, MET, ACVR1B and EGFR. The data is shown in Figure 7. The first column shows the 36 drugs and the next 8 columns are the scaled dissociation constants for the 8 selected Kinases. The scaled dissociation constants ($0 \leq k_d \leq 1$) are obtained from[7] and[14] where a high value reflects that the Kinase is inhibited by the drug. To binarize the inhibition of kinases by a drug, we considered a kinase to be inhibited by the drug if the scaled k_d value is ≥ 0.9. The scaled tumor sensitivities are shown in column 10 in Figure 7 where a value of 1 reflects that the tumor is highly sensitive to the drug. We considered the probabilistic Kinase Inhibition Map method and used leave one out cross validation to measure the effectiveness of the PKIM technique. For leave one out error, we used the data on 35 drugs to build the PKIM and used the generated PKIM to predict the sensitivity for the 36th drug. This is done for all the 36 drugs and the average error is 0.067 which is really low for leave one out validation. Note that the maximum error can be 1 and a fixed prediction of 0.5 will result in an error of 0.437. The leave one out errors for each drug is included in the 11th column of Figure 7.

If we use re-substitution error estimate, then the average error decreases to 0.018. The re-substitution errors for each drug is shown in Column 12 of Figure 7. The decrease in error for re-substitution is expected as we are using the same data for training and testing. For example, we should note that the leave one out error for the drug PI-103 is 0.903. This is because the kinase signature of this drug is not available in any other drug i.e. no other drug targets the PIK3CA pathway as this drug does. Targeting the PIK3CA pathway for this cell line is able to reduce the tumor and this knowledge can only be gained from the data from the drug PI-103 and none of the other 35 drugs considered here. However, when we use the sensitivity of drug PI-103 in training the PKIM, we can predict the sensitivity for the drug PI-103 perfectly from the PKIM as shown in the case of re-substitution error. To be able to accurately predict the sensitivity of a new drug, it is imperative that our training data contain sets of kinase inhibitors that target most of the important pathways such as MAPK, JAK-STAT, mTOR, ERBB etc. The important pathway information can be obtained from public databases such as *KEGG*(http : //www.genome.jp/kegg/) or *pathway commons* (http : // www.pathwaycommons.org). Also drugs targeting both individual and combination of pathways can provide complementary information in estimating the entries of the kinase inhibition map.

The probabilistic KIM that was generated using the experimental data from the 36 drugs is shown in Figure 8. The PKIM was then used to build a circuit representation for cell line CanOS1224. The shaded areas in Figure 8 were considered to be 1 for generating the circuit shown in Figure 9. The circuit representation is easier to visualize and to study effective drug combinations, however the probabilistic Kinase Inhibition Map in Figure 8 should be referred to get the actual prediction. To arrive at a simpler circuit representation, some entries in Figure 8 is considered effective even when they have a low value. For instance, the entry for PIM1 being inhibited alone is 0.26 and still we considered it as effective for drawing our circuit. The PKIM or the circuit representation can be extremely useful in designing combination therapy for diseases. For instance, if our goal is to avoid resistance to drugs evolving in a patient, then we should try to block more than one series of kinase sets that each individually can block the tumor. For instance, in Figure 9, if we block PIK3CA, MRCKA and PIM1 then we will have two independent sets of Kinases that block the tumor. If the cancer becomes resistant to inhibition of PIM1, then the other set of Inhibitors MRCKA and PIK3CA will be the next line of defense against the tumor. From the data in figure 7, Flavopiridol and PI-103 can be used to target the three Kinases PIK3CA, MRCKA and PIM1. If our goal is to maximize chances of success and minimize off target effects, then we will pick the drug or drug combinations that inhibit minimum number of kinases but blocks all the paths of the circuit. To have a measure of toxicity of a drug, we should consider the total number of kinases inhibited by the drug among the 317 kinases tested and not just the 8 kinases considered here.

	Kinases								Sensitivities	Leave one out error	Re-substitution error
drugs	EGFR	PIM1	MRCKA	ACVR1B	PIK3CA	EPHA3	MAP4K5	MET	CanOS		
ABT-869	0	0	0	0	0	0.65	0.91	0.87	0.31	0.00791	0.004051
AMG-706	1	0	0	0	0	0	0	0	0	0	0
AST-487	0.95	0	0	0	0	0.99	0.96	0.56	0.51	0.328909	0.208381
AZD-1152HQPA	0.95	0	0	0	0	0	0.98	0	0	0	0
BIRB-796	0.3	0	0.06	0	0	0.91	0	0	0	0	0
BMS-387032/SNS-	0	0	0	0	0	0	0	0	0	0	0
CHIR-258/TKI-258	0	0	0	0	0	0	0.94	0	0	0	0
CHIR-265/RAF-265	0	0	0	0	0	0	0.92	0	0	0	0
CI-1033	1	0	0	0	0	0.79	0.74	0.44	0.26	0.063798	0.040419
CP-690550	0	0	0	0	0	0	0	0	0	0	0
Dasatinib	0.99	0	0.8	0.97	0	1	1	0	0.95	0.050409	0.027331
EKB-569	1	0	0	0	0	0.6	1	0.38	0	0	0
Erlotinib	1	0	0	0	0	0.76	0	0.62	0	0	0
Flavopiridol	0	0.94	0.05	0	0	0.67	0	0	0.94	0.052976	0.011501
Gefitinib	1	0	0	0	0	0	0	0	0	0	0
GW-2580	0	0	0	0	0	0	0	0	0	0	0
GW-786034	0	0	0	0	0	0	0.7	0	0	0	0
Imatinib	0	0	0	0	0	0	0	0	0	0	0
JNJ-7706621	0	0	0	0	0	0	0.79	0.94	0	0	0
Lapatinib	1	0	0	0	0	0	0	0	0	0	0
LY-333531	0	0.97	0	0	0	0	0	0	0.28	0.2766	0.021387
MLN-518	0.96	0	0	0	0	0	0.92	0	0	0	0
MLN-8054	0	0	0	0	0	0.79	0	0	0	0	0
PI-103	0	0	0.82	0	1	0	0	0	0.91	0.9064	0
PKC-412	0.87	0.94	0	0	0	0	0.89	0.93	0.97	0.025076	0.005444
PTK-787	0	0	0	0	0	0	0	0	0	0	0
SB-202190	0.74	0	0	0.91	0	0	0	0	0	0	0
SB-203580	0.83	0	0.38	0.7	0	0	0	0	0	0	0
SB-431542	0	0	0	0.98	0	0	0	0	0	0	0
Sorafenib	0	0	0	0	0	0.81	0.84	0	0.05	0.0489	0.039753
Staurosporine	0.96	1	0.99	0.93	0	1	1	0.98	1	0.0007	0.000581
SU-14813	0	0	0	0	0	0	0.97	0	0	0	0
Sunitinib	0	0	0	0	0	0.79	1	0	0	0.011252	0.009147
VX-680/MK-0457	0	0	0	0	0	0.85	0.99	0.93	0.26	0.074105	0.037949
VX-745	0	0	0	0	0	0	0	0	0	0	0
ZD-6474	1	0	0.74	0	0	0.8	0.95	0.43	0.37	0.568434	0.263415

Fig. 7. Experimental data and predicton errors using probabilistic KIMs.

7. Conclusions

In this article, we presented an approach to generate abstract representation of cancer pathways that can assist in predicting the sensitivities of a new drug given the kinase inhibitors of the drug. We also extended our algorithm to the probabilistic case to tackle latent kinases and noisy data and to predict non-binarized tumor sensitivities. The algorithms were validated on data obtained after application of targeted cancer drugs on a canine cancer cell line. The approach presented here is a novel way to analyze tumor sensitivity data based on Boolean logic and set theory. We expect that improved results can be obtained when the training set

KINASES		MET	MAP4K5 MET	MAP4K5	EPHA3 MAP4K5	EPHA3 MAP4K5 MET	EPHA3 MET	EPHA3	PIK3CA EPHA3	PIK3CA EPHA3 MET	PIK3CA EPHA3 MAP4K5 MET	PIK3CA EPHA3 MAP4K5	PIK3CA MAP4K5	PIK3CA MAP4K5 MET	PIK3CA MET	PIK3CA
	0	0	0	0	0.01	0.3	0	0	0	0	1	1	0	0	0	0
ACVR1B	0	0	0	0	0.49	1	0	0	0	0	1	1	0	0	0	0
MRCKA ACVR1B	0	0	0	0	0.49	1	0	0	1	1	1	1	1	1	1	1
MRCKA	0	0	0	0	0.07	1	0	0	1	1	1	1	1	1	1	0.91
PIM1 MRCKA	1	1	1	1	1	1	1	1	1	1	1	1	1	1	1	1
PIM1 MRCKA ACVR1B	1	1	1	1	1	1	1	1	1	1	1	1	1	1	1	1
PIM1 ACVR1B	1	1	1	1	1	1	1	1	1	1	1	1	1	1	1	1
PIM1	0.26	0.91	0.91	0.91	1	1	1	0.96	1	1	1	1	1	1	1	1
EGFR PIM1	0.91	0.91	0.98	0.91	1	1	1	1	1	1	1	1	1	1	1	1
EGFR PIM1 ACVR1B	1	1	1	1	1	1	1	1	1	1	1	1	1	1	1	1
EGFR PIM1 MRCKA ACVR1B	1	1	1	1	1	1	1	1	1	1	1	1	1	1	1	1
EGFR PIM1 MRCKA	1	1	1	1	1	1	1	1	1	1	1	1	1	1	1	1
EGFR MRCKA	0	0	0	0	0.64	1	0	0	1	1	1	1	1	1	1	1
EGFR MRCKA ACVR1B	0	0	0	0	0.98	1	0	0	1	1	1	1	1	1	1	1
EGFR ACVR1B	0	0	0	0	0.94	1	0	0	0	0	1	1	0	0	0	0
EGFR	0	0	0	0	0.3	1	0	0	0	0	1	1	0	0	0	0

Fig. 8. Probabilistic Kinase Inhibition Map for CanOS1224. The first row and First column denotes the Kinases being inhibited.

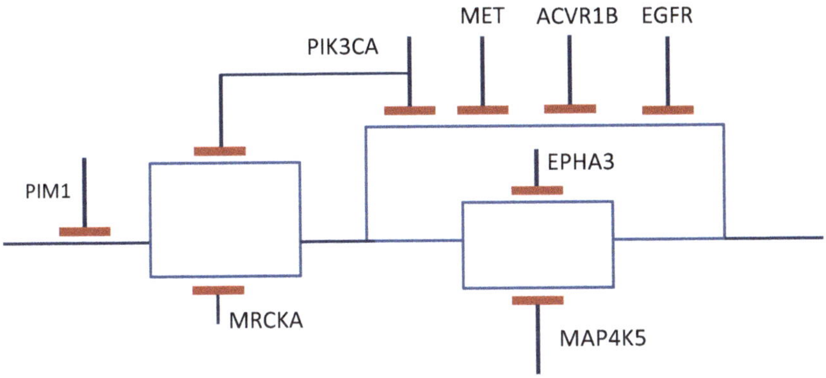

Fig. 9. Circuit representation for CanOS1224 based on PKIM in Figure 8.

will include more drugs with various combination of inhibiting kinases.

8. Acknowledgments

This research was supported by NSF grant CCF 0953366. We would like to thank Charles Keller, Jinu Abraham, Nicolle Hofmann, Jeffrey W. Tyner, Marc M. Loriaux and Brian J. Druker from Oregon Health & Science University and Bernard Seguin from Oregon State University for generating and providing the biological data.

References

1. C. Sawyers, *Nature* **432**, 294 (2004).
2. M. R. Green, *New England Journal of Medicine* **350:21**, 2191 (2004).
3. B. J. Druker, *Oncologist* **9**, 357 (2004).
4. A. Hopkins, J. Mason and J. Overington, *Current Opinion in Structural Biology* **16**, 127 (2006).
5. Z. A. Knight and K. M. Shokat, *Chemistry & Biology* **12**, 621 (2005).
6. G. Manning, D. B. Whyte, R. Martinez, T. Hunter and S. Sudarsanam, *Science* **298**, 1912 (2002).
7. M. W. Karaman, S. Herrgard, D. K. Treiber, P. Gallant, C. E. Atteridge, B. T. Campbell, K. W. Chan, P. Ciceri, M. I. Davis, P. T. Edeen, R. Faraoni, M. Floyd, J. P. Hunt, D. J. Lockhart, Z. V. Milanov, M. J. Morrison, G. Pallares, H. K. Patel, S. Pritchard, L. M. Wodicka and P. P. Zarrinkar, *Nature biotechnology* **26**, 127 (January 2008).
8. Chen, Jianping, Zhang and Xi, *Bioinformatics* **23**, 563 (2007).
9. D. Goldstein and et al., *Nat. Rev. Drug Discov.* **7**, 391 (2008).
10. J. E. Staunton, D. K. Slonim, H. A. Coller, P. Tamayo, M. J. Angelo, J. Park, U. Scherf, J. K. Lee, W. O. Reinhold, J. N. Weinstein, J. P. Mesirov, E. S. Lander and T. R. Golub, *Proceedings of The National Academy of Sciences* **98**, 10787 (2001).
11. J. K. Lee, D. M. Havaleshko, H. Cho, J. N. Weinstein, E. P. Kaldjian, J. Karpovich, A. Grimshaw and D. Theodorescu, *Proceedings of the National Academy of Sciences* **104**, 13086 (August 2007).
12. Z. Szallasi, J. Stelling and V. Periwal, *System Modeling in Cell Biology from Concepts to Nuts and Bolts* (MIT Press, Cambridge, MA, 2006).
13. M. Karnaugh, *Trans. AIEE. pt. I* **72**, 593 (1953).
14. P. P. Zarrinkar, R. N. Gunawardane, M. D. Cramer, M. F. Gardner, D. Brigham, B. Belli, M. W. Karaman, K. W. Pratz, G. Pallares, Q. Chao, K. G. Sprankle, H. K. Patel, M. Levis, R. C. Armstrong, J. James and S. S. Bhagwat, *Blood* **114**, 2984 (2009).

MIXTURE MODEL FOR SUB-PHENOTYPING IN GWAS

DAVID WARDE-FARLEY [3], MICHAEL BRUDNO [2], QUAID MORRIS [2,4], ANNA GOLDENBERG [1*]
[1] *Genetics and Genome Biology, SickKids Research Institute, 101 College Street, Toronto, ON M5G 1L7*
[2] *Department of Computer Science, University of Toronto, 6 King's College Rd., Toronto ON M5S 3G4*
[3] *Département d'Informatique et de Recherche Operationelle, Université de Montréal, CP 6128 succ. Centre-ville, Montréal QC H3C 3J7*
[4] *Donnelly Centre, University of Toronto, 160 College Street, Toronto ON M5S 3E1, Canada*

Genome Wide Association (GWA) studies resulted in discovery of genetic variants underlying several complex diseases including Chron's disease and age-related macular degeneration (AMD). Still geneticists find that in majority of studies the size of the effect even if it is significant tends to be very small. There are several factors contributing to this problem such as rare variants, complex relationships among SNPs (*epistatic effect*), and heterogeneity of the phenotype. In this work we focus on addressing phenotypic heterogeneity. We introduce the problem of identifying, from GWAS data, separate genotypic markers from overlapping mixtures of clinically indistinguishable phenotypes. We propose a generative model for this scenario and derive an expectation-maximization (EM) procedure to fit the model to data, as well as a novel screening procedure designed to identify skew specific to certain phenotypic regimes. We present results on several simulated datasets as well as preliminary findings in applying the model to type 2 diabetes dataset.

Keywords: GWAS; mixture modeling; sub-phenotyping

1. Introduction

With the current revolution in genome sequencing technologies many of the challenges in acquisition of genomic data and discovery of genomic variants have been overcome, supplanted by the challenge of understanding and assigning functions to the discovered variants. Genome Wide Association Studies (GWAS) have shown promise for discovery of disease-related variants, where usually large cohorts are used to find correlations between genomic variants (typically Single Nucleotide Polymorphisms, or SNPs) and disease risk or prognosis. Despite many successes, in majority of studies the size of the effect tends to be very small. Obtaining larger cohorts, which is rather costly, does not necessarily yield much better results. There are several factors contributing to this problem such as rare variants, complex relationships among SNPs (epistasis), and heterogeneity of the phenotype. While current efforts are targeted at collecting rare variant data and there is rich literature in modeling epistatic effect,[1,2] heterogeneity of the phenotype is often overlooked. Misclassification of disease phenotypes, especially when multiple distinct phenotypes are classified as a single disease, where each phenotype could potentially be modulated by different genetic factors, is prominent in cancers and spectrum disorders, such as Autism Spectrum Disorder (ASD).[3] In the presence of such misclassification, strong genotypic effect in a small homogeneous sub-population will appear as very small or even negligible in the whole population. Analyzing richer, continuous phenotypes such as age of onset, as opposed to a binary variable (disease, no disease) may help with discovery of those more homogeneous phenotypic subgroups.

*to whom correspondence should be addressed

A similar situation exists in diabetes: while Types 1 and 2 typically affect children and adults respectively, Maturity-Onset Diabetes of the Young (MODY) is a separate genetic disease that affects patients between 30 and 50 years old. Young MODY patients are often misdiagnosed as having Type 1 diabetes, if a patient is older and obese – as Type 2, and gestational if the patient is pregnant.[4] If we view the age of onset of various types of diabetes together, we observe a mixture, forming distinct but overlapping groups, caused by different genetic (and environmental) conditions. In this paper we develop methodology for simultaneous sub-phenotyping and phenotype-genotype association of such mixtures.

Simultaneous study of quantitative phenotypes and their underlying genotypes also form the basis for identification of expression quantitative trait loci (eQTLs[5,6]). The recent paper of Kim and Xing[7] models the correlation among multiple genes in a Quantitative Trait Network, and identifies SNPs that affect the expression of multiple genes that are tightly connected within the network. Other research[8] stratifies the patient population into groups based on ethnic heritage with assumption that identical phenotypes may be caused by different SNPs in different populations. Our work addresses the case where the phenotype is composed of a set of clinically indistinguishable, but distinct continuous traits, as in spectrum disorders. Both works, ours and,[8] draw additional power by subdividing the population, however our model takes into account information contained in a continuous phenotypic variable and determines genetic variants simultaneously with data partitioning. There have been a few major works that deal with data partitioning according to a phenotypic variable.[9–11] These works perform data stratification (akin a screening step proposed in our paper), mostly deal with discrete traits and do not propose a generative model of the data.

In this work we assume that there are multiple genetic diseases, each caused by a single variant, where each disease corresponds to a Gaussian-distributed quantitative phenotype (e.g., age of onset of the disease), with an unknown mean and variance. The diseases cannot be sub-divided, as they are clinically indistinguishable, hence the phenotype of all patients can be modeled as a mixture of Gaussians. We jointly model the cohesiveness of the sub-phenotypes (likelihood of the Gaussians explaining the phenotypes) and associations (SNPs explaining the sub-phenotypes). We optimize our model using expectation-maximization, and present the results of applying our model to simulated data and to the analysis of Gene-Environment Association Studies (GENEVA) genome-wide association scans for type 2 diabetes.[12]

2. Methods

To address the problem of identifying causal variants from phenotype mixtures we developed an algorithm SNPMix. The algorithm consists of three sub-parts: (1) The screening procedure, where we identify a small set of SNPs that are distributed abnormally in a sub-range of the continuous phenotype; (2) an Expectation Maximization procedure that identifies the sub-phenotypic groups, and the SNPs responsible for these, simultaneously; (3) model selection, that identifies a parsimonious subset of the screened SNPs and corresponding sub-phenotypes.

2.1. *SNP Screening*

Genome Wide Association studies typically genotype millions of SNPs. Only a handful of these are assumed to be associated with the disease of interest. Most approaches that look at multiple SNPs simultaneously *screen* the SNPs in order to identify the subset that is most likely to be associated with the disease for further in-depth analysis. Normally, a statistical test such as χ^2 or Fisher's Exact test is used to find SNPs that have significant deviations in observed allele frequencies between the patient and control populations. In our setting our aim is to identify SNPs that have unlikely frequencies in sub-populations with respect to the control. We thus modify the usual screening procedure: we stratify the population in the case cohort according to their phenotypic value in increasing order and test subsets (halfs, thirds, quarters, etc., halting at eighths) of the stratified population against the control. This way we are able to coarsely capture the modes of our hypothesized Gaussian distributions in one or several of those subsets. We retain the lowest of the p-values from tests made across all subsets ($1 + 2 + 3 + 4 + \ldots + 8 = 36$ tests) for each SNP. We assume that we have N patients and S SNPs. The pseudocode for this algorithm is presented in Table 1.

Table 1: SNP screening procedure

INPUT:	list of SNPs to be screened in $N \times S$ matrix:
	N – patients, S – snps
	test level threshold α
	z - phenotype variable
OUTPUT:	M SNPs – a small subset of the original SNPs

(1) Sort people according to z in ascending order
(2) For each SNP
  (a) calculate the frequencies of 0's, 1's, 2's in the control sample
  (b) for each contiguous (as sorted by phenotype) half, third, fourth, fifth, sixth, seventh and eighth of the patients in the case sample
    i. calculate the frequencies in this portion of the case data
    ii. perform a χ^2 test against the control
  (c) record the minimum p-value over all splits
(3) Rank SNPs according to their p-values

We perform a rough correction based on the Benjamini-Hochberg FDR[13] for the number of SNPs being tested. It is an approximation, since we do not correct for the 36 dependent tests per SNP. The goal of the screening is merely to obtain a superset of the responsible SNPs, and in practice, models fit to real data will include the top K most significant p-values rather than thresholding at a particular significance level.

2.2. *SNPMix Model*

Given the candidate set of screened SNPs, the core of the SNPMix algorithm simultaneously identifies the sub-phenotypic groups and the SNPs responsible for each group by modelling the phenotype and SNPs jointly. We assume that we have M SNPs remaining after the screening procedure, and a continuous phenotype z that follows a mixture-of-Gaussians distribution where each component represents a sub-phenotypic group. For each individual, a given SNP is encoded as a number $j \in \{0, 1, 2\}$, where j is the count of minor alleles: 0 indicates that the given genomic locus is homozygous major, 1 is heterozygous, and 2 is homozygous minor. Thus SNP information is represented as an $N \times M$ table, each entry being 0,1 or 2. Each SNP is modeled as a multinomial distribution. Since we do not know which sub-phenotypic component each patient belongs to, we introduce a latent variable c responsible for the sub-phenotypic class assignment which has as many states as there are components (K). The graphical representation of the model in Figure 1 captures our notion of conditional independence: information about the class de-couples the two, making phenotype and SNP frequencies dependent only on the class and independent of each other. The SNPs that are not corresponding to any sub-phenotypic groups are independent of the phenotype. To summarize, we have N patients,

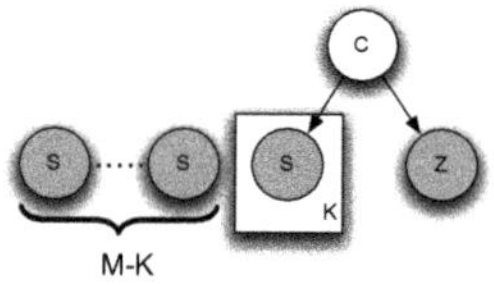

Fig. 1: Graphical model representation. z – phenotype, s – SNP, c – sub-phenotypic class assignment. Our generative model is different from the natural direction of causality, where SNPs cause disease ($s \rightarrow z$, in our notation). What we capture instead is that heterogeneity in the data (sub-phenotypic indicator c) confounds the notion of the dependence of phenotype on genotype. Given this information, we can disambiguate the dependence.

M SNPs, J SNP values ($J = \{j : j \in \{0, 1, 2\}\}$). The phenotype variable z is represented by sets of mean and standard deviation $\{(\mu_k, \sigma_k)\}$ parameters, where $k \in [1 \dots K]$. We can then write our two conditional distributions as follows:

$$p(s_i^n | c = k) = \prod_{j=1}^{J} {p_{ijk}}^{\delta_{j,s_i^n}}, \quad i \in [1, S]$$
$$p(z_n | c = k) = \mathcal{N}(\mu_k, \sigma_k^2)$$

s_i^n indicates the value of SNP $i \in [1 \dots M]$ for person $n \in [1 \dots N]$, δ_{j,s_i^n} is an indicator function which takes the value of 1, if $s_i^n = j$ and is 0 otherwise and $\{(\mu_k, \sigma_k)\}$ are parameters for each of the K mixture components.

Then the joint probability can be written as

$$p(\vec{s}, \vec{z}, c = k^n) = \prod_{k=1}^{K} \left(\alpha_k \mathcal{N}(\mu_k, \sigma_k^2) \prod_{i=1}^{S} \prod_{j=1}^{J} {p_{ijk}}^{\delta_{j,s_i^n}} \right)^{\delta_{k,k^n}} \quad (1)$$

where α_k is a mixture proportion of the kth component, k^n is the class of the person n and δ_{k,k^n} is an indicator function which is 1 if $k = k^n$, $S \in [K \dots M]$ is the current number of SNPs in the model, generalizing to the case where each group can have more than one associated SNP. Let $\Theta = \{\{p_{ijk}\}, \vec{\alpha}, \vec{\mu}, \vec{\sigma}\}$. All of the Θ parameters are estimated as indicated below.

2.3. *Optimization and Model Selection*

We optimize the parameters of the latent variable model developed in the previous section using an expectation-maximization (EM) algorithm by fixing a posterior $Q = p(c|\vec{s}, z; \Theta^{old})$ in the E step and optimizing the *expected complete log likelihood* under Q:

$$\mathbb{E}_Q[\ell(\Theta)] = -\sum_{n=1}^{N}\sum_{k=1}^{K} r_{nk}\left(\log\sigma_k + \frac{(z_n-\mu_k)^2}{2\sigma_k^2}\right) + \sum_{n=1}^{N}\sum_{k=1}^{K} r_{nk}\left(\log\alpha_k + \sum_{s=1}^{S}\sum_{j=1}^{J}\delta_{j,s_i^n}\log p_{ijk}\right) \quad (2)$$

where r_{nk} is the responsibility (i.e. $p(c = k|\vec{s}_n, z_n; \Theta^{old})$) assigned to person n by component k, i.e. the model's estimate of the probability that patient n is from the component k. Note that the expectation now looks very similar to the standard mixture of Gaussians model,[14] where the means and the variances of the Gaussian components depend on the multinomial distributions of the SNPs through the responsibilities.

To obtain the M-step update equations, we differentiate Equation 2 and solve for the parameters which jointly maximize the likelihood given the current estimate of the responsibilities. The updates for μ_k, σ_k and α_k are the standard mixture of Gaussians updates, with p_{ijk} being updated as follows:

$$p_{ijk}^{t+1} = \frac{\sum_{n:s_i^n=j} r_{nk}^t}{\sum_m r_{mk}^t} \qquad p_{ij,k\neq i}^{t+1} = \frac{\sum_{k\neq i}\sum_{n:s_i^{(m)}=j} r_{nk}^t}{\sum_{k\neq i}\sum_m r_{mk}^t}$$

The second equation corresponds to the probability of each SNP's out-of-component distribution, which is the same for all of the components that the SNP is not associated with.

Because the screening procedure results in a selection of a large set of SNPs a subset of which is likely to be responsible for the sub-phenotypic components, we need to perform model selection to eliminate the false positives. In this work, we select $K \leq M$ SNPs that represent sub-phenotypic groups, one per group. We use the greedy model selection procedure described below to select the SNPs, though extensions where many SNPs represent a single component and other model selection techniques, e.g. BIC, are possible. The SNPs that have been selected but are not modeled as associated with the phenotype (indicated by a set of independent SNPs $M' = M - K$ on Figure 1) result in an additive component to our log likelihood:

$$\sum_{n=1}^{N}\sum_{i=K+1}^{M'} \log\binom{2}{s_i^n} + s_i^n\log\gamma_i + (2 - s_i^n)\log(1-\gamma_i) \quad (3)$$

where γ_i and 1 - γ_i here correspond to the usual Hardy-Weinberg p and q values[15] and can be estimated once, in closed form, prior to running EM.

The set of informative SNPs in the model will contain those most relevant to the phenotype, however much of it can be explained by genetic factors. The false positive SNPs found through screening should not explain more than a small proportion of the data. We thus propose a greedy procedure which iteratively removes the SNP that explains the least amount of data. The iterative procedure starts by fitting a component to all of M screened SNPs, and runs the EM algorithm once to obtain the joint model corresponding to the screening result. We remove the SNP where the corresponding component explains the lowest fraction of the data, and iterate with the smaller number of SNPs. The procedure stops when all of the components explain more than a preset threshold, depending on the expected number of components.

3. Results

We will first illustrate our approach on a simulation study and then report our findings on the type 2 diabetes dataset. In all of the experiments below we set $\alpha = 0.05$ and the minimum explained threshold to $\max(1/(K' + 2), .1)$, where K' is the expected number of components (e.g., if the number of components is expected to be 3, the minimum explained threshold is .2, i.e. each component has to explain at least 20% of the data to be accepted).

3.1. *Simulation*

To simulate our data, given $N = 1000$ individuals, $M = 100,000$ SNPs and K Gaussian subphenotypes we

(1) Sample q - the frequency of the minor allele in the population following the empirical distribution observed by the HapMap population as discussed in[16]
(2) Generate two $N \times M$ SNP tables for cases (patients) and controls (healthy population) according to Hardy-Weinberg (HW) principle[15]
(3) Generate mixture proportions uniformly U(a,b) on the interval $[.1, .8]$ (we will use U(.1,.8))
(4) Sample individuals from the mixture to belong to K Gaussians
(5) Sample K SNPs from M to be responsible for each of the K components
(6) For each of the K components
    (a) Generate phenotype z from k^{th} Gaussian with random mean sampled from U(0,K+1) and variance from U(1,2)
    (b) Sample k^{th} SNP values from the empirical multinomial of the control population: with probability .9 inside the Gaussian and with probability .1 – outside.

We refer in the following discussion to "true positives" as those simulated SNPs that are indeed mixture-modulated in the simulated ground truth and are discovered as such by the iterative model selection procedure, and "false positives" as those SNPs assigned a mixture component despite not being generated as such in the simulation.

3.1.1. *Robustness of the model screening procedure*

We first tested the ability of the screening procedure to capture simulated SNPs by generating data from our model for $K = 2$. One would expect to capture subphenotype-modulating SNPs in groups that are not too small and do not have very high overlap. Indeed, the screening procedure contains a sorting step where high overlap between components would imply dilution of the signal for each of the corresponding SNPs. The results are shown in Figure 2. The screening process is able to perfectly recover mixture-dependent simulated SNPs for components with up to half the size of the case cohort with up to 30% overlap. Furthermore, with a slight loss in accuracy we can recover SNPs with up to 70% overlap.

We have then tested the effect of the potentially inaccurate screening procedure on the results of the SNPMix by considering simulated data with 3 to 7 components ($N = 1000, M =$

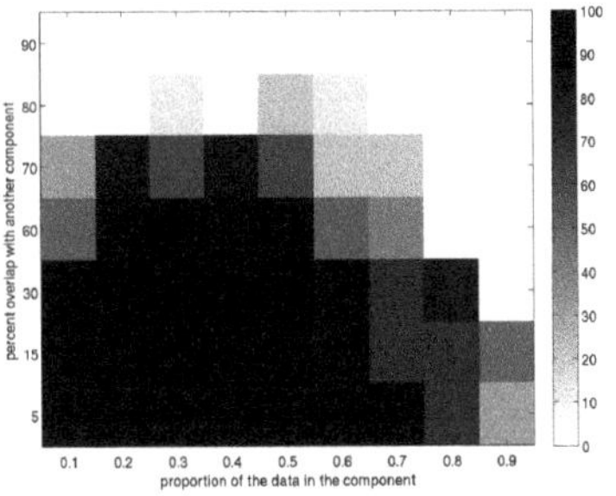

Fig. 2: Heatmap indicating whether we are able to recover true (simulated) SNPs given the size and the overlap of the components

100,000 as before), running 10 trials for each scenario. Figure 3 shows True Positive (TP) rates for SNPMix, on the default set of screened SNPs (those that pass the Benjamini-Hochberg threshold at the time of screening) shown on the left, and a set that includes all of the true positive SNPs in addition to the SNPs found through Screening step whether or not the true SNPs were found by screening procedure (on the right), i.e. if a true SNP was not present in the set of SNPs after screening it was imputed before SNPMix was run. As demonstrated in Figure 3a SNPMix+Screening achieves on average a 40-50% TP rate on the SNPs originally obtained through screening procedure, which compares well with $K/100,000 < 10^{-5}$ baseline prediction accuracy. It appears that the reduced TP rate is due to the inaccuracy of the screening procedure: when all of the true positive SNPs are made available to the model, the TP rate increases to 80-100%, as is illustrated in Figure 3b.

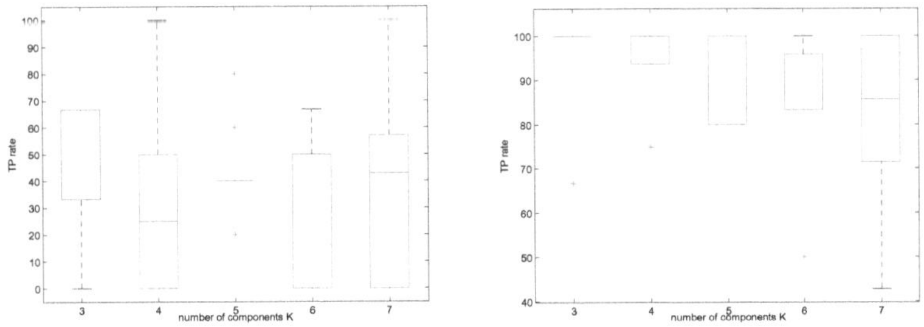

(a) using original results of the Screening step (some true positives might be missing)

(b) using SNP set after Screening augmented to insure the presence of all true positives

Fig. 3: True Positive rate for K=3..7 for the Screen+SNPMix procedure

Figure 4 shows the reduction of false positive rate after running SNPMix, on the same set of simulation data as described above. Comparing Figures 4a and 4b we see that in cases where not all true positives are available, the SNPMix performs slightly worse in reducing FP rate, trying to account for variation in the phenotype with some of the better fitting false positives. To the contrary, in the case where all true positives are available after the Screening step, we observe a 100% reduction in FP rate in most cases.

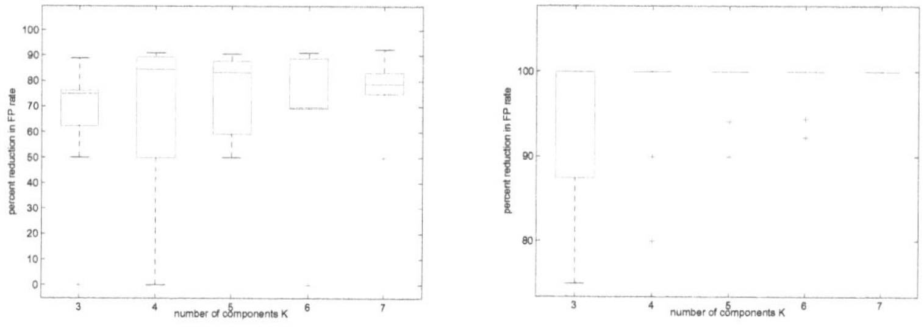

(a) using original results of the Screening step (some true positives might be missing)

(b) using SNP set after Screening augmented to insure the presence of all true positives

Fig. 4: Reduction in false positive rate for K=3..7

3.1.2. *Robustness of the model selection procedure*

Finally, we have attempted to quantify the effectiveness of our model selection procedure. Figure 5 demonstrates how the number of true (TP) and false (FP) positives (represented by red and blue lines on the graph) vary on the simulated data with K=3 SNPs (assuming all true positives are available) as a function of the size of the component (the x-axis on Figure 5 shows the proportion of the minimum number of the patients that have to be in the component for the component to be valid to the total number of patients). If any of the three, in this case, component sizes fall below the allowed threshold of the number of patients per component (as depicted on the x-axis), the component and the corresponding SNP are no longer considered as valid and are added to the pool of SNPs that are not associated with the phenotype.

Figure 5 represents the tradeoff between the number of false and true positives. When the threshold is very stringent, e.g. we require each component in the model to explain at least 33% of the data (for $K = 3$ this means that it is not very likely that all three components will pass the threshold), there are no false positives, but we are not capturing all true positives either. Whereas if we set the threshold too low, the number of false positives might not justify the certainty in retaining the true positives. Fortunately, there is a range (between .15 and .2 for $K = 3$) where the number of false positives is relatively low while all true positives are captured. Such analysis can be performed for varying K to estimate a proper threshold for our model selection procedure prior to running the SNPMix on the real world data.

So far we have not dealt with the case where there is no genotypic variant in our data that could explain the phenotype. This case is illustrated on the Type 2 diabetes dataset.

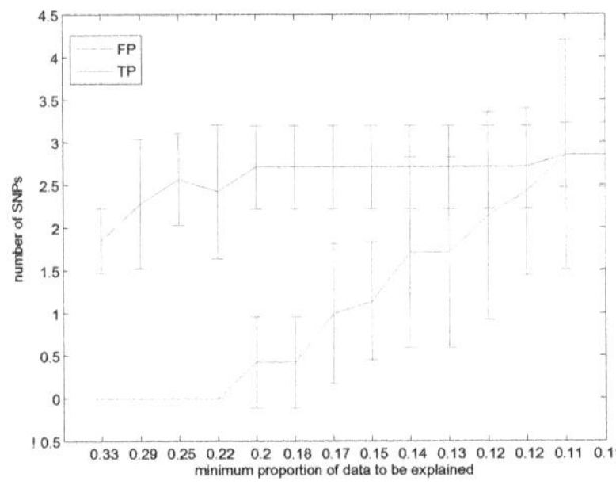

Fig. 5: Average number of SNPs in the model after running a SNPMix procedure for varying minimum-explained thresholds ($K = 3$). Error bars represent 1 std deviation.

3.2. *GENEVA Type 2 diabetes data*

We applied our methods to the Gene Environment Association Studies Initiative (GENEVA)[b] Diabetes.[12] The study contains extensive phenotype and genotype data for 6,000 patient and control individuals. Both phenotype and genotype data was obtained from the dbGAP database.[17] Here we study the 2,502 male subjects (members of the Health Professionals Follow-up Study, HPFS), genotyped by the Affymetrix 6.0 array, including 1,161 diabetes patients and 1,341 controls. By observing the distributions of the various quantitative phenotypes in the study, we chose to model the patient age (at the time of the study) as our quantitative phenotype, as it appeared to be multimodal (Figure 6 hints at the presence of at least two mixture components, supported by fitting a mixture of two Gaussians). For younger patients the age of participation in the study is indicative of the age at diagnosis. Unfortunately, the information on the age at the onset of the disease was not available in this data.

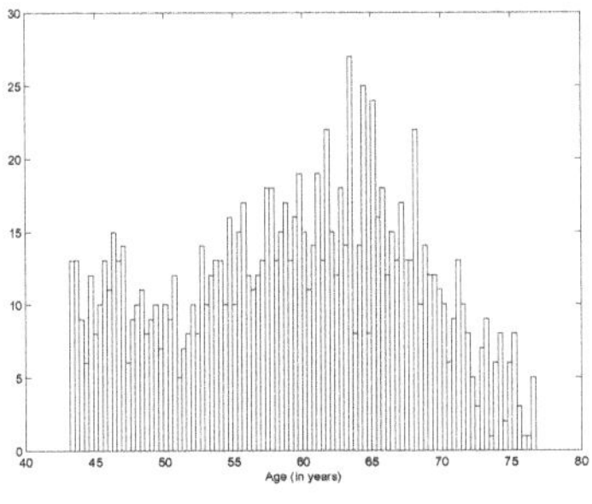

Fig. 6: Age distribution of patients in the GENEVA HPFS type 2 diabetes study

[b]http://www.genevastudy.org/

We first applied the screening procedure and selected the top 25 SNPs. In these 25 SNPs we found 2 regions of linkage disequilibrium (high correlation between SNPs in close proximity on the genome). We chose a single representative SNP for each region, resulting in 20 SNPs total. Figure 7(a) shows the responsibilities after applying the SNPMix procedure, setting the minimum explained threshold at 10% (the approximate size of the younger sub-group). The analysis revealed that the smaller subgroup of younger patients is explained by a single SNP, while a significantly larger number of SNPs are used to explain the larger component. This may be due to a complex relationship among SNPs jointly explaining the larger component (epistasis), not modeled by our method, or to a lack of genetic basis for the larger component. It is also possible that the larger component consists of a very large number of distinct Gaussian distributions. To eliminate the latter possibility, we ran 100 randomized trials of our algorithm, and found that for most individuals in the middle age group the maximum over all trials of the maximum responsibility learned by any of the components is low. At the same time, the model has much higher certainty about the SNPs corresponding to the younger (and older) cohorts, as is illustrated on Figure 7(c). To model a large sub-component with no significant genetic basis we added a "free" Gaussian, that is not explained by any genotype, with mixture proportion based on the fraction of patients for whom the average over 100 randomized trials of the maximum responsibility was below 50% (0.76). We then ran our SNPMix procedure allowing for components that explain at least 10% of the phenotype. The iterative SNPMix procedure converged on 3 Gaussians as can be seen in Figure 7(b), with 2 Gaussians corresponding to SNPs (dbSNP ids rs2002520 and rs4465650), as well as the free Gaussian.

Analyzing the SNPs we found that the minor allele of SNP rs2002520 is rare: in the 934 HapMap individuals genotyped at this position in HapMap, the frequency of the G allele was 0.05, and not a single homozygous minor individual was observed. The allele frequencies, however, were significantly different in the younger sub-group (130 patients), with an allele frequency of 0.09, and 4 individuals carrying the homozygous minor allele. This SNP is located in a region that is annotated as an enhancer by ENCODE based on CHiP-seq of Histone mark. Proximity to a DNase hypersensitivity site indicates a possible regulatory role of the region. The locus is annotated as a human body-mass index QTL by the Rat Genome Database, linked to obesity. Obesity is a common risk factor for type 2 diabetes, so our analysis possibly uncovered a SNP that affects an early onset of diabetes due to this additional risk. None of the genes in immediate proximity of the SNP are linked to diabetes, thus additional analysis is necessary to confirm the role of this variant in the disease.

rs4465650 associated with the older group is harder to justify. It could be a protective SNP that delays the onset of type 2 diabetes in people who have a certain variant of the SNP even if they are predisposed to the disease. However, since we do not have the information on the onset of the disease, we cannot know that these people have not had undetected diabetes for a while. To test whether this SNP is simply related to longevity, we would need to perform and hypothesis test on the controls with matching age distribution, which was unfortunately not available at the time the experiments were performed.

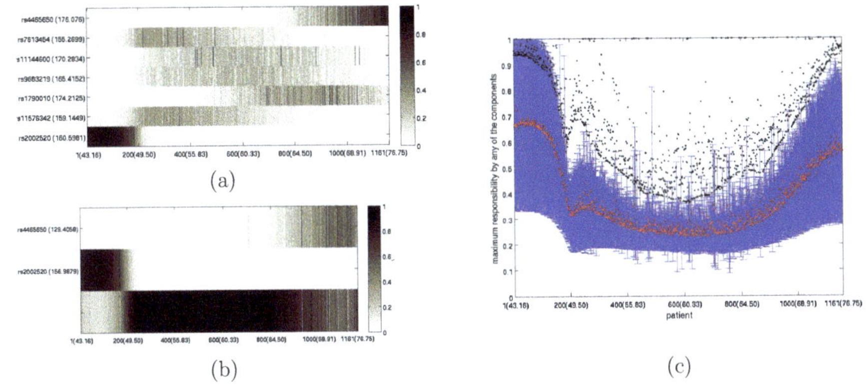

Fig. 7: SNPMix result of modeling age phenotypic variable in the GENEVA type 2 diabetes study starting from 20 independent pre-screened SNPs. (a) Y-axis indicates SNP name corresponding to each component with total responsibility in parenthesis, X-axis indicates patients sorted by age; (b) SNPMix including a phenotype-only (free) component; (c) maximum (in black), mean (in red) and standard deviation (in blue) responsibility for each patient over a 100 runs of SNPMix.

4. Discussion and Conclusion

In this study we consider a novel problem setup, where several clinically indistinguishable phenotypes are assumed to have different underlying associated genotypic variants. This problem is especially inherent in what is known as spectrum disorders, where the case cohort is comprised of several different sub-diseases, that are not easy to identify clinically. The signal for each of the sub-diseases is too weak to be revealed in the full cohort of patients. We believe that it is precisely this reason that led to very weak signal in many known GWA studies. Frustrated with the current state of standard GWAS performance, clinicians are becoming increasingly interested in more complex machine learning approaches, where fewer assumptions are made. Our probabilistic model and the associated expectation-maximization algorithm can recover dependencies on synthetic data where the causal relationships are known. We also present preliminary results on applying the method to a case control study of type 2 diabetes patients. Though the public dataset we used was somewhat simplistic (one could almost read a plausible separation of the Gaussian modes from the graph), it is the simultaneous consideration of phenotype and genetic variants that help us to identify the more accurate borders of the sub-groupings. In the field where there are almost no golden standards of sub-phenotyping, these kind of experiments serve as an indication that the method is doing something reasonable with the data. Further work is needed both to extend our model and to improve model-selection and screening procedures.

Our assumption of a single SNP being associated with each mixture component, while simplifying fitting and model selection, may poorly reflect the reality of complex diseases. The model can be altered to allow a single SNP to flexibly account for perturbations in more than one mixture component, as well as for the assignment of multiple SNPs to a single component, but such a many-to-many relationship greatly expands the space of possible models, and overfitting becomes an increasingly prominent worry. Future work should explore strategies for making principled choices about which dependences to retain, with care taken to appropriately regularize the parameters of richer, more flexible models.

Having narrowed our search space through pre-screening, the discrete choice of which SNPs to model as associated spans the space of all possible subsets of pre-screened SNPs, making exhaustive search prohibitive for even a moderate number of SNPs. Of several techniques we explored, including greedy and beam-search variants of model-pruning using standard model comparison criteria (e.g. the Bayes information criterion), none were entirely satisfactory. The pruning strategy we adopted was among the simplest, yet produced results that offered a parsimonious explanation of the data. Nevertheless, more work is needed on determining the most appropriate model selection criteria and optimization procedures to take full advantage of the class of complex genotype-mixed phenotype models studied in this work.

Acknowledgement

This study was funded by NSERC Discovery Grants to MB and QM.

References

1. H. J. Cordell, *Nat Rev Genet* **10**, 392 (Jun 2009).
2. S. K. Musani, D. Shriner, N. Liu, R. Feng, C. S. Coffey, N. Yi, H. K. Tiwari and D. B. Allison, *Hum Hered* **63**, 67 (2007).
3. P. Lichtenstein, E. Carlström, M. Råstam, C. Gillberg and H. Anckarsäter, *Am J Psychiatry* (Aug 2010).
4. A. Stride and A. T. Hattersley, *Ann Med* **34**, 207 (2002).
5. W. Cookson, L. Liang, G. Abecasis, M. Moffatt and M. Lathrop, *Nat Rev Genet* **10**, 184 (Mar 2009).
6. J. J. Michaelson, S. Loguercio and A. Beyer, *Methods* **48**, 265 (Jul 2009).
7. S. Kim and E. P. Xing, *PLoS Genet* **5** (Aug 2009).
8. K. Puniyani, S. Kim and E. P. Xing, *Bioinformatics* **26**, 208 (Jun 2010).
9. X. Qin, E. R. Hauser and S. Schmidt, *Genet Epidemiol* **34**, 407 (Jul 2010).
10. S. Macgregor, N. Craddock and P. A. Holmans, *Eur J Hum Genet* **14**, 529 (May 2006).
11. A. P. Morris, C. M. Lindgren, E. Zeggini, N. J. Timpson, T. M. Frayling, A. T. Hattersley and M. I. McCarthy, *Genet Epidemiol* **34**, 335 (May 2010).
12. C. Zhang, L. Qi, D. J. Hunter, J. B. Meigs, J. E. Manson, R. M. van Dam and F. B. Hu, *Diabetes* **55**, 2645 (Sep 2006).
13. Y. Benjamini and Y. Hochberg, *Journal of the Royal Statistical Society, Series B* **57**, 289 (1995).
14. A. Dempster, N. Laird and D. Rubin, *Journal of the Royal Statistical Society, Series B* **39**, 1 (1977).
15. A. Edwards, *Foundations of Mathematical Genetics* (Cambridge University Press, 1977).
16. International HapMap Consortium, *Nature* **437**, 1299 (Oct 2005).
17. M. D. Mailman, M. Feolo, Y. Jin, M. Kimura, K. Tryka, R. Bagoutdinov, L. Hao, A. Kiang, J. Paschall, L. Phan, N. Popova, S. Pretel, L. Ziyabari, M. Lee, Y. Shao, Z. Y. Wang, K. Sirotkin, M. Ward, M. Kholodov, K. Zbicz, J. Beck, M. Kimelman, S. Shevelev, D. Preuss, E. Yaschenko, A. Graeff, J. Ostell and S. T. Sherry, *Nat Genet* **39**, 1181 (Oct 2007).

TEXT AND KNOWLEDGE MINING FOR PHARMACOGENOMICS: GENOTYPE-PHENOTYPE-DRUG RELATIONSHIPS

K. BRETONNEL COHEN
Department of Pharmacology, University of Colorado School of Medicine
Aurora, Colorado, USA
Email: kevin.cohen@gmail.com

YAEL GARTEN
Biomedical Informatics Program and Department of Genetics, Stanford University
Stanford, California, USA
Email: ygarten@stanford.edu

NIGAM SHAH
Stanford Center for Biomedical Informatics Research, Stanford University
Stanford, California, USA
Email: nigam@stanford.edu

UDO HAHN
Jena University Language and Engineering Laboratory, Jena University
Jena, Germany
Email: Udo.Hahn@uni-jena.de

Text mining for pharmacogenomics is poised to become a major research topic. This session at the Pacific Symposium for Biocomputing is meant to be a catalyst to push it from potential to realization.

Many research areas are relevant to the general topic of text mining for pharmacogenomics. Minimally, the field calls for work on the following:

- Named entities relevant for the pharmacogenomics literature. This includes drugs, chemicals, diseases, genes, and genomic variations.
- Relations relevant for the pharmacogenomics literature. This includes a wide range of relation types, ranging from simple co-occurrence to is-effective-for, treatment-of, is-associated-with, etc. It is not clear what the full set of relations is, and this presents a challenge for knowledge engineering.
- Infrastructure relevant for mining the pharmacogenomics literature. This includes corpora and gold standards, such as PharmGKB, i2b2 data, the Comparative Toxicogenomics Database, and the Arizona Disease Corpus.
- Evaluation of the state of the art, including both novel evaluations and novel approaches to evaluation.
- Reasoning systems applied over resources like the PharmGKB knowledge base.

The session received a number of relevant submissions, on topics ranging from ontologies; phenotype studies and animal models; drug repurposing; mutation combinations and drug resistance; terminologies; drug-drug interactions; reasoning over syntax and semantics; ranking gene-drug interactions; and genetic variants that affect drug response. A rigorous review process resulted in the selection of a set of papers that arguably represents the best of the state of the art in text and knowledge mining for pharmacogenomics, focusing on genotype-phenotype-drug relationships. Much work remains to be done in the area; we hope that this represents just the beginning of a new flowering in studies of text mining for pharmacogenomics.

THE EXTRACTION OF PHARMACOGENETIC AND PHARMACOGENOMIC RELATIONS – A CASE STUDY USING PHARMGKB

EKATERINA BUYKO, ELENA BEISSWANGER, and UDO HAHN

Jena University Language & Information Engineering (JULIE) Lab
Friedrich-Schiller-Universität Jena, Germany
E-mail: {surname.name}@uni-jena.de
http://www.julielab.de

In this paper, we report on adapting the JREX relation extraction engine, originally developed for the elicitation of protein-protein interaction relations, to the domains of pharmacogenetics and pharmacogenomics. We propose an intrinsic and an extrinsic evaluation scenario which is based on knowledge contained in the PHARMGKB knowledge base. Porting JREX yields favorable results in the range of 80% F-score for *Gene-Disease*, *Gene-Drug*, and *Drug-Disease* relations.

Keywords: Information Extraction, Pharmacogenetics, Pharmacogenomics, PharmGKB

1. Introduction

While molecular biology and clinical medicine have pursued their research agenda in a fairly isolated manner for a long time in the past, this attitude has changed dramatically in the last years and new research themes have been defined which combine the efforts of both scientific camps. This change is mirrored in terms such as 'translational medicine', an area that is explicitly devoted to bringing results from genetic research labs to the bedside in a clinical environment, or 'personalized medicine' which aims at exploiting genetic data from an individual (or a genetically homogeneous cohort) for use in fine-tuned drug dosage recommendations or highly individualized drug development and risk assessments.

Unfortunately, this thematic shift has not found parallels in the field of BioNLP, up until now. Still, the overwhelming amount of work is concerned with proteins and genes (for named entity tagging) and protein-protein interactions (for relation extraction). Although this research has by no means become obsolete or irrelevant (quite the contrary is true given the current performance ceilings of state-of-the-art recognition devices), we here argue for complementing these activities by research efforts that are intended to support the collaborative work of molecular biologists and clinical physicians in terms of suitable information extraction technology. Our special targets are the domains of *pharmacogenetics* (studying the influence of single genes on drug response) and *pharmacogenomics* (studying the influence of multiple genes on drug response, typically using high-throughput techniques).

In order to contribute to this goal, we here consider possible association relations between genes, drugs, and diseases as reported in the scientific literature. Accordingly, we no longer focus on *Protein/Gene-Protein/Gene* relations but extend the scope of our investigations to cover *Gene-Drug*, *Gene-Disease*, and *Drug-Disease* relations. Such a fundamental change in the scope of entities and relations under scrutiny poses new problems. Both, for training (machine learning-based taggers) as well as testing suitable gold standards have to be provided. Since it is not an easy task to set up such metadata, we here build on previous work, the Pharmacogenomics and Pharmacogenetics Research Network and Knowledge Base[a] (PHARMGKB) (see Section 2 for a detailed description). Given this repository, we automatically generate gold standard data out of the knowledge provided by PHARMGKB (see Section 4.1), and use it for two evaluation scenarios, an *intrinsic* and an *extrinsic* one (see Section 5.1). Based on experiments using our relation extraction engine JREX (see Section 4.3), we have evidence that the thematic extension of named entities as well as relations yields favorable results for the three types of relations being gleaned on, in the range of 80% F-score for *intrinsic* settings and up to 77.5% in an *extrinsic* configuration (see Section 5.2).

2. PharmGKB Database

The development of the PHARMGKB repository represents a major step towards an interdisciplinary biomedical information store. PHARMGKB incorporates data on genetic variations and associated phenotypic manifestations. The latter mainly concern the pharmacokinetics of therapeutic drugs (how drugs are absorbed, metabolized and excreted by an organism) and the pharmacodynamics of drugs (how drugs act in an organism). The repository also covers certain non-pharmacological aspects of phenotypes, including susceptibility to disease. Currently (as of May 2011) PHARMGKB contains information about 27,000 pharmacogenes, 2,500 drugs, 3,200 diseases and 23,870 relations between them (amongst others) and it continues to grow as PHARMGKB curators screen the scientific literature for new facts to be inserted. Yet, keeping up with the large amount of articles published every day is virtually impossible. Thus, the development of automatic support tools for the curation process might be a rewarding endeavor. The key task will be to develop a system that is able to recognize genes, drugs and diseases in text and to detect PharmGKB-relevant relations between them.

Other than common relation extraction tasks in the biomedical domain, such as the detection of protein-protein interactions or gene regulatory relations, PharmGKB-relevant relations draw on much broader types of semantic relationships. This may be due to the fact that the focus of interest in pharmacogenetics and pharmacogenomics lies on links between genetic and phenotypic variations. Yet, evidence for these links can originate from different sources, ranging from genetic research papers to clinical studies. Another reason might be that while genes are referred to by a specific name or symbol in text, references to phenotypes are verbalized much more loosely. At the same time, phenotype variation descriptions often mention the drugs and diseases involved. Now, if *Gene-Drug* or *Gene-Disease* relations are extracted from text, this opens up a wide range of interpretations of possible association types involved, from a real physical interaction to correlation with clinical outcomes. In contrast, relations between drugs and diseases often (but not always) adhere to some kind of *treated-by* relation. Given this underspecification of PHARMGKB semantic relationships, we consider PHARMGKB relations as coarse association relations, or merely relations, in the rest of this paper.

The following examples illustrate the association relations we here focus on:

- ***Gene-Drug*** — "*It was found that the genotype of* ***CYP2C19*** *had a significant effect on the N-demethylation of* ***citalopram****.*"

[a]`http://www.pharmgkb.org/`

- ***Gene-Disease*** — *"The urocortin (UCN) gene resides at this interval, and its protein decreases appetite behavior, suggesting that* ***UCN*** *may be a candidate gene for susceptibility to* ***obesity****."*
- ***Drug-Disease*** — *"Decreased expression of BRCA2 mRNA predicts favorable response to* ***docetaxel*** *in* ***breast cancer****."*

The first example contains a mention of the association between the *CYP2C19* gene and the demethylation process of the drug *citalopram*. In the second example, an assumption about the impact of the *UCN* gene on *obesity* is made. The third example contains facts about the role of the gene *BRAC2* in response to the drug *docetaxel* related to *breast cancer*, thus we encounter a relationship between a drug and a disease.

PHARMGKB provides a hierarchically organized category system for different levels of evidence that also classify the significance of the data. The lowest category is "Genotype" (comprising simple genetic variations), the highest "Clinical Outcome" (covering clinically relevant data). In between are the categories "Pharmacodynamics and Drug Response", "Pharmacokinetics", and "Molecular and Cellular Functional Assays". In our relation extraction approach, we do not distinguish between those categories for relations but rather focus on the extraction of generic *Gene-Drug*, *Gene-Disease* and *Drug-Disease* relations for the evaluation of the results on the available PHARMGKB relationship data set.

3. Related Work

In BioNLP, research on (binary) protein-protein interactions (PPIs) was and still is predominant. This is reflected in lots of PPI-annotated corpora (e.g., LLL,[1] AIMED,[2] or BIOINFER[3]), but also the large variety of methodologies to tackle this problem (pattern-based (e.g.,[4]), rule-based (e.g.,[5]), and machine learning-based approaches (e.g.,[6])). Binary PPIs constitute a fairly general abstraction from the complex interactions between genes and proteins, so requests for finer-grained representations were issued. The GENIA event corpus[7] and the BioNLP 2009 Shared Task data[8] contain such detailed annotations of PPIs (amongst others). The BioNLP Shared Task was a first step towards the extraction of specific pathways with precise information about the molecular events involved. The winner system, TURKU,[9] achieved with 51.95% F-score the milestone result in that competition, but was outperformed in the BioNLP 2011 Shared Task with 56% F-score.[10]

Considering relation extraction (RE) in the pharmacogenetics and pharmacogenomics domain, to the best of our knowledge, there are only few studies which deal primarily with phenotype-genotype-drug relations. This may be due to the fact that no large-scale annotated corpora are available, up until now, for phenotype-genotype-drug relationships. Thus, the developed systems mostly focused on named entity recognition (gene, drug, and disease names) and on small-scale evaluations for RE. Rindflesch *et al.* developed the EDGAR system for the extraction of gene and drug names and gene-drug relations relevant for cancer.[11] EDGAR exploits underspecified syntactic parse trees and applies syntactico-semantic rules for the extraction of relationships. An explicit evaluation for the EDGAR system is missing though. Chang and Altman's system recognizes relations between genes and drugs in PubMed abstracts with a co-occurrence-based approach.[12] They further classify the relations into five categories, as specified by PHARMGKB, using a Maximum Entropy based machine learning approach. The relation recognition step is evaluated against a small data set of 215 gene-drug relations manually extracted from a review article, while the classification step is assessed against human-curated articles from PHARMGKB. Evaluation results for all five categories range from 88% recall with 75% precision for the predictions of *Pharmacokinetics*, to 9% recall with 27% precision for the *Clinical Outcome* category. The authors concede that the selected PHARMGKB data set was small, including 325 gene-drug pairs, and that the evaluation results heavily depend upon the size of training data. Chun *et al.* describe a system for disease-gene relation extraction that is based on the co-occurrence of gene and disease name mentions (found via dictionary look-up) and additional filtering of false positives with a machine learning classifier.[13] In the filtering mode, the system achieves 78.5% precision and 87.1% recall on a manually annotated corpus with 1,000 co-occurrences of gene and disease names.

The most recent systems for the extraction of gene-drug relationships are PHARMSPRESSO[14] and GENDRUX.[15] PHARMSPRESSO builds on the TEXTPRESSO tool,[16] a full-text search engine for biological entities and facts such as PPIs. PHARMSPRESSO has been extensively evaluated concerning the detection of gene and drug names, but with respect to relationships it yields only 50% recall on gene-drug *'association'* relations on 45 full-text articles which contain 178 gene name mentions and 142 drug names mentions. GENDRUX is a web-based tool developed for the analysis of documents in the breast cancer domain. Its document collection consists of 4,000 PUBMED abstracts collected using gene and drug name filters. GENDRUX's focus is on the retrieval of documents with gene and drug names related to breast cancer, while RE is based only on the co-occurrence of relevant terms in the titles of documents; an evaluation, however, is lacking. A first large-scale evaluation study was carried out by Coulet *et al.* who extracted PHARMGKB-relevant relationships using a lexicon of key phramocogenomic entities and syntactic parses of 17 million MEDLINE abstracts.[17] The extracted relationships are reported to have a precision up to 87.7%. Still this work does not evaluate the recall of the presented relationship extraction.

With the exception of Coulet *et al.*, we here provide the first large-scale evaluation study of phenotype-genotype-drug relationship extraction for *Gene-Drug*, *Gene-Disease* and *Drug-Disease* relations, using a high-performance relation extractor. We automatically gather a large-scale gold standard based on human-curated texts from the PHARMGKB database. Furthermore, we provide two evaluation scenarios, an *intrinsic* one based on corpus cross-validation and an *extrinsic* one based on PHARMGKB relationship data.

4. Methods

4.1. PharmGKB *as Gold Standard*

Usually, manually annotated gold standard corpora are used to evaluate RE systems. Manual annotation, however, is a time-consuming and costly process. In contrast, we here capitalize on previous curation efforts and derive large-scale gold data for genes, drugs, diseases, and binary relationships among them automatically from the PharmGKB knowledge base.[b] For all relationships, PharmGKB specifies the two participants' names and IDs and zero to many references to PubMed abstracts that provide textual evidence for each relation. In total, references to 5,241 different PubMed abstracts are given. We extracted all available abstracts from the Medline Baseline Repository 2011 to form our initial corpus.[c] Next we limited our focus to those PharmGKB relations that hold between entities of different semantic types (*Gene-Drug*, *Gene-Disease*, *Drug-Disease*). For each entity involved in a relation we compiled a list of its names and all alternative names, plus for genes its symbol and all alternative symbols, as specified in PharmGKB. Finally, for each relation, we matched the names of its participants (case-insensitively) against the abstracts from the referenced PubMed abstracts. Matches covering partial tokens were skipped in this process to avoid false positives. If in a sentence at least one name of the first participant and at least one name of the second participant matched, we marked these sentences as containing a gold PharmGKB relation, and incorporated the abstract as a gold corpus document. The numbers of the unique relations we detected in this way are specified in Table 1.

Table 1. First line: The number of distinct binary *Gene-Drug*, *Gene-Disease*, *Drug-Disease* relations as specified in PharmGKB. Second line: The subset of relations for which at least one PubMed reference is given in PharmGKB. Third line: The fraction of relations with PubMed reference that could be retrieved by our extraction machinery from at least one of the specified PubMed abstracts.

# PharmGKB Relations	*Gene-Drug*	*Gene-Disease*	*Drug-Disease*	Total
all	11,476	8,028	2,639	22,143
with PubMed reference	6,628	7,163	2,634	16,425
retrieved from PubMed abstract(s)	1,686	1,711	673	4,070

The gold corpus we collected using PharmGKB references contains 1,980 PubMed abstracts, where 522 abstracts incorporate *Drug-Disease* relations, 1,414 abstracts hold *Gene-Disease* relations, and 1,262 abstracts contain *Gene-Drug* (see Table 2). In a later step, the collected corpus of 1,980 abstracts, called here gold PharmGKB corpus, was used for the evaluation of our relation extraction approach (see Section 5).

Table 2. Overvuew of gold relations in the gold PhramGKB corpus.

Relation	Abstracts With Gold Relation Annotations	Gold Relation Annotations
Gene-Drug	1,262	9,914
Gene-Disease	1,414	6,626
Drug-Disease	522	3,439
Total	1,980	19,979

[b]The data files genes.zip, drugs.zip and relationships.zip were downloaded on May 17 and diseases.zip on May 18, 2011, from `http://www.pharmgkb.org/resources/downloads_and_web_services.jsp`.
[c]All but six of the referenced PubMed abstracts could be retrieved from the Baseline Repository 2011.

4.2. *Recognition of Relevant Named Entities (Relation Participants)*

For each entity type (gene, drug, disease) we compiled dictionaries from PHARMGKB exploiting preferred names and alternate names of entities, as well as preferred symbols and alternate symbols for genes. The drug dictionary was further extended by terms taken from the *Orange Book Dictionary*[d] extended with MESH[e] term variants, while the disease dictionary was also further extended by headings plus alternate entry terms from the MESH disease branch (starting with the top node "Diseases [C]"). For gene mention recognition, we used GENO[18] and for the remaining NER tasks we applied the LINGPIPE CHUNKER.[f]

4.3. *Extraction of PharmGKB-relevant Relations — JReX*

The relation extraction experiments were run with the relation and event extraction system JREX (Jena Relation eXtractor). Generally speaking, the JREX system classifies pairs of genes in sentences as *interaction pairs* using various forms of syntactic and semantic evidence (see Buyko *et al.*[19] for a deeper account). JREX (under the name of JULIELab) scored on 2nd rank among 24 competing teams in the *BioNLP 2009 Shared Task on Event Extraction*, with 45.8% precision, 47.5% recall and 46.7% F-score. After the competition, this system was further streamlined and now peaks at 57.6% precision, 45.7% recall and 51.0% F-score (,[2021]), and thus considerably narrowed the gap to the winner of the BioNLP'09 Shared Task who scored at 51.95% F-score.[g]

As far as pre-processing is concerned, JREX uses JCORE tools[23] such as JULIELab's sentence splitter and tokenizer. For shallow syntactic analysis it applies the OPENNLP POS Tagger and Chunker, both re-trained on the GENIA corpus. For dependency parsing, the MST parser[24] was retrained on the GENIA Treebank and the parses subsequently converted to the CoNLL'07 representation.[h]

The JREX relation extractor accounts for two major subtasks – first, the structural trimming of dependency graphs, and, second, the identification and ordering of arguments for the relation under scrutiny. Trimming dependency graphs amounts to eliminating informationally irrelevant and to enriching informationally relevant lexical nodes by concept overlays. For example, JREX prunes auxiliary and modal verbs which govern the main verb in syntactic structures such as passives, past or future tense. Accordingly, (see Figure 1), the verb "*activate*" is promoted to the ROOT in the dependency graph and governs all nodes that were originally governed by the modal "*may*". An example of semantic enrichment is also given in Figure 1, where the lexical item "*TNF-alpha*" is turned into the conceptual label *Gene*. This abstraction avoids over-fitting of dependency structures for the machine learning mechanisms on which JREX is based. For a more detailed explanation and evaluation of this approach to syntactic simplification and semantic decoration, see Buyko *et al.*[19] and.[21]

[d]`http://www.accessdata.fda.gov/scripts/cder/ob/default.cfm`
[e]`http://www.nlm.nih.gov/mesh/`
[f]`http://alias-i.com/lingpipe/`
[g]The TURKU system was also improved after the competition and now performs at 52.9 F-score.[22]
[h]We used the GENIA Treebank version 1.0, available from `http://www-tsujii.is.s.u-tokyo.ac.jp`. The conversion script is accessible via `http://nlp.cs.lth.se/pennconverter/`.

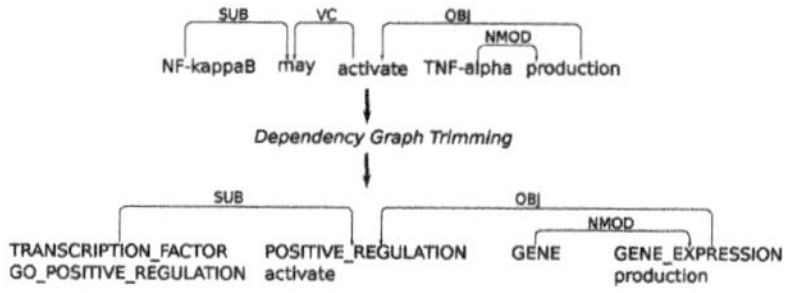

Fig. 1. Trimming of dependency graphs.

The principal relation extraction step is argument detection and ordering. For argument extraction, JREX builds sentence-wise pairs of putative arguments (named entities) and applies two machine learning approaches, one feature-based and the other one kernel-based. In the feature-based classifier, three groups of features are distinguished: (1) *lexical* features (covering lexical items before, after and between the mentions of relation arguments; (2) *chunking* features (concerned with head words of the phrases between two mentions; (3) *dependency parse* features (considering both the selected dependency levels of the arguments, parents and least common subsumer, as well as a shortest dependency path structure between the arguments for *walk* features). For this feature-based approach, the Maximum Entropy (MaxEnt) classifier from MALLET is used.[i]

The graph kernel classifier uses a converted form of dependency graphs in which each dependency node is represented by a set of labels associated with that node. The dependency edges are also represented as nodes in the new graph such that they are connected to the nodes adjacent in the dependency graph. Subgraphs which represent, e.g., the linear order of the words in the sentence can be added, if required. The entire graph is represented in terms of an adjacency matrix which is further processed to contain the summed weights of paths connecting two nodes of the graph (see Airola *et al.*[6] for details). For the graph kernel approach, the LibSVM Support Vector Machine is used as classifier.[j]

5. Experiments and Results

5.1. *Experimental Settings*

We established two evaluation settings, an *intrinsic* and an *extrinsic* one. In the intrinsic, corpus-based scenario we used the gold PHARMGKB corpus introduced in Section 4.1 for a cross-validation of JREX. In the extrinsic scenario we used the PHARMGKB relationship data for evaluation of our relation extraction pipeline.

In the corpus-based evaluation scenario we performed a 10-fold cross-validation using JREX. In order to analyze the corpus settings, we provide here further information on the number of positive and negative instances (see Table 3).[k] The ratio of negative to positive instances is different for genotype/phenotype-drug and for the genotype-phenotype relations.

[i]http://mallet.cs.umass.edu/index.php/Main_Page
[j]http://www.csie.ntu.edu.tw/~cjlin/libsvm
[k]Instances were generated sentence-wise. Positive instances contain pairs of entity mentions (in a sentence), which were annotated to stand in a relation in PHARMGKB, negative instances contain pairs of entity mentions (in a sentence), which were not annotated to stand in a relation in PHARMGKB.

Table 3. Positive and negative training instances in the gold PHRAMGKB corpus.

Relation	Positive Instances	Negative Instances	Total Instances
Gene-Drug	9,914	7,060	16,974
Gene-Disease	6,626	6,664	13,290
Drug-Disease	3,439	2,468	5,907

For example the ratio for *Gene-Drug* and *Drug-Disease* relations is 1.4, while for *Gene-Disease* relations it is about 1.0. These are fairly low numbers in comparison with the ratio of negative to positive instances in the known relationally annotated corpora which provide protein-protein or gene regulatory relations (the ratio here ranges between 5 and 10). This can be explained by the fact that when entities such as drugs, diseases and genes are mentioned in the same sentence they are likely to form a relation as well. Furthermore, our corpus, although annotated with rich dictionaries and GENO tools, contains a high number of gold PharmGKB entities and thus PharmGKB relations. The distribution of positive and negative instances of a corpus clearly influences the overall evaluation figures in terms of F-score. This is the reason why we decided on an additional, extrinsic evaluation of JREX.

In the extrinsic approach, we extracted from the PHARMGKB relationship subset data as presented in Table 1 (4,070 unique relations) about one third of unique relations for each relation type. For this set of relations, we collected the referenced abstracts from our gold PHARMGKB corpus. Thus, for every relation type we generated two splits of the PHARMGKB corpus, the training and the test part as depicted in Table 4. In contrast to the intrinsic evaluation, we focused here on the extraction of unique relations from the test corpus. Furthermore, the training corpus was not allowed to contain the PHARMGKB relations mentioned in the test corpus. The training corpus subset was used for training JREX on the corresponding relation type. The re-trained version of JReX was then used to predict relations on the test part. We generated a list of detected unique relations and compared it against the test relation subset. The figures for the test and training parts in terms of abstracts and unique relations are presented in Table 4.

Table 4. Test and train parts of the gold PHARMGKB corpus (unique relations only).

Relation	Test		Train		Overall	
	Abstracts	Relations	Abstracts	Relations	Abstracts	Relations
Gene-Drug	368	621	1612	1,068	1,980	1,686
Gene-Disease	513	673	1467	1038	1,980	1,711
Drug-Disease	169	198	1811	475	1,980	673

5.2. *Evaluation on the complete gold* PharmGKB *corpus*

The gold PHARMGKB corpus of 1,980 abstracts was used for a 10-fold cross-validation of our relation extraction machinery. As a lower bound to estimate performance, we selected the common co-occurrence approach, i.e., if two entities of interest co-occur in a sentence, they are marked to stand in a relation. In our case, it was particularly interesting to see the baseline results as the ratio between negative and positive instances is lower as, for example, for protein-protein interaction relations and, hence, has a potential to boost co-occurrence figures.

The results are presented in Table 5. The co-occurrence approach allows to extract *Gene-Drug* and *Drug-Disease* relations with about 73.0% F-score, while *Gene-Disease* relations perform at a lower 65.9% F-score.

Table 5. 10-fold cross-validation on PHARMGKB CORPUS

Relation	Co-occurrence			JReX (FB)			JReX (GK)		
	R	P	F	R	P	F	R	P	F
Gene-Drug	100	57.5	73.0	85.1	79.7	**82.3**	77.6	67.4	72.1
Gene-Disease	100	49.2	65.9	76.9	75.2	**76.0**	71.0	65.7	68.2
Drug-Disease	100	58.2	73.5	81.2	77.0	**79.0**	75.3	67.2	71.0
All relations	100	54.5	70.6	82.3	78.0	**80.1**	71.0	65.7	68.2

In the next step, we evaluated JREX in its feature-based (JREX FB) and its graph kernel-based classifier variant (JREX GK). The results are also contained in Table 5. In the feature-based approach, we achieve for *Gene-Drug* relation detection a performance of 82.3% F-score, with 85.1% recall and 79.7% precision. For the *Drug-Disease* relations 79.0% F-score with 81.2% recall and 77.0% precision were determined. The lowest results are achieved for the *Gene-Disease* relation with 76.0% F-score. When JReX learns all relations in one step and performs 3-types classification at once, the overall performance for all three relations settles at 80.1% F-score. This means that learning all three relations in one step is advantageous for achieving a good overall F-score result. The performance of the graph-kernel based classifier (JREX GK) is lower than for the feature-based classifier with up to 10 percentage points difference in F-score results. The feature-based classifier is shown to perform better for all phenotype/genotype-drug relation types than the JREX GK classifier. While the main source of information for the JREX GK classifier are dependency parse trees, the JREX FB classifier considers a range of lexical, morpho-syntactic and syntactic information. It seems that the extraction of phenotype/genotype-drug association relations cannot be captured by considering dependency parse trees only and profits from richer sources of evidence.

Our intrinsic evaluation showed that the JREX based extraction of phenotype/genotype-drug relations outperform significantly the co-occurrence based approach by up to 9.3 percentage points in terms of F-score (see *Gene-Drug* relation evaluation in Table 5). This indicates that this type of relations can effectively be learned by the JREX tool.

5.3. *Evaluation against the* PharmGKB *relation subset*

In Table 4 we presented the figures of test and training parts for each relation type of interest. As the feature-based classifier was shown in the intrinsic evaluation to outperform the graph kernel classifier-based one, for the extrinsic evaluation we used only the feature-based JREX variant. For each relation type, we retrained JREX FB on the training part and performed relation extraction on the test part. The extracted list of unique entities was then compared with the gold PHARMGKB relation list as described previously. In this evaluation, we chose the co-occurrence approach as the baseline, once again. The results are presented in Table 6.

The co-occurrence approach performs with a very low precision of 39.4% for the *Gene-*

Table 6. Evaluation against PHARMGKB relation test sets

Relation	Test Size (Unique)	Co-occurrence			JReX		
		R	P	F	R	P	F
Gene-Drug	621	100	39.4	56.5	90.9	61.9	**73.6**
Gene-Disease	673	100	35.0	52.1	83.3	58.6	**68.8**
Drug-Disease	198	100	45.3	62.3	85.3	71.0	**77.5**
Total	1,492	100	39.9	59.9	86.5	63.8	**73.3**

Drug, 35.0% for *Gene-Disease* and 45.3% for *Drug-Disease* relations. The overall F-score results range between 52.1% for the *Gene-Disease* relation and 62.3% for the *Drug-Disease* relation. The co-occurrence approach is outperformed by JREX in the extrinsic evaluation as in the intrinsic one. JREX peaks for the *Drug-Disease* relation at 77.5% F-score followed by 73.6% F-score for the *Gene-Drug* relation and 68.8% F-score for the *Gene-Disease* relation. In all three extrinsic evaluation results for JREX we see that recall is much higher than precision. This can be explained by the distribution of the positive and negative instances in the training data. The lower the ratio between the positive and the negative mentions is, the higher is the tendency of the classifier to classify an instance as a positive. Still, the precision and thus the overall F-score results show that phenotype/genotype-drug relations can successfully be extracted with JREX, with performance peaking up to 77.5% F-score in a real life extrinsic evaluation on the PHARMGKB relationship data.

5.4. *Caveats – Putting Things in Perspective*

The main concern with the approach above is clearly centred around the reliability and representativeness of the semi-automatically generated gold standard. The collection of PHARMGKB referenced abstracts and the automatic mapping of relationship data in the text may admittedly hide risks of running into bad gold standard data, which is not approved by human annotators. Still, the PHARMGKB references to the abstract texts are a human-curated highly reliable data source for relations that hold between entities of interest.

A manual analysis of selected texts revealed that the annotations seem to reflect the knowledge represented by the PHARMGKB. One student of biology analyzed 200 sentences randomly extracted from the PHARMGKB gold corpus, which should contain gold PHARMGKB relations (100 sentences for *Drug-Disease*, 50 sentences for *Gene-Drug* and 50 sentences for *Gene-Disease* relations). The analysis revealed that 80% of the sentences for *Drug-Disease* relations, and 90% of the sentences for the *Gene-Disease* as well as *Gene-Drug* relations, in effect, contain descriptions of these relations. The figures of the distribution of positive and negative training instances are also an indicator for the reasonable quality of the semi-automatically generated gold standard. Furthermore, numerous previous studies on learning relations from automatically generated corpora in a *distant supervision* mode[l] reinforce the outcomes of our approach (see Mintz *et al.*[25]).

[l]A distant supervision approach produces a large automatically annotated corpus using relation mentions from available databases, it considers all sentences containing those entities to stand in a relation in order to train a relation classifier.[25]

To avoid the intricacies of an intrinsic corpus-based evaluation, we exposed our approach to an extrinsic evaluation similar in spirit to the one we had already carried out on the REGULONDB database.[26] The results, up to 77.5% F-score, are promising and reflect the fact that the PHARMGKB gold standard may indeed be representative for general genotype-phenotype-drug relationships. But as the PHARMGKB curation process seems to involve some automatic preprocessing of the text data for named entity recognition and co-occurrence analysis, the database may lack articles that will not pass the pre-selection step. Furthermore, the PHARMGKB database does not contain references to curated articles which do not contain PHARMGKB-relevant relations (those are dropped after the curation process). These restrictions may explain the low ratio of negative to positive instances in the generated corpus.

6. Conclusions

While the BioNLP community's focus is still almost exclusively gene/protein-centered, both in terms of named entity and relation extraction (PPIs, in particular), this only partially matches the most recent needs at the intersection of molecular biology and clinical/pharmacological research. Our work reported in this paper aims at mitigating that mismatch.

Accordingly, we ported JULIE LAB's high-performance JREX relation extractor from the protein/gene interaction domain proper to the domains of pharmacogenetics and pharmacogenomics. Here, three novel types of relations, namely *Gene-Drug*, *Gene-Disease*, and *Drug-Disease* relations had to be targeted. For these relations we achieved over-all F-scores on the order of 80% in an intrinsic and 73% in an extrinsic evaluation. In both cases co-occurrence-based baselines were clearly outperformed. Our approach crucially relies upon training and test data that we automatically compiled from PHARMGKB.

Despite these encouraging results, our experimental design needs further refinement. First, our triple relation repertoire cannot be matched straightforwardly with the PHARMGKB-specific relation hierarchy (ranging from 'Genotype' to 'Clinical Outcome'). However, such a mapping is needed for directly supporting the curators of PHARMGKB. Second, our internal evaluation is not based on a real gold standard but on an automatically generated substitute only. This requirement opens up the box of Pandora in that not only rarely dealt with entities (diseases, drugs, etc.) but even worse rather underspecified, if not fuzzy relations (such as 'Clinical Outcome') have to be served by massive annotation efforts.

Acknowledgments. This work is funded by a grant from the German Ministry of Education and Research (BMBF) for the *Jena Centre of Systems Biology of Ageing* (JENAGE) (grant no. 0315581D).

References

1. C. Nédellec, Learning Language in Logic: Genic interaction extraction challenge, in *Proceedings of the 4th Learning Language in Logic Workshop*, 2005.
2. R. Bunescu, R. Ge, R. J. Kate, E. M. Marcotte, R. J. Mooney, A. K. Ramani and Y. W. Wong, *Artificial Intelligence in Medicine* **33**, 139 (2005).
3. S. Pyysalo, F. Ginter, J. Heimonen, J. Björne, J. Boberg, J. Jarvinen and T. Salakoski, *BMC Bioinformatics* **8** (2007).
4. C. Blaschke, M. A. Andrade, C. A. Ouzounis and A. Valencia, Automatic extraction of biological information from scientific text: Protein-protein interactions, in *Proceedings of the 7th International Conference on Intelligent Systems for Molecular Biology*, 1999.

5. K. Fundel, R. Küffner and R. Zimmer, *Bioinformatics* **23**, 365 (2007).
6. A. Airola, S. Pyysalo, J. Björne, T. Pahikkala, F. Ginter and T. Salakoski, A graph kernel for protein-protein interaction extraction, in *Proceedings of the BioNLP 2008 Workshop on Current Trends in Biomedical Natural Language Processing*, 2008.
7. J.-D. Kim, T. Ohta and J. Tsujii, *BMC Bioinformatics* **9** (2008).
8. J.-D. Kim, T. Ohta, S. Pyysalo, Y. Kano and J. Tsujii, Overview of BioNLP'09 Shared Task on Event Extraction, in *Proceedings BioNLP 2009. Companion Volume: Shared Task on Event Extraction*, 2009.
9. J. Björne, J. Heimonen, F. Ginter, A. Airola, T. Pahikkala and T. Salakoski, Extracting complex biological events with rich graph-based feature sets, in *Proceedings BioNLP 2009. Companion Volume: Shared Task on Event Extraction*, 2009.
10. S. Riedel, D. McClosky, M. Surdeanu, A. McCallum and C. D. Manning, Model combination for event extraction in BioNLP 2011, in *Proceedings BioNLP 2011 Shared Task Workshop*, 2011.
11. T. C. Rindflesch, L. Tanabe, J. N. Weinstein and L. Hunter, EDGAR: Extraction of drugs, genes and relations from the biomedical literature, in *Proceedings of the 2000 Pacific Symposium on Biocomputing*, 2000.
12. J. T. Chang and R. B. Altman, *Pharmacogenetics* **14**, 577 (September 2004).
13. H.-W. Chun, Y. Tsuruoka, J.-D. Kim, R. Shiba, N. Nagata, T. Hishiki and J. ichi Tsujii, Extraction of gene-disease relations from medline using domain dictionaries and machine learning, in *Proceedings of the 2006 Pacific Symposium on Biocomputing*, 2006.
14. Y. Garten and R. B. Altman, *BMC Bioinformatics* **10** (2009).
15. C. Crasto, D. Luo, F. Yu and D. Chen, *BMC Medical Informatics and Decision Making* **11** (2011).
16. H.-M. Mueller, E. E. Kenny and P. W. Sternberg, *PLoS Biology* **2**, 1984 (2004).
17. A. Coulet, N. H. Shah, Y. Garten, M. A. Musen and R. B. Altman, *Journal of Biomedical Informatics* **43**, 1009 (2010).
18. J. Wermter, K. Tomanek and U. Hahn, *Bioinformatics* **25**, 815 (2009).
19. E. Buyko, E. Faessler, J. Wermter and U. Hahn, Event extraction from trimmed dependency graphs, in *BioNLP 2009. Companion Volume: Shared Task on Event Extraction*, 2009.
20. E. Buyko and U. Hahn, Evaluating the impact of alternative dependency graph encodings on solving event extraction tasks, in *Proceedings of the 2010 Conference on Empirical Methods in Natural Language Processing*, 2010.
21. E. Buyko, E. Faessler, J. Wermter and U. Hahn, *Computational Intelligence* **27** (2011).
22. J. Björne, J. Heimonen, F. Ginter, A. Airola, T. Pahikkala and S. Tapio, *Computational Intelligence* **27** (2011).
23. U. Hahn, E. Buyko, R. Landefeld, M. Mühlhausen, M. Poprat, K. Tomanek and J. Wermter, An overview of JCoRe, the JULIE lab UIMA component repository, in *Proceedings of the LREC'08 Workshop 'Towards Enhanced Interoperability for Large HLT Systems: UIMA for NLP'*, 2008.
24. R. T. McDonald, F. Pereira, K. Ribarov and J. Hajic, Non-projective dependency parsing using spanning tree algorithms, in *Proceedings of the 2005 Human Language Technology Conference and the Conference on Empirical Methods in Natural Language Processing*, 2005.
25. M. Mintz, S. Bills, R. Snow and D. Jurafsky, Distant supervision for relation extraction without labeled data, in *Proceedings of the 2009 Joint Conference of the 47th Annual Meeting of the ACL and the 4th International Joint Conference on Natural Language Processing of the AFNLP*, 2009.
26. U. Hahn, K. Tomanek, E. Buyko, J. J. Kim and D. Rebholz-Schuhmann, How feasible and robust is the automatic extraction of gene regulation events? A cross-method evaluation under lab and real-life conditions, in *Proceedings of the BioNLP 2009 Workshop*, 2009.

LINKING PHARMGKB TO PHENOTYPE STUDIES AND ANIMAL MODELS OF DISEASE FOR DRUG REPURPOSING

ROBERT HOEHNDORF[1,*], ANIKA OELLRICH[2], DIETRICH REBHOLZ-SCHUHMANN[2]
PAUL N. SCHOFIELD[3], GEORGIOS V. GKOUTOS[1]

[1]*Department of Genetics, University of Cambridge*
Downing Street, Cambridge, CB2 3EH, UK

[2]*European Bioinformatics Institute*
Wellcome Trust Genome Campus, Hinxton, Cambridge, CB10 1SD, UK

[3]*Department of Physiology, Development and Neuroscience*
University of Cambridge
Downing Street, Cambridge CB2 3EG, UK, and
The Jackson Laboratory, 600, Main Street
Bar Harbor ME 04609-1500, USA

The investigation of phenotypes in model organisms has the potential to reveal the molecular mechanisms underlying disease. The large-scale comparative analysis of phenotypes across species can reveal novel associations between genotypes and diseases. We use the PhenomeNET network of phenotypic similarity to suggest genotype–disease association, combine them with drug–gene associations available from the PharmGKB database, and infer novel associations between drugs and diseases. We evaluate and quantify our results based on our method's capability to reproduce known drug–disease associations. We find and discuss evidence that levonorgestrel, tretinoin and estradiol are associated with cystic fibrosis ($p < 2.65 \cdot 10^{-6}$, $p < 0.002$ and $p < 0.031$, Wilcoxon signed-rank test, Bonferroni correction) and that ibuprofen may be active in chronic lymphocytic leukemia ($p < 2.63 \cdot 10^{-23}$, Wilcoxon signed-rank test, Bonferroni correction). To enable access to our results, we implement a web server and make our raw data freely available. Our results are the first steps in implementing an integrated system for the analysis and prediction of drug–disease associations for rare and orphan diseases for which the molecular basis is not known.

Keywords: phenotype, drug repurposing, animal models, ontology, pharmacogenetics

1. Background

1.1. *Drug discovery and drug repurposing*

The major challenges currently faced by pharmacological research include the high rates of attrition in the development of new compounds (mainly in Phase II), the increased cost of development itself, and increased regulatory concern about safety and efficacy.[33] Recently, it has been shown that the rate of production of novel drugs has changed little since the 1950s,[28] yet the cost of developing those drugs has soared. As a result, pharmacological research increasingly focuses on *repurposing* or *repositioning* existing drugs for new indications.

Drug repurposing aims to find new indications for existing drugs, and strategies for drug repurposing can be divided into two main types: identification of new targets for known drugs and identification of new indications for a known mechanism of action.[33] Approaches to drug-repurposing include database-driven bioinformatics approaches, *in vivo* and *ex vivo* studies as well as high-throughput screening methods.[33]

Finding new targets for existing drugs depends on our fundamental understanding of the physiology and pathobiology that underlies diseases, their phenotypic manifestations and the molecular sites of action of small-molecule therapeutics. The manifest modularity of disease phenotypes reflects the existence of underlying physiological pathways, and a lesion in any of the components of the same pathway can produce closely overlapping disease phenotypes.[29]

The sequencing of the human genome and that of many animal models, the rapid development of high-throughput phenotyping and genotyping technologies and our ability to create specific mutations in the genomes of model organisms have provided us with a vast amount of information that supports the discovery of meaningful associations between the genotype and phenotype of an organism. This information, in turn, extends our ability to comprehensively characterize the phenotypic manifestations of diseases and generate hypotheses on which the intelligent design of drugs can be built.

With the availability of increasing amounts of information in public and private database, the development of *in silico* approaches that can analyse the data, propose potential drug indications and ultimately reduce the cost and time of drug development is required. Large scale analysis frameworks such as the *PREDICT* framework[14] are examples of such approaches. These frameworks are successfully being applied for drug repositioning and the suggestions of potentially novel drugs.[14] They adopt a variety of methods such as drug response gene expression profiles, drug–drug similarity, drug–disease similarity with respect to known drug–disease associations, text mining of known associations and several further resources. One of the areas though that still remains to be fully exploited is the wealth of phenotypic information that is increasingly becoming available from a variety of studies on animal models of human diseases.

Here, we present an approach for predicting novel associations between drugs and diseases based on the PhenomeNET method for comparing phenotypes across species.[20] We apply our method to the Pharmacogenetics and Pharmacogenomics Knowledge Base (PharmGKB),[16] a central repository containing a wealth of relationships between genetic, genomic, drug-response related phenotype data and clinical information. Our method utilizes predictions of disease–gene associations from the PhenomeNET.[20] These predictions are based on a similarity-based comparative analysis of the phenotypic data associated with humans in the Online Mendelian Inheritance in Men (OMIM) database[1] and the phenotypes associated with animal studies in five major model organisms. We use these gene–disease associations and combine them with the drug–gene associations available from the PharmGKB to suggest new diseases in which a drug may be active.

1.2. *Animal models of human disease*

To gain an in depth understanding of specific drug actions it is often necessary to study and validate their effects as part of a whole system that involves, for example, the organism's physiology including metabolism, absorption, excretion, distribution and toxicity effects. The better we understand the pathophysiology and underlying *in vivo* biology of an organism, the more likely we are to take advantage of the development of new technologies that enable drug discovery.

As such, animal models provide a powerful mechanism for drug discovery since they determine the physiological conditions and complex interdependencies among different cell types and tissues in which the interactions of chemicals with drug targets can be studied. Based on the premise of evolutionarily conserved pathogenetic mechanisms, animal models such as mouse, zebrafish, fruitfly, yeast and worm have been harnessed to provide an in-depth understanding of the biological mechanisms that govern the effect of drug administration. These benefits assign an important role to animal models in the drug discovery process.

The advent of functional genomics allowed for the large scale exploration of gene function based on the systematic comparative analysis of gene activity. The comparison of mutant and "wild-type" phenotypes within a single organism as well as with respect to homologous genes in different organisms allows us to gain a better understanding of human disease. Despite the variations of physiology and pathobiology between species, phenotype information collected from animal models have proved extremely useful in providing new insights into disease mechanisms and etiologies. The analysis and study of phenotypes arising from the various models has direct implications for understanding mammalian physiology in the context of pharmacodynamics and pharmacokinetics studies, in understanding signalling and regulatory networks, in studies that focus on the identification of response regulators, activators and inhibitors, and in chemical genetics.

As a result, mutant strains derived from hypothesis-driven research are now being augmented for several animal models, including the mouse. For example, following the success of phenotype-driven ENU mutagenesis projects,[25] large scale gene knockout programmes have now been established with the ultimate goal of discovering the functions of all of the protein coding genes in the mouse genome.[6]

1.3. *Ontologies and phenotype information*

One of the consequences of all those efforts is the increase in the amount of phenotype information collected around the world and stored in various databases. This increase in phenotype data necessitated the development of computational frameworks that enable the retrieval, comparison and analysis of phenotypes. In response, the biomedical community has developed a plethora of species-specific phenotype ontologies[27,31,34] that are used as controlled vocabularies to annotate phenotypes in several model organism databases.

Ontologies formally specify the meaning of terms in a vocabulary[15] and express this meaning by utilising languages that provide an explicit, formal semantics. Many of the species-specific phenotype ontologies have been augmented with class definition[13,27] based on the Phenotype And Trait Ontology (PATO)[12] and other species-independent ontologies such as the Gene Ontology (GO)[2] and the Chemical Entities of Biological Interest (ChEBI) ontology.[8] Methods have been proposed to formally represent these structured definitions in OWL[19] and utilize them for automated reasoning and verification of annotations.[17,20]

1.4. *PhenomeNET*

PhenomeNET is a network in which nodes represent complex phenotypes resulting from either the phenotype annotations available in model organism databases or the disease descriptions available from the OMIM database. To enable cross-species comparative analyses of phenotypes, PhenomeNET integrates multiple species-specific phenotype ontologies to derive a cross-species phenotype ontology.[20] This integrated ontology is based on the structured definitions of phenotypes that were developed for several species-specific phenotype ontologies,[27] the UBERON mappings between species-specific anatomy ontologies,[37] the Gene Ontology (GO),[2] the PATO ontology[12] as well as several other species-independent ontologies. Efficient automated reasoning over the PhenomeNET is enabled through ontology modularization[18] and design patterns for expressing phenotypes and their links to anatomy and physiology ontologies.[19]

Integrating phenotype ontologies allows for a *direct comparison* of phenotypes of multiple species, and PhenomeNET performs a pairwise comparison of phenotypes using a measure of semantic similarity. As a result, PhenomeNET ranks phenotypes for diseases as well as phenotype annotations from model organism databases. This ranking can predict genes that participate in the same pathway, orthologous genes as well as gene–disease associations based on comparing phenotypes alone.

2. Method

2.1. *Preparation of data*

We obtained the raw PhenomeNET dataset (version 16 September 2011) from `http://phenomeblast.googlecode.com`. The dataset consists of a similarity matrix that represents the pair-wise phenotypic similarity between 87,037 complex phenotypes. These complex phenotypes either represent a phenotype annotation available from one of the model organism databases or a disease phenotype from OMIM.

For our analysis, we filter the PhenomeNET similarity matrix for nodes that represent human genes, human diseases and mouse models. As a result, we obtain a square matrix of similarities between 2,964 OMIM diseases and 26,148 genotypes or genes. Each column and row in this matrix represents either phenotypic descriptions of genes and genotypes in the MGI database[3] or phenotypes associated with OMIM genes and diseases.[1]

We use the MGI report `MGI_PhenotypicAllele.rpt` to map mouse alleles to their corresponding gene. This report includes the MGI allele accession identifier and the corresponding MGI gene identifier. To link mouse genes to their human orthologs, we use the human-mouse orthology mapping available from the MGI (MGI report `HMD_Human4.rpt`).

To link the genes in PhenomeNET to PharmGKB, we use the Gene ID available in MGI report `HMD_Human4.rpt`. OMIM records in PhenomeNET are mapped to their corresponding Gene ID using the `mim2gene` file available from the NCBI (`ftp://ftp.ncbi.nlm.nih.gov/gene/DATA/`).

PhenomeNET contains only heritable diseases that are available in OMIM. The PharmGKB, on the other hand, contains a classification based on UMLS. We use the UMLS to identify the subset of diseases in PharmGKB that can be mapped to OMIM and restrict our analysis to these diseases. The files used for evaluation and mapping were obtained on 9 July 2011.

2.2. *Experimental setup*

PharmGKB provides a list of drugs, genes and diseases, and it associates them when they are discussed together in a published article. This co-occurrence in literature may be indicative of biologically meaningful relations between the associated entities. For example, articles that discuss a drug and a gene together may state that there is some *interaction* between the drug and the gene. Similarly, discussing a drug and a disease together may indicate that a drug can play a pharmacological role in the disease.

Our underlying hypothesis is that we can employ criteria that allow us to combine relations to infer new associations. In particular, we assume that, if a drug is associated with a gene (in PharmGKB), and this gene is involved in a disease, then the drug may play a pharmacological role in the disease. In our analysis, we combine the drug–gene associations available from PharmGKB and the predictions of gene–disease relations from PhenomeNET to infer both known and novel drug–disease associations.

2.3. *Statistical testing*

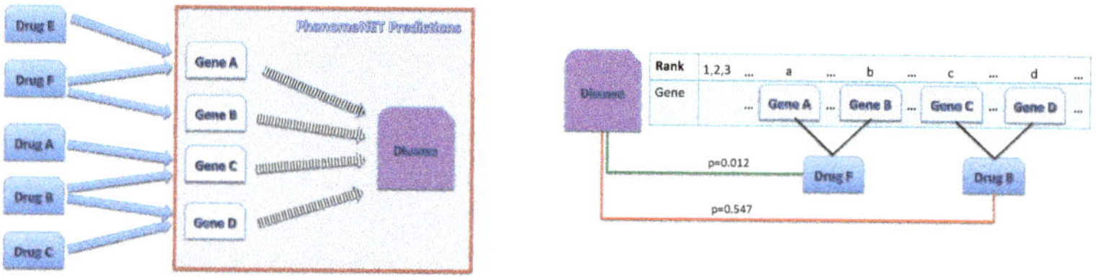

(a) (b)

Fig. 1. **Illustration of parts of our method for establishing drug–disease associations.** Our method uses known drug–gene interactions as input and links them to the gene–disease predictions generated by PhenomeNET (Fig. a). For each disease, PhenomeNET provides a ranked list of genes and genotypes based on their phenotypic similarity to the disease (Fig. b). We identify the subset of the list of genes that is known to interact with a particular drug and performs a Wilcoxon signed-rank test to determine the p-value of observing these genes at this position in the ranked list for that disease. If the genes associated with a drug are distributed uniformly across the ranks, the Wilcoxon signed-rank test will return a relatively high p-value. If the genes associated with a drug are phenotypically very similar to the disease, the test will return a relatively low p-value. Finally, we assign the p-value returned by the Wilcoxon signed-rank test as the value of the association between the drug and the disease.

As a first step in our method, we use the list of drug–gene associations in PharmGKB and the mappings we generate to create a list of PhenomeNET entries for each drug in PharmGKB. As a result, we obtain a list of pairs $(C, < G_1, ..., G_n >)$ such that C is a drug and $G_1, ..., G_n$ are genes or genotypes (in PhenomeNET) with which C interacts.

The second step of our method identifies the distribution of phenotypic similarity for each disease. For each disease that is available in PhenomeNET, we then extract a ranked list of phenotypic similarity of *all* genes and genotypes that we include in our analysis (i.e., all genes and genotypes that have a phenotypic annotation in either the OMIM or MGI database). As a result, for each disease D, we obtain a list of pairs $< G_i, \tau >$ of genes or genotypes G_i and their phenotypic similarity to τ to the disease D. This list includes the genes from the first step of our method, i.e., the genes that have known interaction to particular drugs.

The third step performs a statistical test for significant drug–disease associations. For each disease D, we perform a Wilcoxon signed-rank test for each drug and the its assigned list of gene interactions $(C, < G_1, ..., G_n >)$. The test allows us to identify whether the genes with which C interacts are phenotypically significantly more similar to the disease D than expected by chance. Formally, given a disease D, a list of pairs $< G_i, \tau >$ and a drug C with its interaction partners $< G_1, ..., G_n >$, we perform the Wilcoxon signed-rank test on the distributions of τ_k for $(G_1, \tau_1), ..., (G_n, \tau_n)$ and $< G_i, \tau >$. As a result, we obtain p-values for each pair of a drug and a disease (C, D).

Based on the list of p-values, we perform a correction for multiple testing. We apply two corrections methods to our data set. First, we use the Bonferroni correction to provide a conservative estimate of the p-value for a drug–disease association. However, for many applications it is sufficient to control the false discovery rate in a data set. Therefore, we further apply Benjamini-Hochberg's method (which controls the false discovery rate) to correct for multiple testing.

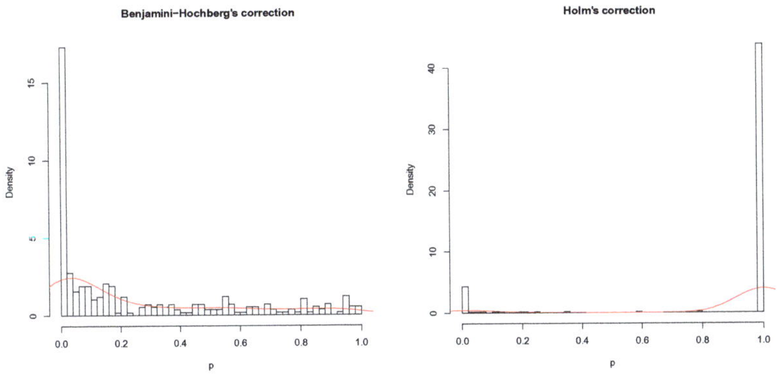

Fig. 2. The figures demonstrate the distribution of p-values for drug–disease associations available in PharmGKB using Bonferroni correction as well as Benjamini-Hochberg's correction for multiple-testing.

3. Results

3.1. *Predicting known drug-disease associations*

The result of our method are p-values for drug–disease associations, and we first identify significant drug–disease associations. In total, using Bonferroni correction, we identify 23,927 drug–disease associations as significant (with $p < 0.1$), and we identify 419,335 drug–disease associations as significant when controlling the false discovery rate using Benjamini-Hochberg's correction. Using a more conservative p-value of 0.0001 as measure of significance, we identify 6,473 significant drug–disease associations when using Bonferroni correction and 55,931 significant associations when using Benjamini-Hochberg's method. Figure 2 shows the distribution of p-values we obtain for the drug–disease associations that are included in PharmGKB.

We then use the manually curated drug–disease assocations available from PharmGKB to evaluate the performance of our method. Within PharmGKB, associations between drugs and diseases are established when both a drug and a disease are discussed together in a scientific article. Such an association does not necessarily indicate that the drug may play a pharmacological role in the disease. Additionally, PharmGKB primarily focuses on genotypes, genes and drugs, and does not provide a comprehensive repository of drug–disease associations. Therefore, we do not only use the drug–disease associations available in PharmGKB in our evaluation, but further use the drug–disease associations from FDA-approved drugs as well as high-confidence (inference score higher than 50) drug–disease associations from the Clinical Toxicogenomics Database (CTD)[7] to evaluate the performance of our method. Our quantitative evaluation is based on analyzing the receiver operating characteristic (ROC) curve for predicting drug–disease associations.

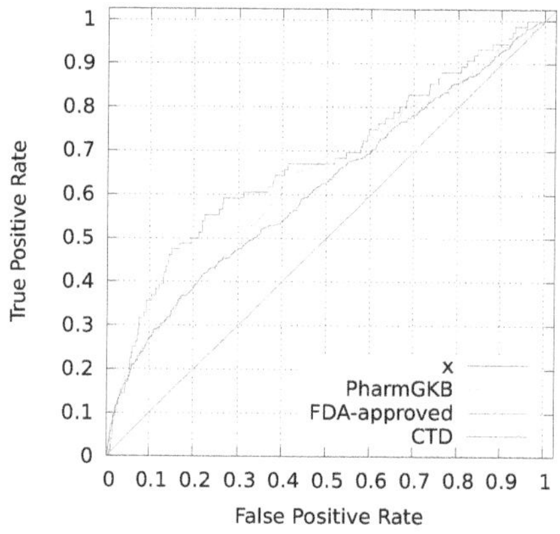

Fig. 3. The ROC curve for predicting known drug–disease associations using our method. We show the ROC curves for comparison with PharmGKB's drug–disease associations, FDA-approved drug indications, and CTD predictions of drug–disease interactions.

A ROC curve is a plot of the true positive rate as a function of the false positive rate. However, while we have a large number of *positive* drug–disease associations (available from PharmGKB), we lack *negative* examples (i.e., drug–disease pairs for which it is *known* that the drug has no effect in the disease). Therefore, we make the simplifying assumption that an *unknown* drug–disease association constitutes a *negative* example. Since our aim is to identify new indications for known drugs, this assumption does not hold for either data set we use for evaluation. Consequently, the result of the ROC analysis represents a *lower boundary* of our method's performance.

The ROC curves resulting from our method are shown in Figure 3. The area-under-curve (AUC) is a measure of the performance of the prediction and equivalent to the probability that a randomly chosen positive example (a known drug–disease association) is ranked higher than a randomly chosen negative example (an unknown drug–disease association).[11] Evaluating against PharmGKB's drug–disease associations yields an AUC of 0.629, against FDA-approved drug indication an AUC of 0.613 and against the CTD dataset an AUC of 0.672.

3.2. *Novel drug-disease associations*

Using the results of our method, we can explore drug–disease association where we predict new uses for existing drugs and potential new drugs for conditions for which they have not been approved or tested. Our criteria for novelty are that the drug-disease relationships are not in the curated knowledge of PharmGKB, not in the list of FDA-approved indications for the drug and not in the high-confidence drug–disease associations available from the CTD.

For example, cystic fibrosis (`OMIM:219700`) is caused by mutations in the *CFTR* gene which is an epithelial apical membrane anion channel regulated by nucleotides and phosphorylation.[30] The disease affects multiple tissues: pancreas, liver, reproductive tract, and the heart, with progressive lung disease accounting for most of the morbidity and mortality. It is an autosomal recessive disease with a mean prevalence of 0.737/10,000 in the EU.[10] The drugs that are significantly associated with cystic fibrosis using our method are levonorgestrel, tretinoin and drospirenone/ethinyl estradiol ($p < 2.65 \cdot 10^{-6}$, $p < 0.002$ and $p < 0.031$, Wilxocon signed-rank test, Bonferroni correction).

Tretinoin (retinoic acid) is already implicated as a potential therapy for cystic fibrosis in the scientific literature. For example, a case report in 2006 showed dramatic improvement in cystic fibrosis patients treated with isotretinoin for acne.[4] Another study shows that retinoic acid can restore the levels of nucleoside diphosphate kinase, which are reduced in CF, and reduce membrane transgutaminase activity, suggesting that retinoic acid could be a novel therapy for some aspects of CF.[36]

Inspection of the genes associated with the functionally related drugs levonorgestrel and estradiol in PharmGKB shows that mouse phenotypes associated with mutations of these genes have a high representation of reproductive phenotypes. Recent reports show that β-estradiol directly rescues F508CFTR functional expression in human cystic fibrosis airway cells[9] and that 17b-estrodiol potentiates activity of the cardiac isoform of CTFR. This may indicate that a potential role for sex steroids in cystic fibrosis therapy may be directly on CFTR itself and not only through action in the reproductive system. CTFR has been known for some time to be up-regulated by estrogens[32] and down-regulated by progesterone.[26]

Instead of looking for potential drugs for a disease, we can also explore the drug–disease associations by aiming to identify indications for a known drug. For example, the drug *ibuprofen* is associated with four diseases in PharmGKB (hemorrhage, pain, myocardial infarction and stomach neoplasm) and FDA-approved indications of ibuprofen are rheumatoid arthritis, osteoarthritis, pain and primary dysmenorrhea. Our method strongly suggests chronic lymphocytic leukemia ($p < 2.63 \cdot 10^{-23}$, Wilcoxon signed-rank test, Bonferroni correction) as an indication in which ibuprofen may be active. Ibuprofen has been shown to inhibit proliferation and induce apoptosis in promyelocytic cells,[21] and intake of ibuprofen has been linked with significantly decreased risk of leukemia.[22]

3.3. *Interface and availability*

We make our results freely available on our project's website at `http://code.google.com/p/phenomeblast/wiki/PhenomeDrug`. In addition, we enable access to the results of our method through a web server. The webserver enables search for either drugs (from PharmGKB) or diseases (from the OMIM database) and lists the drug–disease pairs as well as their p-values.

4. Discussion

We developed and implemented a method to suggest novel drug–disease associations based on analyzing information about phenotypes available from animal studies. Our initial results demonstrate that we can identify known drug–disease associations and suggest new ones, although our method does not perform as well as comprehensive drug repurposing and discovery frameworks such as PREDICT, which achieve an AUC of over 0.9 for predicting known drug–disease associations.[14] However, these frameworks are commonly based on machine learning and rely on the availability of information about the molecular mechanisms underlying disease (i.e., known gene–disease or drug–disease interactions that can be used to identify the features based on which a classifier can be trained). Our approach does not utilize machine learning, and neither PhenomeNET nor our method for drug-repurposing require any prior knowledge about gene–disease or drug–disease associations. Since our approach relies only on the knowledge of drug targets and the comparisons of phenotypes to establish a link between drug and disease, it can be applied to suggest drugs for orphan diseases of which the molecular basis is unknown.

A major goal for future research is to refine the analysis and evaluation of our method. At the moment, PharmGKB provides associations between drugs and genes based on co-occurrence in scientific articles. Discussing a drug and a gene together may indicate that the drug targets the gene, but can also be indicative of further biological relations. Since our method crucially depends on curated information about drug–gene relations, it may be improved when more specific relations between drugs and genes (or genotypes) become available within PharmGKB. Furthermore, we intend to apply our method to other databases such as DrugBank[23] and the Comparative Toxicogenomics Database[7] which also contain drug–gene interactions. We also plan to include phenotypes of further model organism available in PhenomeNET, such as yeast, fly, worm and fish, to develop an integrated system for the analysis and prediction of drug–disease associations for rare and orphan diseases based on comparisons of phenotypes.

Performing enrichment analyses over both the diseases that are associated with a drug, using disease-relevant ontologies such as the Human Disease Ontology,[24] and enriching the drugs that are associated with a disease using the chemical classifications of the ATC, MeSH or ChEBI, are areas of further research that may improve the performance of our method.

5. Conclusions

We use the PhenomeNET cross-species phenotype network to provide ranked lists of gene–disease associations. Since the PhenomeNET method performs a comparison of phenotypes directly, it can prioritize genes for orphan diseases of which the molecular basis is unknown. Using this information in our study, we can predict drugs for diseases when the disease mechanisms are not known and therefore suggest potentially new drugs for rare and orphan diseases. The use of PhenomeNET further provides direct links to animal models that can be used to investigate the drug and disease mechanisms as well as the drug's role in the disease.

Animal models of disease play an important role in the investigation of the pathobiology of disease[5] and of drug mechanisms.[35] Our method can identify relevant animal models and thereby improve the speed and reduce the cost required to test and investigate a novel candidate drug. PhenomeNET contains not only phenotype description associated with genes, but rather with the *genotype* of a model organism that belongs to a particular *strain* and within a certain *environment*. The link between PharmGKB and the PhenomeNET resource will therefore allow us to target particular strains and genotypes, and utilize the results from large-scale phenotyping projects such as the IMPC.[3,6] The link to the environmental conditions present in some model organism databases may further improve identification of targets for testing drugs within an *in vivo* environment.

Acknowledgments

Funding for RH was provided by the European Commission's 7th Framework Programme, RICORDO project, grant number 248502. Funding for PS was provided by an NIH grant, number R01 HG004838-02. Funding for GG was provided by BBSRC grant BBG0043581. Funding for AO and DRS was provided by the European Bioinformatics Institute.

References

1. J. Amberger, C. Bocchini, and A. Hamosh. A new face and new challenges for online mendelian inheritance in man (OMIM). *Hum Mutat*, 2011.
2. M. Ashburner, C. A. Ball, J. A. Blake, D. Botstein, H. Butler, M. J. Cherry, A. P. Davis, K. Dolinski, S. S. Dwight, J. T. Eppig, M. A. Harris, D. P. Hill, L. I. Tarver, A. Kasarskis, S. Lewis, J. C. Matese, J. E. Richardson, M. Ringwald, G. M. Rubin, and G. Sherlock. Gene ontology: tool for the unification of biology. *Nature Genetics*, 25(1), May 2000.
3. J. A. Blake, C. J. Bult, J. A. Kadin, J. E. Richardson, J. T. Eppig, and the Mouse Genome Database Group. The Mouse Genome Database (MGD): premier model organism resource for mammalian genomics and genetics. *Nucleic Acids Research*, 39(suppl 1):D842–D848, 2011.

4. J. L. Buckley, M. A. Chastain, and R. L. Rietschel. Improvement of cystic fibrosis during treatment with isotretinoin. *SKINmed: Dermatology for the Clinician*, 5(5):252–225, 2006.
5. C. J. Bult, J. T. Eppig, J. A. Kadin, J. E. Richardson, and J. A. a. Blake. The mouse genome database (mgd): mouse biology and model systems. *Nucleic acids research*, 36(Database issue), January 2008.
6. F. S. Collins, R. H. Finnell, J. Rossant, and W. Wurst. A new partner for the international knockout mouse consortium. *Cell*, 129(2):235, 2007.
7. A. P. Davis, B. L. King, S. Mockus, C. G. Murphy, C. Saraceni-Richards, M. Rosenstein, T. Wiegers, and C. J. Mattingly. The comparative toxicogenomics database: update 2011. *Nucleic Acids Research*, 2010.
8. K. Degtyarenko, P. Matos, M. Ennis, J. Hastings, M. Zbinden, A. McNaught, R. Alcantara, M. Darsow, M. Guedj, and M. Ashburner. ChEBI: a database and ontology for chemical entities of biological interest. *Nucleic Acids Research*, 2007.
9. T. Fanelli, R. A. Cardone, M. Favia, L. Guerra, M. Zaccolo, S. Monterisi, T. De Santis, S. M. Riccardi, S. J. Reshkin, and V. Casavola. Beta-oestradiol rescues deltaf508cftr functional expression in human cystic fibrosis airway cfbe41o- cells through the up-regulation of nherf1. *Biol Cell*, 100(7):399–412, 2008.
10. P. M. Farrell. The prevalence of cystic fibrosis in the european union. *Journal of Cystic Fibrosis*, 7(5):450 – 453, 2008.
11. T. Fawcett. An introduction to ROC analysis. *Pattern Recognition Letters*, 27(8):861 – 874, 2006. ROC Analysis in Pattern Recognition.
12. G. V. Gkoutos, E. C. Green, A.-M. M. Mallon, J. M. Hancock, and D. Davidson. Using ontologies to describe mouse phenotypes. *Genome biology*, 6(1), 2005.
13. G. V. Gkoutos, C. Mungall, S. Dolken, M. Ashburner, S. Lewis, J. Hancock, P. Schofield, S. Kohler, and P. N. Robinson. Entity/quality-based logical definitions for the human skeletal phenome using PATO. *Annual International Conference of the IEEE Engineering in Medicine and Biology Society.*, 1:7069–7072, 2009.
14. A. Gottlieb, G. Y. Stein, E. Ruppin, and R. Sharan. PREDICT: a method for inferring novel drug indications with application to personalized medicine. *Molecular Systems Biology*, 7, June 2011.
15. N. Guarino. Formal ontology and information systems. In N. Guarino, editor, *Proceedings of the 1st International Conference on Formal Ontologies in Information Systems*, pages 3–15. IOS Press, 1998.
16. T. Hernandez-Boussard, M. Whirl-Carrillo, J. M. Hebert, L. Gong, R. Owen, M. Gong, W. Gor, F. Liu, C. Truong, R. Whaley, M. Woon, T. Zhou, R. B. Altman, and T. E. Klein. The pharmacogenetics and pharmacogenomics knowledge base: accentuating the knowledge. *Nucleic acids research*, 36(Database issue), Jan. 2008.
17. R. Hoehndorf, M. Dumontier, A. Oellrich, D. Rebholz-Schuhmann, P. N. Schofield, and G. V. Gkoutos. Interoperability between biomedical ontologies through relation expansion, upper-level ontologies and automatic reasoning. *PLOS ONE*, 6(7):e22006, July 2011.
18. R. Hoehndorf, M. Dumontier, A. Oellrich, S. Wimalaratne, D. Rebholz-Schuhmann, P. Schofield, and G. V. Gkoutos. A common layer of interoperability for biomedical ontologies based on OWL EL. *Bioinformatics*, 27(7):1001–1008, April 2011.
19. R. Hoehndorf, A. Oellrich, and D. Rebholz-Schuhmann. Interoperability between phenotype and anatomy ontologies. *Bioinformatics*, 26(24):3112 – 3118, 10 2010.
20. R. Hoehndorf, P. N. Schofield, and G. V. Gkoutos. Phenomenet: a whole-phenome approach to disease gene discovery. *Nucleic Acids Research*, 2011.
21. J. Jakubikova, T. Duraj, X. Takacsova, L. Hunakova, B. Chorvath, and J. Sedlak. Non-steroidal anti-inflammatory agent ibuprofen-induced apoptosis, cell necrosis and cell cycle alterations in

human leukemic cells in vitro. *Neoplasma*, 48(3):208–213, 2001.
22. C. M. Kasum, C. K. Blair, A. R. Folsom, and J. A. Ross. Non-steroidal anti-inflammatory drug use and risk of adult leukemia. *Cancer Epidemiology Biomarkers & Prevention*, 12(6):534–537, June 2003.
23. C. Knox, V. Law, T. Jewison, P. Liu, S. Ly, A. Frolkis, A. Pon, K. Banco, C. Mak, V. Neveu, Y. Djoumbou, R. Eisner, A. C. Guo, and D. S. Wishart. Drugbank 3.0: a comprehensive resource for omics research on drugs. *Nucleic Acids Research*, 2010.
24. P. LePendu, M. Musen, and N. Shah. Enabling enrichment analysis with the human disease ontology. *Journal of Biomedical Informatics*, 2011. In press.
25. H. Morgan, T. Beck, A. Blake, H. Gates, N. Adams, G. Debouzy, S. Leblanc, C. Lengger, H. Maier, D. Melvin, H. Meziane, D. Richardson, S. Wells, J. White, J. Wood, T. E. Consortium, M. H. de Angelis, S. D. M. Brown, J. M. Hancock, and A.-M. Mallon. EuroPhenome: a repository for high-throughput mouse phenotyping data. *Nucleic Acids Research*, 38(suppl 1):D577–D585, 2010.
26. A. Mularoni, L. Beck, R. Sadir, G. L. Adessi, and M. Nicollier. Down-regulation by progesterone of cftr expression in endometrial epithelial cells: A study by competitive rt-pcr. *Biochemical and Biophysical Research Communications*, 217(3):1105 – 1111, 1995.
27. C. Mungall, G. Gkoutos, C. Smith, M. Haendel, S. Lewis, and M. Ashburner. Integrating phenotype ontologies across multiple species. *Genome Biology*, 11(1):R2+, 2010.
28. B. Munos. Lessons from 60 years of pharmaceutical innovation. *Nature reviews. Drug discovery*, 8(12):959–968, Dec. 2009.
29. M. Oti and H. G. Brunner. The modular nature of genetic diseases. *Clinical Genetics*, 71:1–11, 2007.
30. J. R. Riordan. CFTR function and prospects for therapy. *Annual review of biochemistry*, 77:701–726, 2008.
31. P. N. Robinson, S. Koehler, S. Bauer, D. Seelow, D. Horn, and S. Mundlos. The human phenotype ontology: a tool for annotating and analyzing human hereditary disease. *American journal of human genetics*, 83(5):610–615, 2008.
32. L. Rochwerger and M. Buchwald. Stimulation of the cystic fibrosis transmembrane regulator expression by estrogen in vivo. *Endocrinology*, 133(2):921–30, 1993.
33. S. H. Sleigh and C. L. Barton. Repurposing strategies for therapeutics. *Pharmaceutical Medicine*, 24(3):151–159, 2010.
34. C. L. Smith, C.-A. W. Goldsmith, and J. T. Eppig. The mammalian phenotype ontology as a tool for annotating, analyzing and comparing phenotypic information. *Genome Biology*, 6(1):R7, 2004.
35. S. Tickoo and S. Russell. Drosophila melanogaster as a model system for drug discovery and pathway screening. *Current Opinion in Pharmacology*, 2(5):555 – 560, 2002.
36. K. J. Treharne, O. Giles Best, and A. Mehta. Transglutaminase 2 and nucleoside diphosphate kinase activity are correlated in epithelial membranes and are abnormal in cystic fibrosis. *FEBS Letters*, 583(17):2789–2792, 2009.
37. N. L. Washington, M. A. Haendel, C. J. Mungall, M. Ashburner, M. Westerfield, and S. E. Lewis. Linking human diseases to animal models using ontology-based phenotype annotation. *PLoS Biol*, 7(11):e1000247, 11 2009.

INTEGRATING VA'S NDF-RT DRUG TERMINOLOGY WITH PHARMGKB: PRELIMINARY RESULTS

JYOTISHMAN PATHAK, PhD
Department of Health Sciences Research, Mayo Clinic
200 1st Street SW, Rochester, MN, USA
Email: pathak.jyotishman@mayo.edu

LAURA C. WEISS*
Bethel University
3900 Bethel Drive, St. Paul, MN, USA
Email: laura-weiss@bethel.edu

MATTHEW J. DURSKI, BS
Department of Health Sciences Research, Mayo Clinic
200 1st Street SW, Rochester, MN, USA
Email: durski.matthew@mayo.edu

QIAN ZHU, PhD
Department of Health Sciences Research, Mayo Clinic
200 1st Street SW, Rochester, MN, USA
Email: zhu.qian@mayo.edu

ROBERT R. FREIMUTH, PhD
Department of Health Sciences Research, Mayo Clinic
200 1st Street SW, Rochester, MN, USA
Email: freimuth.robert@mayo.edu

CHRISTOPHER G. CHUTE, MD, DrPH
Department of Health Sciences Research, Mayo Clinic
200 1st Street SW, Rochester, MN, USA
Email: chute@mayo.edu

Abstract

Biomedical terminology and vocabulary standards play an important role in enabling consistent, comparable, and meaningful sharing of data within and across institutional boundaries, as well as ensuring semantic interoperability. The Veterans Affairs (VA) National Drug File Reference Terminology (NDF-RT) is a federally recommended standardized terminology resource encompassing medications, ingredients, and a hierarchy

* This work was done while the author was a summer intern at Mayo Clinic.

for high-level drug classes. In this study, we investigate the drug-disease relationships in NDF-RT and determine how PharmGKB can be leveraged to augment NDF-RT, and vice-versa. Our preliminary results indicate that with additional curation and analyses, information contained in both knowledge resources can be mutually integrated.

1. Introduction

Standardized biomedical terminologies play an important role in enabling consistent representation and interoperability of healthcare data and information systems. Within the realm of pharmaceutical drugs and medications, NDF-RT[2] created and maintained by the U.S. Department of Veterans Affairs is one of the publicly available federal medication terminologies. The goal of NDF-RT is to allow various clinical information systems using different drug nomenclatures to share and exchange medication data efficiently[3]. NDF-RT includes information about drugs and ingredients, provides a way to link and map standard clinical drug names to other drug terminologies (e.g., RxNorm), and is updated on a monthly schedule. Additionally, NDF-RT contains a multi-axial hierarchical knowledge structure that classifies ingredients and drug products based on their therapeutic intent, mechanism of action, pharmacokinetics and other aspects (see Section 2) It also provides several ways to create "meaningful groups" for the analysis of medication data. Several research efforts in the recent past, including our own prior work[4,5], have studied different aspects of NDF-RT or classification of medication data and their applications in information exchange[6], linkage[7] and querying[8].

However, to the best of our knowledge, there are no existing studies investigating and comparing drug-disease relationships in NDF-RT and Pharmacogenomics Knowledge Base (PharmGKB[9]), which is a comprehensive resource on pharmacogenes (i.e., genes involved in modulating the response to drugs), their variations, pharmacokinetics, pharmacodynamic pathways, and their effects on drug-related phenotypes. In particular, it is not known if drug-disease relationships that are manually curated in PharmGKB can be leveraged for augmenting the information contained in NDF-RT, and vice versa. Such a mutual integration of information from two related and publicly available knowledge resources could identify areas for additional curation, facilitate exchange of information as well as make such information sources more robust, increasing their utility for research and clinical applications.

To this end, in this paper we report our preliminary findings in comparing and mapping the drug-disease relationships in PharmGKB with those in NDF-RT. In particular, we studied the pharmacodynamics (PD) and pharmacokinetics (PK) relationships between drugs and diseases in PharmGKB, and compared them with the drug-disease relationships in NDF-RT. Our results indicate that while both sources contain related information, additional curation and analyses is required to enable mutual information integration.

2. Background

2.1. *Veterans Affairs National Drug File Reference Terminology*

The National Drug File Reference Terminology (NDF-RT)[2] is created and maintained by the Department of Veteran Affairs (VA). It includes information about drugs and ingredients, but also contains a multi-axial hierarchical knowledge structure that classifies various ingredients and drug products. In particular, NDF-RT uses a description logic-based formal reference model that groups drug products into the high-level drug classes for Chemical Structure (e.g., Acetanilides), Mechanism of Action (e.g., Prostaglandin Receptor Antagonists), Physiological Effect (e.g., Decreased Prostaglandin Production), drug-disease relationship describing the Therapeutic Intent (e.g., Pain), Pharmacokinetics describing the mechanisms of absorption and distribution of an administered drug within a body (e.g., Hepatic Metabolism), and legacy VA-NDF classes for Pharmaceutical Preparations (VHA Drug Class; e.g., Non-Opioid Analgesic). Figure 1 shows NDF-RT's content model.

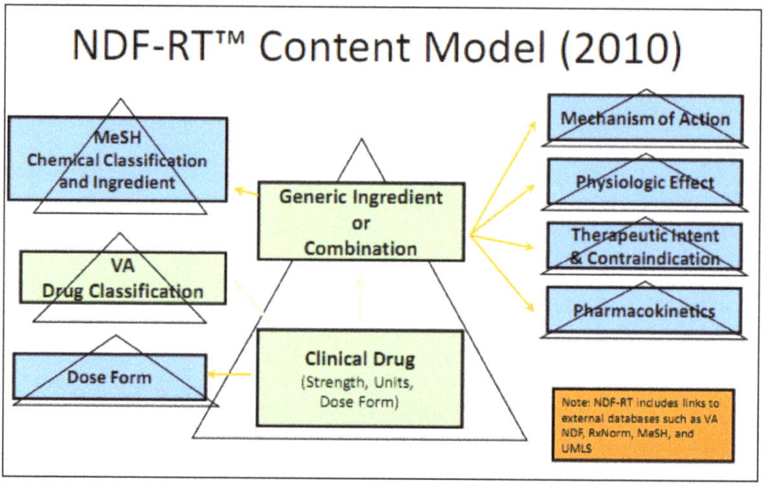

Figure 1 NDF-RT content model (Source: http://evs.nci.nih.gov/ftp1/NDF-RT)

This figure shows the structure of NDF-RT: triangles denote hierarchies of related concepts, each hierarchy being categorized in the rectangles within the triangles. Taxonomic or ISA relationships (upward-pointing green arrows) unify NDF-RT clinical drug concepts into a polyhierarchy, classified both by their VA drug class and their generic ingredient(s). Various named role relationships (sideways-pointing amber arrows) define the central drug concepts (green) from which they originate in terms of the reference hierarchy concepts (blue) pointed to. Role relationships are also inherited into subsumed clinical drug concepts. Note that NDF-RT also augments a "legacy" classification system (called VA-NDF[1]) which classified drug products into groupings developed by VA (denoted by VHA Drug Class) to support organization and decision support for medication usage in clinical care settings.

2.2. *The Pharmacogenomics Knowledge Base*

The Pharmacogenomics Knowledge Base (PharmGKB)[9] is a comprehensive resource for pharmacogenes (i.e., genes involved in modulating the response to drugs), their variations, pharmacokinetics, pharmacodynamic pathways, and their effects on drug-related phenotypes. Its overarching goal is to support the integration, aggregation and curation of the information contained in various life sciences and biological databases that contain information about genetic variation and associated phenotypes. The data in PharmGKB is curated from the literature and other databases that report genetic variations in known pharmacogenes. To facilitate indexing and retrieval of the information, the PharmGKB data sets are annotated with genes and/or drugs based on five different categories: clinical outcomes, pharmacodynamics and drug response, pharmacokinetics, molecular and cellular functional assays, and genotype (see Figure 2).

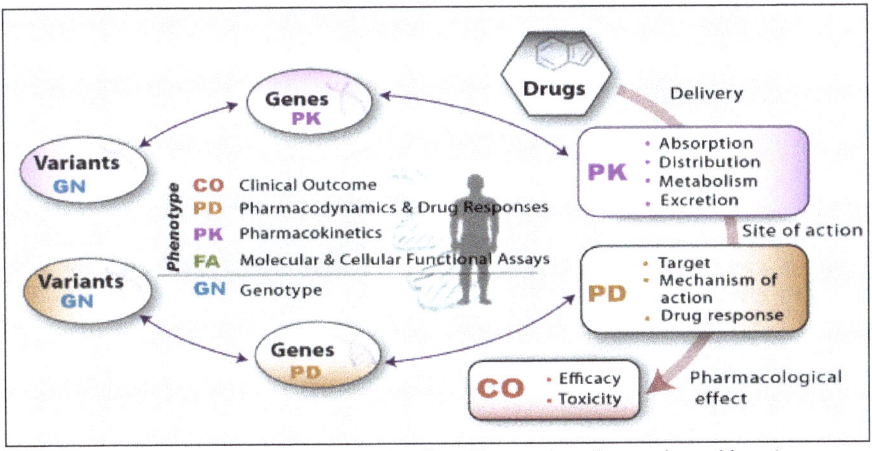

Figure 2 Pharmacogenomics information flow (Source: http://www.pharmgkb.org)

3. Materials and Methods

3.1. *Materials*

The primary materials used in this study are the following:

- The July 6, 2011 PharmGKB relationships data set, available for download via http://www.pharmgkb.org/resources/downloads_and_web_services.jsp. The full data set contains a total of 24,213 relationships between the genes, drugs and diseases in PharmGKB (see Table 1). For this study, our investigative efforts focused exclusively on the 2,697 drug-disease relationships in PharmGKB. This data set included all four types of drug-disease relationships: both PD and PK, PD only, PK only, and neither PD nor PK (see Table 1). For example, the drug *Imatinib* has several types of relationships with different diseases: both PD and PK with the disease *Gastrointestinal Stromal Tumors*, only PD with the disease *Glioma*, only PK with the disease *Neoplasms*, and neither PD nor PK with the disease *Leukemia*.

	Type of Relationship						
	Drug-Drug*	Disease-Disease*	Gene-Gene*	Drug-Disease	Gene-Disease	Gene-Drug	Total
PD Only	56	0	0	1,786	102	2,989	4,933
PK Only	230	0	0	137	1	1,280	1,648
Both PD and PK	47	0	0	446	5	986	1,484
No PD and No PK	303	145	735	328	8,166	6,471	16,148
Total	636	145	735	2,697	8,274	11,726	24,213

Table 1 Relationships in PharmGKB, before analysis (highlighted cells indicate relationships studied in this work)
*The numbers for these relationships include duplicates.

- The June 2011 NDF-RT ontology of drugs and diseases, accessible via http://evs.nci.nih.gov/ftp1/NDF-RT/. NDF-RT is organized as a relationship-based terminology source of medications and related entities (i.e. drugs, diseases, ingredients, etc.). Each unique term can be described via its relationship with some other term in the ontology. This identification process is further illustrated in NDF-RT's content model (see Figure 1), which is a visual representation of an individual term's organizational structure. For this study, we investigated the information contained in the *Role Relationships* section of a particular drug. This feature of NDF-RT connects a specific drug with a variety of other factors, such as an ingredient, mechanism of action, or disease (see Table 2). Similar

Type of Relationship	Total
has_ingredient	23,244
has_MoA (Mechanism of Action)	14,749
has_PE (Physiologic Effect)	25,370
has_PK (Pharmacokinetic)	750
has_TC (Therapeutic Category)	76
may_diagnose	954
may_treat	47,636
may_prevent	5,968
Total	118,747

Table 2 Relationships in NDF-RT, before analysis (highlighted cells indicate relationships studied in this work)

to the restrictions for PharmGKB, in this study we were concerned only with the relationships between a drug and disease. Therefore, we limited our scope to include four relationship types from NDF-RT: 'has_PK', 'may_treat', 'may_prevent', and 'may_diagnose'. It should be noted that while NDF-RT contains a 'has_PE' relationship, due to ambiguity in the NDF-RT documentation about its relevance to the pharmacodynamics of a given drug, it was excluded from our evaluation.

3.2. *Methods*

For this study, we investigated the 2,697 drug-disease relationships from PharmGKB in comparison with NDF-RT. Of these 2,697 relationships in PharmGKB, we removed all relationships between drug classes and diseases (n=363) from analysis for uniformity with NDF-RT. The reason for doing this is because in order for a drug-disease relationship to exist in NDF-RT, the term must be an individual drug within a drug class, rather than the drug class itself. In other words, NDF-RT does not provide relationships between drug classes and diseases, but rather between individual drugs (that belong to a drug class) and diseases. Hence, our comparison process began with the remaining 2,334 drug-disease relationships from PharmGKB.

For each unique drug-disease pair in the PharmGKB data set, we searched for an identical drug-disease pair in NDF-RT. If an appropriate match was identified, we recorded the match and type of relationship in the data set. If no match was identified in NDF-RT, the match was designated as "missing". For example, the drug *Fluorouacil* has a PD relationship with *Colonic Neoplasms* in PharmGKB. A relationship between this drug and disease was identified in NDF-RT as a "may_treat" relationship. This identified match and its corresponding classification were recorded in the data set. Throughout our comparison process, we discovered several additional drug-disease relationships that existed in NDF-RT, but not in PharmGKB. The type of relationships for these pairs was also recorded and they were added to the data set. To continue with the example above, *Fluorouracil* had a 'may_treat' relationship with *Keratosis* in NDF-RT. An identical relationship of any type did not exist in PharmGKB; thus an entry for the relationship in NDF-RT was added to the data set for analysis.

This process was repeated for all 2,334 drug-disease relationship pairs in PharmGKB, adding entries from NDF-RT when necessary. The drug ID and disease ID was recorded for each pair in the list from each knowledge base, as well as the type of relationship in both PharmGKB (PD, PK, both, or neither) and NDF-RT (has_PK, may_treat, may_prevent, may_diagnose, or a combination of the four). Several of the drug-disease pairs in our final analysis data set have multiple types of relationships for the same pair. In PharmGKB, many pairs have both a PD and a PK relationship (e.g., drug *Adalimumab* and disease *Arthritis, Rheumatoid*). Additionally, there were a large number of entries from NDF-RT that had both a 'may_treat' and 'may_prevent' relationship (e.g., drug *Aspirin* and disease *Pain*).

Throughout our comparison, there were a total of 57 entries in which the drug was matched with a synonym for the drug. For example, PharmGKB's *Salbutamol* was identified in NDF-RT as

Albuterol, a synonym for *Salbutamol*. Similarly, there were 22 diseases that were matched to a synonymous disease name. Furthermore, 77 drug entries from PharmGKB were matched with a drug in NDF-RT that did not have a *Role Relationships* section available (e.g., *Nitrazepam*). For these, the PharmGKB drug-disease pairs were recorded as not having a match in NDF-RT and no additional entries were added to the data set from NDF-RT. Finally, the drugs in 168 entries from PharmGKB did not have any appropriate match in NDF-RT and could not, therefore, be matched with a suitable drug in NDF-RT (e.g., *Orbofiban*). These outlying cases were all included in the final analysis, as they have potential to augment and clarify both PharmGKB and NDF-RT.

4. Results

The comparison between PharmGKB and NDF-RT began with the 2,334 drug-disease relationships in PharmGKB. Of these 2,334 relationships, 2,039 existed only in PharmGKB, while 295 were common to both PharmGKB and NDF-RT. By the end of the comparison, a total of 1,499 relationships identified in NDF-RT were added to the original data set for analysis. These additional entries were drug-disease pairs that existed only in NDF-RT, and not in any form in PharmGKB. See Table 3 for a summary of these results.

Relationship	Number of Relationships
Common to both PharmGKB and NDF-RT	295
Exist only in PharmGKB	2,039
Total in PharmGKB (starting number for analysis)	2,334
Exist only in NDF-RT (added to data set for analysis)	1,499
Total from PharmGKB and NDF-RT for analysis	3,833

Table 3 Relationships from PharmGKB and NDF-RT studied in this work (highlighted cells indicate relationships studied in this work)

In order to augment the relationships in PharmGKB using NDF-RT, and vice versa, each drug-disease relationship must be classified according to its type. The only relationship type that is common to both PharmGKB and NDF-RT in their current releases is PK. Therefore, throughout our investigation, we were primarily focused on any PK relationship from one knowledge base that could be augmented in the other. For example, PharmGKB's PK relationship between the drug *Cocaine* and the disease *Apnea* could potentially be added to NDF-RT. As demonstrated in Table 4, there are a total of 537 PK relationships in PharmGKB that are currently not represented in NDF-RT under the relationship type has_PK, although they may be classified under some other relationship type (e.g., 'may_prevent'). For example, PharmGKB has an identified PK relationship between the drug *Lozartan* and the disease *Hypertension*. While this relationship is captured in NDF-RT, the drug-disease pair is represented as a may_treat relation, instead of has_PK. As for the has_PK relationships identified in NDF-RT, all 34 do not exist in PharmGKB under any relationship type.

	Number of relationships, per classification				
	PK	PD	may_treat	may_prevent	may_diagnose
Total in PharmGKB (not represented in NDF-RT)	537 (537)	1,942 (1,942)	N/A	N/A	N/A
Total in NDF-RT (not represented in PharmGKB)	34 (34)	N/A	1,662 (1,662)	183 (183)	8 (8)

Table 4 Comparison between drug-disease relationships in PharmGKB and NDF-RT, according to classification

Unlike the PK relationship, PharmGKB's PD classification does not currently exist in NDF-RT. Therefore, all PD relationships in PharmGKB (1,942) are unique to that data set and could not be added to NDF-RT for evaluation. The same can be said for NDF-RT's 'may_treat', 'may_prevent', and 'may_diagnose' relationships. As these classifications do not currently exist in PharmGKB, a relationship of the same type could not be identified. For example, NDF-RT has a 'may_prevent' relationship between the drug *Flunisolide* and the disease *Asthma.* Although this relationship does in fact exist in PharmGKB with a relationship type of PD, a 'may_prevent' classification is not yet available. Therefore, we conclude that all relationships of these types in NDF-RT do not exist under the same classification in PharmGKB.

Due to PharmGKB not having the 'may_treat', 'may_prevent', and 'may_diagnose' relationships in its current release, it is not possible to compare these types of relationships in PharmGKB and therefore they have been marked as "not applicable" in Table 4. Similarly, there were no PD relationships identified in NDF-RT due to the absence of this relationship in NDF-RT's current release. See Table 4 for a summary of these results.

5. Discussion

The principal goal of this study was to identify and analyze the relationships in NDF-RT that could be used to augment PharmGKB, and vice versa. As the PK relationship is common to both knowledge bases, this relationship is the most immediately applicable finding for our purposes. A total of 537 PK relationships were identified in PharmGKB and 34 has_PK relationships were identified in NDF-RT. Interestingly, of these 571 drug-disease PK pairs, none were common to both knowledge bases. One plausible reason for this is the lack of a consistent definition of 'has_PK', 'may_treat', and 'may_prevent' relationships in NDF-RT with the PK relation in PharmGKB. Another reason for non-overlap might be due to the focus and scope of these knowledge resources: NDF-RT is primarily developed and maintained for encoding clinical data, whereas the main focus of PharmGKB is for biological and life sciences. However, in spite of these issues, these results indicate that there is a potential for unification and curation of the PK relationships in the two knowledge sources. Such an effort should encourage the development of a common set of gene-drug, gene-disease, and drug-disease relationships so that information is more easily comparable and integrable.

The comparative process described in this work began with the drug-disease pairs in PharmGKB that had both a PD and PK relationship. Throughout this work, we emphasized the fact that our study originated with these specific entities, and expanded to include additional relationships after comparison with NDF-RT. Some of these additional relationships that were added from NDF-RT were found to be present in PharmGKB, though without *both* a PD and PK relationship. For this reason, a number of relationships that have just a PD, or just a PK, or neither in PharmGKB have been included in our analysis.

From this study, we also determined that on several occasions the relationship type (i.e., PD or PK) is not specified between a drug-disease pair in PharmGKB. While we acknowledge that this under-specification of the relationship type could be due to limited scope of curation (i.e., only selected journals are curated by PharmGKB curators), not having a proper designation of the type of relationship between the drug and disease entities that exists in the PharmGKB can lead to confusion. This reiterates the goal of this study to investigate publicly available knowledge resources, such as NDF-RT, for further evaluation and mutual information integration. For example, it might be worth investigating the incorporation of NDF-RT relationship types, such as 'may_treat', 'has_MOA' etc., within the PharmGKB knowledge base. The same could be said about NDF-RT's potential to include a PD relationship. Furthermore, one might consider investigating Structured Product Labels to perform a similar analysis between drugs and adverse drug reactions.

Finally, this study focused exclusively on drug-disease relationships, as they are common to both knowledge bases. However, PharmGKB encompasses gene-disease, drug-drug and gene-drug relationships, as well. Integrating these relationships in NDF-RT could be beneficial in addition to the drug-disease relationships that currently exist in NDF-RT. We plan to pursue an extended analysis encompassing such relationships in the future.

6. Conclusion

In this study, we investigate the drug-disease relationships in two publicly available knowledge resources, namely PharmGKB and NDF-RT. Our results indicate that with additional curation and authoring of relationships, both resources can be mutually integrated.

7. Acknowledgments

This work is supported by the PGRN Pharmacogenomic Ontology Network Resource (PHONT; U01GM061388).

References

1. Nelson SJ, Brown SH, Erlbaum MS, et al. A Semantic Normal Form for Clinical Drugs in the UMLS: Early Experiences with the VANDF. *AMIA Annual Symposium.* 2002:557-561.
2. Brown S, Elkin P, Rosenbloom T, et al. VA National Drug File Reference Terminology: A Cross-Institutional Content Coverage Study. *MedInfo: Studies in Health Technology and Informatics.* 2004:477-781.
3. Bodenreider O. Biomedical Ontologies in Action: Role in Knowledge Management, Data Integration and Decision Support. In: Geissbuhler A aKC, ed. *IMIA Yearbook of Medical Informatics.* Vol 47: International Medical Informatics Association; 2008:67-79.
4. Pathak J, Chute C. Analyzing categorical information in two publicly available drug terminologies: RxNorm and NDF-RT. *Journal of American Medical Informatics Association.* 2010;17(4):432-439.
5. Pathak J, Chute C. Further revamping VA's NDF-RT drug terminology for clinical research. *Journal of American Medical Informatics Association.* 2011;18(3):347-348.
6. Bouhaddou O, Warnekar P, Parrish F, et al. Exchange of Computable Patient Data between the Department of Veterans Affairs (VA) and the Department of Defense (DoD): Terminology Mediation Strategy. *Journal of the American Medical Informatics Association.* 2008;15(2):174-183.
7. Burton MM, Simonaitis L, Schadow G. Medication and Indication Linkage: A Practical Therapy for the Problem List? *AMIA Annual Symposium.* 2008:86-90.
8. Palchuk M, Klumpennar M, Jatkar T, Zottola R, Adams W, Abend A. Enabling Hierarchical View of RxNorm with NDF-RT Drug Classes. *AMIA Annual Symposium.* Washington, DC2010:577-581.
9. Altman RB. PharmGKB: a logical home for knowledge relating genotype to drug response phenotype. *Nat Genet.* 2007;39(4):426-426.

DISCOVERY AND EXPLANATION OF DRUG-DRUG INTERACTIONS VIA TEXT MINING

BETHANY PERCHA
Biomedical Informatics Program, Stanford University, Stanford, CA 94305, USA

YAEL GARTEN
Biomedical Informatics Program and Department of Genetics, Stanford University, Stanford, CA 94305, USA

RUSS B. ALTMAN*
Departments of Bioengineering, Genetics and Medicine, Stanford University, Stanford, CA 94305, USA
**Corresponding author.*

Drug-drug interactions (DDIs) can occur when two drugs interact with the same gene product. Most available information about gene-drug relationships is contained within the scientific literature, but is dispersed over a large number of publications, with thousands of new publications added each month. In this setting, automated text mining is an attractive solution for identifying gene-drug relationships and aggregating them to predict novel DDIs. In previous work, we have shown that gene-drug interactions can be extracted from Medline abstracts with high fidelity - we extract not only the genes and drugs, but also the type of relationship expressed in individual sentences (e.g. *metabolize, inhibit, activate* and many others). We normalize these relationships and map them to a standardized ontology. In this work, we hypothesize that we can combine these normalized gene-drug relationships, drawn from a very broad and diverse literature, to infer DDIs. Using a training set of established DDIs, we have trained a random forest classifier to score potential DDIs based on the features of the normalized assertions extracted from the literature that relate two drugs to a gene product. The classifier recognizes the combinations of relationships, drugs and genes that are most associated with the gold standard DDIs, correctly identifying 79.8% of assertions relating interacting drug pairs and 78.9% of assertions relating noninteracting drug pairs. Most significantly, because our text processing method captures the semantics of individual gene-drug relationships, we can construct mechanistic pharmacological explanations for the newly-proposed DDIs. We show how our classifier can be used to explain known DDIs and to uncover new DDIs that have not yet been reported.

1. Introduction

Americans are living longer than ever before, and with that increased age comes a greater reliance on pharmaceuticals. For example, recent estimates by the Kaiser Family Foundation indicate that the average 70-year-old American fills over 30 prescriptions per year.[1] The chance of an adverse drug reaction increases exponentially as each new drug is added to an individual's regime. Because clinical trials for new drugs do not typically test for drug-drug interactions (DDI) directly, serious DDIs are often not discovered until a drug is already on the market. In addition, a patient who is unaware that a symptom he experiences is due to a DDI may attribute it to other factors. Many DDIs, therefore, probably go unreported.

Biologically, many DDIs are the result of conflicting or synergistic interactions between a pair of drugs and similar genes or molecular pathways within the human body.[2,3] Therefore, what we observe as drug-drug interactions often take the form of drug-gene-drug interactions. Unfortunately, while lists of known DDIs are widely available and commonly-used in clinical practice, drug-gene interactions are not as widely known. In addition, genes and drugs can interact in a variety of ways, and it is unclear which interaction types are most predictive of a drug's tendency to interact with other drugs. Furthermore, no complete databases exist that concisely describe the exact mechanisms by which drugs and genes interact; most of these interactions are only described in papers buried deep within the scientific literature.

In this environment, text mining presents a solution to the problem of uncovering novel DDIs.[4–6] Our work extends a growing body of research that has sought to classify DDIs and better understand gene-drug relationships using text mining; for example, Tari et al[7] developed a method that combined text mining and automated reasoning to extract novel DDIs. Other authors have built text-based networks of biological entities and used reasoning techniques to uncover new biologically-relevant relationships among them.[8,9] Previous work from our own group[10,11] has established methods for using a syntactical parser to identify and characterize drug-gene relationships. The end result was a semantic network of drug-gene relationships in which the edges consisted of several hundred interaction types and subject/object context terms normalized to concepts in an ontology. All of these approaches have sought to infer novel relationships among biological entities by combining known facts expressed in scientific text.

Our current work extends this line of research by using our semantic network - in particular, paths through the network that connect pairs of drugs - to infer the types of drug-gene relationships that can predict drug-drug interactions. An advantage of our method is the fact that it makes almost no *a priori* assumptions about the nature of these relationships, instead using a machine learning algorithm (a random forest) to identify the kinds of gene-drug relationships that best predict DDIs. Besides learning which textual features are most relevant for predicting DDIs, the method can also be used to predict novel DDIs and to "explain" these predictions through suggested mechanisms of interaction; this explanatory process is a built-in component of the algorithm. In this paper, we describe the main features of our algorithm, show how it can be used to predict possible mechanisms of interaction for a known DDI, and describe how it can be used in the future to predict novel DDIs.

2. Methods

2.1. *Extracting Drug-Gene Interactions*

This project builds on an earlier method for text mining Medline to extract drug-gene interactions.[11] Briefly, that method works as follows:

(1) *Create two lexicons of terms, one for gene names and one for drug names.* We used two custom lexicons. The first consisted of a set of 731 known pharmacodynamic and pharmacokinetic genes identified by the PharmGKB database curators.[12] The second consisted of 2,910 unique drug and drug-class names, also from PharmGKB. The gene lexicon also included all common synonyms for each gene; we required the drug name to be in its

generic form (rather than a brand name) to be included.[a]

(2) *Obtain a corpus of Medline article abstracts.* The Helix Group at Stanford University maintains a corpus of all Medline abstracts published before 2009. The corpus contained about 17.5 million abstracts and 88 million sentences.

(3) *Retrieve all sentences in Medline that mention both a drug and a gene of interest.* (For the purposes of this project, the drug and gene entities of interest will be known as *seeds.*) We accomplished this using the two aforementioned lexicons and running 100 search processes in parallel on Stanford's BioX2 cluster.[13]

(4) *Represent sentences as dependency graphs using the Stanford Parser.*[14,15] The dependency graphs are rooted, oriented, and labeled graphs, where the nodes are words and the edges are dependency relations between words. If two seeds were not located in the same sentence clause, that sentence was removed from consideration. In addition, if a graph contained more than one clause and there was a clause that did not contain either seed, that clause was removed from consideration. A sample dependency graph for one Medline sentence of interest is shown in Figure 1.

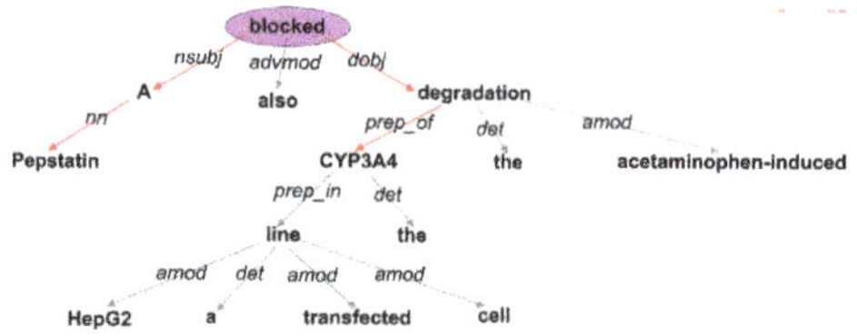

Fig. 1. Dependency graph for the sentence "Pepstatin A also blocked the acetaminophen-induced degradation of the CYP3A4 in a transfected HepG2 cell line" (PMID: 15078344). The red arrows show the path through the graph that connects the seeds Pepstatin A (a drug) and CYP3A4 (a gene). Because this path contains a verb - in this case, "blocked" - this is a sentence of interest.

(5) *Identify and normalize composite entities.* A seed does not usually occur in isolation in a sentence, but as part of a larger composite entity that includes the surrounding context. For example, a gene name like *CYP3A4* will usually occur as part of a larger entity, such as *CYP3A4 degradation* or *CYP3A4 elimination.* We used a previously-established algorithm[10] to identify the context terms surrounding each seed and normalize them. The normalization process involved mapping context terms with similar semantics but different syntax, such as *degradation of CYP3A4* and *CYP3A4 degradation*, to the same concept (*Elimination*) using a previously-constructed ontology.[11]

(6) *Extract relations between composite entities.* Relations describe the nature of the interaction between the two entities in a given sentence. They take the form $R(a, b)$, where a and b represent the locations of the two entities in the dependency graph, and R is a node that connects a and b and indicates the nature of their relationship. For a sentence to progress past this stage of the analysis, the relation connecting the gene and drug entities must

[a]Note that throughout this analysis, we use the term "gene" interchangeably with "gene product" or "protein"; it is actually the protein product of a gene that interacts with a drug.

(a)

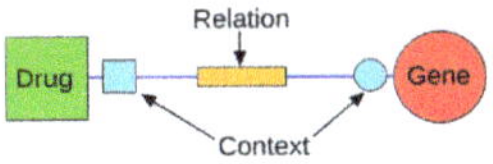

(b)

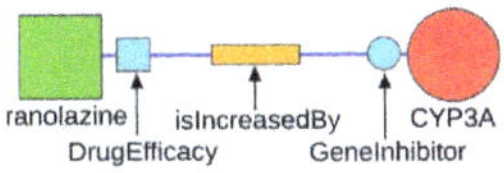

Fig. 2. A single drug-gene edge in the semantic network. A composite entity consists of a drug or gene and its surrounding [normalized] context terms. (a) The general form of an edge. (b) A specific example.

have been a verb (e.g. *associated*) or a nominalized verb (e.g. *association*).

(7) *Normalize relations.* The extracted relations, like the context terms surrounding each seed, were normalized. During normalization, the raw relations were mapped onto a much smaller set of normalized relationships taken from the ontology. For example, the verbs *associated* and *related* both map to the ontological entity *isAssociatedWith*. In addition, less-common terms like *augment* were mapped to more common synonyms, like *increase*.

The overall goal of the normalization process for both composite entities and relations was to collapse statements with the same semantic meaning but different word choice or syntax to the same basic relationship, reducing the number of features that needed to be considered later when building the DDI classifier. When tested on a smaller set of drug-gene relationships extracted from Medline, our ontology was able to properly normalize approximately 80% of all relation types mentioned in the literature. Nonetheless, by including only those sentences where the relation could be normalized, we necessarily excluded some true facts about drug-gene relationships from our network. It is important to note that only sentences for which the *relation* could not be normalized were thrown out; sentences for which the *context terms* could not be normalized were still included - the context was simply normalized to *Thing*, as described further in Section 3.1.

2.2. *Semantic Network*

When applied to the entire Medline corpus, the relation extraction and normalization process yielded 76,784 different normalized gene-drug relationships of the form shown in Figure 2. We eliminated all relationships in which the verb could not be normalized (i.e. was not one of the relations contained in the ontology), which left us with 53,208 relationships [b]. We then put all relations in active voice, collapsing passive/active pairs of normalized verbs such as *isMetabolizedBy* and *metabolizes* into a single feature. This left 49,021 normalized relations. However, many of these normalized relations were duplicates of each other because a given drug-gene relationship could be reported in similar ways many different times throughout the biomedical literature. We chose to eliminate duplicate paths of this nature. After collapsing the duplicate edges, we were left with 24,155 unique edges, which we used to construct a

[b]Examples of relations that could not be normalized included *protects, mimicked, oxidize, encode,* and *seen.* We are in the process of expanding our ontology to include some of these less common relations.

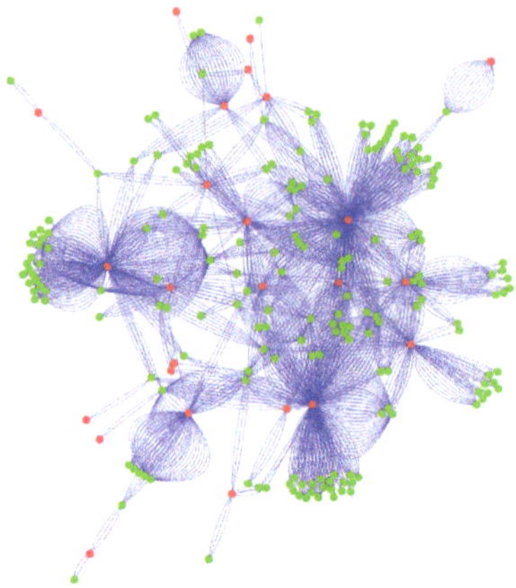

Fig. 3. A subset of the semantic network (selected to enhance visual clarity), including only the 43 most pharmacogenomically-important genes from PharmGKB and 600 drugs that were known to interact with at least one other drug. The green nodes represent drugs and the pink nodes genes. The context terms and relations are not shown in this picture, but are present on every edge, as shown in Figure 2. Multiple edges between the same gene-drug pair in this figure represent different textual relationships found between that gene and drug in the literature.

semantic network, a subset of which is shown in Figure 3. Each edge in the semantic network had the form shown in Figure 2, but is simplified in Figure 3 for clarity.

2.3. *Feature Extraction*

The feature extraction phase of this project relied on one central assumption: that the shortest textual path linking two drugs in the network represented the simplest explanatory mechanism of their interaction (if any such mechanism existed). The set of relevant features then consisted of all the genes, relations, and context terms found on the shortest path. To find the shortest path between any two drugs D_1 and D_2, we performed a breadth-first search for D_2, starting at D_1. Breadth-first search is guaranteed to yield the optimal (shortest) path between two points on a graph.[16] The shortest possible path between any two drugs in the network has the form shown in Figure 4. For the purposes of building our training set, we considered only drug pairs that had one or more shortest paths of the form in Figure 4; if the shortest path was longer than this, the drug pair was not included in the training set. We made this decision because we wanted to explore only those drug pairs for which the mechanistic explanation provided by the shortest path could be interpreted easily.

By assigning each feature a numeric index, we could easily convert the lists of normalized terms found on the shortest paths into a matrix of numbers, with each row corresponding to a single path and the columns corresponding to the number of occurrences of each feature on the path. If multiple shortest paths were found, we included a separate row in our training matrix for each unique path.

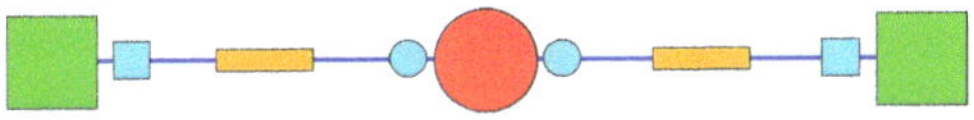

Fig. 4. The minimum-length path between two drugs in the network. It is two edges long. The colors and symbols in this figure are identical to those in Figure 2: green squares represent drugs, and the pink circle represents a gene. The yellow rectangles represent relations, and the blue circles and squares represent context terms for genes and drugs, respectively.

2.4. *Classification*

The next step was to train a supervised machine learning classifier to recognize interacting drug pairs based on the textual features of their connecting paths. We randomly sampled 5000 drug pairs from a list of known interacting pairs provided by DrugBank,[17] then selected 5000 additional drug pairs randomly from the drug lexicon, ensuring that none of them were on DrugBank's list of interactions[c]. For each of the 5000 pairs in our positive and negative training sets, we found all of the paths between the two drugs in the pair that took the form shown in Figure 4 and recorded the features observed along the paths. Each path between an interacting drug pair became a positive training example, and each path between a noninteracting drug pair became a negative training example.

We used a random forest,[18] specifically the implementation found in the R library `randomForest`, to perform the final classification for all of the drug pairs in our training set. The random forest is an ensemble method in which many uncorrelated decision trees "vote" to classify data points; it outperformed both logistic regression and a support-vector machine classifier used in the early stages of this project. Each tree in the random forest uses only a subset of the features for classification, which ensures that votes from different trees are uncorrelated. We found that the overall classification error stabilized when approximately 200 trees were included in the forest.

2.5. *Performance Evaluation*

The standard metric of performance for the random forest is the out-of-bag (OOB) estimate of the error, which is similar to leave-one-out cross-validation.[18] Each tree in the random forest is constructed using only about 2/3 of the available training data; the rest of the data points are referred to as the "out of bag" data for that tree. Thus it is possible to build the entire forest, then reclassify each training example using only those trees for which it was OOB. The generally-accepted rule is to use a voting cutoff of 50% to classify a training point as positive; this means that for the random forest to assign the label "interacting" to a path, 50% or more of the trees in the forest had to classify that path as interacting. We used the standard OOB estimate of the error to evaluate the random forest's performance on our training data.

One interesting feature of the random forest is that it provides a natural measure of its classification certainty for each training example - namely, the fraction of trees that voted "interact" for that example. By ranking the paths for a particular drug pair based on the number of "yes" votes each received from the random forest, we can determine which path(s)

[c]DrugBank obtains its list of drug interactions from a variety of sources, including the Physician's Desk Reference, e-Therapeutics, MedicinesComplete, Epocrates RX, and Drugs.com (which in turn uses Cerner Multum).

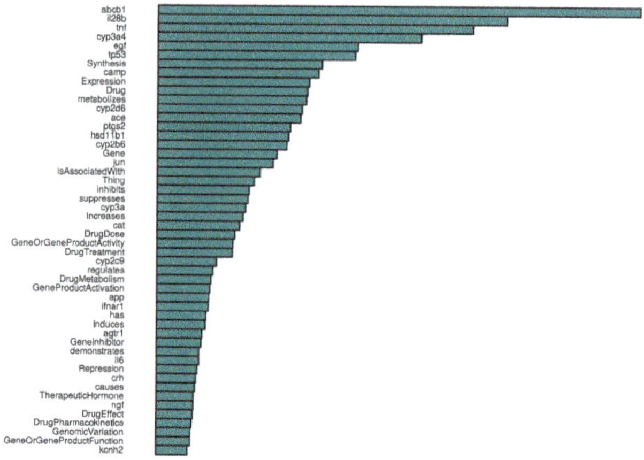

Fig. 5. The 50 most important features to the random forest classifier, ordered according to a permutation metric;[18] the numeric values of importance are not as informative as the relative sizes of the bars.

represent the most likely mechanism(s) of interaction for that pair.

3. Results

3.1. *Feature Extraction*

A total of 1806 entities were represented in our network: 1061 drugs, 532 genes, 172 context terms, and 41 relations. There were 172,271 negative training examples (paths between 5000 noninteracting drug pairs) and 182,534 positive training examples (paths between 5000 interacting drug pairs) in our training set, all of which had the form shown in Figure 4.

The random forest uses a permutation method to provide an estimate of which textual features were most important to the classification process; the 50 most important features are shown in Figure 5. Among the most important features are the genes *ABCB1, IL28B, TNF, CYP3A4, EGF, CAMP*, and *CYP2D6*, the context terms *Synthesis, Expression, DrugDose, GeneOrGeneProductActivity, DrugTreatment, DrugMetabolism, GeneProductActivation, GeneInhibitor, Repression*, and *DrugEffect*, and the relations *metabolizes, isAssociatedWith, inhibits, suppresses, increases, regulates*, and *induces*.

Three context terms on this list appear strange at first glance: *Drug, Gene*, and *Thing*. Context is normalized to the term *Drug* or *Gene* if the drug/gene seed is itself the subject or object of the verb or nominalized verb in the sentence, as in the example sentence *CYP2C9 metabolizes warfarin*. In this sentence, the gene CYP2C9 would be normalized to "CYP2C9 Gene" and the drug warfarin would be normalized to "warfarin Drug". Context is normalized to the term *Thing* if the real context is some property of the drug or gene that cannot be otherwise normalized. For example, in one sentence, the authors used the term "polymorphism" incorrectly as a modifier of a drug name, referring to "polymorphisms in (drug name) drug dose...". Because the drug context in that sentence was *polymorphisms* but the seed was a drug name and not a gene name, *polymorphisms* was not found in the ontology among the acceptable context terms for the drug and the context was normalized to *Thing*. One can therefore think of *Thing* as a marker for cases where normalization of a context term was not possible (using the current version of the ontology), but normalization of the relation proceeded normally.

Table 1. The final contingency table for the random forest classifier.

True Class	Random Forest Classification 0	1	Class-wide Error
No Interaction	135,842	36,429	0.211
Interaction	36,915	145,619	0.202

3.2. *Classification and Performance Evaluation*

The final contingency table for the random forest classifier is shown in Table 2. The random forest correctly assigned 281, 461 out of 354, 805 training paths (79.3%; 135, 842 non-interacting and 145, 619 interacting paths) to the correct class. It said 36, 429 paths represented interactions when the drug pair involved did not appear on the list from DrugBank (false positives), and claimed that 36, 915 paths did not represent interactions when in fact the drug pair did appear on the list from DrugBank (false negatives).

We can get a sense of the significance of this result by considering what would happen if we simply flipped a coin to assign the label "interacting" or "noninteracting" to each path. Roughly 50% of the paths in our training set corresponded to interacting drug pairs, and the other 50% to noninteracting drug pairs. Therefore, if we assigned the labels "interacting" and "noninteracting" entirely at random, we would expect to correctly classify about 50% of paths (with the false positive error rate approximately equal to the false negative error rate). Our method thus represents an improvement in accuracy of nearly 30% over simple guessing.

3.3. *Predicting Mechanisms of Interaction*

In addition to classifying interacting and noninteracting drug pairs with nearly 80% accuracy, our method provides valuable insight into the possible mechanisms by which drugs interact. By choosing a path from one drug to the other through a particular gene, we obtain one potential mechanism for how the two drugs could interact. For example, Figure 6 shows a selection of the highest-ranking paths for a known interacting drug pair: verapamil and atorvastatin. Table 2 shows the Medline sentences corresponding to the edges that comprise these paths. All of these paths received at least 90% "yes" votes from the random forest.

Most of the edges connecting verapamil and ABCB1 in Figure 6 seem to indicate that verapamil inhibits the activity of ABCB1. The edges connecting atorvastatin and ABCB1 indicate that atorvastatin upregulates the production of P-glycoprotein, the protein product of ABCB1. The two drugs' effects on ABCB1 therefore interfere with each other. Following an-

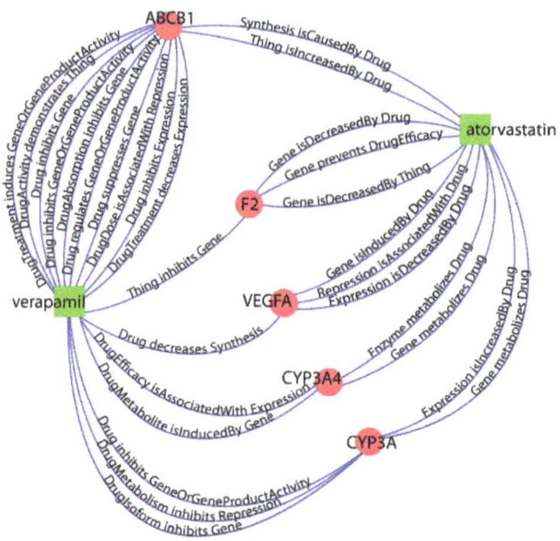

Fig. 6. A selection of the highest-ranking paths between verapamil and atorvastatin. The total number of connecting paths between verapamil and atorvastatin in the network was 293. All of the paths shown here received more than 90% "interact" votes from the random forest.

other path, this time through the gene CYP3A4, we see that CYP3A4 induces the breakdown of verapamil into its metabolites, specifically by N-dealkylation and N-demethylation of the drug. Since CYP3A4 is a major metabolizing enzyme for atorvastatin, we might expect that coadministration of the two drugs could lead to heightened levels of one or both of them in the body, leading to toxicity. These represent two different possible mechanisms of interaction.

4. Discussion

4.1. *Predicting Interactions and Mechanisms*

These suggested mechanisms are useful because they provide summaries of what the scientific community knows about pharmacogenomically-mediated interactions between drug pairs of interest. The drug-gene relationships that form the basis of these mechanisms are all existing knowledge; however, our method provides a novel way to connect disparate facts from across the biomedical literature to provide mechanistic explanations of drug-drug interactions.[19] In the case of drug pairs that are already known to interact, using this approach provides a list of potential mechanisms of interaction, which may help us uncover new mechanisms that are not yet part of common medical knowledge. By looking at known interacting drug pairs with similar mechanisms of interaction, we can also begin to predict what the phenotypic effects of our newly-predicted interactions might be.

Once the random forest has been trained on a set of known interacting drug pairs, it can be used to predict whether any other drug pair will interact. In particular, it can be applied to a novel test set that does not include drugs from the original training set. Provided a drug pair is connected by at least one path of the form shown in Figure 4 in the semantic network,

Table 2. The raw sentences from Medline abstracts that correspond to the edges shown in Figure 6. Each path between verapamil and atorvastatin consists of two edges (i.e. two sentences).

PMID	Normalized Relation	Sentence
Relationships involving f2 (thrombin)		
2611956	verapamil Thing inhibits Gene f2	Ilexonin A and verapamil markedly inhibited the thrombin induced Ca2+ influx.
12921859	atorvastatin DrugEfficacy prevents Gene f2	In addition, thrombin induced NF-kappaB translocation and membrane translocation of RhoA in smooth muscle cells which were both prevented by pre-treatment of the cells by atorvastatin.
12921859	atorvastatin Drug decreases Gene f2	How atorvastatin could limit the pro-inflammatory response to thrombin was studied in cultured rat aortic smooth muscle cells.
15792677	atorvastatin Drug decreases Thing f2	Atorvastatin reduces thrombin generation after percutaneous coronary intervention independent of soluble tissue factor.
Relationships involving ABCB1 (P-glycoprotein, MDR1)		
16996216	atorvastatin Drug causes Synthesis ABCB1	Atorvastatin at 10 and 20 microM up-regulated ABCB1 expression resulting in a significant 1.4-fold increase of the protein levels.
16996216	atorvastatin Drug increases Thing ABCB1	Treatment of HepG2 cells with 20 microM atorvastatin caused a 60% reduction on mRNA expression (p<0.05) and a 41% decrease in ABCB1-mediated efflux of Rhodamine123 (p<0.01) by flow cytometry.
9607955	verapamil DrugTreatment induces GeneOrGeneProductActivity ABCB1	Previous drug exposure of the cells showed that verapamil, celiprolol, and vinblastine induced the P-gp expression, while metkephamid (MKA) decreased the P-gp expression level as compared to the control.
9636053	verapamil DrugActivity demonstrates Thing ABCB1	P-gp proteoliposomes from P. pastoris showed a strong verapamil- and valinomycin-stimulated ATPase activity, with characteristics (KM, Vmax) similar to those measured in mammalian cells.
9535788	verapamil Drug inhibits Gene ABCB1	In addition, the DNA-damaging agent was found to enhance in a dose-dependent manner cellular efflux of the P-gp substrate rhodamine 123, which was inhibited by the P-gp inhibitor verapamil, thus providing evidence that exposure to MMS led to increased P-gp-related drug transport in rat liver cells.
7769842	verapamil Drug inhibits GeneOrGeneProductActivity ABCB1	When P-gp function was assessed by Rhodamine 123 (Rh123) efflux kinetics, we found that only KG1a and KG1 cells, which have an early (immature) CD34+ CD33- CD38- phenotype, and to a lesser extent TF1, with an intermediate (CD34+ CD33+ CD38+) phenotype, displayed significant P-gp activity which could be inhibited by both verapamil and SDZ PSC 833.
16457995	verapamil DrugAbsorption inhibits Gene ABCB1	While cyclosporine and verapamil significantly increased the absorption of methylprednisolone and vinblastine through potent inhibition of intestinal P-gp, tacrolimus failed to achieve this.
17936633	verapamil Drug regulates GeneOrGeneProductActivity ABCB1	The results displayed that only compound 3c was P-gp inhibitor as Elacridar, while compound 3a and reference compounds Cyclosporin A and Verapamil modulated P-gp activity saturating the efflux pump as substrates.
16260035	verapamil Drug suppresses Gene ABCB1	Depsipeptide-resistant KU812 cells expressed P-glycoprotein (P-gp) and their resistance was abolished by co-treatment with verapamil.
15257901	verapamil DrugDose isAssociatedWith Repression ABCB1	DL-PPMP and verapamil were found to inhibit MDR1 gene expression in KBV(200) cells at the mRNA level, and complete inhibition occurred after a 48-hour DL-PPMP treatment at 25 micromol/L.
15257901	verapamil Drug inhibits Expression ABCB1	The inhibition of GCS and mdr1 gene expressions is positively correlated with the concentrations of DL-PPMP and verapamil, which can reverse MDR by inhibiting synthesis of GCS and mdr1 gene, indicating the positive correlation between the expression of GCS gene and MDR in KBV(200) cells.
7749215	verapamil DrugTreatment decreases Expression ABCB1	The level of mdr1 mRNAs is decreased in the presence of verapamil (with a maximum effect obtained at the 24th hour), which suggests that the mechanism of action of verapamil is transcriptional and/or post-transcriptional.
Relationships involving VEGFA		
14615256	verapamil Drug decreases Synthesis VEGFA	Verapamil (100 microM) decreased IL-6 and VEGF production (P¡0.03 and P¡0.005, respectively) in central keloid fibroblasts cultures at 72 h.
16701707	atorvastatin Drug induces Gene VEGFA	We observed that atorvastatin significantly stimulated VEGF release in a dose-dependent manner.
17389519	atorvastatin Drug isAssociatedWith Repression VEGFA	Atorvastatin effectively inhibited laser-induced CNV in mice and was associated with downregulation of CCL2/MCP-1 and VEGF and reduced macrophage infiltration into the RPE/choroid.
12084593	atorvastatin Drug decreases Expression VEGFA	Atorvastatin therapy reduced VEGF plasma levels in CAD patients (from 31.1 +/- 6.1 to 19.0 +/- 3.6 pg/ml; p ¡ 0.05).
Relationships involving CYP3A4		
15001968	verapamil DrugEfficacy isAssociatedWith Expression CYP3A4	Values for the maximum rate of metabolism (V(max)) of verapamil N-dealkylation (formation of D-617) and N-demethylation (formation of norverapamil) activities correlated with the CYP3A4 protein content in both organs.
11907487	CYP3A4 Gene induces DrugMetabolite verapamil	Consistent with expression data, formation of verapamil metabolites catalyzed by CYP3A4 and CYP2C was shown.
11005703	CYP3A4 Enzyme metabolizes Drug atorvastatin	Atorvastatin, cerivastatin, lovastatin and simvastatin are predominantly metabolised by the CYP3A4 isozyme.
11061579	CYP3A4 Gene metabolizes Drug atorvastatin	Atorvastatin is metabolized solely by CYP3A4, and pravastatin metabolism is not well defined.
Relationships involving CYP3A		
16513446	verapamil Drug inhibits GeneOrGeneProductActivity CYP3A	Verapamil inhibited CYP3A activity, with a maximum effect occurring within 10 days.
16013069	verapamil DrugMetabolism inhibits Repression CYP3A	The above data suggested that the metabolism of verapamil and the formation of norverapamil was inhibited by naringin possibly by inhibition of CYP3A in rabbits.
14744949	verapamil DrugIsoform inhibits Gene CYP3A	The present study showed that verapamil enantiomers and their major metabolites [norverapamil and N-desalkylverapamil (D617)] inhibited CYP3A in a time- and concentration-dependent manner by using pooled human liver microsomes and the cDNA-expressed CYP3A4 (+b5).
12433810	atorvastatin Drug increases Expression CYP3A	Treatment of 2- to 3-day-old human hepatocyte cultures with 3 x 10(-5) M lovastatin, simvastatin, fluvastatin, or atorvastatin for 24 h increased the amounts of CYP2B6 and CYP3A mRNA by an average of 3.8- to 9.2-fold and 24- to 36-fold, respectively.
16258024	CYP3A Gene metabolizes Drug atorvastatin	Atorvastatin (ATV) is primarily metabolized by CYP3A in the liver to form two active hydroxy metabolites.

the random forest can vote on each connecting path and rank it based on the probability that it represents a mechanism of interaction. This provides us with a powerful tool for predicting mechanisms of interaction that are not yet known. In the future, we hope to use the trained random forest to predict the most likely mechanisms of interaction between drug pairs that are often prescribed together but whose interaction status is not yet known.

4.2. *Study Limitations*

There are several limitations to the present approach that we hope to address in subsequent iterations of this work. One major limitation is that we only searched Medline for 731 known pharmacodynamic and pharmacokinetic genes, many of which were liver cytochromes and other enzymes known for their involvement in drug metabolism[d]. While interactions involving these drugs are interesting, most are already known, and we also tend to miss more specific interactions, such as drug pairs that share the same pharmacologic target. In the future, we plan to expand our data set to encompass a much wider variety of genes - there are $26,216$ genes in the full lexicon from PharmGKB - but the increase in computational time required to search for $26,216 \times 2,910 = 76,288,560$ drug-gene pairs is substantial.

On a similar note, because we included only pharmacokinetic and pharmacodynamic genes in our analysis, we were unable to capture physical or chemical interactions that were not the result of two drugs interacting directly with the same gene. Examples of such missed interactions might include a drug that increases the pH of the stomach, reducing absorption of another drug, or two drugs that have similar, relatively nonspecific phenotypic effects (such as reducing inflammatory responses throughout the body). We also miss interactions in which two drugs interact with components of the same metabolic pathway but not the same gene, or those in which one drug interacts with a transcription factor that controls the activity of an enzyme responsible for metabolizing another drug. All of these are valid interactions that could be captured if we expanded the semantic network to include gene-gene interactions, as well as interactions of both drugs and genes with certain disease states and phenotypes, all of which we plan to do in the future.

A final limitation of our model is its inability to resolve anaphoras. An example of an anaphora is a two-sentence combination like *CYP2C9 is a gene. It metabolizes warfarin.* Our model would not pick up the relationship between CYP2C9 and warfarin because the two entities are found in separate sentences. As we refine this work, we would like to find ways to resolve anaphoras, perhaps by considering pairs of entities that are mentioned in the same abstract, not just the same sentence.

5. Conclusion

We have described a method for predicting and explaining drug-drug interactions based on automated extraction of relevant pharmacogenomic facts from the biomedical literature. The method classifies known drug-drug interactions with nearly 80% accuracy using only textual

[d]Readers interested in drug interactions mediated by this class of genes are encouraged to visit http://medicine.iupui.edu/clinpharm/ddis/, a valuable source of information on DDIs mediated by liver cytochromes.

features from descriptions of drug-gene relationships, and provides reasonable mechanistic explanations for its classification decisions. Its success opens many doors to the future use of similar techniques in text mining, perhaps to predict gene-gene interactions, uncover interactions of drugs and genes with diseases, and generate testable hypotheses about the relationships between drugs, genes, and phenotypes.

6. Acknowledgements

This work was supported by the National Institutes of Health (grants LM05652 and GM61374) and the National Library of Medicine (contract HHSN276201000025C).

References

1. http://www.statehealthfacts.org/ Accessed Wednesday, June 8, 2011. This statistic includes both new prescriptions and refills.
2. Katzung BG, Masters SB, Trevor AJ (2009) *Basic and Clinical Pharmacology* (McGraw-Hill: New York). Chapter 2: Drug Receptors and Pharmacodynamics.
3. Li A, Ed. (1997) *Drug-Drug interactions: scientific and regulatory perspectives.* (Academic Press: San Diego).
4. Cohen AM, Hersh WR (2005) A survey of current work in biomedical text mining. *Brief Bioinform* 6:57-71.
5. Garten Y, Coulet A, Altman RB (2010) Recent progress in automatically extracting information from the pharmacogenomic literature. *Pharmacogenomics* 11:1467-89.
6. Plake C, Schroeder M (2011) Computational polypharmacology with text mining and ontologies. *Curr Pharm Biotechnol* 12:449-57.
7. Tari L, Anwar S, Liang S, Cai J, Baral C (2010) Discovering drug-drug interactions: a text-mining and reasoning approach based on properties of drug metabolism. *Bioinformatics* 26:i547-53.
8. Chen H, Sharp BM (2004) Content-rich biological network constructed by mining PubMed abstracts. *BMC Bioinformatics* 5:147.
9. Plake C, Shiemann T, Pankalla M, Hakenberg J, Leser U (2006) AliBaba: PubMed as a graph. *Bioinformatics* 22:2444-5.
10. Coulet A, Altman RB, Musen MA, Shah N (2010) Integrating heterogeneous relationships extracted from natural language sentences. In *Proc Bio-ontologies SIG, ISMB 2010* Boston, USA.
11. Coulet A, et al. (2010) Using text to build semantic networks for pharmacogenomics. *Journal of Biomedical Informatics* 43(6):1009-19.
12. Klein TE, et al. (2001) Integrating Genotype and Phenotype Information: An Overview of the PharmGKB Project. *The Pharmacogenomics Journal* 1:167-170.
13. The Bio-X^2 cluster at Stanford University was funded by the National Science Foundation, NSF award CNS-0619926.
14. Klein D, Manning CD (2003) Accurate Unlexicalized Parsing. *Proceedings of the 41st Meeting of the Association for Computational Linguistics* 423-430.
15. De Marneffe MC, MacCartney B, Manning CD (2006) Generating Typed Dependency Parses from Phrase Structure Parses. In *LREC 2006.*
16. Russell SJ, Norvig P (2010) *Artificial intelligence: A modern approach.* (Prentice Hall: Upper Saddle River).
17. Wishart DS, et al. (2008) DrugBank: a knowledgebase for drugs, drug actions and drug targets. *Nucleic Acids Res* 36:D901-6.
18. Hastie T, Tibshirani RJ, Friedman J (2009) *The Elements of Statistical Learning: Data Mining, Inference, and Prediction* (Springer: New York). Chapter 16: Random Forests.
19. Other authors have already pointed out that by finding new ways to connect seemingly-unrelated facts from throughout the scientific literature, we can often generate interesting, and novel, hypotheses. For an example, see Swanson DR (1990) Medical literature as a potential source of new knowledge. *Bull Med Libr Assoc* 78:29-37.

RANKING GENE-DRUG RELATIONSHIPS IN BIOMEDICAL LITERATURE USING LATENT DIRICHLET ALLOCATION

YONGHUI WU*

Department of Biomedical Informatics, Vanderbilt University
Nashville, TN 37203, USA
**E-mail: yonghui.wu@Vanderbilt.Edu*

MEI LIU

Department of Biomedical Informatics, Vanderbilt University
Nashville, TN 37232, USA
E-mail: mei.liu@Vanderbilt.Edu

W. JIM ZHENG

Department of Biochemistry, Medical University of South Carolina
Charleston, SC 29425, USA
E-mail: zhengw@musc.edu

ZHONGMING ZHAO

Department of Biomedical Informatics, Vanderbilt University
Nashville, TN 37232, USA
E-mail: zhongming.zhao@Vanderbilt.Edu

HUA XU

Department of Biomedical Informatics, Vanderbilt University
Nashville, TN 37232, USA
E-mail: hua.xu@Vanderbilt.Edu

Drug responses vary greatly among individuals due to human genetic variations, which is known as pharmacogenomics (PGx). Much of the PGx knowledge has been embedded in biomedical literature and there is a growing interest to develop text mining approaches to extract such knowledge. In this paper, we present a study to rank candidate gene-drug relations using Latent Dirichlet Allocation (LDA) model. Our approach consists of three steps: 1) recognize gene and drug entities in MEDLINE abstracts; 2) extract candidate gene-drug pairs based on different levels of co-occurrence, including abstract level, sentence level, and phrase level; and 3) rank candidate gene-drug pairs using multiple different methods including term frequency, Chi-square test, Mutual Information (MI), a reported Kullback-Leibler (KL) distance based on topics derived from LDA (LDA-KL), and a newly defined probabilistic KL distance based on LDA (LDA-PKL). We systematically evaluated these methods by using a gold standard data set of gene-drug relations derived from PharmGKB. Our results showed that the proposed LDA-PKL method achieved better Mean Average Precision (MAP) than any other methods, suggesting its promising uses for ranking and detecting PGx relations.

Keywords: Gene-drug Relation; Latent Dirichlet Allocation; Ranking; Pharmacogenomics.

1. Introduction

There exists striking variability in individual responses to drug therapy as exemplified by severe adverse drug reactions which have been ranked as among the commonest causes of death in hospitalized patients.[1,2] Pharmacogenomic (PGx) research is to impact this problem by linking the inherited differences to variable drug responses. However, much of our current understanding in pharmacogenomics has been dispersed across numerous journals. Hence, it is extremely important, as well as, immediately needed, to extract important facts across publications into a comprehensive knowledge base. The Pharmacogenomics Knowledge Base (PharmGKB),[3] sponsored by National Institutes of Health (NIH), is such a database annotated manually by a team of curators. Compared to the vast amount of relations implicitly exist in published scientific literatures, the number of relations annotated in the PharmGKB is still limited. Thus, there is a great interest in developing automated methods to accurately detect PGx relations, such as gene-drug relations from biomedical literature.

In this paper, we present a study on detecting and ranking gene-drug relations in MEDLINE abstracts by using LDA model. Starting with MEDLINE abstracts, we first recognized gene and drug entities using existing Natural Language Processing (NLP) tools such as MetaMap. Then we extracted candidate gene-drug pairs based on different levels of co-occurrence, including abstract level (text from both titles and abstracts), sentence level, and phrase level. In order to find the most related gene-drug relations, we finally ranked candidate gene-drug pairs using different methods including three baseline methods: frequency, Chi-square test, and Mutual Information (MI), and two LDA-based methods: a previously reported Kullback-Leibler (KL) distance based on topics derived from LDA LDA-KL, and a newly defined probabilistic KL distance based on LDA LDA-PKL. The evaluation using a manually annotated data set from PharmGKB indicated that our LDA-PKL method outperformed others. To our best knowledge, this is the first attempt that applied LDA models to rank gene-drug relations for building PGx knowledge bases from biomedical literature.

2. Background

Different methods have been proposed to detect pharmacogenomic entity relations from MEDLINE abstracts, including co-occurrence based, rule based and machine learning based methods.[4] The co-occurrence method is a commonly used method in literature mining, building on the assumption that two entities co-occurring in the scope of the same abstract, single sentence or single phrase,[5] are likely to be related. On the other hand, the rule based methods extract relations using predefined or automatically derived patterns. Machine learning based methods train a classifier on a set of annotated relations to classify the candidate relations. Since entities co-occurring in a specific scope of text do not always define a relation, the simple co-occurrence method would generate false positives, which makes it difficult to validate the correct relations from the large numbers of candidates to integrate them into knowledge bases. Thus, other methods such as syntactic rule based methods (rule based methods using syntactic patterns) and machine leaning methods were proposed to further analyze the co-occurred relations to reduce the false positives. The syntactic rule based methods assume that the entities co-occurring in certain syntactic patterns (such as subj and obj relations) are more likely to be related.

A number of systems have been developed to extract relations among gene, drug, and diseases. Most of them, such as XplorMed, iHOP and CoPub Mapper, tried to extract relations between gene, drug and disease from abstracts of biomedical articles.[6–9] Jenssen et al.[9] used a co-occurrence based method to build a gene-gene relationship network by extracting all the gene-gene pairs co-occurring in the same MEDLINE abstract. The relations were then weighted by the number of times the two genes co-occurred in the same abstract. Their method achieved a precision of 60% and recall of 50%. Pharmspresso[10] and Textpresso[11] investigated methods to extract relationships from the full text articles, rather than abstracts or sentences. Tari et al.[12] applied a rule based method using loose patterns defined by a wildcard operator ("_") to describe syntactic relations between two entities. Recent rule-based methods, such as RelEx[13] and OpenDMAP[14] used dependency parse trees to detect protein-protein interactions. Coulet et al.[15] presented their research on detecting PGx relations using dependency graphs generated by the Stanford Parser.[16] Through manual evaluation of the randomly selected 220 raw relations, they reported precisions within the range of 70-87.7%. These automatically detected relations were normalized into a knowledge base to guide the annotation of PharmGKB.

There are also studies that have focused on detecting gene-drug relations, which is an important category of pharmacogenomic relations. For example, Chang et al.[17] used a machine learning method called Maximum Entropy to classify the gene-drug relations detected by a co-occurrence based method into five categories defined in PharmGKB. Garten et al.[18] created a gene-drug network from the sentence level co-occurrence over full text using Pharmspresso. Then, a logistic regression classifier was trained using these automatically extracted gene-drug relations and a group of manually curated relations. The evaluation result showed that the classifier trained from automatically detected relations was as good as, and sometimes even better than, the classifier trained from manually curated knowledge bases.

Semantic is another type of useful information to analyze co-occurred entities. The semantic meanings implicitly existed in literatures can be automatically uncovered using probabilistic models. Latent Dirichlet Allocation (LDA)[19] is a widely used model to identify semantic topics from large document collections. Wang et al.[20] described a method that used a variation of LDA model, named Bio-LDA model, to detect entity relationships from MEDLINE abstracts. They reported that the LDA models could detect the relationships between two bio-terms even if they did not co-occur in the same text.

In this study, we applied LDA model to rank candidate gene-drug relations. Instead of directly applying LDA to relation detection, we used it as a ranking method to prioritize candidate gene-drug relations derived from co-occurrence methods based on different levels of scope of text. Ranking is an effective way to help researchers to focus on the most related relations from large numbers of candidates mixed with false positives. We compared the LDA-based ranking methods with traditional ranking algorithms such as frequency, Chi-Square and Mutual Information, and demonstrated its usability for ranking PGx relations.

3. Methods

In this study, 249,181 MEDLINE abstracts listed in the gene2pubmed file from NCBI (National Center of Biotechnology Information) were used as the study corpus. Our approach consists of three steps: 1) recognize gene and drug entities in MEDLINE abstracts using natural language processing (NLP) tools; 2) extract candidate gene-drug pairs based on different levels of co-occurrence, including abstract level, sentence level, and phrase level; and 3) rank candidate gene-drug pairs using five different methods including frequency, Chi-square test, MI, LDA-KL, and LDA-PKL. These five ranking methods were then systematically evaluated using the Mean Average Precision (MAP) on a manually annotated data set derived from PharmGKB. Figure 1 shows an overview of the study design.

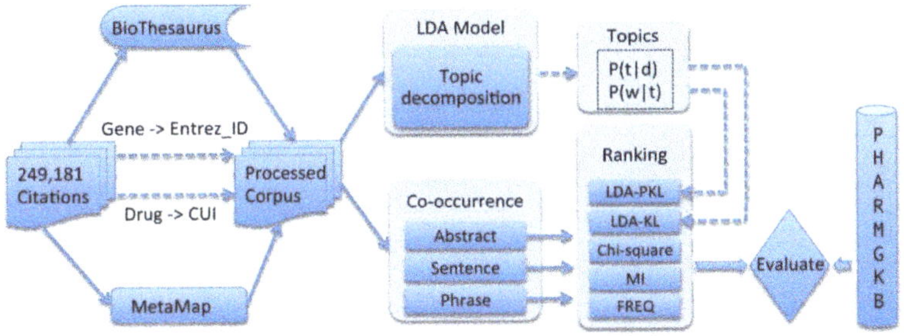

Fig. 1. An overview of study design.

3.1. *Data set and gene/drug recognition*

In this study, we started with MEDLINE articles listed in the gene2pubmed file from NCBI, which ensures that each article in our corpus mentions at least one gene. Articles that mention many genes are often about high-throughput technologies; therefore they were removed from our study, resulting in a corpus containing 249,181 MEDLINE articles.

For each article in the corpus, its title and abstract were downloaded and processed by two NLP tools: the MetaMap program[21] and a gene lookup program based on Biothesaurus.[22] Instead of running MetaMap by ourselves, we downloaded the corpus of *"2011 MetaMapped Medline Baseline Results"*, which is a MEDLINE corpus processed by MetaMap by National Library Medicine (NLM). MetaMap outputs were used to identify drug entities - CUIs (Concept Unique Identifiers from UMLS[23]) from the MetaMap output with semantic types of phsu (Pharmacologic Substance) and antb (Antibiotic) were identified as drugs. One issue with this approach was that very general drug terms such as "drug" and "medicines" were also labeled as drugs. Therefore we manually reviewed top 100 frequent drugs terms from MetaMap outputs and manually removed general drug CUIs (total 15 CUIs).

To recognize genes in abstracts, we developed a gene lookup program using lexicons from Biothesaurus. To ensure the high performance of gene name recognition, we only annotated gene names that were listed in gene2pubmed and were recognized by the lookup program. If a gene name within an article was identified by the lookup program, but it was not annotated for that article according to gene2pubmed, we would ignore that gene name. An example of two processed sentences is shown in Table 1. For sentence 1, the gene entity "abcb1" as well as its alternative names: "mdr1" and "glycoprotein" were mapped to the same Entrez ID "5243". But the ambiguous one, "impact" in sentence 2, was not considered since it could not be matched by Biothesaurus. Using sentence level co-occurrence, the candidate relation (5243, C0012265) is detected from sentence 1.

Table 1. Example for pre-processed data.

Sentences from MEDLINE abstract 17652833: 1. Effect of ABCB1 (MDR1) 3435C >T and 2677G >A,T polymorphisms and P-glycoprotein inhibitors on salivary digoxin secretion in congestive heart failure patients. 2. Evaluation of the impact ...
1. effect/noun of/prep <GENE><NAME>abcb1</NAME><CUI>5243</CUI><ST>bioth</ST> </GENE> <GENE><NAME>mdr1</NAME><CUI>C0376622,5243</CUI><ST>gngm,bioth</ST> </GENE> 3435c/noun t/noun and/conj 2677g/noun a-t/noun polymorphisms/noun and/conj p/noun <GENE><NAME>glycoprotein</NAME><CUI>5243</CUI><ST>bioth</ST></GENE> inhibitors /noun on/prep salivary/adj <DRUG><NAME>digoxin</NAME><CUI>C0012265</CUI><ST>carb ,phsu ,strd</ST></DRUG> secretion/noun in/prep congestive-heart-failure patients/noun ./punc 2. evaluation/noun of/prep the/det <GENE><NAME>impact</NAME><CUI>C1825598</CUI><ST> gngm</ST></GENE>

In order to evaluate the ranking of gene-drug relations using existing annotations from PharmGKB, we further filtered extracted gene and drug entities to those that appeared in PharmGKB. For genes, PharmGKB and our gene lookup program both used Entrez Gene IDs; so it was straightforward to limit extracted genes to those in PharmGKB. However, it is more challenging to map drug entities – PharmGKB uses its own IDs for drug entities and there is no direct mapping between UMLS CUIs and PharmGKB Drug IDs. We employed KnowledgeMap,[24] a general UMLS concept extraction system, to bridge the gap between UMLS drug CUIs from MetaMap and ParmGKB drug IDs. Each drug term in PharmGKB was processed by KnowledgeMap and its corresponding UMLS CUIs were identified. There were a few issues associated with the automated mapping method. First, not all drug terms in PharmGKB were mapped to UMLS CUIs by KnowledgeMap - among the 3,004 drug entities in the PharmGKB database, 2,474 of them were mapped to CUIs. Second, multiple PharmGKB drug terms might be mapped to one UMLS CUI, because of different levels of granularity between PharmGKB and UMLS drug names. For example, three PharmGKB drug terms "PA164712641 : Corticosteroids and mydriatics in combination ", "PA164712644: Corticosteroids, Combinations With Antiseptics", and "PA164712645 : Corticosteroids, Dermatological Preparations", were mapped to one UMLS CUI "C0001617 Corticosteroid". As the focus of this study was the ranking methods instead of entity recognition, we removed un-mapped drug terms in PharmGKB and only kept mapped UMLS CUIs in the final list of drug entities for our evaluation. After this, a total of 9,700 distinct gene entities and 1,115 distinct drug entities were recognized from the corpus and used for our evaluation.

3.2. *Detecting candidate gene-drug relation pairs*

If a gene and a drug entity occur in the same scope of text, we define them as a candidate gene-drug pair. In this study, three different levels of scope were investigated: 1) abstract level, where a gene and a drug occur within the same abstract (including title); 2) sentence level, where a gene and a drug occur within the same sentence of an article (sentence boundary was determined by MetaMap program); 3) phrase level, where a gene and a drug occur within the same phrase. A phrase was defined as a fragment between any two punctuations, within one sentence.[5]

3.3. *Ranking candidate gene-drug relations*

Five different ranking methods were used to rank candidate gene-drug pairs derived from different levels of co-occurrence patterns. Three of them were baseline methods: frequency (FREQ), Chi-square test, and Mutual Information (MI). Two of them were LDA-based methods: LDA-KL (a previously reported measure) and LDA-PKL (a newly defined measure in this study). Details of each ranking algorithm were listed below.

(1) FREQ - the frequency based method simply ranked all the relations according to the number of times that a gene and a drug entity co-occurred in the same scope of text.
(2) The Chi-square test score and MI score for a gene-drug relation (w_g, w_d) were calculated as shown in equation 1 and equation 2, respectively.

$$\chi^2 = \frac{N(O_{11}O_{22} - O_{12}O_{21})^2}{(O_{11}+O_{12})(O_{11}+O_{21})(O_{12}+O_{22})(O_{21}+O_{22})} \tag{1}$$

$$MI(w_g, w_d) = log\frac{D(w_g, w_d)}{D(w_g)D(w_d)}. \tag{2}$$

Where O_{11} is the number of abstracts w_g and w_d co-occurred; O_{12} is the number of abstracts only w_g appeared; O_{21} is the number of abstracts only w_d appeared; O_{22} is the number of abstracts neither w_g or w_d appeared. $D(w_g, w_d)$ denotes the number of abstracts containing this relation, and $D(w_g)$ denotes all the number of abstracts talking about w_g.
(3) LDA-KL, shown in equation 4, is a score defined using Kullback-Leibler (KL) distance over all the topics derived from the LDA model, which was originally used to detect complex "bio-terms" relations from literature.[20] The relations with a lower LDA-KL score are more likely to be related.

$$KL(w_g, w_d) = \sum_{i=1}^{T} p(w_g|t_i)log(\frac{p(w_g|t_i)}{p(w_d|t_i)}) \tag{3}$$

$$LDA_{KL}(w_g, w_d) = KL(w_g, w_d) + KL(w_d, w_g). \tag{4}$$

(4) LDA-PKL - a new distance measure by combining the KL distance over different topics and the posterior probabilities of topics over MEDLINE abstracts, which is shown in equation 6. Where $p(w_g|t_j)$ is the probability that gene wg appeared in topic j; T is the total number of topics automatically determined by the LDA model (T = 100 in this

study); $p(t_j|d_i)$ is the posterior probability of topic j to MEDLINE abstract i; D is the set of MEDLINE abstracts containing relation (w_g, w_d).

$$PD(w_g, w_d) = \sum_{i=1}^{D} p(w_g|d_i)p(w_d|d_i)$$
$$= \sum_{i=1}^{D}[(\sum_{j=1}^{T} p(w_g|t_j)p(t_j|d_i)) \times (\sum_{j=1}^{T} p(w_d|t_j)p(t_j|d_i))] \quad (5)$$
$$LDA_{PKL}(w_g, w_d) = \frac{PD(w_g, w_d)}{LDA_{KL}(w_g, w_d)}. \quad (6)$$

In order to rank the relations using LDA latent topics, we run the LDA model with a fixed topic number of 100 on the data set containing 249,181 MEDLINE abstracts using the c-version of LDA by Blei et al.[19] The data set was pre-processed before running. The gene entity and drug entity were mapped to Entrez ID and drug CUI using BioThesaurus and MetaMap, respectively. A list of 570 English stop words[a] was removed from the vocabulary. To further reduce noise, we removed 160,560 words whose frequency were less than three. The final data set contains 162,590 words.

3.4. *Evaluation*

To evaluate the performance of multiple ranking methods, we created a gold standard of gene-drug relations by leveraging annotations from PharmGKB. We downloaded PharmGKB files on 6/20/11 from the official website of PharmGKB. There were 11,607 gene-drug relations, associated with 3,283 MEDLINE articles. The overlap between our corpus (249,191) and PharmGKB corpus (3,283) were 898 MEDLINE articles. These 898 MEDLINE articles were associated with 1,530 gene-drug relations, according to PharmGKB.

If all the gene and drug entities were recognized correctly by our approach, candidate gene-drug relations detected at the abstract level co-occurrence should be able to cover all 1,530-candidate relations from 898 MEDLINE articles (i.e., a recall of 100%). However, there will always be errors in gene and drug entity recognition. Moreover, gene-drug relations annotated in PharmGKB were from review of full text articles; while we used only abstracts. As the purpose of this study was to evaluate relation ranking algorithms instead of entity recognition methods, we limited our evaluation data set to articles where gene/drug entities reported by PharmGKB were correctly identified by our programs. Therefore, if a PharmGKB gene-drug relation from one article contained entities (either gene or drug) that were not recognized by MetaMap and Biothesaurus from that article, we removed the relation from our evaluation. This resulted in 831 gold standard gene-drug relations from 722 MEDLINE abstracts, which served as the final evaluation data set for this study.

The candidate relations detected from the 722 MEDLINE abstracts using different co-occurrence levels were evaluated using the 831 gold standard relations. Precision and recall for detecting candidate gene-drug relations were calculated using following formulas:

[a]The stop word list is available at ftp://ftp.cs.cornell.edu/pub/smart/english.stop

$$Precision = \frac{\text{Number of relations validated by 831 gold standard relations}}{\text{Number of detected relations}} \quad (7)$$

$$Recall = \frac{\text{Number of relations validated by 831 gold standard relations}}{\text{Number of gold standard relations(831)}}. \quad (8)$$

To evaluate the ranking result of different algorithms, we reported the Mean Average Precision (MAP)[25] as well as the precision-recall curve. The MAP score is commonly used as the standard evaluation method for ranked results from information retrieval tasks. Equation 9 shows the equation for MAP where Q is the query set, which is 1 in our case, R_{jk} is the set of ranked relations until you get K true relations, and $\{r_1, \ldots, r_{mj}\}$ is the set of true relation in R_{jk}.

$$MAP(Q) = \frac{1}{|Q|}\sum_{j=1}^{Q}\frac{1}{m_j}\sum_{k=1}^{m_j} Precision(R_{jk}). \quad (9)$$

4. Result

4.1. *Detecting gene-drug relations*

Using different levels of co-occurrence methods, the number of gene-drug relations detected from entire corpus and from the 722 gold standard MEDLINE abstracts were (79,885, 1,707), (43,309, 1,006), and (30,960, 692) for abstract level, sentence level and phrase level, respectively. Table 2 shows the performance of co-occurrence methods at different levels (abstract, sentence, phrase) evaluated on the 831 gold standard relations.

Table 2. Precision and recall for detected gene-drug relations using different levels of co-occurrence.

Method	# of extracted relations	# of extracted and correct relations	# of relations in gold standard	Precision	Recall
Abstract	1,707	831	831	48.6%	100%
Sentence	1,006	589	831	58.5%	70.8%
Clause	692	447	831	64.5%	53.7%

Since the 831 gold standard relations only included the PharmGKB gene-drug relations between gene entity and drug entity that could be recognized using MetaMap and Biothesaurus from 722 abstracts, the abstract level co-occurrence yielded a precision of 48.6% and recall of 100%. And as expected, lower recall but higher precision was observed when we strengthen the constraint on the context level. For instance, the most strict method at the phrase level achieved a higher precision (64.5%) at the cost of lower recall (53.7%).

4.2. *Ranking relations*

Figure 2 shows the MAP score and precision-recall curve for all the ranking algorithms over different levels of co-occurrence (Fig.2a and Fig.2b for abstract level, Fig.2c and Fig.2d for sentence level, Fig.2e and Fig.2f for phrase level). Among all three levels of co-occurrence based methods, our proposed LDA-PKL ranking method outperformed all others in terms of MAP score after the curves became steady. Similar results were also observed on the precision-recall curves.

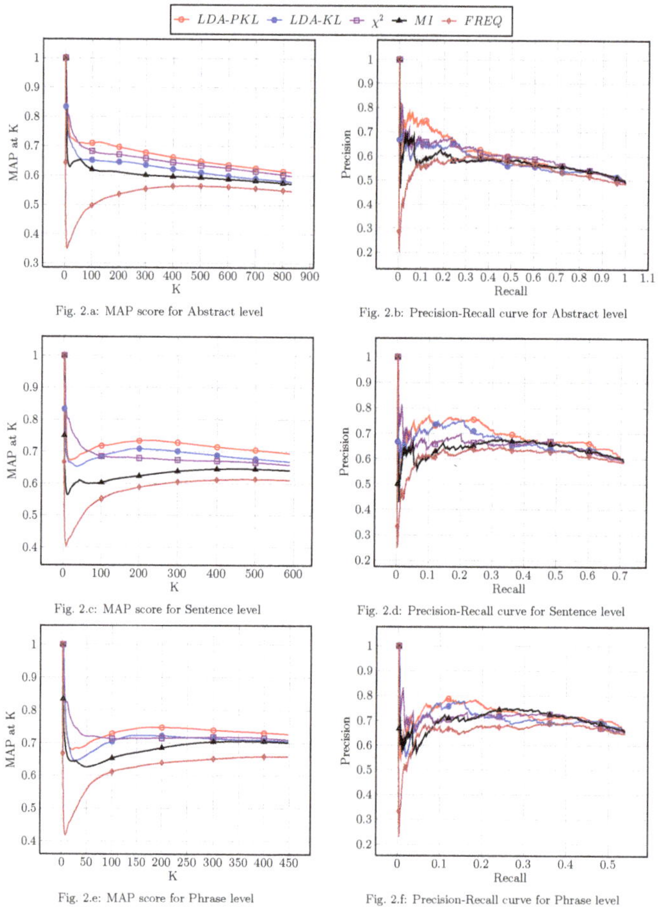

Fig. 2. MAP score and precision-recall curve for all ranking algorithms.

5. Discussion

We conducted a study on ranking candidate gene-drug relations that were extracted from MEDLINE abstracts based on simple co-occurrence methods using LDA model. Different levels of co-occurrence were used to extract candidate gene-drug relations from 249,181 MEDLINE abstracts. These candidate relations derived from literature were evaluated using a gold standard set of gene-drug relations derived from manually curated PharmGKB knowledge base. We reported the precision and recall for gene-drug relation detection at different co-occurrence levels: 48.6% and 100% for abstracts, 58.5% and 70.8% for sentences, and 64.5% and 53.7% for phrases. These results were lower than previously reported results on co-occurrence based relation detection methods for general biomedical terms by Ding et al.[5] where they had 57.1% and 100% for abstracts, 63.8% and 84.9% for sentences, and 74.3% and 62.1% for phrases, respectively. This indicated gene/drug entities and their relations were more difficult to extract than general biomedical terms. We analyzed the errors and found that some gene-drug relations were missed because 1) they were not recognized by the gene and drug entity recognition programs; or 2) there were no mapping between UMLS drug CUIs and PharmGKB drug entity IDs using the KnowledgeMap.

Five different ranking methods were used to prioritize the candidate gene-drug relations detected using different levels of co-occurrence. The MAP score of proposed LDA-PKL method outperformed other methods over all the position "K" after curves became steady, which suggests that the semantic topics derived from LDA model could improve the ranking of gene-drug relations. The proposed LDA-PKL method outperformed the LDA-KL method, which only considered the KL distance over all topics, by further weighting the KL distance using the posterior probabilities of topics over MEDLINE abstracts. The Chi-square method also outperformed LDA-KL method on abstract level and part of phrase level. When considering the statistical information from the entire corpus, the Chi-square and MI methods outperformed the FREQ method that simply ranks relations by their co-occurrence frequency. The advantage of LDA-PKL on abstract level and sentence level is better than the phrase level, since the semantic topics used in ranking were derived from the abstract level using LDA model.

Although the proposed LDA-PKL algorithm outperformed other methods on MAP score after the curves became steady, in some cases from precision-recall curves, the Chi-square and MI outperformed LDA-PKL on precision. This suggested that the semantic topics derived from LDA model also introduced some noise when inferring relationships between entities (In LDA model, the PGx entities could co-occur in all topics with a probability, even if they do not co-occur in any abstract).

There are limitations in this initial study. First, better Named Entity Recognition (NER) tools are needed for identifying gene/drug entities. We analyzed the errors and found that some gene-drug relations were missed because the gene and drug entities were not recognized correctly. Although this research is not focused on NER, better gene/drug entity recognition would help detect more relations. Second, we used a fixed number of topics 100 for topic decomposition of the LDA model. It would be interesting to further analyze the performance of LDA-PKL under different numbers of topics.

The ranking algorithm proposed in this paper, LDA-PKL, could help researchers to focus on the most likely related gene-drug relations from large numbers of candidates mixed with false positives. Since LDA-PKL is not specific to gene-drug relations, our method could be extended to other types of relations between PGx entities. Our future work includes: developing more effective NER methods to accurately recognize more gene/drug entities, applying this method to mining other relations among PGx entities, and further evaluating the false positives to discover new relations that are not currently in knowledge databases.

6. Conclusion

In this paper, we presented our research on ranking gene-drug relations extracted from biomedical literature using LDA model. Our proposed method, LDA-PKL, outperformed other existing ranking methods including Chi-square, MI, and a previously reported LDA-KL method, at all co-occurrence levels, suggesting that appropriate uses of the semantic topics derived from LDA model could help analyze candidate relations that co-occur in a piece of text.

7. Acknowledgments

The datasets used were obtained from NLM. The evaluation data were obtained from PharmGKB website: http://www.pharmgkb.org. The authors would like to thank Dr. Josh Denny at Vanderbilt University for providing KnowledgeMap, Dr. Hongfang Liu at Mayo Clinic for providing Biothesaurus, Dr. Alan Aronson at NLM for making MetaMapped corpus available, and the PharmGKB team for the gold standard PGx relations.

This study was supported in part by grants from the NHLBI 5U19HL065962, the NLM R01LM010681, and the NCI R01CA141307.

References

1. D. M. Roden, R. B. Altman, N. L. Benowitz, D. A. Flockhart, K. M. Giacomini, J. A. Johnson, R. M. Krauss, H. L. McLeod, M. J. Ratain, M. V. Relling, H. Z. Ring, A. R. Shuldiner, R. M. Weinshilboum and S. T. Weiss. Pharmacogenomics: challenges and opportunities, *Ann Intern Med* **145**, 749 (2006).
2. K. M. Giacomini, R. M. Krauss, D. M. Roden, M. Eichelbaum, M. R. Hayden and Y. Nakamura. When good drugs go bad, *Nature* **446**, 975 (2007).
3. T. E. Klein, J. T. Chang, M. K. Cho, K. L. Easton, R. Fergerson, M. Hewett, Z. Lin, Y. Liu, S. Liu, D. E. Oliver, D. L. Rubin, F. Shafa, J. M. Stuart and R. B. Altman. Integrating genotype and phenotype information: an overview of the pharmgkb project. pharmacogenetics research network and knowledge base, *Pharmacogenomics J* **1**, 167 (2001).
4. Y. Garten, A. Coulet and R. B. Altman. Recent progress in automatically extracting information from the pharmacogenomic literature, *Pharmacogenomics* **11**, 1467 (2010).
5. J. Ding, D. Berleant, D. Nettleton and E. Wurtele. Mining medline: abstracts, sentences, or phrases?, *Pac Symp Biocomput* , 326 (2002).
6. C. Perez-Iratxeta, P. Bork and M. A. Andrade. Xplormed: a tool for exploring medline abstracts, *Trends Biochem Sci* **26**, 573 (2001).
7. R. Hoffmann and A. Valencia. Implementing the ihop concept for navigation of biomedical literature, *Bioinformatics* **21 Suppl 2**, ii252 (2005).
8. B. T. Alako, A. Veldhoven, S. van Baal, R. Jelier, S. Verhoeven, T. Rullmann, J. Polman and G. Jenster. Copub mapper: mining medline based on search term co-publication, *BMC Bioinformatics* **6**, p. 51 (2005).

9. T. K. Jenssen, A. Laegreid, J. Komorowski and E. Hovig. A literature network of human genes for high-throughput analysis of gene expression, *Nat Genet* **28**, 21 (2001).
10. Y. Garten and R. B. Altman. Pharmspresso: a text mining tool for extraction of pharmacogenomic concepts and relationships from full text, *BMC Bioinformatics* **10 Suppl 2**, p. S6 (2009).
11. H. M. Muller, A. Rangarajan, T. K. Teal and P. W. Sternberg. Textpresso for neuroscience: searching the full text of thousands of neuroscience research papers, *Neuroinformatics* **6**, 195 (2008).
12. L. Tari, J. Hakenberg, G. Gonzalez and C. Baral. Querying parse tree database of medline text to synthesize user-specific biomolecular networks, *Pac Symp Biocomput* , 87 (2009).
13. K. Fundel, R. Kuffner and R. Zimmer. Relex–relation extraction using dependency parse trees, *Bioinformatics* **23**, 365 (2007).
14. L. Hunter, Z. Lu, J. Firby, J. Baumgartner, W. A., H. L. Johnson, P. V. Ogren and K. B. Cohen. Opendmap: an open source, ontology-driven concept analysis engine, with applications to capturing knowledge regarding protein transport, protein interactions and cell-type-specific gene expression, *BMC Bioinformatics* **9**, p. 78 (2008).
15. A. Coulet, N. H. Shah, Y. Garten, M. Musen and R. B. Altman. Using text to build semantic networks for pharmacogenomics, *J Biomed Inform* **43**, 1009 (2010).
16. D. Klein and C. D. Manning. Accurate unlexicalized parsing, *Proceedings of the 41st Annual Meeting on Association for Computational Linguistics* **1**, 423 (2003).
17. J. T. Chang and R. B. Altman. Extracting and characterizing gene-drug relationships from the literature, *Pharmacogenetics* **14**, 577 (2004).
18. Y. Garten, N. P. Tatonetti and R. B. Altman. Improving the prediction of pharmacogenes using text-derived drug-gene relationships, *Pac Symp Biocomput* , 305 (2010).
19. D. M. Blei, A. Y. Ng and M. I. Jordan. Latent dirichlet allocation, *J. Mach. Learn. Res.* **3**, 993 (2003).
20. H. Wang, Y. Ding, J. Tang, X. Dong, B. He, J. Qiu and D. J. Wild. Finding complex biological relationships in recent pubmed articles using bio-lda, *PLoS One* **6**, p. e17243 (2011).
21. A. R. Aronson. Effective mapping of biomedical text to the umls metathesaurus: the metamap program, *Proc AMIA Symp* , 17 (2001).
22. H. Liu, M. Torii, Z.-z. Hu and C. Wu. Mapping gene/protein names in free text to biomedical databases, *Data Mining Workshops, International Conference on* **0**, 101 (2007).
23. O. Bodenreider. The unified medical language system (umls): integrating biomedical terminology, *Nucleic Acids Res* **32**, D267 (2004).
24. J. C. Denny, P. R. Irani, F. H. Wehbe, J. D. Smithers and r. Spickard, A. The knowledgemap project: development of a concept-based medical school curriculum database, *AMIA Annu Symp Proc* , 195 (2003).
25. M. Sanderson and J. Zobel. Information retrieval system evaluation: effort, sensitivity, and reliability, *Proceedings of the 28th annual international ACM SIGIR conference on Research and development in information retrieval* , 162 (2005).

THE STRUCTURE AND FUNCTION OF CHROMATIN AND CHROMOSOMES

WILLIAM STAFFORD NOBLE[1,2], C. ANTHONY BLAU[1,3], JOB DEKKER[4], ZHI-JUN DUAN[3] AND YI MAO[1]

[1]*Department of Genome Sciences, University of Washington*
[2]*Department of Computer Science and Engineering, University of Washington*
[3]*Department of Hematology, University of Washington*
[4]*Department of Biochemistry and Molecular Pharmacology, University of Massachusetts Medical School*

1. Background

One of the striking features of the eukaryotic nucleus is that chromosomes adopt preferred conformations that vary across different tissues and developmental stages. Many lines of evidence suggest the functional significance of genome architecture, including the existence of chromosomes territories [1], the correlation between radial positioning of chromosomes with gene density, transcriptional activity, replication timing and chromosome size [2-5], the tissue-specific spatial configurations of chromosomes relative to one another [6, 7], the co-localization of disparate DNA elements into functionally defined aggregates or "factories" [9-13], and the predisposition to human disease, including cancer by predisposition to acquisition of karyotypic defects [14-17]. Collectively, these various lines of evidence indicate an intimate relationship between genome structure and function. These findings also emphasize the need to characterize both local and global chromosome structure to understand the underlying regulatory mechanisms of various genome functions.

In broad terms, three types of molecular biology tools are currently available to characterize chromosome structure.

1. Microscopy-based imaging, including live cell imaging [18], DNA FISH [19], and tomography [20] are established but relatively low throughput methods.
2. DNA adenine methyltransferase identification (DamID) [21] can be used to characterize genome structures associated with nuclear landmarks or other proteins. This method requires the use of genetically engineered cell lines.
3. Finally, chromatin conformation capture (3C) [24] and related methods, including 4C, 5C, ChIA-PET [22, 23] and Hi-C [35], are relatively simple and have been widely adopted for studies of long-range chromatin interactions.

3C and its derivatives are now the most commonly used tools for characterizing chromosome structure [24-34]. During the past two years, three 3C-based methods have been developed to identify chromatin interactions genome-wide, revealing the folding principles of the human genome [35] and enabling the construction of 3D models of the genomes of haploid budding yeast [36] and fission yeast [37].

The regulation of eukaryotic gene expression depends on transcription factors and cofactors acting on functional DNA elements such as promoters, enhancers, silencers and insulators, which are situated in a three-dimensional space. While systematic efforts such as those of the NIH

ENCODE Consortium have led to high throughput strategies for rapidly annotating functional elements in the human genome, how these elements interact in three dimensions is not well understood. The focus of this workshop is the development and application of computational techniques that link genome structure to function.

2. Workshop Contributions

The workshop includes two invited talks and five talks that were selected on the basis of submitted abstracts.

- **Leonid A. Mirny** is a Professor in the Division of Health Sciences and Technology and Department of Physics at the Massachusetts Institute of Technology. His invited lecture, "Higher-order chromatin architecture: bridging physics and biology", will describe computational simulations of human chromatin based on Hi-C data, which support the hypothesis that, on a megabase scale, human chromatin adopts a so-called *fractal globule* configuration. The resulting model can be used to test mechanisms responsible for chromatin structure formation during the cell cycle, to investigate cell-to-cell variability in chromatin structure, and to establish a possible link between chromosomal abnormalities in cancer and high-order chromatin structure.
- **Yijun Ruan** is a Senior Group Leader and the Associate Director of Genomic Technologies at the Genome Institute of Singapore. His invited lecture, "Chromatin interaction and transcription regulation," will describe the successful use of Chromatin Interaction Analysis with Paired-End-Tag sequencing (ChIA-PET) to map long-range chromatin interactions associated with CTCF and RNA polymerase II in human cells. These analyses suggest a central role for CTCF-mediated looping in the general landscape of chromosomal organization. The results also show that many genes are transcribed cooperatively in a process heavily influenced by chromatin organization.
- Damian Wójtowicz, Fedor Kouzine, Arito Yamane, Wolfgang Resch, Craig J. Benham, Rafael Casellas, David Levens and Teresa M. Przytycka. "Genome-wide mapping and computational analysis of non-B DNA structures *in vivo*." This talk describes a new experimental method, called ssDNA-seq, that identifies occurrences of single-stranded DNA (ssDNA) across a genome. Applied to mouse B cells and human Raji cells, the assay indicates an enrichment of ssDNA near transcription start sites, suggesting a role for ssDNA in gene regulation. The data also provides insights into existing computational methods for predicting non-canonical local DNA structure.
- Zackary Scholl, Jianling Zhong and Alexander Hartemink. "Mapping transcription factories in *Saccharomyces cerevisiae*." In this project, the authors integrate published yeast data sets related to genome architecture [36], transcription localization [38, 39] and polymerase II occupancy [40] to show statistical evidence for the presence of transcription factories.
- Thomas Asbury, Matt Mitman, William Arndt, Jijun Tang and Jim Zheng. "Modeling, Analyzing and Visualizing Human Genome in 3D." This talk describes an ongoing effort

to develop next-generation software for interactively visualizing eukaryotic genomes in three dimensions. Using simulated data, the authors demonstrate the ability to visualize the 3D human genome in real time and at scales ranging from 1bp resolution to the entire genome.

- Geet Duggal, Hao Wang, Michelle Girvan, Sridhar Hannenhalli and Carl Kingsford. "Graph-theoretic approaches for testing embeddability and spatial functional enrichment in chromosome conformation data." In this project, the authors apply graph rigidity theory to detect regions of the yeast genome in which observed genome architecture data [36] is sufficient to identify a rigid embedding in three-dimensional space. This procedure yields a single, large component comprising the entire genome, with sub-components corresponding to each chromosome. The authors also propose a statistical test for co-localized sets of genes.

3. Outlook

Similar to the rest of the genomics field, progress in elucidating genome structure and relating structure to biological function is driven largely by technological innovations. As such, future directions for the field are somewhat difficult to guess. Nonetheless, given current technologies, we can identify a few key areas in which concerted effort by computational biologists will likely lead to significant insights.

3.1. *Statistical Confidence Estimation*

Large-scale genome structure assays such as Hi-C generate as their primary output a huge collection of sequence reads. Relatively robust, user-friendly tools exist for assigning these reads in an efficient fashion to a reference genome; however, such tools are currently lacking for the subsequent steps of aggregating reads with respect to, for example, restriction enzyme fragments and of assigning statistical confidence estimates to observed interactions. Existing statistical confidence estimation methods use straightforward empirical or analytic null models, and these models have not been extensively validated [35-37].

3.2. *Inference of 3D Structure*

Given information about the frequency of pairwise interaction between pairs of DNA loci, as produced by medium- or high-throughput methods such as 4C, 5C or Hi-C, a natural next step is to attempt to infer properties of the 3D conformation of the DNA in the nucleus. Several successful attempts have been made along these lines, including a relatively straightforward, direct optimization approach [36, 37], as well as more sophisticated statistical sampling methods [41]. The latter approach is clearly preferable, because it has the potential to model cell-to-cell variability in 3D structure. Clearly, any 3D inference algorithm must take into account physical constraints imposed by the spatial organization of the nucleus itself as well as the physical properties of the DNA polymer. Furthermore, any solution to the 3D inference problem will likely need to be capable of integrating multiple types of data, including 3C-based measurements as well as information from FISH, DamID or ChIA-PET. Finally, a particular challenge for 3D inference is appropriately handling multiploid genomes, and the fact that large populations of cells are analyzed, in which the observed count of interactions between any given pair of loci in the cell and in the cell population is actually a mixture of counts from all possible pairs of homologous copies of those loci. For more details on this general problem, see the recent excellent review by Marti-Renom and Mirny [42].

3.3. *Visualization of Genomic Structure*

Widely used genome browsers, such as the Santa Cruz Genome Browser [43], the Ensembl Genome Browser [44] or JBrowse [45], are designed to work with data that is arrayed linearly along the genome. As such, representing pairwise interaction data within these frameworks is difficult. The Genome3D viewer [46], which is described in one of our submitted talks, is an excellent first step in this direction, but a fully functional platform that integrates the wide variety of available 1D and 3D data requires further effort.

3.4. *Statistical Methods for Co-localization Analysis*

As pointed out above, one primary hypothesis driving the current interest in genome structure is that structure is causally linked to genome function in some non-trivial fashion. Accordingly, there is a growing demand for statistical methods that can bridge the gap between our current knowledge of gene function, as represented, for example, by Gene Ontology annotations or various functional genomics assays, and empirical observations of genome architecture. For example, recent evidence suggests that co-regulated yeast genes tend to be co-localized [47]. In this workshop, the contributed talk by Scholl *et al.* represents a novel approach to this type of problem. Ultimately, the field is aiming toward developing a physical model that suggests how genome form affects function.

References

1. T. Cremer and M. Cremer. Chromosome territories. *Cold Spring Harb Perspect Biol.* **2**(3), a003889 (2010). PMCID: 2829961.
2. S. Boyle, S. Gilchrist, J. M. Bridger, N. L. Mahy, J. A. Ellis and W. A. Bickmore. The spatial organization of human chromosomes within the nuclei of normal and emerin-mutant cells. *Hum Mol Genet.* **10**(3), 211-219 (2001).
3. J. A. Croft, J. M. Bridger, S. Boyle, P. Perry, P. Teague and W. A. Bickmore. Differences in the localization and morphology of chromosomes in the human nucleus. *J Cell Biol.* **145**(6), 1119-1131 (1999). PMCID: 2133153.
4. F. Grasser, M. Neusser, H. Fiegler, T. Thormeyer, M. Cremer, N. P. Carter, et al. Replication-timing-correlated spatial chromatin arrangements in cancer and in primate interphase nuclei. *J Cell Sci.* **121**(Pt 11), 1876-1886 (2008). PMCID: 2687722.
5. T. Misteli. Beyond the sequence: cellular organization of genome function. *Cell.* **128**(4), 787-800 (2007).
6. L. A. Parada, P. G. McQueen and T. Misteli. Tissue-specific spatial organization of genomes. *Genome Biol.* **5**(7), R44 (2004). PMCID: 463291.

7. I. Solovei, M. Kreysing, C. Lanctot, S. Kosem, L. Peichl, T. Cremer, et al. Nuclear architecture of rod photoreceptor cells adapts to vision in mammalian evolution. *Cell.* **137**(2), 356-368 (2009).
8. H. Tanabe, S. Muller, M. Neusser, J. von Hase, E. Calcagno, M. Cremer, et al. Evolutionary conservation of chromosome territory arrangements in cell nuclei from higher primates. *Proc Natl Acad Sci U S A.* **99**(7), 4424-4429 (2002). PMCID: 123664.
9. M. Thompson, R. A. Haeusler, P. D. Good and D. R. Engelke. Nucleolar clustering of dispersed tRNA genes. *Science.* **302**(5649), 1399-1401 (2003).
10. G. L. Zhou, L. Xin, W. Song, L. J. Di, G. Liu, X. S. Wu, et al. Active chromatin hub of the mouse alpha-globin locus forms in a transcription factory of clustered housekeeping genes. *Mol Cell Biol.* **26**(13), 5096-5105 (2006). PMCID: 1489176.
11. C. S. Osborne, L. Chakalova, J. A. Mitchell, A. Horton, A. L. Wood, D. J. Bolland, et al. Myc dynamically and preferentially relocates to a transcription factory occupied by Igh. *PLoS Biol.* **5**(8), e192 (2007). PMCID: 1945077.
12. F. Bantignies, V. Roure, I. Comet, B. Leblanc, B. Schuettengruber, J. Bonnet, et al. Polycomb-dependent regulatory contacts between distant Hox loci in Drosophila. *Cell.* **144**(2), 214-226 (2011).
13. C. H. Eskiw, N. F. Cope, I. Clay, S. Schoenfelder, T. Nagano and P. Fraser. Transcription factories and nuclear organization of the genome. *Cold Spring Harb Symp Quant Biol.* **75**, 501-506 (2010).
14. T. Misteli. Higher-order genome organization in human disease. *Cold Spring Harb Perspect Biol.* **2**(8), a000794 (2010).
15. J. J. Roix, P. G. McQueen, P. J. Munson, L. A. Parada and T. Misteli. Spatial proximity of translocation-prone gene loci in human lymphomas. *Nat Genet.* **34**(3), 287-291 (2003).
16. C. Lin, L. Yang, B. Tanasa, K. Hutt, B. G. Ju, K. Ohgi, et al. Nuclear receptor-induced chromosomal proximity and DNA breaks underlie specific translocations in cancer. *Cell.* **139**(6), 1069-1083 (2009). PMCID: 2812435.
17. R. S. Mani, S. A. Tomlins, K. Callahan, A. Ghosh, M. K. Nyati, S. Varambally, et al. Induced chromosomal proximity and gene fusions in prostate cancer. *Science.* **326**(5957), 1230 (2009). PMCID: 2935583.
18. F. Hediger, F. R. Neumann, G. Van Houwe, K. Dubrana and S. M. Gasser. Live imaging of telomeres: yKu and Sir proteins define redundant telomere-anchoring pathways in yeast. *Curr Biol.* **12**(24), 2076-2089 (2002).
19. A. Bolzer, G. Kreth, I. Solovei, D. Koehler, K. Saracoglu, C. Fauth, et al. Three-dimensional maps of all chromosomes in human male fibroblast nuclei and prometaphase rosettes. *PLoS Biol.* **3**(5), e157 (2005). PMCID: 1084335.
20. D. P. Bazett-Jones, R. Li, E. Fussner, R. Nisman, and H. Dehghani. Elucidating Chromatin and Nuclear Domain Architecture with Electron Spectroscopic Imaging. *Chromosome Res.* **16**, 397-412 (2008).
21. L. Guelen, L. Pagie, E. Brasset, W. Meuleman, M. B. Faza, W. Talhout, et al. Domain organization of human chromosomes revealed by mapping of nuclear lamina interactions. *Nature.* **453**(7197), 948-951 (2008).
22. M. J. Fullwood, M. H. Liu,Y. F. Pan, J. Liu, H. Xu, Y. B. Mohamed, et al. An oestrogen-receptor-alpha-bound human chromatin interactome. *Nature.* **462**(7269), 58-64 (2009). PMCID: 2774924.

23. L. Handoko, H. Xu, G. Li, C. Y. Ngan, E. Chew, M. Schnapp, et al. CTCF-mediated functional chromatin interactome in pluripotent cells. *Nat Genet.* **43**(7), 630-638 (2011).
24. J. Dekker, K. Rippe, M. Dekker, N. Kleckner. Capturing chromosome conformation. *Science.* **295**(5558), 1306-1311 (2002).
25. J. Dostie, T. A. Richmond, R. A. Arnaout, R. R. Selzer, W. L. Lee, T. A. Honan, et al. Chromosome Conformation Capture Carbon Copy (5C): a massively parallel solution for mapping interactions between genomic elements. *Genome Res.* **16**(10), 1299-1309 (2006). PMCID: 1581439.
26. S. Horike, S. Cai, M. Miyano, J. F. Cheng and T. Kohwi-Shigematsu. Loss of silent chromatin looping and impaired imprinting of DLX5 in Rett syndrome. *Nat Genet.* **37**(1), 31-40 (2005).
27. J. Q. Ling, T. Li, J. F. Hu, T. H. Vu, H. L. Chen, X. W. Qiu, et al. CTCF mediates interchromosomal colocalization between Igf2/H19 and Wsb1/Nf1. *Science.* **312**(5771), 269-272 (2006).
28. S. Lomvardas, G. Barnea, D. J. Pisapia, M. Mendelsohn, J. Kirkland and R. Axel. Interchromosomal interactions and olfactory receptor choice. *Cell.* **126**(2), 403-413 (2006).
29. M. Simonis, P. Klous, E. Splinter, Y. Moshkin, R. Willemsen, E. de Wit, et al. Nuclear organization of active and inactive chromatin domains uncovered by chromosome conformation capture-on-chip (4C). *Nat Genet.* **38**(11), 1348-1354 (2006).
30. V. K. Tiwari, L. Cope, K. M. McGarvey, J. E. Ohm and S. B. Baylin. A novel 6C assay uncovers Polycomb-mediated higher order chromatin conformations. *Genome Res.***18**(7),1171-1179 (2008). PMCID: 2493406.
31. Z. Zhao, G. Tavoosidana, M. Sjolinder, A. Gondor, P. Mariano, S. Wang, et al. Circular chromosome conformation capture (4C) uncovers extensive networks of epigenetically regulated intra- and interchromosomal interactions. *Nat Genet.* **38**(11), 1341-1347 (2006).
32. D. Bau, A. Sanyal, B. R. Lajoie, E. Capriotti, M. Byron, J. B. Lawrence, et al. The three-dimensional folding of the alpha-globin gene domain reveals formation of chromatin globules. *Nat Struct Mol Biol.* **18**(1), 107-114 (2011). PMCID: 3056208.
33. S. Schoenfelder, T. Sexton, L. Chakalova, N. F. Cope, A. Horton, S. Andrews, et al. Preferential associations between co-regulated genes reveal a transcriptional interactome in erythroid cells. *Nat Genet.* **42**(1), 53-61 (2010).
34. B. Tolhuis, R. J. Palstra, E. Splinter, F. Grosveld and W. de Laat. Looping and interaction between hypersensitive sites in the active beta-globin locus. *Mol Cell.* **10**(6), 1453-1465 (2002).
35. E. Lieberman-Aiden, N. L. van Berkum, L. Williams, M. Imakaev, T. Ragoczy, A. Telling, I. Amit, B. R. Lajoie, P. J. Sabo, M. O. Dorschner, R. Sandstrom, B. Bernstein, M. A. Bender, M. Groudine, A. Gnirke, J. Stamatoyannopoulos, L. A. Mirny, E. S. Lander, and J. Dekker. Comprehensive mapping of long-range interactions reveals folding principles of the human genome. *Science.* **326**(5950), 289–293 (2009).
36. Z. Duan, M. Andronescu, K. Schutz, S. McIlwain, Y. J. Kim, C. Lee, J. Shendure, S. Fields, C. A. Blau, and W. S. Noble. A three-dimensional model of the yeast genome. *Nature.* **465**, 363–367 (2010). PMC2874121.
37. H. Tanizawa, O. Iwasaki, A. tanaka, J. R. Capizzi, P. Wickramasighne, M. Lee, Z. Fu, and K. Noma. Mapping of long-range associations throughout the fission yeast genome reveals global genome organization linked to transcriptional regulation. *Nucl Acids Res.* (2010).

38. C. Miller, B. Schwalb, K. Maier, D. Schulz, S. Dümcke, B. Zacher, A. Mayer, et al. Dynamic transcriptome analysis measures rates of mRNA synthesis and decay in yeast. *Mol Syst Biol*. 7, 485(2011).
39. L. S. Churchman and J. S. Weissman. Nascent transcript sequencing visualizes transcription at nucleotide level. *Nature*. **469**(7330), 368-373 (2011).
40. A. J. Jasiak, H. Hartmann, E. Karakasili, M. Kalocsay, A. Flatley, E. Kremmer, K. Strässer, D. E. Martin, J. Söding and P. Cramer. Genome-associated RNA Polymerase II Includes the Dissociable Rpb4/7 Subcomplex. *J Biol Chem*. 283(39), 26423-26427 (2008).
41. D. Bau, A. Sanyal, B. R. Lajoie, E. Capriotti, M. Byron, J. B. Lawrence, J. Dekker, and M. A. Marti-Renom. The three-dimensional folding of the α-globin gene domain reveals formation of chromatin globules. *Nature Structural Molecular Biology*. **18**(1), 107–114 (2011).
42. M. A. Marti-Renom and L. Mirny. Bridging the resolution gap in structural modeling of 3D genome organization. *PLoS Computational Biology*. **7**(7), e100002125 (2011).
43. W. J. Kent, C. W. Sugnet, T. S. Furey, K. M. Roskin, T. H. Pringle, A. M. Zahler, and D. Haussler. The Human Genome Browser at UCSC. *Genome Res*. **12**, 996–1006 (2002).
44. P. Flicek, M. R. Amode, D. Barrell, K. Beal, S. Brent, et al. Ensembl 2011. *Nucleic Acids Research*. **39**, D800-D806 (2010).
45. M. E. Skinner, A. V. Uzilov, L. D. Stein, C. J. Mungall and I. H. Holmes. JBrowse: a next-generation genome browser. *Genome Res*. **19**(9), 1630-1638 (2009).
46. T. M. Asbury, M. Mitman, J. Tang, and W. J. Zheng. Genome3D: a viewer-model framework for integrating and visualizing multi-scale epigenomic information within a three-dimensional genome. *BMC Bioinformatics*. **11**, 444 (2010).
47. Z. Dai and X. Dai. Nuclear colocalization of transcription factor target genes strengthens coregulation in yeast. *Nucl Acids Res*. (2011).

LAW, BIOETHICS AND THE CURRENT STATUS OF OWNERSHIP, PRIVACY, INFORMED CONSENT IN THE GENOMIC AGE

GREG HAMPIKIAN and ERIC M. MESLIN

The rise of biobanks including the related databases containing increasing amount of health information are now commonplace – found on every continent of the globe including Antarctica. Common though they may be, legal and ethical issues for collecting, storing and using the associated health/genetic information continue to perplex clinicians, researchers, law enforcement, and the lay public alike. For example, one's image or likeness is copyrightable, but not one's genome. Yet billions of dollars are being made from patented genes, patented species, private databases, and human tissues. Several health care systems now collect human DNA samples from patients through "opt-out" programs, and vast collections of DNA are being assembled without compensation to donors. Now that whole genome sequencing is becoming more affordable, the bioinformatics community is rightly asking about its enduring obligations and responsibilities vis a vis genetic ownership, residual rights, privacy, and informed consent. And that's just in the USA. Many issues are magnified when material or data are intended to cross national borders. This workshop is designed to unpack many of the issues in pragmatic detail – hopefully dispelling certain myths while offering cautionary reminders about legal and ethical obligations. The two co-chairs will lead in-depth discussion and facilitate small group exercises to emphasize these issues. A particular highlight will be a mock "House of Commons" debate where workshop participants will be asked to take positions, ask questions, and vote with their feet. In other words, this workshop is meant to be provocative, user-friendly and skill enhancing.

Greg Hampikian is a professor of Biology, and Criminal Justice, at Boise State University. He is also director of the Idaho Innocence Project, and spends a lot of time in US and foreign courtrooms.

Eric M. Meslin is Director of the Indiana University Center for Bioethics, Associate Dean (Bioethics) IU School of Medicine, and Professor of Medicine; Medical & Molecular Genetics; Public Health and Philosophy. He needs a larger business card.

SYSTEMS PHARMACOGENOMICS–BRIDGING THE GAP

P-STAR
Marylyn Ritchie, The Pennsylvania State University
Nancy Cox, University of Chicago
Cheng Cheng, St. Jude Children's Research Hospital

PGRN Systems Biology Group
Scott Weiss, Channing Laboratory, Harvard Medical School

PharmGKB
Teri Klein and Russ Altman, Stanford University

Biological systems are quintessentially complex. Normal function requires the constant orchestration of numerous pathways on widely varying spatial and temporal scales. To understand such a complex system, data has been collected on the molecular, cellular, organ, and population scales and across various -omes such as the genome, proteome, and metabolome. Unfortunately, it is often the case that each data set is analyzed *in vacuo*, without regard to information on other scales or -omes. Multi-scale integration of data types to answer fundamental and practical questions in biomedicine is a significant challenge to the experimental and computational biology communities. This is a topic that is being tackled in a significant way in Systems Pharmacogenomics – which is exploring systems biology approaches for drug treatment response.

Systems researchers in the academic sector are more biology driven. The focus tends to be on developing models and solutions to various questions that lead to new hypotheses to test and explore back at the bench. Whereas the pharma sector tend to be more data driven, where the goal is to find predictive models that can be used for drug treatment development or prediction, but may not be intuitive in terms of the underlying biology. It seems as though approaches and techniques developed by each sector would be enhanced by coming together to discuss the strengths and weakness of their respective approaches.

We will host the second Systems Pharmacogenomics – Birds of a Feather workshop at PSB 2012 with invited speakers from both academic and industry sectors. This event will be sponsored by the PGRN Systems Biology Working Group, PharmGKB, and P-STAR (the PGRN Statistical Analysis Resource). We will have a series of talks following by a panel discussion aimed at bridging the gap between academic and pharma systems pharmacogenomics.

Our invited speakers include: Pankaj Agarwal, Director, Computational Biology, Molecular Discovery & Development, GlaxoSmithKline; Stephane Bourgeois, Geneticist, Wellcome Trust Sanger Institute; Trey Ideker, Division Chief of Medical Genetics at UCSD School of Medicine; and Nicholas J. Schork, Ph.D., Director of Biostatistics and Bioinformatics, The Scripps Translational Science Institute, Professor, Molecular and Experimental Medicine The Scripps Research Institute.

This event will introduce a variety of current approaches for systems pharmacogenomics and potential avenues for future research. Special thanks to the PGRN for their sponsorship and to the speakers for participating in this very exciting event.

www.ingramcontent.com/pod-product-compliance
Lightning Source LLC
LaVergne TN
LVHW050548311026
851422LV00009B/35
9789814596374